FUNDAMENTALS OF RESIDENTIAL CONSTRUCTION

FUNDAMENTALS OF RESIDENTIAL CONSTRUCTION

Edward Allen
and
Rob Thallon

Featuring the Drawings *of*
Joseph Iano

JOHN WILEY & SONS, INC.

Photo on page ii by Rob Thallon.

Library of Congress Cataloging-in-Publication Data:

Allen, Edward, 1938–
    Fundamentals of residential construction / Edward Allen and Rob Thallon
        p. cm.
    Includes index.
    ISBN 0-471-38687-1 (cloth : alk. paper)
        1. House construction. I. Thallon, Rob. II. Title.

    TH4811 .A463 2001
    690'.837—dc21

                                                                2001045402

Printed in the United States of America.
10 9 8 7 6 5 4 3 2

CONTENTS

PART THREE
WOOD LIGHT FRAME CONSTRUCTION 157

PREFACE

When *Fundamentals of Building Construction, Materials and Methods* was first published more than 15 years ago, it was adopted immediately by hundreds of colleges and universities as a text for general courses in construction technology. It also precipitated immediately the first of a growing stream of requests from teachers for a companion volume that would concentrate on residential construction while retaining the qualities of the parent book. We are pleased to respond to those requests with the book that you hold in your hands.

From its parent, this book inherits several important traits: straightforward, readable writing; clear drawings; and extensive photographic illustrations. These elements are blended on spacious, attractive pages, and, for the reader's convenience, nearly every illustration is on the same two-page spread as the referencing text. Retained, too, is the concern for both technical and aesthetic matters, because we believe that both are important for the quality of buildings and the lives of the people who inhabit them.

Although both authors teach in schools of architecture, we are not mere ivory-tower academicians. Between us, we are the architects of well over 200 constructed houses and innumerable remodeling projects. Both of us have spent countless hours on construction sites, working with residential builders, developers, contractors, and craftspeople on the day-to-day minutiae of getting houses built. We have both constructed houses with our own hands, from excavation to finishes. Both of us are authors of prior books on construction that have found enthusiastic acceptance by the building professions.

To make this book inclusive of regional differences in construction practices, we have found it extremely helpful that one of us works in wintry New England and the other in the damp but mild Pacific Northwest. To extend the boundaries of our own experiences, we also consulted frequently with colleagues in other regions of the United States and Canada.

ACKNOWLEDGEMENTS

Edward Allen is grateful to be teamed with Rob Thallon, a gifted teacher, accomplished author, and award-winning architect. He thanks Rob for sharing his vision of excellence and for his tireless work. He is grateful to Joseph Iano, whose ideas and innovations for the parent book still show in this book. And he thanks Mary M. Allen for her support and encouragement.

Rob Thallon is especially thankful to Edward Allen, his mentor and friend, for selecting him to work on this important project. He thanks his co-author for having been his keenest critic and most fervent champion during his previous writing projects. He thanks the talented illustrators who helped develop and render the new drawings for this book: Mu-Yun Chang, Lisa Ferretto, Laura Houston, and especially Dave Bloom, who unfailingly assisted during the entire process. He also thanks Lynne Clearfield and Greg Thomson, who diligently tracked down new photographs for this publication.

Finally, he particularly thanks his family for their understanding, support, and good humor during the many months he has been preoccupied with the quality of this book.

Together, Edward Allen and Rob Thallon are both grateful to all the people of John Wiley & Sons, who gave so much of themselves in producing this book. We thank especially Executive Editor Amanda Miller, her assistant Nyshie Perkinson, Associate Editor Matt Van Hattem, Designer Karin Kincheloe, and Associate Managing Editor Donna Conte. The quality of a book, like that of a house, is proportional to the skill and dedication of the individuals on the production team.

Edward Allen
South Natick, Massachusetts

Rob Thallon
Eugene, Oregon

DISCLAIMER

The drawings, tables, descriptions, and photographs in this book have been obtained from many sources, including trade associations, suppliers of building materials, governmental organizations, and architectural firms. They are presented in good faith, but the author, illustrator, and publisher, while they have made every reasonable effort to make this book accurate and authoritative, do not warrant, and assume no liability for, its accuracy or completeness or its fitness for any particular purpose. It is the responsibility of users to apply their professional knowledge in the use of information contained in this book, to consult the original sources for additional information when appropriate, and to seek expert advice when appropriate.

Fundamentals of Residential Construction

CONTEXT

THE CONTEXT FOR CONSTRUCTION

1

People have been building houses for thousands of years. These houses have provided shelter, afforded privacy, defined territory, enhanced status, and, in some cases, provided defense. The earliest houses were opportunistic uses of naturally sheltered places like caves and were more like nests than houses. As time passed, people learned to assemble materials collected from nature to make simple freestanding structures. In many cultures, these structures evolved into highly crafted houses that are elegant expressions of cultural patterns and values (Figures 1.1 and 1.2). In the past 150 years, technology has afforded us conveniences such as electricity, plumbing, and automatic heating and air conditioning that have made houses, in the words of the famous architect Le Corbusier, "machines for living." Most recently, housing of the industrialized world has emphasized energy conservation and efficient production. Today, residential designers have a rich history from which to draw, and residential builders have the best tools and most complete palette of quality materials with which to build that have ever existed. The challenge for this new generation of designers and builders is to improve the built environment in the face of decreasing natural resources and increasing population.

Throughout history, the forms of houses have differed from region to region. House form varies primarily in relation to climate, to available building materials and tools, and to the culture of the people being housed. The influence of climate on house form is dramatically demonstrated by the comparison of the igloo from polar regions with the open-sided palm-thatched structure from tropical zones. The forms of houses in the same climate can vary also, however, because of the use of different building materials. In Mexico, for example, the introduction of reinforced concrete has spawned a collection of flat-roofed houses that contrast sharply with the traditional sloped roof made of timber covered with clay tiles. The culture of the people being housed also has considerable influence on house form. Native American tribes who were nomadic built dwellings such as tepees that were easily folded and transported, while rooted tribes from the same region built stationary houses of earth, stone, and wood.

The modern North American house has evolved largely from 16th-century timber-framed houses that had been developed in response to the climate, materials, and culture of northern Europe (Figure 1.3). Early pioneers landing on the eastern shores of the continent found a new homeland rich with timber that had to be cleared to make way for development, so it was logical to use wood for the construction of new houses. The settlers soon discovered, however, that the European tradition of exposed timber frame was inadequate in the harsher climate of the New World, so they developed an exterior skin of clapboards to protect the frame (Figure 1.4). This wooden structure and its details evolved over the years in response to changes in tools, transportation, and social norms. Other building materials and systems such as brick and stone masonry developed simultaneously, but were never as prevalent in North America as the clapboard-clad timber-framed building.

Then, in the 1840s, after more than 300 years of development, the heavy timber frame yielded its preeminence almost overnight to a new system of construction, the *wood light frame*. The emergence of the light frame was made possible by two technological developments: the mass production of the inexpensive wire nail, and the ability of water-powered sawmills to cut large quantities of consistently dimensioned lumber. These developments allowed the large timbers and complicated connections of the traditional timber-framed buildings to be replaced with numerous small structural pieces, simply connected (Figure 1.5). The advantages of the wood light frame over its predecessor were so numerous and compelling that it has dominated residential and other small-scale construction for the 150 years since its introduction, and it still shows no sign of giving way to other systems. Today, the wood light frame accounts for over 90 percent of all new site-built residential construction and is the basis for most factory-built housing as well.

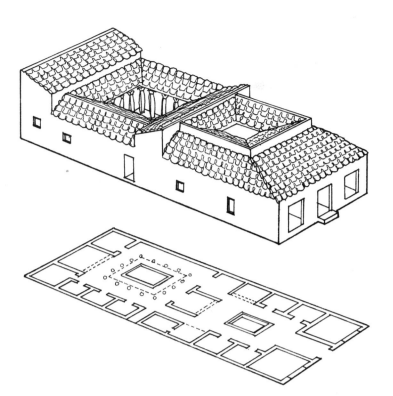

FIGURE 1.1
The Roman domus, developed more than 2000 years ago, had individual rooms for common daily functions and was built around a central courtyard that helped to cool the rooms naturally. (*From Peter Hodge,* The Roman House: Aspects of Roman Life, *Longman Group, Ltd., London, 1975*)

FIGURE 1.2
A traditional Japanese house from the Nara period, 710–784 A.D. The house had both open and closed spaces, and the enclosed indoor space had no permanent partitions. Houses such as this influenced traditional Japanese house design and construction to the present day. (*From Nishi Kazuo and Hozumi Kazuo,* What Is Japanese Architecture? *Kodanshu International, Ltd., Tokyo, 1985*)

FIGURE 1.3
European timber house forms generally followed a progression of development from crude pit dwellings, made of earth and tree trunks, to cruck frames, to braced frames.

Cruck

PIT DWELLING

CRUCK FRAME

BRACED FRAME

FIGURE 1.4
The North American climate was more severe than the European climate, so early pioneers found a way to wrap the wooden frame with cladding, protecting it more securely from the weather than the exposed half-timbers of European houses. This example, built in Essex County, Massachusetts, is still standing. (*Photo courtesy Library of Congress, Prints and Photographs Division, Historic American Buildings Survey, Reproduction Number, HABS, MASS, 5-TOP, 1-6*)

FIGURE 1.5
The wood light frame uses fewer materials and less labor to construct than does its predecessor, the timber frame. For lateral stability, light framing relies on sheathing such as plywood applied to the exterior of the frame.

WOOD LIGHT FRAME

A BUILDING CULTURE

Houses are built within the context of the many individuals and institutions that affect their design and construction. In primitive and vernacular societies, the context is relatively local and involves few people. The head of a household might acquire a piece of land through the family, formulate a simple design based on local traditions, consult with a local builder about schedule and cost, arrange for the purchase of local building materials, and work together with the builder using traditional methods to build the house. The building of a house today in North America involves a much more complex process and many more participants. Nonetheless, all these participants are instrumental to the success of the project, and all are connected to what can be called a residential building culture — a network of people and institutions, which we will call "subcultures," that are directly or indirectly dedicated to the production of houses. The principal subcultures are discussed in the following paragraphs and in later chapters of this book.

Contractors and Subcontractors

At the center of today's residential building culture are the contractors and subcontractors whose job it is to construct houses. These people — carpenters, plumbers, masons, electricians, and myriad others — devote their professional lives to assembling materials in concert with one another to make houses. Their work depends on direct contributions from many other sectors of the building culture such as designers, material suppliers, and code enforcement agencies. Indirect contributions from realtors, financial institutions, educators, and publishers also play an important role in their work. Contractors and subcontractors are discussed extensively in Chapter 2.

Builders and Developers

Builders bring together and coordinate the numerous parts of an entire building project for the purpose of offering it for sale or rent. The builder purchases a building lot, obtains financing, hires the designers and other consultants to produce plans, hires the contractor to do the construction, markets the project, and sells or rents it. Builders can work on one house at a time or can build large tracts of houses or large multifamily structures.

Whereas builders are the entrepreneurs who produce houses for sale, developers are entrepreneurs who produce building lots. Developers purchase large tracts of land, contract for the design of roads and utilities, obtain the necessary governmental permissions to develop the land, contract for the installation of roads and utilities, and sell the divided land as building lots. Developers often expand their operation to become builder/ developers, and builders likewise can expand in the other direction. Non-profit builder/developers produce affordable housing for rent or sale to low-income families or individuals. Builders and developers are discussed further in Chapter 2.

Designers

The members of the building culture most responsible for communication are the designers. This group includes architects, building designers, engineers of several kinds, landscape architects, landscape designers, and interior designers. They are responsible for being knowledgeable about current building practices, understanding and interpreting the various codes and laws that regulate building design, having a current understanding of the availability and performance of building materials, and integrating all these factors into designs that are appreciated by their clients. These various participants in the role of residential design are discussed in Chapter 3.

Material Manufacturers and Distributors

There are thousands of companies, large and small, that manufacture and sell the materials and assemblies used to construct houses. The manufacturers generally sell their goods wholesale to retail stores, which, in turn, sell to contractors and to the general public. Contractors, because they are fre-

FIGURE 1.6
Large retail outlets such as this provide one-stop shopping for professional builders and homeowners alike. Because of the large volume of building materials, tools, and books sold at these outlets, prices are usually competitive, and professionals receive an additional discount. (*Photo by Rob Thallon*)

quent customers who often buy in volume, usually are offered a discount at the retail outlets. Product information in both printed and electronic form is distributed to contractors and designers and is disseminated to the general public via commercial advertising in periodicals.

Building material manufacturers have also formed a large number of organizations that work toward the development of technical standards and the dissemination of information with relation to their respective products. The Western Wood Products Association, for example, is made up of producers of lumber and wood products. It carries out programs of research on wood products, establishes uniform standards of product quality, certifies mills and products that conform to its standards, and publishes authoritative technical literature concerning the use of lumber and related products. Associations with a similar range of activities exist for virtually every material and product used in building. All of them publish technical data relating to their fields of interest, and many of these publications are indispensable references for the architect or engineer. A considerable number are incorporated by reference into various building codes and standards.

Realtors

Realtors are the salespeople of the building culture and play a critical role in marketing houses built for sale. They are responsible for knowing what the buying public wants in a house and for selling or renting houses as they are built. Because realtors have direct contact with consumers and are in a position to learn their desires, they are frequently queried by resourceful builders who are trying to discover new design features that will make their houses more marketable.

New speculative houses are typically advertised and sold by realtors via a *listing agreement* under which real-

FIGURE 1.7
Lumberyards play an important role in residential construction. Based on a set of building plans, an employee of the yard will estimate the quantity of lumber that is required to build a project and will furnish a competitive bid for the entire package of lumber, delivered to the building site. Yards prefer doing business with contractors who are organized so that deliveries can be concentrated into five or six truckloads for an average-sized house. (*Photo by Rob Thallon*)

tors assume numerous responsibilities including negotiating the price of the house, the terms of the sale, and the conditions of the contract, with particular attention paid to the aspect of financing. For this service, realtors are generally paid a percentage of the cost of each house sold. Large builder/developers will often create their own real estate company for the purpose of marketing and selling their own houses.

Regulatory Agencies

Building design and construction are regulated by zoning ordinances and building codes written for the purpose of providing safe and healthy built environments. *Zoning ordinances* are local laws that divide the locality into zones and regulate such things as what kinds of buildings may be built in each

zone and to what uses these buildings may be put. For example, these regulations restrict the use of buildings within residential zones so that dangerous or obnoxious activities do not get mixed in with houses. Within residential zones, the minimum size of lots, the distance a house must be from the property line, requirements for off-street parking, and maximum fence heights are typically regulated. *Building codes* are designed to ensure structural safety and a healthy living environment within the house itself. The sizes of structural members, minimum standards for plumbing and wiring, minimum ceiling heights, the design of stairs and handrails, and provision for emergency escape are all examples of the regulations found in the building codes. Zoning ordinances and building codes are further discussed later in this chapter.

Financial Institutions

Most residential construction projects require financial resources beyond the immediate means of the owner. Banks and other financial institutions provide capital for the projects in the form of long-term loans to qualified owners. The ability to resell a house if the owner defaults on payments is a primary concern of lending institutions, so they are less inclined to loan money for the purchase of houses that appear to be very different from the norm. Financial institutions are further discussed later in this chapter.

Educational and Research Institutions

Most designers and builders have some formal training. Architects are required to have at least a 5-year professional degree, and most plan service drafters have taken courses in drafting and residential construction. Many courses in both the business and the physical skills required of their specialties are offered to contractors and subcontractors. Some contractors and subcontractors are required to be licensed, and there are sanctioned courses offered by different institutions for this purpose. The training of residential designers is explored in Chapter 3, and that of builders and contractors is discussed in Chapter 2.

Associations

There are many associations that relate to the design and construction of houses. The American Institute of Architects (AIA) and the National Association of Home Builders (NAHB) are two of the largest such associations, and there are numerous other organizations of manufacturers, building trades, and other groups within the building culture. Hybrid groups that include members from several disciplines also exist. For example, the International Code Council, which is responsible for writing building codes, includes architects, builders, and building code officials.

Publishers

The publishing industry has long been an integral part of the residential building culture. For hundreds of years, periodicals have advertised the latest building materials, tools, and other products (Figure 1.8). Popular magazines such as *Better Homes and Gardens, Sunset,* and *Home* have carried articles about design, while others such as *Builder* and *Fine Homebuilding* have focused on construction. Books and journals are devoted to a variety of related topics. Recently, there has been a proliferation of how-to books for the do-it-yourself market. Whether the motive is advertising or education, the most successful published materials come from sources with strong connections to the building culture and especially to the design and construction processes.

METALLIC SHINGLES

make the most durable and ornamental roof in the world. The only shingle manufactured from metal that makes an absolutely tight roof. Send for full descriptive circular and new prices to

ANGLO-AMERICAN ROOFING CO.,
22 Cliff Street **NEW YORK.**

FIGURE 1.8
Ads such as this one from the year 1882 have appeared in popular journals for as long as the journals have existed. Many modern ads refer to Web pages and/or offer free demonstration videos or CDs. (*From* **Builder and Wood-Worker,** *Vol. XVIII, Chas. D. Lakey, New York, 1882*)

CONSTRUCTION SYSTEMS

For the past 150 years, most houses in North America have been built using wood light frame construction, which is the most flexible of all building systems. There is scarcely a shape it cannot be used to construct, from a plain rectilinear box to cylindrical towers to complex roofs with dormers of every description (Figure 1.9). Since it first came into use, wood light framing has served to construct buildings in styles ranging from reinterpretations of nearly all the historical fashions to uncompromising expressions of every architectural philosophy of the last hundred years. It has assimilated without difficulty a succession of technical improvements in building: gas lighting, electricity, indoor plumbing, central heating, air conditioning, thermal insulation, prefabricated components, and electronic communications.

Wood light frame buildings are easily and swiftly constructed with a minimal investment in tools. Many observers of the building industry have criticized the supposed inefficiency of light frame construction, which is carried out largely by hand methods on the building site, yet it has successfully fought off competition from industrialized building systems of every sort, partly by incorporating their best features, to remain the least expensive form of durable construction.

However, wood light frame construction has its deficiencies: If ignited, it burns rapidly; if exposed to dampness, it decays. It expands and contracts by significant amounts in response to changes in humidity, sometimes causing chronic difficulties such as cracking plaster, sticking doors, and buckling floors. The framing itself is so unattractive to the eye that it is seldom left exposed in a

FIGURE 1.9
The Carson House, built in 1885 in the Queen Anne style for a lumber baron in Eureka, California, qualifies as one of the most elaborate residential forms ever built and stands as a testament to the flexibility of the wood light frame. (*Courtesy of University of Oregon Visual Resources Collection. Original photography by Michael Shellenbarger*)

building. These problems can be controlled, however, by clever design and careful workmanship, and there is no arguing with success: Frames made by the monotonous repetition of wooden joists, studs, and rafters are likely to remain the number one system of building in North America for a long time to come. The wood light frame system is described in detail in Chapters 7 to 19.

If 90 percent of all site-built residential construction consists of light wood frame, the remaining 10 percent is divided among several other residential construction systems. In some regions of the South, loadbearing masonry is the dominant system. Throughout the continent, other systems such as timber frame, light-gauge steel frame, insulating concrete forms, insulated masonry, and panelized construction are used in significant numbers of dwellings. These systems are important for their roles in developing new materials and building methods and for inducing innovation in the dominant wood light frame system. These less common systems are discussed in Chapters 20 to 24.

The *manufactured housing* industry factory-builds entire houses as finished boxes, often complete with furnishings, and trucks them to prepared foundations where they are set in place and made ready for occupancy in a matter of hours (Figure 1.10). If the house is 14 feet (4.27 m) or less in width, is constructed on a rubber-tired frame, and is completely finished in the factory, it is known as a *mobile home*. If wider than this, or more than one story high, it is built in two or more completed sections that are joined at the site, and is known as a *sectional home* or *modular home*. Mobile homes are sold at a fraction of the price of conventionally constructed houses. This is due in part to the economies of factory production and mass marketing, and in part to the use of components that are lighter and less costly and, therefore, of substantially shorter life expectancy. At prices that more closely approach the cost of conventional on-site construction, however, many companies manufacture modular housing to the same standards as conventional construction. Manufactured housing is an important component of the housing industry, but highly specialized. Because the units are made in a factory rather than at the site, the designs are strongly driven by considerations of production and transport, and their construction process is significantly different from site-built housing. For these reasons, manufactured housing can be given only limited coverage in this book.

FIGURE 1.10
This manufactured house was trucked to the site in sections, which were joined as they were placed on the site-built concrete foundation. The garage will be built at the site because garages, having no framed floors, are difficult to transport and are economical to frame on site. Manufactured houses account for approximately 25 percent of all new housing in the United States. (*Photo by Rob Thallon*)

FIGURE 1.11
Manufactured housing is typically single-story construction, but some companies experiment with two-story models. This house was set on the foundation within a matter of hours, but it took weeks for the site crew to add the porch, finish the trim, connect the utilities, and complete the painting. (*Courtesy of Fischer SIPS, Louisville, Kentucky*)

(*a*)

(*b*)

(*c*)

TYPES OF RESIDENTIAL DEVELOPMENT

At the present time, houses are built in the United States at the rate of about 1.6 million new units per year. Of this total, 75 percent are single-family detached (freestanding) dwellings, 21 percent are units within large structures (5 or more units), and the balance are in buildings with 2 to 4 units (Figure 1.12). Most new housing is built at the site, but about

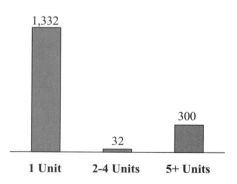

FIGURE 1.12
1999 U.S. housing production measured in thousands of units, broken down by the number of separate dwelling units per building. (*Source: U.S. Census Bureau, Housing Completions, December 1999*)

370,000 manufactured houses (representing about 23 percent of the total) are built in factories and shipped to the site each year (Figure 1.13). Remodeling of existing houses is more difficult to quantify because it includes projects that range in scope from a new window to an addition larger than the original house. However, it is clear that remodeling is a substantial component of the residential construction industry. The U.S. Census Bureau estimates that residential remodeling in the United States in 1999 accounted for $99 billion in economic activity, about 28 percent of the value of all new residential construction (Figure 1.14).

U.S. Housing Starts and Shipments (in thousands)

	1992	1993	1994	1995	1996	1997	1998	1999
Total	**1,241**	**1,380**	**1,502**	**1,416**	**1,524**	**1,487**	**1,644**	**1,681**
New site-built single-family housing starts	1,030	1,126	1,198	1,076	1,161	1,134	1,271	1,332
Percentage of total	83	82	80	76	76	76	77	79
Manufactured homes shipped	211	254	304	340	363	353	373	349
Percentage of total	17	18	20	24	24	24	23	21

FIGURE 1.13
New site-built single-family housing starts vs. manufactured-home shipments. (*Source: Manufactured Housing Institute*)

U.S. Annual Value of U.S. Residential Construction

	1995 Billions of Dollars	1995 Percentage of Total	1999 Billions of Dollars	1999 Percentage of Total
Total residential construction	247	100	349	100
New residential units	171	69.3	250	71.5
Residential improvements	76	30.7	99	28.4

FIGURE 1.14
Annual value of construction put in place: new housing units vs. improvements. An additional $43 billion was spent in 1999 on repairs to existing residential buildings. Residential construction accounted for 45.6 percent of all construction in the United States in 1999. (*Source: U.S. Census Bureau, Document C30, Table 1: Annual Value of Construction Put in Place in the U.S.: 1995–1999*)

Privately Owned Housing Units

	1990 No. of Units (1,000s)	1990 Percentage of Total	1995 No. of Units (1,000s)	1995 Percentage of Total	1999 No. of Units (1,000s)	1999 Percentage of Total
All units	1,308	100	1,313	100	1,636	100
1 unit total	966	74	1,066	81	1,307	80
For sale	594	46	682	52	913	6
Contractor built for owner	199	15	204	16	208	13
Owner built	147	11	146	11	144	9
For rent	26	2	33	2	43	2
2 units or more total	342	26	247	19	329	20
For sale	76	6	51	4	54	3
For rent	266	20	196	15	275	17

FIGURE 1.15
Statistics for privately owned housing units completed in three separate years. (*Source: U.S. Census Bureau: Characteristics of New Housing, 1999*)

In Canada in the year 2000, more than 151,000 units were built, 61 percent of which were single-family detached houses.

Houses are built for a number of different reasons, depending principally on who pays for their construction (Figure 1.15):

• Many are built for personal use and are financed from start to finish by the future owner. Houses in this category are virtually always single-family detached dwellings, although a few are condominium units within larger buildings.

• Houses may also be built for profit, either to be sold or to be rented by entrepreneurial housing developers. This activity is called *speculative building* and accounts for more than half of all housing units built each year. Speculative houses built for sale are most likely to be detached dwellings, whereas speculative rental housing is usually consolidated into large buildings.

• Finally, low-cost houses intended for low-income families are built for the public good by government or nonprofit agencies. Like houses built for profit, *affordable housing* can be detached or part of a large structure, for sale or for rent.

The most popular form of residence in North America has always been the single-family detached house. In 1999, 1.25 million detached units were built, representing 75 percent of all site-built residential construction activity (Figure 1.12). Symbolic of independence, family life, and a connection with nature, the single-family detached house has evolved through numerous styles, including Colonial, Federal, Victorian, Bungalow, and Ranch (Figure 1.16). In a survey of prospective buyers conducted by *Professional Builder* magazine, all respondents indicated a strong preference for the detached unit when offered a choice between this and an attached house such as a townhouse

GEORGIAN—New England 1715–1780

QUEEN ANNE 1880–1910

CRAFTSMAN 1900–1930

MODERNE 1920–1940

FIGURE 1.16
The single-family detached house has always been the most popular type of residence. It has been constructed in a variety of styles throughout its evolution in North America. (*Reproduced with permission from John Milnes Baker, AIA,* **American House Styles: A Concise Guide,** *New York, W. W. Norton, 1994*)

Type of Home	Buyer Type			
	First Time	Move-up	Empty Nester	Retiree
Detached production house	50.0	46.1	48.7	42.4
One-of-a-kind custom house	22.5	42.1	22.0	24.6
Specialty home	6.6	4.8	1.1	8.8
Modular, panelized, package home	5.3	2.6	16.2	8.6
Townhouse	10.1	0.6	3.7	5.2
Mobile home	1.6	1.8	1.2	2.8
Condominium in low-rise building	1.4	—	5.9	3.0
Duplex	2.1	0.7	1.2	4.0
Condominium in high-rise building	0.5	1.3	—	0.7

FIGURE 1.17
In a consumer survey, several types of potential buyers were asked, "Which one type of home described is the type you would attempt to purchase if you were buying a home at the present time?" Results of surveys such as this influence the construction and design of all types of housing. (*Source: Professional Builder Consumer Survey on Housing, 1998, Cahners Business Information*)

FIGURE 1.18
Tracts of identical or similar houses built at the edges of existing development are largely responsible for the sprawl of the suburbs. Production line repetition and a dearth of landscape features contribute most significantly to the lack of character in these instant neighborhoods.
(*Photo © Bill Owens*)

or condominium (Figure 1.17). This preference for detached housing has been largely responsible for the proliferation of suburbs since the end of World War II.

The largest number of single-family houses are built in tracts of many units where developer/builders repeat house plans in order to reduce construction costs by means of production line repetition (Figure 1.18). In tracts, a considerable amount of time and money must be invested to obtain governmental approvals and install infrastructure such as roads, storm drainage, sewers, water lines, electricity, and telecommunications before any houses can be built. This large initial investment limits the development of large and medium sized tracts to experienced developer/builders to whom financial institutions will loan the large sums of money required for such endeavors.

The design of housing tracts must conform to zoning ordinances that stipulate minimum street widths, off-street parking requirements, minimum lot sizes, minimum distances of buildings from lot lines, maximum building heights, and many other constraints. These regulations are designed to avoid infringement of any homeowner's rights and property values by the activities of other homeowners. A tract that is developed to comply completely with these regulations is called a subdivision, whether the houses are all the same or are unique (Figure 1.19). Most municipalities also have laws that allow a residential tract to be developed as a planned unit development (PUD). In a PUD, the houses are designed simultaneously with a coordinated site plan to assure privacy, individuality, visual harmony, and a pleasant neighborhood environment. A PUD generally achieves the qualities that are sought by zoning ordinances, but often does so without literally complying with them. For example, in an area where zoning ordinances call for half-acre lots, a PUD might achieve this overall density by clustering houses in tight groupings, each with a small private yard or garden, and providing generous communal open spaces between the clusters. The concept of the planned unit development is that, in recognition of the quality of design that can be achieved when the entire project is designed by a coordinated team of design professionals, the literal enforcement of zoning ordinances may be relaxed (Figure 1.20).

Many new single-family detached houses are built on individual lots, independently of the construction of other units. These houses may be speculative projects offered for sale or may be built for or by the owner of the lot. Speculative houses on individual lots tend to be constructed by small-scale developers or by developers who prefer the variety of experience this type of project affords. The construction of a new residence on

FIGURE 1.19

A subdivision has streets and building lots designed to accommodate houses that vary considerably in design from one lot to the next. For the sake of efficiency, builders usually repeat several house designs again and again in each subdivision, but each design can be built in both its original configuration and its mirror image, and small cosmetic features can be introduced for variety. (*Photo by Rob Thallon*)

FIGURE 1.20

The street in this PUD is narrower than normal and the garages are located close to the street in order to create an intimate feeling for the residents. Adjustments like this to zoning regulations are possible in PUDs because the entire tract is designed at the same time by a professional design team. (*Photo by Rob Thallon*)

individual lots by owners for their own occupancy is also a very common occurrence, accounting for as much as 20 percent of all new house construction (Figure 1.21). In this case, owners either hire a general contractor to manage the construction or act as the general contractor themselves. When acting as general contractor, owners may do some or most of the work themselves with the remaining work performed under their direction by subcontractors.

Large multiunit structures less than four stories in height are usually built with the same materials and methods as single family residences, but invariably require a larger, more highly capitalized contractor to do the work. Multiunit structures are almost always built either for profit or for the public good and tend to be sited as densely as possible. In 1999, new residential structures with 5 or more units contained an average of 15 units. Because cities, counties, and other jurisdictions have regulations about how many automobile parking spaces must be provided for each unit, the number of units that can be located on a given site often depends entirely on the number of parking spaces that the site can accommodate. Without building a parking garage, the greatest density can be achieved by covering as much of the street level of a site as possible with parking and placing the living units on the second floor and above (Figure 1.22). When units are located at ground level, private outdoor yards can be obtained at the expense of density.

Multiunit, multistory structures are designed and built essentially the same as single-family detached houses but have special problems:

1. The site planning process is more involved, requiring more neighborhood meetings, necessitating more permits and approvals, and taking more time than a single residence. Parking and the movement of vehicles onto and through the site strongly influence the building design.

FIGURE 1.21
Custom-designed houses, built by or for owners on their own lot, account for about one-fifth of all residential construction. Custom-designed houses may also be built on speculation. In either case, the builder/developers of these unique houses tend to be small-volume entrepreneurs who value the challenge of variety more than that of efficient production. (*Photo by Rob Thallon*)

FIGURE 1.22
For rental housing, the design goal is most often to maximize density within the guidelines allowed by governmental regulations. The more units on the site, the more income for the owner. (*Photo by Rob Thallon*)

2. Code requirements are more stringent with regard to accessibility by emergency vehicles and emergency egress by occupants.

3. Building codes require that individual units be separated by walls or floor/ceiling assemblies that are resistant to the passage of fire. Most residences in multiunit buildings are acoustically insulated as well.

4. A passenger elevator is required in most cases where buildings are three or more stories tall.

5. Where cars are parked below living units, a garage made of concrete or concrete masonry with a concrete slab ceiling is required as a way of protecting the dwellings from vehicle fires.

Zoning Ordinances, Building Codes, and Other Legal Constraints

Zoning Ordinances

The legal restrictions on buildings begin with local zoning ordinances. The most basic purpose of zoning ordinances is to designate areas of land in a town or county for particular kinds of uses. In a town or city, some areas are designated for commercial use, some for civic, some for schools, others for residential use, and so forth. In rural areas, zones are set aside for agriculture, for commerce, for residential development, and for other uses. Residential zones within urban boundaries are usually divided into low-, medium-, and high-density areas. The device of zoning prevents such things as automobile repair shops or slaughterhouses from being located in residential neighborhoods and creates a mix of residential densities. In residential zones, ordinances usually define minimum lot sizes, off-street parking requirements, maximum building heights, and *setback requirements,* which dictate how far buildings must be from each of the property lines. Zoning ordinances often contain other provisions such as tree cutting restrictions, erosion control measures, fencing restrictions, solar setback requirements, and sidewalk specifications. Copies of the zoning ordinance for a municipality are available for purchase or reference at the office of the building inspector or the planning department, or they may be consulted at public libraries.

Building Codes

Building codes were originally developed to protect the occupants of a building from careless or unscrupulous builders. In ancient Babylon, building codes held a builder accountable for his work to the extent that he was slain if a house he had built should fall and kill the householder. Engineers of ancient Roman arches were required to stand beneath each arch as the formwork was removed. Modern building codes still make builders and designers responsible for structural safety, but place more emphasis on checking plans and verifying workmanship during construction and less on penalizing the builder or designer for failures.

Codes that regulate the design and construction of residences came into existence in the United States in response to disastrous fires and unhealthy and unsafe living conditions. One of the first codes was written in the 1630s when the governor of the colony of Massachusetts issued a proclamation forbidding the construction of wooden chimneys or the use of thatch for roofing. In 1867, the New York Tenement House Act called for fire escapes, windows for ventilation, running water within each building, and handrails on all stairways. In the 1920s, fire insurance companies were successful in setting fire safety standards for all major construction throughout the United States, and it wasn't long before other interests followed suit. The first national code intended for adoption by local and state governments, the Uniform Building Code, was first published in 1927.

Since that time, there has been a proliferation of codes that prescribe minimum standards for building design and construction as well as specialized codes for plumbing, electrical wiring, fire safety, mechanical equipment, and energy efficiency. These codes are called *model codes* because they are prepared by national organizations of local building code officials. Their purpose is to provide models for adoption by local jurisdictions such as states, counties, and cities. In recent decades, there have been several competing model building codes. In the western United States and parts of the Midwest, most codes are modeled after the Uniform Building Code (UBC). In the East and other areas of the Midwest, the BOCA National Building Code (BOCA) is the model. The Standard Building Code (SBC) has been adopted by many southern and southeastern states. The specialized codes such as plumbing and electrical codes have followed a similar pattern.

In recent years, the major building code organizations have published residential versions of their model codes. These specialized editions of the model codes were intended only for single-family dwellings and duplexes (two families in one building), while larger residential structures were subject to the standards in the complete codes. The residential codes were replaced in the 1990s with a single, national model code created by the *Council of American Building Officials (CABO),* an organization with representation from all major regional code associations. The CABO One- and Two-Family Dwelling Code, published in 1992 and 1995, combines into one document relevant standards from the model building codes as well as standards from the national model electrical and plumbing codes.

Beginning in the year 2000, local code jurisdictions throughout the United States began adopting the new *International Residential Code (IRC),* which applies not only to one- and two-family dwellings, but also to *townhouses,* which are multiple single-family dwellings with separate means of *egress* (emergency escape). The IRC is drawn largely from its predecessor, the CABO Code, and includes standards for electrical wiring, plumbing, and energy conservation. Large multifamily buildings with common means of emergency egress are increasingly governed by the International Building Code (IBC), the first fully unified model code in U.S. history. Canada publishes its own model code, the National Building Code of Canada.

While these model codes for residential structures differ in detail, they are similar in approach and intent. Emergency exit requirements generally include minimum size and maximum sill height dimensions for bedroom windows to allow occupants to escape and firefighters to enter. An automatic fire alarm system, including smoke detectors, is almost universally required in residential buildings to awaken the occupants and get them moving toward the exits before the building becomes fully engulfed in flames. Automatic fire sprinkler systems are also sometimes required in residential construction, and these requirements are likely to become more widespread as simpler, less expensive sprinkler systems are developed. A typical code also establishes standards for natural light; ventilation; structural design; floor, wall, ceiling, and roof construction; chimney construction; plumbing; electrical wiring; and energy efficiency. Codes will be discussed further in each chapter as they apply.

Other Legal Constraints

The U.S. Occupational Safety and Health Act (OSHA) sets safety standards for construction operations. Fire insurance organizations (Underwriters Laboratories, for example) exert a major influence on construction standards through their testing and certification programs and through

their rate structures for building insurance coverage, which offer strong financial incentives for more fire-resistant construction.

In addition, an increasing number of states have placed legal limitations on the quantities of volatile organic compounds in paints and construction adhesives that building products can release into the atmosphere. Most states and localities also have conservation laws that protect wetlands and other environmentally sensitive areas from encroachment by buildings.

Multiple-unit residential buildings must adhere to legal restrictions that go beyond the building codes that apply to single-family dwellings. Access standards regulate the design of entrances, stairs, doorways, elevators, and toilet facilities for a small percentage of dwellings in multifamily residential buildings to assure that they are accessible by physically handicapped members of the population. The Americans with Disabilities Act (ADA) makes accessibility to buildings by disabled persons a civil right of all Americans.

BUILDING COSTS AND FINANCING

Every building project has a budget, which plays a crucial role in its design and construction. Costs for a typical single-family residence include the initial cost of the land, the costs of site improvements (such as a driveway, utilities, and landscaping), materials and labor to construct the building, building plans or design fees, building permit(s), and the cost of financing. There are rules of thumb such as cost-per-square-foot figures by which one may estimate these costs for the purpose of determining project feasibility.

Once a project begins, there is a series of conversations among the owner, the designer, and the contractor in order to establish precise costs and allocate resources appropriately. In some cases, the owner buys generic plans and deals directly with contractors to establish prices and maintain quality. A construction cost can be established either by negotiating with a single contractor or by means of a competitive bidding

Item	Area	Unit Cost	Cost
Main floor	1853	$120	$222,360
Second floor	1294	$120	$155,280
Basement	631	$ 30	$ 18,930
Porches (first floor)	270	$ 50	$ 13,500
Porch (second floor)	112	$ 60	$ 6,720
Garage	528	$ 60	$ 31,680
Connector	300	$ 50	$ 15,000
Buildings total			$463,470
Architectural fees			$ 42,000
Engineering			$ 2,000
Permits			$ 3,000
Systems development fees			$ 3,000
Site			
Septic			$ 3,000
Road			$ 5,000
Landscape			$ 20,000
Total			$544,470

FIGURE 1.23

A cost estimate done by an architect for a client in the year 1999. The cost of the house in this type of estimate is based on cost per square foot and is done early in the design process. A contractor will later determine the actual cost of the house based on bids from subcontractors.

process among several contractors. In other cases, the owner employs an architect or designer who makes early estimates of construction costs based on rules of thumb and later consults with contractors during the negotiating or bidding process to establish more accurate cost projections (Figure 1.23). In all cases, it is ultimately the owner who makes the decisions that affect the cost and the design of a project.

There are a number of variables that must be considered when determining the cost of a residential building project. The most important of these are the overall size, the complexity of the design, and the quality of materials. Larger houses are generally more expensive than smaller houses, but the cost per square foot tends to be lower for larger houses because of the economy of scale. Design complexity can have a significant effect on construction cost. Keeping the overall building form simple and respecting the modular sizes of standard materials are key principles in projects where affordability is a primary objective. Material selection can also have a huge impact on budget because material costs vary so considerably. The cost of a simple residence made with the most affordable materials can easily be doubled if the same residence is built with more luxurious materials.

Owners must often grapple with the difference between *first cost* and long-term value. The first cost is the initial cost of construction, and the tendency of owners is to keep this cost as low as possible. Unfortunately, a low first cost frequently leads to more expenditure in the long run. More insulation in the walls of a house, for example, can lead to lower heating and cooling bills that can recoup the initial extra investment in a short time (Figure 1.24). Higher quality materials can require less maintenance and add to the overall value of a house when it is sold. Numerous examples such as this have led to the practice of *life-cycle cost analysis,* which is a long-term analysis of construction, operation, and maintenance costs, and is often employed in large-scale projects.

Large residential subdivision projects and multifamily structures follow the same cost analysis procedures as a single-family house except that the cost of site development is a much larger percentage of the cost of the overall project, so more evaluation is required in this area. The amount of building construction is also more extensive in large projects, so the stakes are higher and much more attention is given to repetition of house or apartment designs in order to gain the advantages of the assembly line.

Financing

Building a house is such an expensive endeavor that almost no one can afford to pay for it out of pocket. Banks and savings and loan institutions are in the business of lending money for this significant investment in return for interest, which is a rental cost for the money expressed as a percentage of the amount loaned. In order to make a loan on the construction of a house, a bank needs to be convinced that the owner has the financial capacity to repay the loan over time and that the house can be resold to pay off the loan if the owner should fail to make the loan payments. The bank assesses the owner's ability to make loan payments on the basis of investigation of the owner's credit record and the evaluation of recent tax returns. Owners often will prequalify for a loan up to a certain amount in order to establish a project budget before the house is designed. The projected *resale value* of a house has a tremendous impact on how much financing a bank will provide or if financing will be provided at all. In order to establish resale value, an official of the bank appraises the building based on plans and material specifications and adds the value of the lot to arrive at a total *appraised value* of the developed property as proposed. To minimize their risk, banks usually require that the owner invests at least 20 percent of the appraised value of a project.

The traditional mechanism used to finance the construction of a house is the combination of a *construction loan* while the house is being built and a *permanent mortgage* for the period thereafter. The construction loan is generally limited to a period of 9 months or less and has an interest rate that is higher than the permanent mortgage. As construction proceeds, progress is verified by an agent of the bank, and funds from the construction loan are disbursed to the contractor. The permanent mortgage goes into effect when con-

Monthly Expenses

	Mortgage	Fuel	Total
Base case home	$1057	$124	$1181
Passive home	$1082	$ 74	$1156
Total	$ +25	$ -50	$ -25

FIGURE 1.24
A first-cost vs. life-cycle analysis comparing two 2124-square-foot houses in New York State, one with an attached sunspace and additional insulation. The first cost of the sunspace and insulation was $880, which translated to an additional $25 monthly mortgage payment. With fuel savings, however, the total monthly expenditures for the passive home were $25 less than the base home, and the initial additional outlay was recovered in less than 3 years (and has additional amenity). (*Source:*** The Passive Solar Design and Construction Handbook,** *Steven Winter Associates, Michael J. Crosbie, Editor, John Wiley & Sons, New York, 1998***)**

struction is complete and runs for a period that may range from 15 to 30 years with payments due to the bank every month. These two loans can be consolidated into a single *all-in-one loan* that is negotiated before construction begins at a single prevailing interest rate and avoids the duplication of closing costs and other costs associated with dual-loan agreements.

GREEN BUILDING: THE ENVIRONMENTAL MOVEMENT

As expanding populations and diminishing resources begin to produce global-scale environmental impacts, there are many who see the need for a more environmentally aware approach to the way we design and build our houses. The construction of well over a million new houses per year represents the consumption of unimaginable amounts of raw materials, the emission of uncountable tons of toxic chemicals, and huge long-term energy requirements. Even a relatively small shift in the way these houses are designed and built can have a significant impact on resources and on the lives of those directly affected by their construction (Figure 1.25).

Those concerned with the environmental consequences of new construction have organized into various groups that promote the cause of what they commonly call "green building." The green building movement generally has a comprehensive approach to improving the situation that includes three strategies:

1. Improve the energy efficiency of buildings by designing and building them to use less energy to heat and cool.

2. Improve the resource efficiency of buildings by utilizing fewer and better materials in their construction.

3. Lower the toxicity of new buildings to minimize health risks to workers and occupants.

Many of these strategies increase the cost of constructing a new house, but proponents of environmental responsibility see no alternative in the long run and point out that consumers are increasingly aware of the issues and expect some response by those who design and build the houses they buy. As the volume of green building activity increases, its cost premiums will decrease.

Energy Efficiency

Environmentally responsible designers can have a large impact on the long-term energy consumption of a house. Reduction of energy consumption starts by making the house no larger than needed and by increasing thermal insulation and other weatherizing elements beyond code-mandated minimums. The shape of the building and its orientation can be modified to provide possibilities for passive solar heating and passive cooling (Figure 1.26). In addition, the selection of an efficient mechanical system can make a big difference in the amount of energy consumed for heating and cooling. Low-volume plumbing fixtures conserve water and lower the volume of sewage that needs to be treated.

Resource Efficiency

The first step toward resource efficiency is to make the house as small as is practical, and beyond this, a number of smaller steps can also be taken. The use of lumber that has been harvested from a certified *sustainable yield forest* protects timber as a resource and lessens the pressure to harvest from natural forests with volatile ecosystems. The use of materials with minimal *embodied energy*, which is a measure of all the energy it takes to produce and transport a building product, contributes to overall energy conservation. The use of materials with *recycled content* such as carpet made of recycled beverage containers and building panels made of recycled newspapers reduces the

demand for natural resources and diminishes the volume of landfill. The detailing of the house to eliminate structural framing where it is not required and designing to take advantage of the standard dimensions of building materials can save material, reduce labor costs, and minimize construction waste. Finally, the use of materials that can be recycled easily when the house is ultimately remodeled or demolished will save energy and materials in the future.

Low Toxicity

Environmentally conscious builders use materials that do not emit dangerous chemicals over their lifetime. Carpeting is often a big offender in this regard because of the adhesives used to bond some types of fibers. Certain panel products used for subflooring and cabinet cores have been notorious for emitting harmful formaldehyde gas, but formaldehyde-free panels are now manufactured and need only be specified. Paint used to have a high lead content, but this situation has been corrected through the efforts of concerned citizens. However, many paints still contain high levels of *volatile organic chemicals (VOCs)*, which are extremely unhealthy for painters, pose health risks for future occupants, and pollute the atmosphere. Paint manufacturers have been developing more and more water-based coatings that are equally as durable as solvent-based coatings but have much lower VOC levels. The issue of toxic building materials was first raised by people with extremely high sensitivity who were falling ill due to the various emitted gases. The health dangers from these materials were communicated to the general public by environmental advocates and alert journalists, and manufacturers and builders have responded, but the response has been slow and much work remains for the green building movement to accomplish.

Material	Weight (lbs)
Solid sawn wood	1600
Engineered wood	1400
Drywall	2000
Cardboard	600
Metals	150
Vinyl (PVC)	150
Masonry	1000
Containers	50
Other	1050
Total	8000

FIGURE 1.25
Typical construction waste for a new 2000-square-foot house. Siding material was assumed to be vinyl on three sides and brick on the street side. (*Source: NAHB, Residential Construction Waste: From Disposal to Management*)

FIGURE 1.26
This passive solar house, designed by architect Michael Utsey and built in 1976, employs south-facing glass, interior thermal mass, and extra insulation to reduce heating and cooling costs. Still popular as a strategy, passive solar heating and cooling has been joined in recent years by the green building movement, whose goals relate more generally to the environment and to human health. (*Photo © Jane Lidz*)

BUILDING A HOUSE: THE TYPICAL PROCESS

The construction of a house involves a series of steps that vary somewhat from project to project, but that typically follow a similar path (Figure 1.27). This path is outlined in the paragraphs that follow and described in detail for a typical wood light frame house in Chapters 7 to 19. For large subdivision or multiple-family projects, the first steps are typically considerably more complicated, but once framing begins, the only significant difference between a large project and a single-family project is one of scale.

Building Permit

After the construction documents are complete, the first step in the construction process is the application for a building permit. Plans for the building are submitted to the city or county building department where they are reviewed for compliance with local zoning ordinances and building codes. When the review is complete, any necessary corrections have been made, and all fees have been paid, a permit for construction is issued. As construction proceeds, inspections are requested by the contractor at designated stages of construction, in response to which the building department sends an inspector to verify the quality of the work and its correspondence with the approved plans. In rural locations where a septic system is required, approval of the design of the system must generally be obtained before a building permit can be issued.

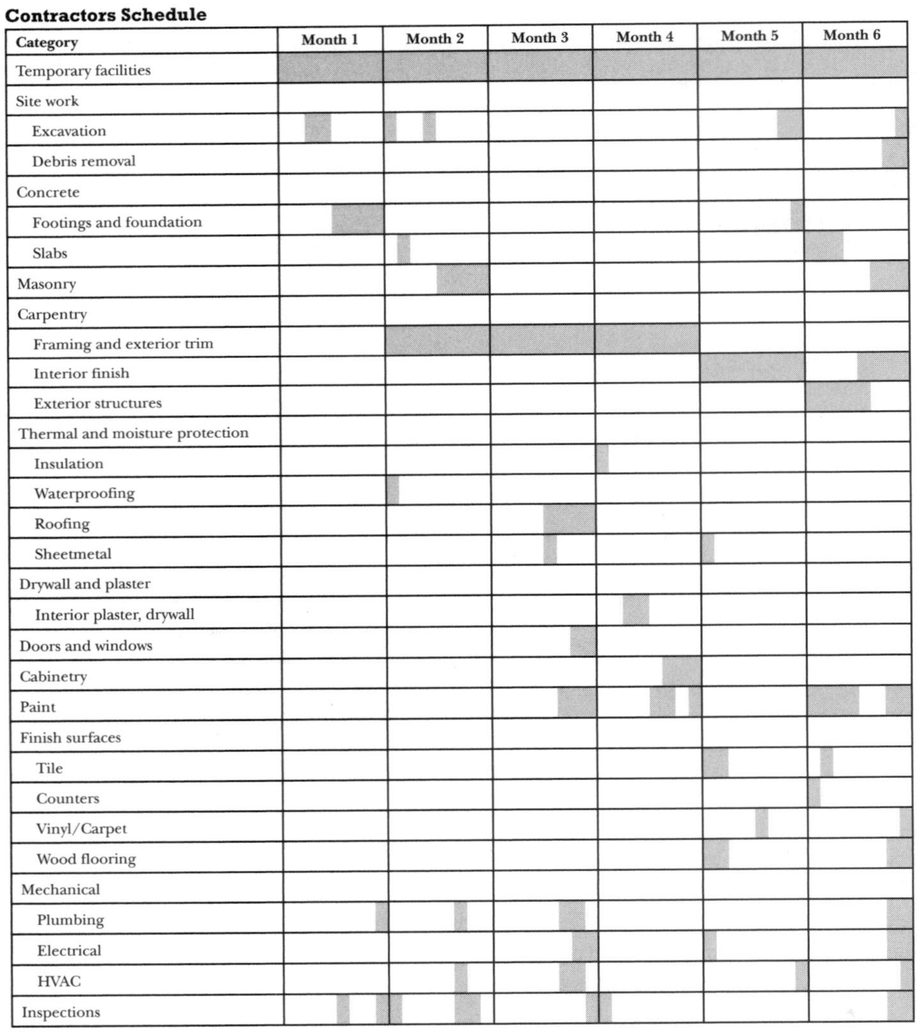

Building Site Preparation

Preparation of the building site for construction is the first physical step in the building process. The area of the site to be occupied by the house is cleared, the building is located and its corners staked, and a hole is excavated for the foundation. In addition, utilities such as electrical power, water, gas, and phone lines are extended to the building site. Sewage and storm drain lines are installed.

The Foundation

Construction of a foundation to transmit the weight of the building into the earth is the next step in the construction process. Foundations can range from a concrete slab poured on top of the ground to a basement excavated deeply into the site (Figure 1.28). Foundations for large multi-family buildings often incorporate parking garages and thus involve much more complicated construction.

Structural Framing

When the foundation is in place, the framing can proceed. With a slab foundation that acts as a floor, framing begins with the walls. When there is no slab, a wooden floor structure is constructed on the foundation. A framing crew builds the walls, floors, and roof with lumber and structural wooden panels (Figure 1.29).

Contractors Schedule

Category	Month 1	Month 2	Month 3	Month 4	Month 5	Month 6
Temporary facilities	▓	▓	▓	▓	▓	▓
Site work						
Excavation	▓	▓			▓	▓
Debris removal						▓
Concrete						
Footings and foundation	▓				▓	▓
Slabs		▓				▓
Masonry		▓				
Carpentry						
Framing and exterior trim		▓	▓	▓		
Interior finish					▓	▓
Exterior structures					▓	
Thermal and moisture protection						
Insulation		▓				
Waterproofing		▓				
Roofing			▓			
Sheetmetal			▓		▓	
Drywall and plaster						
Interior plaster, drywall				▓		
Doors and windows			▓			
Cabinetry				▓		
Paint			▓	▓	▓	▓
Finish surfaces						
Tile					▓	
Counters				▓		
Vinyl/Carpet					▓	
Wood flooring					▓	
Mechanical						
Plumbing		▓	▓	▓	▓	
Electrical		▓			▓	
HVAC			▓		▓	
Inspections	▓	▓	▓	▓	▓	

FIGURE 1.27
A typical schedule of the work for an average-sized house. Notice how many different trades are at the site during the last week of construction. (*Courtesy of Treeborn Carpentry, Eugene, Oregon*)

FIGURE 1.28
A typical perimeter foundation around a crawlspace. Underfloor plumbing, heating/cooling ducts, and first-floor framing have been completed, and soil has been backfilled around the foundation. Next the floor deck will be laid, and wall framing will begin. (*Photo by Rob Thallon*)

FIGURE 1.29
The house in the foreground stands silhouetted against a framed house. The roof of the framed house has been covered with building paper and the windows have been installed to protect the inside so that interior work can proceed protected from the weather. (*Photo by Rob Thallon*)

FIGURE 1.30
Large-scale residential projects often employ the same wood light frame technology as single-family residences. Separation between the units to inhibit the rapid spread of fire is achieved with gypsum board panels. (*Photo by Rob Thallon*)

Roofing and Siding

Exterior finish materials are applied to the completed framing to make the building weathertight and durable. Roofing is applied to the roof, and provisions are made to gather rainwater from the roof and channel it away from the building. Windows and doors are installed, and the walls are covered with siding. Any required exterior painting is done. With roofing and siding in place, the building is "tight to the weather," and the overall project is about 50 percent complete (Figure 1.31).

Utilities and Insulation

While the exterior of the house is being finished, the interior utility systems are installed. First the plumbing, then the heating and cooling ductwork, and finally the electrical wiring are installed within the cavities between the studs, joists, and other framing members (Figure 1.32). When these systems are all in

FIGURE 1.31
The roofing, windows, doors, siding, gutters, and downspouts complete the weather envelope at the exterior of the structural frame. These workers are applying the last of the siding. (*Photo by Rob Thallon*)

FIGURE 1.32
The plumbing, heating/cooling system, and electrical wiring are installed inside the house before the walls and ceiling are insulated and covered with a finish surface. (*Photo by Greg Thomson*)

place and fastened to the framing, thermal insulation is placed into the cavities between the studs of exterior walls and ceiling joists to form a continuous thermal envelope around the perimeter of the house.

Interior and Exterior Finish

Upon completion of the thermal insulation, the studs and other framing are covered with interior finish materials, which are immediately painted. Next the flooring and cabinets are installed and trim is fastened around floors, doors, and windows to cover the gaps between the various materials (Figure 1.33). To complete the interior of the house, the finish hardware, plumbing fixtures, and electrical trim are installed, and all remaining unfinished surfaces receive a protective coat

FIGURE 1.33
Finishing the interior of a house involves more than half of the building trades, some of whom are returning to the job for the second or third time. Drywall hangers and tapers, floor installers, tilesetters, cabinetmakers, finish carpenters, painters, plumbers, electricians, and others are required to complete the work. (*Photo by Rob Thallon*)

FIGURE 1.34
The W. G. Low House, built in Bristol, Rhode Island, in 1887 to the design of architects McKim, Mead, and White, illustrates both the essential simplicity of wood light framing and the complexity of which it is capable. (*Photo by Wayne Andrews*)

of paint, stain, or varnish. Simultaneously with the finishing of the interior of the house, the site work is completed with the installation of paving, decks and terraces, irrigation system, fencing, and planting.

After the house is occupied, adjustments are often needed, such as balancing the heating/cooling system according to the occupants' preference, adjusting the operation of doors and windows, touching up the paint, and explaining the operating and care instructions for equipment. In most regions, the contractor must warranty the quality of the construction for 1 year after initial occupation of the house.

Selected References

1. Gideon, Sigfried, and Bobenhausen, William. *American Building: The Environmental Forces That Shape It.* New York: Oxford University Press, 1999.

Connects the environment, the act of building, and human nature. The classic discussion of the evolution of American architecture at all scales.

2. Davis, Howard. *The Culture of Building.* New York: Oxford University Press, 1999.

Elegantly written and beautifully illustrated, this book describes a global building culture that is evaluated for its ability to produce a vital built environment. North American building culture is explored in depth and can be more profoundly understood in the context of examples from other parts of the world.

3. Wright, Gwendolyn. *Building the Dream: A Social History of Housing in America.* Cambridge, MA: MIT Press, 1983.

A scholarly classic describing the reasons behind the form of most American housing. Not so much about building as it is about social history.

4. International Code Council, Inc. *2000 International Residential Code for One- and Two-Family Dwellings.*

This is the model code for all aspects of residential construction, including detached houses, duplexes, and townhouses. Updated regularly.

5. Periodicals are an excellent source for learning about current thinking in the residential building culture. *Fine Homebuilding* and *The Journal of Light Construction* focus on construction techniques and also have articles about tools, materials, and design. *Builder* and *Professional Builder* focus on marketing and business and also have articles about materials and design.

6. *Environmental Building News,* a periodical published monthly for the past 15 years, is the most respected voice of the environmental building movement. Articles range from site selection to evaluation of the environmental impact of roofing materials.

Key Terms and Concepts

wood light frame
contractor
subcontractor
builder
developer
designer
manufacturer
realtor
listing agreement
American Institute of Architects (AIA)
National Association of Home Builders (NAHB)
manufactured housing
mobile home
sectional home

modular home
speculative building
affordable housing
single-family detached house
multiunit housing
tract
planned unit development (PUD)
zoning ordinance
setback
building code
model building code
Council of American Building Officials (CABO)
International Residential Code
townhouse
egress

Occupational Safety and Health Act (OSHA)
Americans with Disabilities Act (ADA)
first cost
long-term value
life-cycle cost analysis
resale value
appraised value
construction loan
permanent mortgage
all-in-one loan
green building
sustainable yield forest
embodied energy
recycled content
volatile organic chemical (VOC)

REVIEW QUESTIONS

1. Who are the professionals involved with the production of a typical speculative house from conception to occupation by the owner? How do the participants change if the house is built for the owners on their own property?

2. What are the differences between site-built and manufactured housing?

3. Compare a PUD with a tract development.

4. What are the advantages and disadvantages of multiunit development from the point of view of the developer?

5. Explain the differences between zoning ordinances and building codes. Give examples of specific provisions of each.

6. Explain the general concept of green building. What are the particular steps a builder can take to implement the principles of green building?

7. Describe the typical construction process.

EXERCISES

1. Have each class member interview a residential builder to determine the type of work he or she typically undertakes. Discuss the range of responses in class.

2. Repeat the preceding exercise for residential architects and building designers.

3. Obtain copies of your local zoning ordinances and building code (they may be in your library). Look up the applicable provisions of these documents for a house on a specific site. What setbacks are required? How tall can the house be? How many cars must be able to be parked on the site? What are the requirements for emergency egress from the house?

4. Examine ads in your local phone directory or inquire at the local chapter of the NAHB to find builders who specialize in green building. Interview the green builder(s) to determine how their work differs from common practice? What different measures would they like to institute that they are not able to?

2

THE CONSTRUCTION COMMUNITY: BUILDERS, CONTRACTORS, AND DEVELOPERS

HISTORY

Builders in North America have always been industrious and inventive. The colonists, arriving in a land without permanent buildings or a labor force to construct them, had to do all the work themselves, so they were driven to devise labor-saving processes. Thus, North America had water-powered lumber mills even before the mother country England did, and brickyards and foundries were created in the colonies despite discouragement by the British crown. In the beginning, American builders adapted imported European building methods to their more severe climate, but later, they invented an entirely new system that was more flexible, used materials more efficiently, and was easier to construct. This new system, the Wood light frame, has been the predominant system of residential construction for over 150 years and remains so today, even as creative builders experiment with new systems (Chapters 20 to 24).

There has been specialization in the building industry since it was established early in the life of the American colonies. The first houses were constructed by carpenters who made the basic structure of wood and masons who laid foundations and made the fireplaces of stone or brick. The specialized trades of window and door making, roofing, plastering, and painting soon evolved. In the late 1800s, when utilities were introduced into the house, plumbers, gas fitters, and electricians were added to the list of specialists. Today, there are between 15 and 20 specialized trades that work on the average new residence (Figure 2.1).

Standardization of building components and the building process increases productivity, which translates into increased profit for professional builders. Standard sized bricks were typically used rather than random sizes of stone for masonry in early buildings because the regular bricks could be laid faster with a repetitive motion. Standard sized doors, windows, and other components were adopted at an early date because they made construction easier and communication between builders and material suppliers simpler. The concept of standardization of the entire building process as an assembly line, introduced in the past century, was the ultimate step to speed on-site production. Nowhere were the benefits of assembly line production demonstrated more dramatically than at Levittown, a suburban New York development built in the late 1940s, where the developer

FIGURE 2.1
In the late 19th century, builders had more responsibility for and control over the design of a house because there were virtually no codes or residential designers. (*Photo by Edward Allen*)

organized the construction process into 27 distinct steps and prefabricated every possible part of the house in a shop (Figure 2.2). At the peak of production, the builders of Levittown built 30 houses per day. Since that time, many labor-saving devices such as the backhoe, the concrete pump, and the pneumatic nailer have further increased the speed of residential construction. The introduction of factory-assembled components such as engineered roof trusses and prefabricated fireplaces has reduced the on-site labor required to construct a new house even more. Construction efficiency has reached the point that, in 1999, the largest residential builder in the United States produced over 100 houses per day (Figure 2.3).

BUILDERS AND CONTRACTORS

A *builder* is someone who engages in the activity of constructing buildings. A *contractor,* on the other hand, is someone who enters into a legal relationship with an owner of a property to build the owner's building or perform other work. Under these definitions, not all builders are contractors, because some builders construct houses for sale and are not under contract to an owner. Neither are all contractors builders, because some, such as landscape contractors, do not engage in the act of making buildings. Further confusion arises because the term "builder" is also used loosely to mean both builder and contractor and specifically to refer to a speculative builder. In the discussions that follow, we will use the term "builder" in the general sense to refer to residential builders and will qualify the term when referring to a specific group.

FIGURE 2.2
Capitalizing on the high demand for housing after World War II, Bill Levitt streamlined the residential construction process to produce Levittown, New York. The largest development ever constructed by a single builder, Levittown has 17,400 houses and about 82,000 residents. As originally built, there were five variations of the Cape Cod–style house, each with about 750 square feet, 2 bedrooms, 1½ bathrooms, and a fireplace. A similar Levittown of 16,000 houses was later built in Pennsylvania. (*Courtesy of Nassau County Museums Collection, Long Island Studies Institute at Hofstra University*)

Types of Builders and Contractors

There is a significant distinction between builders who produce new residences for sale on the open market and those who work for a property owner to build a new house or remodel or repair an existing one. The former take on more responsibility than other builders — procuring land for construction, arranging construction financing, obtaining a design, and selling the completed dwelling. These builders are entrepreneurs and assume responsibility for making decisions about a product that must be attractive to consumers. If the product does not sell at a fair price, the builder will fail to make a profit. This endeavor involves a certain amount of speculation, and therefore these builders are called *speculative builders*. The latter category of builder, called a *contract builder* or contractor, is under contract to an owner and takes the risk that no profit will be made if the construction work under contract takes more time or costs more than anticipated. It obviously behooves both types of builders to run their businesses efficiently in order to make a profit, but the speculative builder has the potential to recoup losses from inefficiencies by raising the selling price of the finished house.

Another important difference between builders relates to whether they do *new construction* or *remodeling* and/or repair to existing houses. Both speculative builders and contractors can specialize in either type of construction, and many do both. Remodeling contractors generally work on smaller projects and have smaller companies than those who do new work because all the work is custom, precluding large-volume production. Likewise, contractors who specialize in building new custom-designed houses have limited production because each project is unique (Figure 2.5). Large-volume production of new houses is achieved by working on large tracts of repetitive units, and contractors achieve success in this arena by managing labor and materials on the building site as they would a housing production line in a factory (Figure 2.6).

Sizes of Companies

The National Association of Home Builders (NAHB) has categorized building companies into three groups according to size. Those building fewer than 25 units per year are small builders, those building 26 to 100 units per year are medium builders, and those building more than 100 units per year are large builders (Figure 2.7). In 1994, small builders

FIGURE 2.3
Large-volume residential builders compete in the market by carefully managing the production of houses that are designed with ease of construction as a high priority. The same designs are repeated over and over to maximize production line efficiency. (*Photo by Rob Thallon*)

FIGURE 2.4
Labor-saving equipment speeds the construction process considerably. At an efficiently organized site, the same boom truck that delivers trusses to the roof also lifts a tub/shower enclosure into place on the second floor. (*Photo by Rob Thallon*)

FIGURE 2.5
Custom-designed houses such as this take much longer to build than do tract houses designed for efficient construction. Custom-designed houses tend to be constructed by small-volume builders who complete only a few projects in a year. (*Architect: David Edrington. Photo by Rob Thallon*)

FIGURE 2.6
Foundation in the foreground, framed house across the street, and finished shell in the distance. Large-volume builders organize house construction as a portable production line. By keeping subcontractors busy without having to travel between jobs, builders can get their work for less and be competitive with the prices of houses offered for sale. (*Photo by Rob Thallon*)

FIGURE 2.7
Large-volume builders consume materials so rapidly that they prefer to store them adjacent to the houses being constructed. Here lumber and other materials are stored on developed building lots for use in houses being constructed across the street. Notice the job shack, equipment and material storage trailer, and backhoe in the background. (*Photo by Rob Thallon*)

accounted for 79 percent of all NAHB member builders, having increased their share of the total by 21 percent over the previous 25 years (Figure 2.8). Although the small builders far outnumber medium and large builders, they produce the fewest houses (14 percent) because many small builders produce only one or two houses per year.

Financial success in the residential construction industry is not necessarily proportional to the number of units produced. The mass-produced houses built by large builders generally have lower sale prices than houses built by small builders (Figure 2.9). This is because numerous small builders with low overhead costs are engaged in building large custom houses, while large and medium builders have high overhead costs and generally concentrate on the production of lower priced repetitive units in tracts. So it is not uncommon for a small builder who completes only one expensive custom-designed house per year to make as much money as a medium builder who finishes more than 25 units in a year.

Contractors and Subcontractors

A *general contractor* is responsible for all construction work from start to finish. A residential general contractor signs a contract with the owner to build an entire house, or group of houses, or to complete an entire remodeling or repair project. The general contractor, in turn, hires *subcontractors* such as plumbers, plasterers, and painters to complete specialized parts of the work. The subcontractors are not employees of the general contractor, but each has an agreement with the general contractor to complete specified portions of the work. General contractors frequently hire their own carpentry crew to do the framing, siding, and finish trim. In addition, general contractors often employ their own laborers to move materials on the site, to prepare for the various subcontractors, and to perform other categories of work that are typically performed by subcontractors (Figure 2.11).

Size of Builder	Single Family				Multifamily			
	Number of Builders	Percentage of Total	Starts per Year	Percentage of Total	Number of Builders	Percentage of Total	Starts per Year	Percentage of Total
Small volume (1–24 starts per year)	73,552	93.0	331,970	38.8	3,329	67.5	30,268	9.5
Medium volume (25–99 starts per year)	4,338	5.6	182,812	21.3	831	18.8	43,216	13.6
Large volume (100+ starts per year)	1,212	1.5	341,841	39.9	772	15.7	244,746	76.9

FIGURE 2.8
Builder size as related to number and type of housing starts in the United States. A small number of large builders account for the greatest percentage of residential building starts. (*Source: National Association of Home Builders, Economics Department, 1999*)

Price of House	Size of Builder			
	Small Volume (<25 units/year)	Medium Volume (25–99 units/year)	Large Volume (100+ units/year)	Total
< $75,000	4%	10%	7%	7%
$75,000–$99,999	10%	19%	22%	19%
$100,000–$149,999	22%	31%	33%	30%
$150,000–$199,999	36%	19%	22%	24%
$200,000–$249,999	9%	11%	10%	10%
$250,000–$499,999	14%	8%	5%	8%
$500,000 and over	5%	2%	—	2%
Median	$169,444	$133,890	$140,000	—

FIGURE 2.9
Percentages of houses built within each price range according to size of builder. Prices are of single-family detached houses built in 1994 and include land. (*Source: Gopal Ahluwalia*, Home Builders and Their Companies, *Housing Economics, NAHB, 1995*)

FIGURE 2.10
One of the most important tools for the contractor is the pickup truck. Equipped with a rack to carry lumber, ladders, and other long objects; a lockable box for tools; and a bed to carry large or heavy items; the pickup serves to get workers, equipment, and materials to the job; and the cab is good for coffee breaks and lunch on cold days. Many subcontractors prefer vans, large enclosed trucks, or trailers, according to the nature of their specialties. (*Photo by Rob Thallon*)

Jobs Subcontracted	Always (%)	Sometimes (%)	Never (%)
Foundations	88	8	4
Framing	66	19	15
Roofing	82	13	5
Doors and windows	72	9	20
Exterior siding	73	16	11
Plumbing	96	3	1
Electrical wiring	97	2	2
Fireplace	88	8	5
Drywall	89	10	2
Interior doors	66	13	21
Wood flooring	84	11	5
Carpeting	96	3	1
Kitchen cabinets	74	14	12
Kitchen countertops	79	15	6
Painting	84	11	5
Concrete flatwork	87	10	3

FIGURE 2.11
Portions of residential construction projects that are subcontracted. These data indicate that most general contractors subcontract most of the work and are therefore principally managers of the construction process. (*Source: National Association of Home Builders, Economics Department*)

Licensing of Builders

In most states, contractors and subcontractors are required to be *licensed, bonded,* and *insured.* The license signifies that the contractor has passed a government-sanctioned test that measures proficiency in the area of his or her specialty. For example, a general contractor's license indicates that the licensee has a basic understanding of contracts, building codes, materials, and construction procedures, whereas a plumber's license indicates a specialized comprehension of plumbing. Bonding of contractors is a form of insurance paid for by the contractor to protect the owner against nonperformance or fraud. If a contractor should fall ill while building a project, for example, the bond would pay the additional cost beyond the original contract sum to hire a substitute contractor to complete the work. Contractor's insurance protects the owner financially from damage to the owner or the owner's property caused by the contractor.

OBTAINING WORK

Because they generate their own work, speculative builders do not depend for their livelihood on being hired by an owner. In this sense, speculative builders have an advantage over contractors because they do not have to spend time bidding against other contractors for their work, nor do they depend on being hired by a continuous succession of owners. However, speculative builders do depend on selling their completed projects for a profit in order to finance the next speculative project.

There are two basic mechanisms by which contractors and owners come to agreement about contracts: *direct hiring* and *competitive bidding.* In both cases, it is an advantage to the contractor to have established a reputation based on honesty, professionalism, and quality work. When the contractor is hired directly, a good reputation is essential because there is no competition to control the cost of the work. When bidding for a job, a good reputation is useful because the owner and architect together usually decide on a list of contractors who are invited to bid.

Competitive Bidding

Contractors often compete for a construction contract by submitting competitive bids. Typically, the owner and designer select three or more general contractors with experience that indicates they are qualified to do the work. The contractors are given several weeks to evaluate the plans and specifications, secure bids from subcontractors, and submit comprehensive bids for the entire project. The bid price includes the cost of all labor, material, subcontractors, and contractor's profit and overhead. A proposed schedule of the work is usually submitted with the bid. Each bid is evaluated against the bids of the other contractors, and one contractor is selected to do the project.

Producing a bid involves a considerable investment of time and money on the part of each general contractor. Hours are spent reading the plans to understand in detail what is proposed to be built. Each general contractor distributes copies of the plans to as many as 20 or 30 subcontractors. Each of these subcontractors must prepare a separate bid for the portion of the work in which he or she specializes. The general contractor must be sure that each subcontractor understands the scope of the work and includes it all in his or her contract without overlapping with the contracts of other subcontractors. If the general contractor is planning to perform some of the work, estimates for labor and material for this portion of the work must be prepared as well. During this process of preparing a bid, the general contractor or subcontractors may contact the designer for clarification of the construction documents and may make suggestions to the designer for changes or substitutions.

Direct Hiring

There are several reasons for an owner to hire a contractor directly, without competitive bidding. The job may be small and not worth the expense and the time of the bidding process. Or the work may not be clearly defined, so that bids would not be precise and would have to be high to cover unforeseen circumstances. Or the quality of the work or the construction schedule may be more important than the cost. Perhaps the owner has a favorite contractor with whom she or he wishes to work. These are but a few examples of the myriad situations in which the owner would want to hire a contractor based on reputation rather than on a bid for the cost of the work.

If a contractor is hired directly, however, it does not mean that contractor is working with a blank check. Hourly rates for the contractor are agreed upon before the project begins, and if there are subcontractors who must be managed, a fee for this work is also established. Often when a contractor is hired directly, his or her estimates and cost accounting are furnished to the owner so both parties can search for potential savings and keep track of costs.

Contract Formats

For a small remodeling or repair job, the contract may be just a handshake. Often, it is a simple handwritten agreement signed by both contractor and owner, or it may be a more formal contract. Larger projects, ranging from the construction of a new single-family house to a tract or multifamily building, typically require a more comprehensive legal document. The American Institute of Architects (AIA) publishes a set of contracts that are commonly used for this purpose and specify clearly and unambiguously the responsibilities of the contractor, the owner, and the architect.

There are two basic formats for the contract between contractor and owner — the *stipulated sum contract*

(fixed fee) and the *cost-plus contract*. A contract resulting from competitive bidding can take a variety of forms, but the most common is the stipulated sum contract, under which the contractor agrees to build the house (or houses) for a fixed amount. Changes in the scope of the work such as an added window or deleted fireplace are kept track of by means of a *change order*. The change order is an agreement between the owner and the contractor about an adjustment to the contract amount due to a change in the work. The most serious shortcoming of the stipulated sum contract is that all bidding contractors must augment their bids somewhat to protect themselves from the possibility that an unforeseen circumstance such as a dramatic jump in material prices may occur. If nothing occurs during the course of construction to add to the contractor's cost of the work, the owner still pays the full contract sum.

When the contractor is hired directly, without competitive bidding, the agreement between owner and contractor is typically spelled out in a cost-plus contract, under which the owner is to pay the contractor the cost of the work plus a fee for managing the project. Receipts for materials and services (including those of subcontractors) establish the cost of the work. Thus, in the cost-plus contract, if an unexpected cost occurs, the owner pays for it. If some aspect of the work costs less than expected, the owner is the beneficiary. The contractor's fee can be a fixed figure, or it can be a percentage of the construction cost.

The contracts between the general contractor and the subcontractors are similar to those between the general contractor and the owner. When the main contract is a stipulated sum contract, subcontractors must generally sign a stipulated sum contract as well. When the general contractor has a cost-plus agreement with the owner, however, the subcontractor may work for the general contractor on either a fixed-fee or a cost-plus basis. Each general contractor typically has a list of preferred subcontractors with whom he or she has worked. In some cases, contractors and subcontractors form a loose alliance, agreeing to work with each other whenever possible. In other cases, they behave as a team, always working together.

Responsibilities of the Contractor

When a contractor signs an agreement with an owner, he or she assumes a number of responsibilities regarding the execution and completion of the project. These responsibilities are detailed in a section of the AIA contract documents entitled "General Conditions." The contractor, whether a general contractor or a subcontractor, is usually responsible for the following:

1. Scheduling of all aspects of construction related to the contractor's specialty. In the case of the general contractor, the scheduling includes coordination of all subcontractors and their work.

2. Purchasing of materials and transporting them to the site and, once

they are on the site, protecting them from theft, weather, and damage.

3. Installing the materials into the house in a professional fashion to complete the work as described in the contract documents.

4. Complying with the building codes and calling for inspections by building officials at the appropriate times during construction.

5. Keeping records of all costs related to the project — for the contractor's own records in the case of a stipulated sum contract and for the owner in the case of a cost-plus contract.

6. Warranting the work against defects in workmanship, generally for a period of 1 year after construction is complete.

THE RESIDENTIAL DEVELOPER

The Developer

The business of constructing houses has expanded over the years to include the development of land upon which to build. The *residential developer* is the

person (or company) who procures land, divides the land into legally defined building lots, and installs improvements such as roads and utilities to serve the lots (Figure 2.12). This is usually a lengthy process, involving multiple approvals by local (and regional) zoning authorities and coordination with utility companies. Nonetheless, by purchasing in volume and controlling development costs, developers can generate residential building lots inexpensively and make a significant profit on their sale.

Some developers stop at this stage and sell bare lots to smaller developers, builders, or individuals, while others continue with the development process to profit from the construction and sale of houses as well. The problem with selling bare lots is that it is difficult to control the quality of the neighborhood, so lots sold in this way are often encumbered with *deed restrictions* that set minimum floor areas or mandate particular architectural styles for houses built upon them. Small developers often buy groups of lots in these tracts for their own speculative projects.

FIGURE 2.12
Site construction at a subdivision. The concrete curbs and gutters have been laid at the edges of the streets, and a compacted gravel base has been installed between them. The underground utilities are being laid in the trench at the right as gravel backfill is being placed over them to make a smooth site. The complexity and cost of site development on sloping sites lead developers to prefer flat ground. (*Photo by Rob Thallon*)

The Developer/Builder

Developers who construct houses for sale on the lots they develop are called *developer/builders* (Figure 2.13). Like the speculative builder of individually custom-designed houses, the developer/builder takes a risk that the completed project will not produce a profit or, worse yet, that it will not sell at all. Thus, all the actions of the developer/builder are calculated to be efficient and to entice buyers. Serious consideration is given to ensure that the location for a new development will be attractive to potential buyers — that schools and shopping are convenient and that the neighborhood is safe and aesthetically pleasing. The design of the new houses (including their sites) is usually tailored to meet current preferences among average buyers, perpetuating the status quo and often making it difficult for marginal sectors of the house-buying population to find new housing to meet their needs (Figure 2.14).

There is a wide range of scales of residential development — from developer/builders who build one or two houses per year to those responsible for thousands of units. In 1999, Pulte Corporation of Bloomfield

FIGURE 2.13
Construction at the edge of a large subdivision produced by a developer/builder. The design and construction of the streets, the lots, and the houses are all under the control of the same company. (*Photo by Rob Thallon*)

FIGURE 2.14
A typical speculative house offered for sale. This house was built in the year 2000 by Centex Corporation, one of the largest homebuilders in the United States. Its design was based on market trends in the area where it is built. (*Photo by Rob Thallon*)

Hills, Michigan, the largest developer/builder in the United States, sold 26,662 houses, accounting for almost 2 percent of all residential construction in the country (Figure 2.15). The largest developer/builders invariably have all the expertise needed to complete a project with their own staffs, or they have subsidiary companies that do this work for them. They employ land planners and engineers to develop buildable lots, architects and landscape architects to produce designs for development, contractors to implement the designs, and realtors to market and sell the product. Small developers rarely have any of these experts on their staffs except that many act as their own building contractors.

These small developers rely on the advice of colleagues, realtors, and consultants hired for specific tasks.

Financing

To procure financing for prospective projects, speculative builders and developer/builders must themselves first invest their own "seed money" in the project. For large-scale projects, land use consultants, planners, engineers, and architects must all be paid from this seed money to study the feasibility of a proposed project and to develop schematic plans. For single-lot projects, most financial institutions require that the speculative builder own the lot before a construction loan will be considered.

Builders as Designers

When looking for a design for a speculative house, a speculative builder is primarily concerned with making a profit when the house is sold. Major factors that are considered include the ability to construct the house efficiently, the suitability of the design for the area in which it is to be built, and the inclusion of features desired by the potential buyers. Construction efficiency is largely related to simplicity of form and use of common materials and methods. Builders developing large tracts gain efficiency by repeating the same design and achieve variety with mirror-image right- and left-hand versions of the same design and minor

FIGURE 2.15
With projects like this all over the country, the largest residential developer/builder in the United States completes more than 100 houses every working day. (*Courtesy of Pulte Corporation. Photo © Dunn Aerial Photography*)

FIGURE 2.16
Large builder/developers attempt to achieve a sense of variety from house to house but incorporate as few actual differences as possible. A few different floor plans are repeated, usually with mirror-image versions. Simple changes such as different paint colors, different siding on the street side of the house, or different porch designs add to variety. (*Photo by Rob Thallon*)

FIGURE 2.17
A house showcased in the 2000 Parade of Homes, Denver, Colorado. Promotional tours such as this are common in large and mid-sized U.S. cities and allow local builders to demonstrate their skills to a large audience. Tours typically feature a range of houses from small, affordable residences to large, expensive ones such as this. (*Courtesy of Laureate Homes, luxury home division of US Home Corporation*)

modifications to the facades (Figure 2.16).

Finding the right combination of site and design for a residential development involves a balance between the location of the project and the perceived needs and desires of the target population. A house designed for a family with children, for example, will be most attractive to potential buyers if it is located in a neighborhood within a good school district and is on a site with sufficient room for a yard in which children can play. Small or remote sites are most suitable for target populations of individuals or couples without children. Speculative builders must be careful to develop or choose a design that is of appropriate size, quality, and style for the neighborhood in which it is built, because potential buyers are generally looking for long-term value, which would be compromised by a house that starts its life as a white elephant.

Some speculative builders choose to start the process of matching site and design by selecting a target market and searching for the right site on which to build. Others start with the site and proceed to determine the best kind of house to build on it. In whichever sequence the choices are made, builders are aided by working with realtors and by observing the successes and failures of other builders in the area. In fact, much of the inspiration for what to build next comes from scrutinizing the products of one's competitors. This is done at annual "tour of homes" and "street of dreams" shows sponsored by local builders' associations or by merely driving around to see what has been recently placed on the market (Figure 2.17). The National Association of Home Builders tabulates buyers' preferences and publishes them frequently, both in print and on the Internet (Figure 2.18).

Designs for houses built by speculative builders may come from a variety of sources, depending primarily on the size of the builder and the nature of the target market. The largest builders and some of medium size employ in-house architects, landscape architects, and interior designers to generate house plans and site plans. This team of design professionals usually works simultaneously on several large projects at different stages of development. Independent architects and landscape architects are often hired by smaller builders of multiple-unit projects or by builders of expensive single-residence projects. Small builders often use stock plans or low-cost residential designers who modify stock plans to suit the circumstance.

Marketing and Selling

Most new houses are sold by realtors who know how to market them and

Characteristics of New Single-Family Homes: 1988-1999

	1988	1989	1990	1991	1992	1993	1994	1995	1996	1997	1998	1999
Total completed (000s)	1,085	1,026	966	838	964	1,039	1,160	1,065	1,129	1,116	1,160	1,307
Central A.C. installed	75	77	76	75	77	78	79	80	81	82	83	84
2-car garage or more	66	70	72	71	75	77	78	76	78	78	79	81
1 fireplace or more	65	65	66	62	64	63	64	63	62	61	61	62
2½ baths or more	42	44	45	44	47	48	49	48	49	50	52	55
2 stories or more	49	49	49	47	47	48	47	48	47	49	50	51
1 story	46	46	46	48	48	48	49	49	49	49	48	48
Slab	41	43	40	38	38	40	41	42	44	45	45	47
Full or partial basement	39	37	38	40	42	40	39	39	37	37	37	37
4 bedrooms or more	26	28	29	28	29	30	30	30	31	31	33	34
Brick exterior	17	17	18	19	21	21	21	20	21	21	21	21
No garage or carport	18	17	16	17	15	14	13	14	13	13	12	12
2,400 ft.^{2} or more	25	26	29	28	29	29	28	28	30	31	32	34
1,200 ft.^{2} or less	12	13	11	12	10	9	9	10	9	8	7	7
Average ft.^{2}	1,995	2,035	2,080	2,075	2,095	2,095	2,100	2,095	2,120	2,150	2,190	2,250
Median ft.^{2}	1,810	1,850	1,905	1,890	1,920	1,945	1,940	1,920	1,950	1,975	2,000	2,030
Median lot size ft.^{2}	9,225	9,500	10,000	9,750	9,600	9,600	9,500	9,375	9,100	9,000	8,800	8,750
Average lot size ft.^{2}	14,220	13,845	14,680	14,275	14,060	13,440	13,645	13,665	13,705	13,290	12,870	12,910

FIGURE 2.18
Characteristics of new single-family homes: 1988–1999. With the exception of the first row and the last four rows, numbers are percentages. Statistics such as these help speculative builders determine what features to include in their houses. (*Source: U.S. Census Bureau, compiled by NAHB Economics Department*)

are familiar with the necessary legal paperwork. Large builder/developers often own a real estate subsidiary, which markets and sells exclusively their own residences. Small developers usually work with a local real estate firm with an agreement to give exclusive rights to market their new houses in exchange for discounted fees.

TRAINING THE CONSTRUCTION COMMUNITY

On-the-Job Training

Construction is an occupation that one can learn by doing, and many successful builders have started their careers as assistants and laborers. Builders and many of the building trades that operate as subcontractors often hire young workers eager to start making a living, and these beginnings develop into vocations. Family construction businesses regularly hire young family members as a way of preparing the next generation for running the family business. In addition, on-the-job training is the best way to learn the physical aspects of how buildings are assembled. There is no substitute for feeling the weight and texture of a material in your hand, sensing its resistance to being cut, fitting it into place, and attaching it to the building. No laboratory or classroom can impart the understanding of the construction process that is gained from being in the middle of it day after day. Yet on-the-job training can be slow, because job site efficiency demands that new workers perform simple, repetitive tasks. In addition, neither the theoretical principles of construction practice nor the business acumen necessary to be financially successful are easily gleaned from co-workers intent on productivity and camaraderie.

Apprenticeships

Some trades formalize on-the-job training by requiring those entering their trade to complete an apprenticeship program. The licensed trades such as plumbers and electricians typically have apprenticeship programs, and some unlicensed trades such as carpenters often do as well. An apprentice plumber, for example, must be employed for 4 years by a licensed plumber while performing in a prescribed range of situations in the field. The apprentice must attend two courses per year and must pass a written examination at the end of the apprenticeship period in order to obtain a journeyman plumber's license.

Trade Schools

The shortcomings of on-the-job training may be overcome in trade schools, where theory is taught in parallel with practical skills. Emphasis is placed on exposing students to a wide range of experiences, including situations where their own skills will rub up against those of other trades. Some trade schools are independent, but most are incorporated into community colleges throughout the country. Programs in trade schools are designed to benefit trades both with and without apprenticeship programs.

Builder Associations

There are many associations that disseminate information and promote the exchange of ideas among builders. The largest of these is the *National Association of Home Builders (NAHB)*, which was founded in 1942 and now has over 200,000 members, including builders (about one-third of its members), material manufacturers, realtors, and mortgage lenders. The NAHB is headquartered in Washington, DC, and has state and local chapters whose members meet monthly. The stated goals of the association are to inform members about "changes in building products and techniques, consumer preferences, marketing, finance, regulation, and legislation that affect the building industry."

There is an annual national convention at which participants network, attend classes, and are introduced to the newest products and processes. The NAHB also maintains a Web site with a wealth of information about the industry, including current statistical data about new construction by region. In Canada, the comparable organization is the *Canadian Home Builders Association (CHBA)*.

There are also numerous smaller associations that support specific materials and trades. Examples include the National Concrete Masonry Association (NCMA), the National Roofing Contractors Association (NRCA), and so forth. Each of these associations is generally composed of members who are contractors and material suppliers.

Building professionals have joined with material manufacturers to form organizations that work toward the development of technical standards and the dissemination of information concerning their respective fields of specialization. The Construction Specifications Institute (CSI), whose MasterFormat™ standard is described in the following section, is one example. It is composed of independent building professionals such as architects and engineers, of builders, and of members from industry.

MasterFormat

The *Construction Specifications Institute (CSI)* of the United States and *Construction Specifications Canada (CSC)* have evolved over a period of many years a standard outline called MasterFormat for organizing information about construction materials and components. MasterFormat is used as the outline for construction specifications for nearly all large construction projects in the two countries, including most multiunit residential projects. It also forms the basis on which trade associations' and manufacturers' technical literature is cataloged and filed. Its 16 primary divisions are as follows:

Division 1   General Requirements

Division 2   Site Work

Division 3   Concrete

Division 4   Masonry

Division 5   Metals

Division 6   Wood and Plastics

Division 7   Thermal and Moisture Protection

Division 8   Doors and Windows

Division 9   Finishes

Division 10   Specialties

Division 11   Equipment

Division 12   Furnishings

Division 13   Special Construction

Division 14   Conveying Systems

Division 15   Mechanical

Division 16   Electrical

Within these broad divisions, several levels of subdivision are established to allow the user to reach any desired degree of detail. A five-digit code, in which the first two digits correspond to the division numbers listed previously, gives the exact reference to any category of information. Within Division 7, Thermal and Moisture Protection, for example, some standard reference codes are

07210	Building Insulation
07300	Shingles and Roofing Tiles
07600	Flashing and Sheet Metal
07620	Sheet Metal Flashing and Trim
07650	Flexible Flashing

Most chapters of this book give the major MasterFormat designations for the information presented, to help the reader know where to look in construction specifications and in the technical literature for further information. The full MasterFormat system is contained in the volume referenced at the end of this chapter.

Publications

Much of the education of those involved in the building industry, and much of their communication, occurs through the pages of books, journals, and other publications. Of the current journals marketed broadly to the building industry, *Professional Builder* is the oldest, having begun in 1936 as *Practical Builder* magazine. Another magazine with similar objectives is *Builder*, which is the publication of the NAHB. Both of these publications have articles on residential design, construction materials and techniques, marketing, and business, and both are filled with advertisements for building products and tools. Other magazines target different audiences related to residential construction. *Fine Homebuilding*, started in 1981, is addressed to custom home builders and is also widely read by do-it-yourselfers. *The Journal of Light Construction* is both comprehensive and practical in its approach and is marketed to serious builders and designers.

Magazines targeted for remodelers and subcontractors are also widely available. *Remodeling* and *Professional Remodeler* from the Remodelers Council of the NAHB are the principal publications for remodeling contractors. Each subcontracting specialty has several journals from which to choose, ranging from *Electrical News* to *Reeves Journal/Plumbing-Heating-Cooling, Woodshop News*, and *Tools of the Trade*.

FIGURE 2.19
Images such as this appear in newspaper advertisements all over the country. This same image was featured in both Pennsylvania and Texas. (*Courtesy of Pulte Corporation*)

The Monroe Model as photographed at Smithfield Estates.

SELECTED REFERENCES

1. Eichler, Ned. *The Merchant Builders.* Cambridge, MA: MIT Press, 1982.

Chronicles the history of residential development since World War II. Clearly written by an industry insider.

2. Urban Land Institute. *Community Builder's Handbook: Residential Development Handbook.* Washington, DC: Urban Land Institute, 1978.

3. Construction magazines are the best source for up-to-date information on techniques, tools, and materials for builders. The most popular general periodicals are *Fine Homebuilding* and the *Journal of Light Construction.*

4. Management and marketing strategies for builders appear monthly in *Builder* and *Professional Builder* magazines. Both publications also maintain Web sites.

5. The Construction Specifications Institute and Construction Specifications Canada. *MasterFormat — Master List of Section Titles and Numbers.* Alexandria, VA, and Toronto, 1995.

This book gives the full set of numbers and titles under which construction information is filed and utilized.

KEY TERMS AND CONCEPTS

builder
contractor
speculative builder
contract builder
new construction
remodeling
general contractor
subcontractor

licensed
bonded
insured
direct hiring
competitive bidding
stipulated sum contract
cost-plus contract
change order

residential developer
developer/builder
deed restriction
National Association of Home Builders (NAHB)
Canadian Home Builders Association (CHBA)
MasterFormat
Construction Specifications Institute (CSI)
Construction Specifications Canada (CSC)

REVIEW QUESTIONS

1. What is the difference between a builder and a contractor? Between a general contractor and a subcontractor?

2. What are the advantages and disadvantages to the owner of direct hiring as opposed to contract bidding? What are the advantages and disadvantages to the contractor?

3. Compare the typical activities of a developer to those of a developer/builder.

4. Discuss the relationship between residential builder company size and the type and size of houses built.

5. Why do builders need to be licensed, bonded, and insured?

EXERCISES

1. Have class members contact a variety of residential subcontractors and ask them about their training, their methods of obtaining work, and their interactions with other trades. Conduct a class discussion to compare notes.

2. Repeat the preceding exercise with general contractors who are asked to describe their process of building a house from obtaining the job to handing the key to the owner.

3. Visit a local home show and take notes about the latest products offered for sale. Have a class discussion about how the most interesting of these products are likely to affect the construction process in the future.

4. Visit a residential building site and document the activities at the site over the period of time in which your class meets. Translate your notes into a graphic format that describes the construction schedule for the period in which you study the site. Extend this schedule to include activities at the beginning and end of construction.

THE DESIGN COMMUNITY

3

- **History**
- **Residential Designers**
  - Architects
  - Building Designers
  - Owner/Designers
- **Consultants**
  - Engineers
  - Landscape Architects
  - Land Planners
  - Interior Designers

- **The Design Process**
  - Stock Plans
  - Custom Design
  - Drafting Services
  - Design/Build
- **Design Sources**
  - Speculative House Resources
  - Custom House Resources
  - Technical Resources
  - Design Associations
- **Communication Between Designer and Builder**

HISTORY

Hundreds of years ago, when a new house was constructed, there were no designers to organize the rooms and tailor them to the needs of the owner. Instead, the design of the house was copied from one of several traditional models and was adjusted in modest ways to accommodate the owner's family and to fit the specific conditions of the site where it was to be built. This process of arriving at a design was almost instinctive and usually involved only the owner and the builder. The plans for the house, if any, were drawn simply for the purpose of a contractual agreement, and the details were left to the builder. Styles throughout North America were strongly related to the simple form of the timber frame (Figure 3.1).

Until the 19th century, the making of a new building was an integrated set of events in which considerations of aesthetics, planning, structure, and construction were all considered to be part of an overall, unified effort. In some cases, houses were constructed under the direction of a single master builder, who was expert in all these areas. In other cases, the homeowners built their new houses themselves, relying on a building culture that had evolved to the degree that the entire population understood basic design principles and the qualities and capabilities of materials.

Differentiation of the roles of residential designer and builder began in the mid-19th century when architects, who were already designing churches and civic buildings, were hired to act as designers for complex multifamily housing projects. Architects had also been designing large elaborate houses for wealthy clients, but this constituted a very small portion of the residential construction market. In the 1920s, new building codes related to ordinary single-family dwellings created the need for a new kind of expert, and architects stepped forward to fill that role. The two world wars involved architects further with mass housing because of the need for efficiency in the use of both space and materials. Today, most new house construction involves either architects or other residential designers whose role is quite distinct from that of the builder (Figures 3.2 and 3.3).

With the differentiation of the roles of designers and builders, the construction of buildings now follows a much more fragmented process in which a number of specialists are each responsible for a small, isolated part of the overall activity. Of these specialists, builders have the most complete understanding of how materials go together from a practical point of view, and designers are trained to understand a building from structural, programmatic, and aesthetic points of view. It is true that houses are much more complex than they were only a hundred years ago, and for this reason it is rational to expect that their design and construction will require input from specialists. Beautiful buildings, however, can be made only by integrating all aspects of design and construction so that they reinforce one another. Therefore, it is now the responsibility of the designer to acquire a technical understanding equal to that of the builder in order to design buildings that resonate with the human spirit.

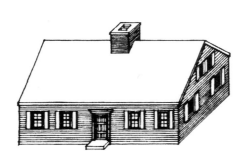

Cape Cod
A style of cottage developed mainly on Cape Cod, Massachusetts, in the 18th and early 19th centuries, typically a rectangular, one- or one-and-a-half-story, wood-frame house with white clapboarded or shingle walls, a gable roof with low eaves and usually no dormer, a large central chimney, and a front door located on one of the long sides.

saltbox
A type of wood-framed house found esp. in New England, generally two full stories high in front and one story high in back, the roof having about the same pitch in both directions so that the ridge is well toward the front of the house.

Dutch Colonial
Of or pertaining to the domestic architecture of Dutch settlers in New York and New Jersey in the 17th century, often characterized by gambrel roofs having curved eaves over porches on the long sides.

FIGURE 3.1
Early American house styles had simple forms that were strongly related to the timber frame structure that supported them.
(*Reprinted with permission from Frank Ching,* A Visual Dictionary of Architecture, *John Wiley & Sons, New York, 1996*)

FIGURE 3.2
Designers of speculative housing must interpret the speculator's marketing strategy into physical form. This exercise all too often involves merely the adjustment of existing house plans. Designers of speculative housing are usually instructed by their employer to pay particular attention to construction efficiency. (*Photo by Rob Thallon*)

FIGURE 3.3
Designers of custom houses meet directly with the future owners of the house. This gives them the best opportunity of any residential designer to match the particular needs and lifestyle of the future occupants with the spaces and features of the house. Aesthetics, style, and the concept of beauty are discussed most frequently with this type of residential design project. (*Architect: SALA Architects, Inc., Minneapolis, Minnesota. Photo by Dale Mulfinger*)

RESIDENTIAL DESIGNERS

Architects

Architects are the most highly trained of all the professional house designers. Architects must complete a minimum of 5 years in an accredited professional program plus 3 years of internship before they can sit for their professional licensing examination. The examination is a nationally standardized test that is administered by each state. Passage of the examination gives a person the right to call him- or herself an architect, and it is illegal for anyone who has not passed the exam to use the word "architect" or any derivation of that word in his or her title or company name. The appearance of the initials A.I.A. after one's name indicates that the person is an architect who is a member of the *American Institute of Architects,* but not all licensed architects choose to join this organization.

The education of architects in professional schools is not related only to housing because curricula are designed to give students a broadly based design education. Studio-based design training usually gives a student experience in the design of museums, libraries, and other building types in addition to housing. Technical training is similarly general in that schools emphasize the principles of construction, structural design, and environmental control rather than specific details that relate to housing or any other building type or system. Architecture students with a primary interest in housing can gain specialized training in this area, however, by their selection of coursework and especially by their selection of an architectural firm with whom to intern after they graduate.

FIGURE 3.4
Designers of multifamily housing must deal with the difficult issues of privacy, security, fire separation, and connection to the outdoors. Almost all multifamily housing is designed by registered architects. (*Architects: Bergsund & Associates and David Edrington, Eugene, Oregon. Photo by Rob Thallon*)

FIGURE 3.5
The motives of the client have a significant effect on the design of rental housing. The student housing on the left is privately owned and designed to maximize profit. The university-owned student housing on the right was designed with student comfort and a contribution to the community as high priorities. (*University Architects: CES/T&E, Berkeley, California, and Eugene, Oregon. Photo by Rob Thallon*)

Of the more than 1.5 million dwelling units built each year in the United States, licensed architects are directly involved on only about 20 percent. State laws generally require that large multifamily residential buildings be designed by architects and that architects be included in the design team for large tracts and subdivisions. Nonprofit agencies and institutions that produce affordable housing usually hire architects (sometimes by a competitive process) to take advantage of their expertise in planning, space utilization, code interpretation, and efficient material usage. Together, these instances in which architects are hired to design new housing account for about 15 percent of all

new units. The remaining 5 percent of new houses for which architects are directly employed are one-of-a-kind houses for individual clients. Some architects also produce stock plans that are mass-marketed to customers whom the designer never meets.

Building Designers

Building designers are a very diverse group with a wide range of training. Some have degrees in architecture but have either not qualified for or not passed the architectural licensing exam. Others have only minimal academic preparation but have entered the field of residential design through practical experience in drafting or

construction. There are no legal or educational requirements for assuming the title of building designer, but a certification program administered by the *American Institute of Building Design (AIBD)* leads to the status of *Certified Professional Building Designer (CPBD)*, allowing a building designer to display a credential of competence.

Building designers are responsible for more residential design than all other categories of residential designers combined. Residential designers are also responsible for most stock plans that are published. However, most of the work of building designers occurs in local plan shops and drafting services that serve both speculative builders and owner/builders. The more talented residential designers compete with architects for the middle-income and wealthy clients who desire custom-designed houses.

Owner/Designers

It is surprising, given the complexity of the process, the size of the investment, and the significance of the results, how many owners with little or no training design houses for themselves. This would not be so unexpected, however, if the emotional connection people have with their houses were considered at the same time. There is a prevailing attitude of independence in North America, and perhaps as many as 10 percent of new houses are designed by their owners. The word "designed" does not have quite the same meaning in this context as it does with architects or building designers because owners rarely have the expertise to carry their ideas through to working drawings adequate for a building permit. Instead, most owner/designers take their ideas to a drafting service, a building designer, or an engineer to draw up their plans (Figure 3.6).

CONSULTANTS

Today's houses are much more complicated than they were when master builders single-handedly orchestrated their design and construction. Furthermore, expanding and ever more complex building codes make house design more challenging each year. It is not at all uncommon today for a residential designer to hire consultants for the portions of the design work that require specialized training, such as structural engineering or landscape design. The best designs occur when the consultant works in collaboration with the architect or building designer from the beginning of a project.

Engineers

In custom-designed houses and multifamily residences, structural complexities frequently occur that are more efficiently resolved by an engineer than by an architect or building designer. Such tasks as the design of structural stiffening to resist lateral loads imposed by earthquakes and high winds, the framing of complex

roofs, and the design of foundations for steep sites or unconsolidated soils are typically performed by the structural engineer. The principal designer of the house hires the engineer as a consultant to resolve these issues during the design process and passes the cost on to the owner.

Like architects, *structural engineers* are licensed professionals who are legally obligated to be familiar with building regulations and to take legal responsibility for their work. Every engineer must earn a professional degree and pass a licensing exam. In many jurisdictions, buildings or groups of buildings that require the stamp of a professional may bear the stamp of either an architect or an engineer.

Landscape Architects

Whereas architects and engineers are professionals trained to design buildings, *landscape architects* are trained to design the outdoor spaces around buildings (Figures 3.7 and 3.11). Large projects such as multifamily residences, subdivisions, PUDs, and custom houses for affluent owners generally involve landscape architects.

Landscape architects are licensed in most states much like architects. Licensure requires a degree from an accredited 5-year program, an internship with a licensed firm, and successful completion of a national examination. Licensing and continuing education are administered by the *American Society of Landscape Architects (ASLA)*. Many landscape firms and individuals practice without a license as landscape designers.

Land Planners

Land planners are typically employed as consultants only on large housing projects such as subdivisions, PUDs, and multifamily structures. The role of a land planner is to analyze the site and prepare a design for its division into building lots and its required improvements. The land planner is familiar with governmental approval processes and prepares the documents

FIGURE 3.6
Although most owner/designers succeed in producing houses of acceptable quality, and some even produce delightful results, the potential is high for these largely untrained designers to make fundamental design mistakes. This example, located in a northern rainy climate, has no roof over the front door. Or is the front door the one on the porch overlooking the street, accessible only by walking past the "side" door? (*Photo by Rob Thallon*)

necessary for review by governmental agencies with respect to roadways, parking, lot lines, utilities, drainage, tree preservation, and other site-related issues.

At present, states do not require that land planners be licensed, but there is a voluntary training program through the *American Planning Association (APA)* that prepares the planner to take a uniform national examination. In order to qualify for the exam, a candidate must have a 4-year degree in planning from an accredited planning program plus 2 years of internship under a certified planner. Upon passing the national exam, a planner becomes certified as a member of the *American Institute of Certified Planners (AICP)*. Certification is valued more highly in the public sector than the private sector and is a requirement for some governmental positions.

Interior Designers

Interior designers are trained to bring life to interior spaces. They design lighting; select colors, cabinetry, finish surfaces, floor and window coverings; and coordinate furnishings with the built space (Figure 3.8). At the residential scale, most are employed directly by owners after completion of the construction process or at the very end of construction. Architects occasionally work with interior designers as consultants during the design phase of their work, and builders frequently hire interior designers to coordinate the finishes and furnishings of interior spaces for model homes.

The practice of interior design is regulated in approximately one-third of the states. In these states, in order to call oneself an *interior designer,* a practitioner must have completed a degree in a professional school and this must be followed by a period of internship and passage of an examination. *Interior decorators,* on the other hand, have no requirement for formal training and are hired based on their good taste and/or their business acumen.

FIGURE 3.7
Landscape architects are trained to design outdoor spaces that complement the indoor spaces of a house. For subdivisions and PUDs, a landscape architect can also design streets and open spaces to enhance both community and privacy. (*Architect: SALA Architects, Inc., Minneapolis, Minnesota. Photo by Dale Mulfinger*)

THE DESIGN PROCESS

Stock Plans

Most houses in the United States and Canada are built from *stock plans*—plans that are selected by the owner from a catalog containing hundreds of predesigned houses. The first popular catalog of plans to appear in print was the book *Cottage Residences* by the architect A. J. Downing, which was published in 1842 (Figure 3.9). Previous to this, house plans were found primarily in pattern books that were used by carpenters to lay out and build traditional houses. Downing's

FIGURE 3.8
Interior designers and interior architects are trained to design interior spaces. Cabinetry, lighting design, material and color selections, window treatments, and furnishings all fall within the realm of their expertise. (*Architect: ABS Architects, New York. Photo by Alison Snyder*)

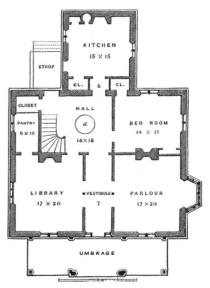

FIGURE 3.9
Plans such as this from the architect A. J. Downing's *Cottage Residences*, published in 1842, were widely admired and used across the United States. Plan books have kept pace with public taste throughout the decades and remain a major source of residential designs.

book was so well received that it went through four editions and was followed by numerous other books of plans by him and others. In 1846, *Godey's Lady's Book* magazine started publishing house plans with the offer that a more complete set of plans suitable for building could be obtained free of charge by writing to the magazine. *Godey's* plans were the first true stock plans and were used for the construction of thousands of residences before the magazine finally ceased publication in 1898. The stock plan tradition came to be assumed by building supply retailers, who benefited by distributing house plans free of charge.

In the early 1900s, a number of enterprising architects and builders recognized the potential to sell the stock plans that had previously been distributed for free. A number of stock plan companies emerged and flourished during the first half of the century. The production of these plans was simplified by the fact that the residential building profession had developed strong traditions, allowing details for each design to be sparse. Most of these companies went out of business during World War II when there was essentially no private residential building, but new companies, fueled by the postwar building boom, soon took their place.

Today, the stock plan business is a $70 million per year industry with about 250 companies mass-marketing their plans in the United States and Canada. Stock plans are commonly marketed in catalogs that can be found on the magazine racks in supermarkets, newsstands, and building supply outlets. They are often organized thematically under such titles as "Country House Plans," "Vacation Homes," or "Top-Selling 100 Plans." Plans are also syndicated in newspapers, published in popular home magazines, and offered on line. Typically, each design is formatted onto a single, magazine-sized page for the purpose of giving prospective customers an overview. This one-page advertisement usually includes a rendering of the house, a floor plan, a list of floor areas, and often a written description of the spaces (Figure 3.10). The most sophisticated marketers of stock plans offer compact discs (CDs) with three-dimensional walk-throughs of the most popular models and computerized plan selection based on criteria such as floor area, number of floors, number of bedrooms, and architectural style.

The consumer may order plans through the mail, over the phone, or on line. For $250 to $450, the customer receives a full set of construction drawings, including plans, elevations, section(s), foundation plan, interior elevations, and outline specifications. Additional copies are available at greatly reduced prices, and most companies offer a "construction package" of five to eight sets for distribution to contractors, subcontractors, and the building department. The plan services also offer plans in the form of reproducible transparencies that may be altered to customize the design.

Wraparound Delight

- This Country-style home enjoys an airy wraparound porch. A few perfectly placed rocking chairs translates into delightful evenings sipping a cold drink and chatting with family members.
- Inside, the living area's openness gives the feeling of a much larger home. Visit with friends for a spell in the living room before moving to the dining room to enjoy a delectable meal. Ample windows flood the space in natural light, and a door accesses the porch.
- The kitchen boasts a pantry closet, space for the washer and dryer, and lots of counter space for meal preparation. A window above the sink brightens daily chores.
- The master bedroom offers bright windows, a walk-in closet and access to the adjacent full bath. Two more bedrooms accommodate children or guests.
- The blueprints for this design include plans for a detached, one-car garage with a fully enclosed storage area.

Plan DD-1006	
Bedroom: 3	Baths: 1
Living Area:	
Main floor	1,005 sq. ft.
Total Living Area:	**1,005 sq. ft.**
Detached garage	224 sq. ft.
Storage	62 sq. ft.
Exterior Wall Framing:	**2x4**
Foundation Options:	
Crawlspace	
Slab	
(All plans can be built with your choice of foundation and framing. A generic conversion diagram is available.)	
BLUEPRINT PRICE CODE:	A

Figure 3.10

Today's plan books arrive mostly in the form of periodicals available at the supermarket. Typical information includes a perspective drawing from the street side, a plan with the sizes of rooms, and a verbal description. Full sets of plans suitable for construction can be ordered through the periodical. (*Courtesy of Homestyles, Inc., St. Paul, Minnesota*)

FIGURE 3.11
Fallingwater, designed by Frank Lloyd Wright, is perhaps the most famous example of the integration of house and landscape. (*Photo by Corey Saft*)

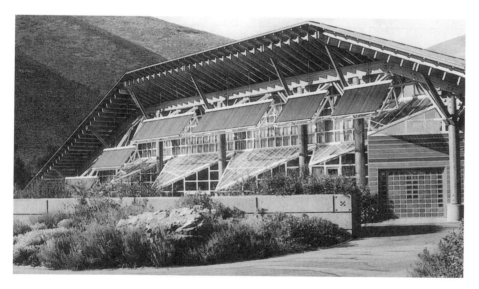

FIGURE 3.12
Sensitive design work by skilled professionals can yield superb results in a custom-designed house. (*Architect: Arne Bystrom, Seattle. Landscape Architect: Bob Murase, Seattle. Photo by Rob Thallon*)

posed setbacks from the property line, the slope of the site, and the locations of utilities, the driveway, and the septic tank if there is one.

A major drawback of stock plans is that they are designed without reference to any particular site. This problem is illustrated by two houses with the same floor plan, one built on the north and one on the south side of the same street. One of these houses will have a sunny front yard, the other a sunny back yard, yet the rooms and connections to the outdoors all too often are not adjusted for this fact.

Custom Design

The hiring of an architect or building designer to custom-design a house is a much more time-consuming and expensive process than purchasing stock plans, but it can produce far superior results. Principal among the advantages of a *custom-designed house* are the ability to integrate the house with the specific features of the site and the potential to tailor the house without compromise to the vision of the owners (Figure 3.11).

The process of designing a custom house varies considerably according to the type of project and the preferences of the participants. When an architect designs a house, the process can take 6 months or more and the design fees usually equal 10 to 15 percent of the construction cost of the house. The process is very meticulous and is intended to produce a house that meets the client's wishes in every detail. At a lower level of detail and customization, a building designer can produce a new design for the builder of a moderate-income speculative tract development in less than 2 weeks for a fee much lower than that of the architect, but still more costly than stock plans. Obviously, there are innumerable situations between these two extremes, and the more time that can be devoted to a project, the more refined the final product can be (Figure 3.12).

The stock plans offered for sale today are much more detailed than their predecessors and can generally be used, often with minor modifications, to procure a building permit. Because codes differ from jurisdiction to jurisdiction, stock plan services cannot guarantee that their plans will meet local codes, but they do claim that all plans are drawn to meet the requirements of national model codes. Modifications are commonly needed to meet structural requirements, which vary considerably from region to region. In order to apply for a building permit, a *site plan* must be added to the stock plan to inform the building department of the pro-

The designer of a custom house who is providing full services to a client will perform at least the following phases of work, each of which is described in the AIA literature:

1. *Schematic design.* The designer analyzes the site and distills the needs and desires of the owner through a series of meetings. This information is synthesized into a design that is approximate in terms of its size, its form, and the arrangement of the rooms (Figure 3.13).

2. *Design development.* The schematic design is refined so that the dimensions are precise and materials are described.

3. *Construction drawings.* The developed design is drawn in its final form to be distributed to the building department for permits, to contractors for bidding, to the bank for appraisal, and to the owner for review and for the purpose of making a contract with a general contractor. A written specification, which describes the expected quality of the work and lists the materials in detail, is produced to accompany the drawings.

4. *Contract negotiations.* The designer assists the owner with the competitive bidding process or in negotiating a contract for construction with a general contractor. In speculative work, this phase is often eliminated because the owner and contractor are one and the same.

5. *Construction observation.* The designer acts as the owner's representative at the construction site, assuring that the work is done according to the contract and helping to make decisions about constructional and aesthetic issues.

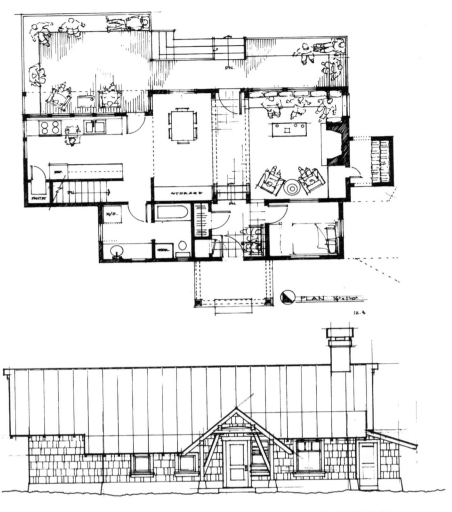

FIGURE 3.13
In custom-design work, the house can be tailored to a sloping site, and unique requests from the client can be incorporated without compromise. Schematic design sketches such as these form the basis of early communication between designer and client.

Although these phases are described as distinct sequential events, the first three or four often overlap. Usually, only licensed architects carry out the full complement of these phases. Building designers tend to work mostly in the second and third phases but may do more. Interaction with consultants can and does occur during any of the phases, but the best work is produced when consultants begin to collaborate with the architect or designer very early in the process.

Drafting Services

Many houses are designed by a builder or owner who takes an ad for a stock plan, sketches modifications to the design on the back of an envelope, and takes this to a *drafting service* or plan shop where it is "drafted up." Not much designing takes place at the drafting service, save for adding a bedroom or two or shifting the garage to the opposite side of an existing stock plan. Drafting services are run by building designers who hire drafters and specialize in turning simple ideas into a form suitable for obtaining a building permit and/or bids for construction. Drafting services also do some custom design, but their primary function is to provide low-cost plans to local speculative builders and owners looking for an inexpensive house to build on their own lot.

In most areas of the country, drafting services account for a whopping 70 to 80 percent of all single-family plans turned in for a permit. Drafting services have an advantage over national stock plan services because the proprietor, usually a building designer, is attuned to local building codes and practices. Drafting services usually have their own supply of stock plans from which prospective clients can choose, and much of their work consists of adjusting these or stock plans from other sources.

Design/Build

Design/build firms both design and build houses. Designers in design/

FIGURE 3.14
Models are used to demonstrate three-dimensional relationships. This model is a study of the house illustrated in Figure 3.3. (*Architect: SALA Architects, Inc., Minneapolis, Minnesota. Photo by Dale Mulfinger*)

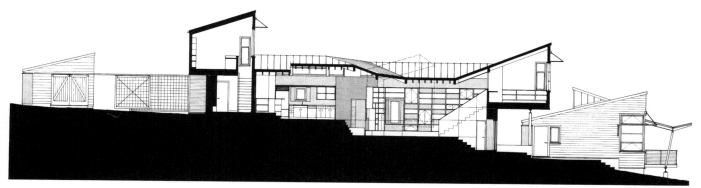

FIGURE 3.15
A computer-generated section/elevation study of a custom-designed house. (*Courtesy of Fernau & Hartman Architects, Inc., Berkeley, California*)

build firms may be either architects or building designers, and the builders often have an interest in design. Many firms consist of a single master builder. In larger design/ build firms, the communication between designer and builder is improved significantly because the same people work with each other on a consistent basis. Designers in a design/build firm have the advantage of being able to ask questions of the builders during the design process. Builders have opportunities to influence the design to make it more efficient to construct.

Because of the nature of the business, most of the work taken on by design/build firms is custom designed. Clients of design/build firms tend to want a custom-designed house or the remodeling of an existing house, but many design/build firms work for public or nonprofit agencies or produce their own speculative houses.

DESIGN SOURCES

Speculative House Resources

Builder magazines conduct annual polls to gauge public preferences with regard to the type, size, configuration, and style of house (Figure 3.16). The data from these surveys are used extensively by the designers of speculative houses and housing. Similar kinds of data are available to the designer through statistics showing the characteristics of recently purchased housing (Figure 2.18). Equally important to speculative builders and their designers are ideas from other successfully built projects in the area, which are readily incorporated into new designs in the hope that they will spawn a fresh success.

Custom House Resources

The recent work of other designers in the field is a great resource for designers of custom houses. Innovative

designs are fervently sought for publication by the popular media and are therefore available both to designers and to the lay public who become their clients. In fact, the house and garden magazines and coffee table books that feature beautiful and unique residences are one of the few components of the building culture that does not encourage the status quo. They are one of the main tools used by clients to inform their designers of their preferences, and they are used extensively by designers to illustrate design possibilities to their clients (Figure 3.17).

Technical Resources

Books and CDs on topics ranging from framing details to passive solar design to standard furniture sizes line the shelves of residential designers' offices (Figure 3.18). Produced independently by material manufacturers, trade associations, or testing

agencies, these references are consulted regularly during the design process. The building code is, of course, an indispensable reference in generating a design and resolving technical issues. When answers to technical questions are beyond the expertise of the designer, advice is sought from a consultant such as a structural engineer.

Design Associations

There are several professional associations that support the education of and exchange of information among residential designers. Architects may choose to join the American Institute of Architects (AIA), which has local, regional, and national meetings to exchange information about recent technical innovations and code interpretations and to share ideas about recently completed buildings. Building designers may join the American Institute of Building Design (AIBD),

First-Time Homebuyers		Move-Up Homebuyers	
Feature	**Percentage**	**Feature**	**Percentage**
High ceilings	86	Separate tub and shower	89
Bay windows	82	Walk-in pantry	78
Refrigerator	75	Grilltop range	75
Fireplace in family room	74	His-and-her closets	62
Ceiling fan in kitchen	72	Matching cabinets/appliances	62
Separate laundry room	70	Double oven	58
Built-in microwave oven	69	Formal dining room	56
Ceramic tile in master bath	68	Storage room	55
Stainless steel kitchen sink	67	Front porch	52
Single oven	66	Bookshelves	51
Separate tub and shower	65	Family room French doors	48
Walk-in closets	63	Attic	39
Walk-in pantry	54	Hardwood entry flooring	33
		Home office	36
		Game room	30

FIGURE 3.16
The features expected most frequently by various homebuyer groups. Consumer surveys such as this influence builders who instruct their designers to include features perceived to be attractive. (*Source:* Builder, *July 1992, Hanley-Wood, Inc.*)

FIGURE 3.17
Architects and other designers often disseminate their ideas to the general public through publication in the popular media. This house was commissioned by Life Magazine for its "Life Dream House®, 1999." (*Architect: SALA Architects, Inc., Minneapolis, Minnesota. Photo © Karen Melvin Photography*)

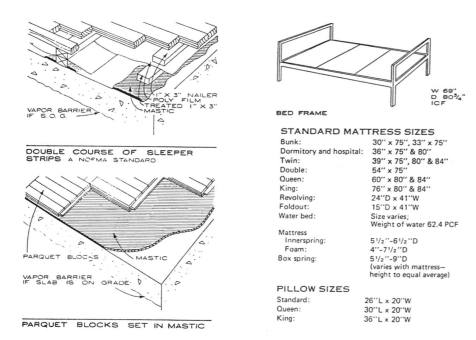

FIGURE 3.18
Reference books list the sizes of furniture and equipment, illustrate construction details, and explain systems. (*Reprinted with permission from Ramsey/Sleeper*, Architectural Graphic Standards, *10th ed., John Wiley & Sons, New York, 2000*)

which sponsors educational events and programs concentrated in the areas of residential design and planning. The AIBD also enforces a certification program for building designers to show evidence of their competency by demonstrating minimum levels of education and experience. Both the AIA and the AIBD offer extensive on-line support and information systems.

COMMUNICATION BETWEEN DESIGNER AND BUILDER

The designer conveys the design for a new house to the builder through a set of *construction documents*. The documents include *architectural drawings* done on large sheets (usually 18 by 24 inches or 24 by 36 inches) (A2 or A1 in Canada) and typed *specifications* on 8½ × 11 sheets (A4).

The drawings show configurations, locations, and sizes of things, and how things are assembled (Figures 3.19 to 3.22). Architectural drawings typically include

1. Site plan @ ¹⁄₁₆ inch per foot scale or smaller (1 : 200)

2. Floor plans @ ¼ inch per foot scale (1 : 50)

3. Exterior elevations @ ¼ inch per foot scale (1 : 50)

4. Foundation plan @ ¼ inch per foot scale (1 : 50)

5. Building sections @ ¼ inch per foot scale (1 : 50)

6. Floor and roof framing plans @ ¼ inch per foot scale or smaller (1 : 50)

7. Details of roof and wall edges and trim @ 1 inch per foot scale (1 : 10) or larger

8. Interior elevations of cabinets and special walls @ ¼ inch per foot scale (1 : 50)

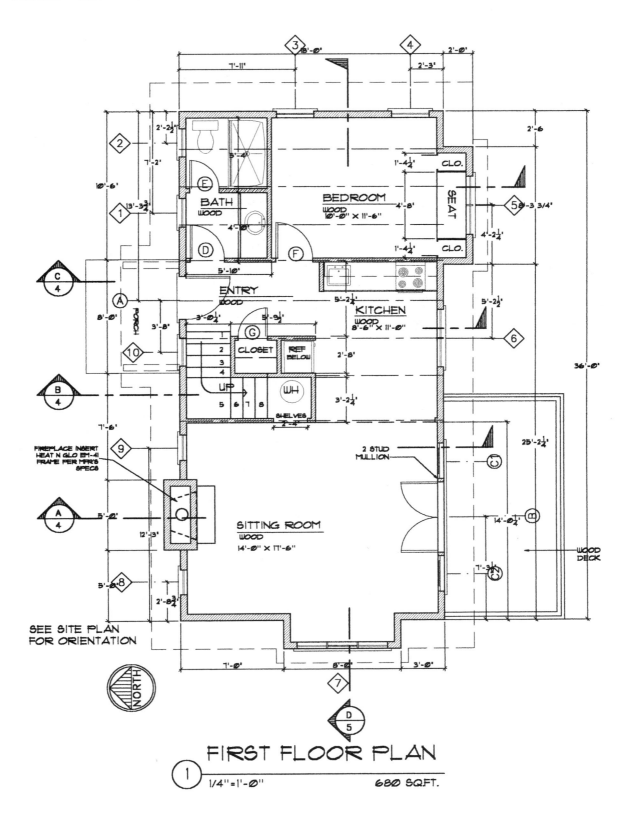

FIRST FLOOR PLAN

1 — 1/4"=1'-0"   680 SQ.FT.

FIGURE 3.19

Architectural plan drawings show essential information needed to construct a new residence. Rooms are labeled for reference, horizontal dimensions locate all walls, windows and doors are referenced to charts that describe them, and section lines (e.g., A4, B4, C4, and D5) show where section drawings occur. (*Courtesy of David Edrington, Architect*)

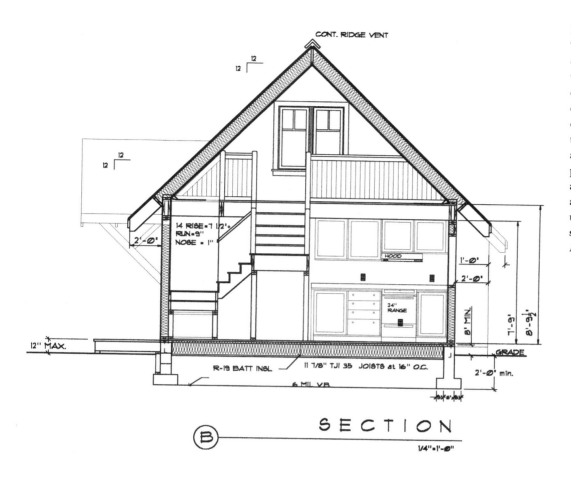

FIGURE 3.20
The section drawings give critical vertical dimensions that, together with the dimensions on the plan drawings, describe precisely the size and shape of the house. Section drawings also typically show roof pitch and insulation values and location. Eave details and foundation details are usually drawn at a larger scale. (*Courtesy of David Edrington, Architect*)

FIGURE 3.21
The simple building section and wall section are the only sections included in the construction drawings for a tract house. (*Courtesy of Dietz Design, Eugene, Oregon*)

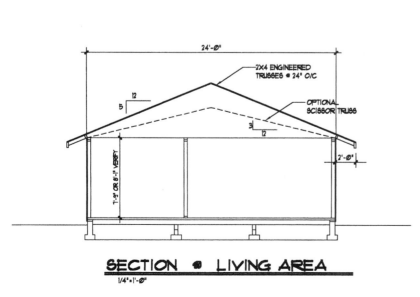

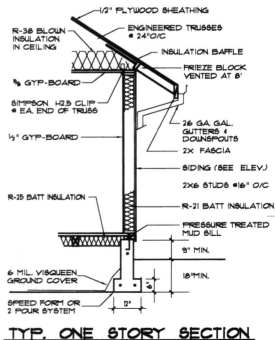

The specifications describe the parts of the design not easily illustrated in drawings such as the quality of materials, finishes, and workmanship. Typical sections cover the strength of the concrete in the foundation, the species and grade of lumber used for the frame and finishes, types and thicknesses of thermal insulation, type and quality of roof shingles and siding, and myriad other details down to the number of coats of paint on each surface of the house. In addition, specifications often describe contractual items such as who pays for electricity during construction and what is to be done with excavated soil. The amount of detail contained in the specifications depends on the type of project, with documents for single speculative houses usually being the least detailed and those for large multifamily projects the most.

After the construction documents have been completed, the further involvement of the designer may range from very minimal to extensive, depending on the type of project. In mass-produced speculative work, the designer is rarely consulted after the drawings have been completed unless there is a serious problem. The design itself has been worked out to facilitate uninterrupted work flow by subcontractors, and input by designers during the construction process only impedes progress and delays the schedule. In custom-designed work, on the other hand, designers often allocate as much as 20 percent of their fee for construction observation. Especially in remodeling projects, the problems and opportunities for integrating new work with old cannot be visualized until the existing building has been demolished to expose its construction. Even in new work, designs are typically so complex that opportunities to make adjustments for site conditions and for the construction process are frequent. Taking advantage of these opportunities often adds quality to the building. For example, the slight adjustment of a window for a particular view, the selection of finish materials such as tile or paint in the light of the newly constructed space, and the detailing of trim are decisions best made by the designer on the site (Figure 3.23). Effective communication between builder and designer is mandatory to allow design decisions to be made in this way.

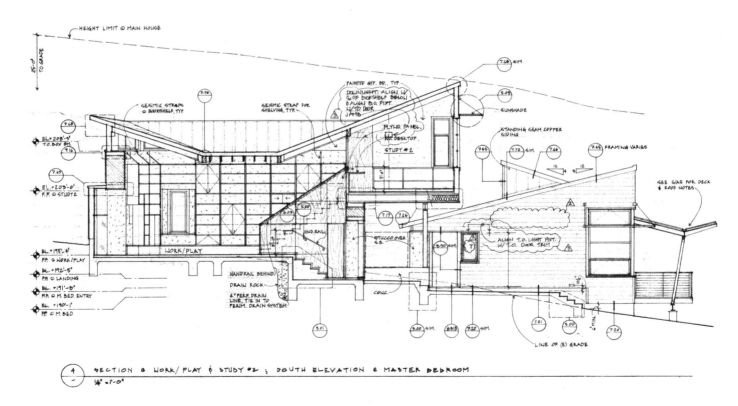

FIGURE 3.22
This detailed section drawing is only one of several included in the construction drawings for a custom-designed house on a sloping site.
(*Courtesy of Fernau & Hartman Architects, Inc., Berkeley, California*)

FIGURE 3.23
Many design decisions are made at the site, especially for custom-designed houses. When not controlled, this practice can lead to cost overruns, but in the best of cases, it can incorporate into the work subtle adjustments to the site and insights from craftsmen. In this case, the base of the stair was designed as a collaboration between architect and finish carpenter. Notice the return air grill integrated into the baseboard below the railing. (*Architect: Thallon Architecture. Photo by Rob Thallon*)

SELECTED REFERENCES

1. Gutman, Robert. *The Design of American Housing.* New York: Publishing Center for Cultural Resources, 1985.

This interesting and scholarly work clearly portrays the architect's role in the design of housing, but does not seriously investigate the role of others.

2. Walker, Lester. *American Shelter.* Woodstock, NY: Overlook Press, 1996.

A chronicle of American domestic architecture from Native American shelters of the 6th century A.D. to the present day. Each style is documented with drawings and the materials and logic of its form are discussed.

3. Moore, Charles, Lyndon, Donlyn, and Allen, Gerald. *The Place of Houses.* Berkeley: University of California Press, 2000.

This is a thought-provoking inquiry into the nature of houses and the art of their design.

4. Alexander, Christopher, et al. *A Pattern Language.* New York: Oxford University Press, 1977.

This widely used and often-quoted work outlines a design philosophy and strategy based on positive human values. Only slightly more than half of the book deals with design at a residential scale.

5. Wentling, James W. *Housing by Lifestyle.* New York: McGraw–Hill, 1995.

A well-written and clearly illustrated book of design suggestions for the builders and developers of contemporary houses and suburbs.

6. Hunter, Christine. *Ranches, Row Houses, and Railroad Flats.* New York: Norton, 1999.

Following a lucid discussion of housing requirements and the means to satisfy them, design aspects of the three basic types of houses—freestanding houses, attached houses, and apartments—are considered.

7. Hoke, John Ray, Jr., Ed. *Architectural Graphic Standards* (10th ed.). New York: John Wiley & Sons, 2000.

The standard reference for architects and designers describing the sizes of things and roughly how they are assembled. The book deals with buildings of all scales.

KEY TERMS AND CONCEPTS

architect
American Institute of Architects (AIA)
building designer
American Institute of Building Design (AIBD)
Certified Professional Building Designer
structural engineer
landscape architect
American Society of Landscape Architects (ASLA)

land planner
American Planning Association (APA)
American Institute of Certified Planners (AICP)
interior designer
interior decorator
stock plan
site plan
custom-designed house
schematic design

design development
construction drawings
contract negotiations
construction observation
drafting service
design/build
construction documents
architectural drawings
specifications

REVIEW QUESTIONS

1. What is the difference between an architect and a building designer?

2. Discuss the roles of the landscape architect and the land planner in the process of developing a multifamily residential project such as a PUD.

3. What is the role of the interior designer as compared to the architect?

4. How are stock plans different from plans for a custom-designed house?

5. List the five principal phases of the architect's work in a residential project.

6. What are the advantages and disadvantages of the design/build organizational structure from the point of view of the professional? From the point of view of the owner/client?

7. Describe the difference between construction drawings and construction documents.

EXERCISES

1. Have class members visit the offices of design professionals involved in residential work—an architect, a building designer, a landscape architect, a structural engineer, and an interior designer. At each office, try to determine the profile of the typical client, the amount of time allocated to a typical project, and the nature and frequency of interactions with the builder, the client, and other professionals. Conduct a class discussion to compare notes.

2. Obtain used sets of both stock plans and architect-designed custom plans. Display these in class and discuss their differences.

3. Visit a range of new housing in your area and compare the designs in relation to probable design intentions.

MATERIALS

THE MATERIAL WOOD

4

Wood is probably the best loved of all the materials that we use for building. It delights the eye with its endlessly varied colors and grain patterns. It invites the hand to feel its subtle warmth and varied textures. When it is fresh from the saw, its fragrance enchants. We treasure its natural, organic qualities and take pleasure in its genuineness. Even as it ages, bleached by the sun, eroded by rain, worn by the passage of feet and the rubbing of hands, we find beauty in its transformations of color and texture.

Wood earns our respect as well as our love. It is strong and stiff, yet by far the least dense of the materials used for the beams and columns of buildings. It is worked and fastened easily with small, simple, relatively inexpensive tools. It is readily recycled from demolished buildings for use in new ones, and, when finally discarded, it biodegrades rapidly to become natural soil. It is our only renewable building material, one that will be available to us for as long as we manage our forests with an eye to the perpetual production of wood.

However, wood, like a valued friend, has its idiosyncrasies. A piece of lumber is never perfectly straight or true, and its size and shape can change significantly with changes in the weather. Wood is peppered with defects that are relics of its growth and processing. Wood can split; wood can warp; wood can give splinters. If ignited, wood burns. If left in a damp location, it decays and harbors destructive insects. However, the skillful designer and the seasoned carpenter know all these things and understand how to build with wood to bring out its best qualities, while neutralizing or minimizing its defects.

TREES

Because wood comes from trees, an understanding of tree physiology is essential to knowing how to build with wood.

Tree Growth

The trunk of a tree is covered with a protective layer of dead *bark* (Figure 4.1). Inside the dead bark is a layer of living bark composed of hollow longitudinal cells that conduct nutrients down the trunk from the leaves to the roots. Inside this layer of living bark lies a very thin layer, the *cambium*, which creates new bark cells toward the outside of the trunk and new wood cells toward the inside. The thick layer of living wood cells inside the cambium is the *sapwood*. In this zone of the tree, nutrients are stored and sap is pumped upward from the roots to the leaves and distributed laterally in the trunk. At the inner edge of this zone, sapwood dies progressively and becomes *heartwood*. In many species of trees, heartwood is easily distinguished from sapwood by its darker color. Heartwood no longer participates in the life processes of the tree but continues to contribute to its structural strength. At the very center of the trunk, surrounded by heartwood, is the *pith* of the tree, a small zone of weak wood cells that were the first year's growth.

An examination of a small section mof wood under a low-powered microscope shows that it consists primarily of tubular cells whose long axes are parallel to the long axis of the trunk. The cells are structured of tough *cellulose* and are bound together by a softer cementing substance called *lignin*. The direction of the long axes of the cells is referred to as the *direction of grain* of the wood. Grain direction is important to the designer of wooden buildings because the properties of wood parallel to grain and perpendicular to grain are very different.

In temperate climates, the cambium begins to manufacture new sapwood cells in the spring, when the air is cool and groundwater is plentiful, conditions that favor rapid growth. Growth is slower during the heat of the summer when water is scarce. *Springwood* or *earlywood* cells are therefore larger and less dense in substance than the *summerwood* or *latewood* cells. Concentric bands of springwood and summerwood make up the annual growth rings in a trunk that can be counted to determine the age of a tree. The relative proportions of springwood and summerwood also have a direct bearing on the structural properties of the wood a given tree will yield because summerwood is stronger and stiffer than springwood. A tree grown under continuously moist, cool conditions grows faster than another tree of the same species grown under warmer, drier conditions, but its wood is not as dense or as strong.

Softwoods and Hardwoods

Softwoods come from coniferous trees and *hardwoods* from broad-leafed trees. The names can be deceptive because many coniferous trees produce harder woods than many broad-leafed trees, but the distinction is nevertheless a useful one. Softwood trees have a relatively simple microstructure, consisting mainly of large longitudinal cells (*tracheids*) together with a small percentage of radial cells (*rays*), whose function is the storage and radial transfer of nutrients (Figure 4.2). Hardwood trees are more complex in

FIGURE 4.1
Summerwood rings are prominent and a few rays are faintly visible in this cross section of an evergreen tree, but the cambium, which lies just beneath the thick layer of bark, is too thin to be seen, and heartwood cannot be distinguished visually from sapwood in this species. (*Courtesy of Forest Products Laboratory, Forest Service, USDA*)

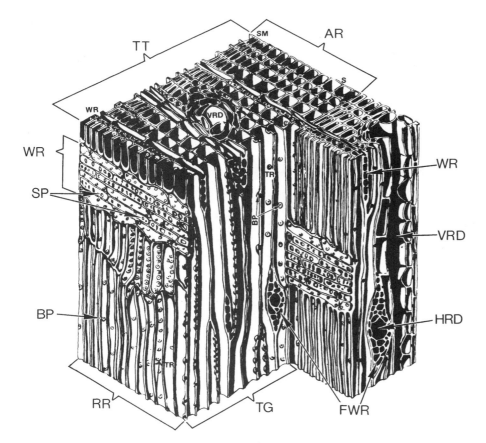

Cell structure of a softwood

FIGURE 4.2
Vertical cells (tracheids, labeled TR) dominate the structure of a softwood, seen here greatly enlarged, but rays (WR), which are cells that run radially from the center of the tree to the outside, are clearly in evidence. An annual ring (labeled AR) consists of a layer of smaller summerwood cells (SM) and layer of larger springwood cells (S). Simple pits (SP) allow sap to pass from ray cells to longitudinal cells and vice versa. Resin is stored in vertical and horizontal resin ducts (VRD and HRD) with the horizontal ducts centered in fusiform wood rays (FWR). Border pits (BP) allow the transfer of sap between longitudinal cells. The face of the sample labeled RR represents a radial cut through the tree, and TG, a tangential cut. (*Courtesy of Forest Products Laboratory, Forest Service, USDA*)

structure, with a much larger percentage of rays and two different types of longitudinal cells: small-diameter *fibers* and large-diameter *vessels* or *pores*, which transport the sap of the tree (Figure 4.3).

When cut into lumber, softwoods generally have a coarse and relatively uninteresting grain structure, while many hardwoods show beautiful patterns of rays and vessels (Figure 4.4). Most of the lumber used today for the frames of buildings comes from softwoods, which are comparatively plentiful and inexpensive. For fine furniture and interior finish

details, hardwoods are often chosen. A few softwood and hardwood species widely used in North America are listed in Figure 4.5, along with the principal uses of each. However, it should be borne in mind that literally thousands of species of wood are used in construction around the world and that the available species vary considerably with geographic location. The major lumber-producing forests in North America are in the western and eastern mountains of both the United States and Canada, and the southeastern United States, but other regions also produce significant quantities.

(a)

FIGURE 4.4
The grain figures of two softwood species (left) and two hardwoods (right) demonstrate the difference in cellular structure between the classes of woods. From left

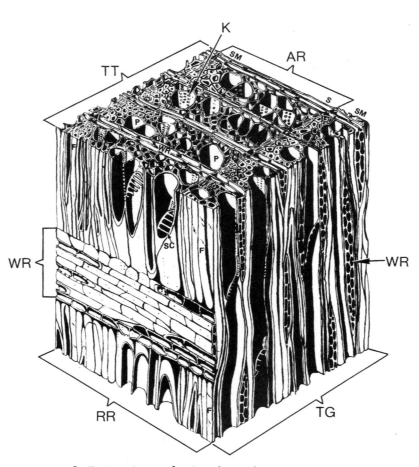

Cell structure of a hardwood

FIGURE 4.3
Rays (WR) constitute a large percentage of the mass of a hardwood, as seen in this sample, and are largely responsible for the beautiful grain figures associated with many species. The vertical cell structure is more complex than that of a softwood, with large pores (P) to transport the sap, and smaller wood fibers (F) to give the tree structural strength. Pore cells in some hardwood species end with crossbars (SC), while those of other species are entirely open. Pits (K) pass sap from one cavity to another. (*Courtesy of Forest Products Laboratory, Forest Service, USDA*)

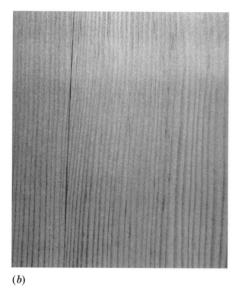  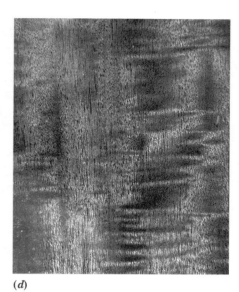

(b) (c) (d)

to right: (*a*) The cells in sugar pine are so uniform that the grain structure is almost invisible except for scattered resin ducts. (*b*) Vertical-grain Douglas fir shows very pronounced dark bands of summerwood. (*c*) Red oak exhibits large open pores amid its fibers. (*d*) This quarter-sliced Mahogany veneer has a pronounced "ribbon" figure caused by varying light reflections off its fibers. (*Photos by Edward Allen*)

Softwoods

Used for Framing, Sheathing, Paneling

Alpine fir
Balsam fir
Douglas fir
Eastern hemlock
Eastern spruce
Eastern white pine
Englemann spruce
Idaho white pine
Larch
Loblolly pine
Lodgepole pine
Longleaf pine
Mountain hemlock
Ponderosa pine
Red spruce
Shortleaf pine
Sitka spruce
Southern yellow pine
Western hemlock
White spruce

Used for Moldings, Window and Door Frames

Ponderosa pine
Sugar pine
White pine

Used for Finish Flooring

Douglas fir
Longleaf pine

Decay-Resistant Woods. Used for Shingles, Siding, Outdoor Structures

California redwood
Southern cypress
Western red cedar
White cedar

Hardwoods

Used for Moldings, Paneling, Furniture

Ash
Beech
Birch
Black walnut
Butternut
Cherry
Lauan
Mahogany
Pecan
Red oak
Rosewood
Teak
Tupelo gum
White oak
Yellow poplar

Used for Finish Flooring

Pecan
Red oak
Sugar maple
Walnut
White oak

FIGURE 4.5
Some species of wood commonly used in construction in North America, listed alphabetically in groups according to end use. All are domestic except Lauan (Asia), Mahogany (Central America), rosewood (South America and Africa), and teak (Asia).

LUMBER

Sawing

The production of *lumber,* lengths of squared wood for use in construction, begins with the felling of trees and the transportation of the logs to a sawmill (Figure 4.6). Sawmills range in size from tiny family operations to giant, semiautomated factories, but the process of lumber production is much the same regardless of scale. Each log is first passed repeatedly through a large *headsaw,* which may be either a circular saw or a bandsaw, to reduce it to untrimmed slabs of lumber (Figure 4.7). The *sawyer* (with the aid of a computer in the larger mills) judges how to obtain the maximum marketable wood from each log and uses hydraulic machinery to rotate and advance the log in order to achieve the required succession of cuts. As the slabs fall from the log at each pass, a conveyor belt carries them away to smaller saws that reduce them to square-edged pieces of the desired widths (Figure 4.8). The sawn pieces at this stage of production have rough-textured surfaces and may vary slightly in dimension from one end to the other.

Most lumber intended for use in the framing of buildings is *plainsawed,* a method of dividing the log that produces the maximum yield and therefore the greatest economy (Figure 4.9). In plainsawed lumber, some pieces have the annual rings running virtually perpendicular to the faces of the piece, some have the rings on various diagonals, and some have the rings running almost parallel to the faces. These varying grain orientations cause the pieces to distort differently during seasoning, to have very different surface appearances from one another, and to erode at different rates when used in applications such as flooring and exterior siding. For uses in which any of these variations will cause a problem, especially for finish flooring, interior trim, and furniture, hardwoods are often *quartersawn* to produce pieces of lumber that have the annual rings running more nearly perpendicular to the face of the piece. Boards with this *vertical grain* orientation tend to remain flat despite changes in moisture content. The visible grain on the surface of a quartersawn piece makes a tighter and more pleasing *figure.* The wearing qualities of the piece are improved because there are no broad areas of soft springwood exposed in the face, as there are when the annual rings run more nearly parallel to the face.

Seasoning

Wood in a living tree contains a quantity of water that can vary from about 30 percent to as much as 300 percent of the oven-dry weight of the wood. After a tree is cut, this water slowly starts to evaporate. First to leave the wood is the *free water* from the cavities of the cells. When the free water is gone, the wood still contains about 26 to 32 percent moisture, and the *bound water* held within the cellulose

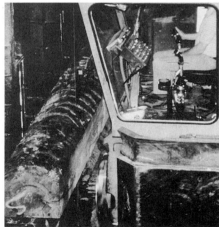

FIGURE 4.7
In a large, mechanized mill, the operator controls the high-speed bandsaw from an overhead booth. (*Courtesy of Western Wood Products Association*)

FIGURE 4.6
Loading Southern pine logs onto a truck for their trip to the sawmill. (*Courtesy of Southern Forest Products Association*)

FIGURE 4.8
Sawn lumber is sorted into stacks according to its cross-sectional dimensions and length. (*Courtesy of Western Wood Products Association*)

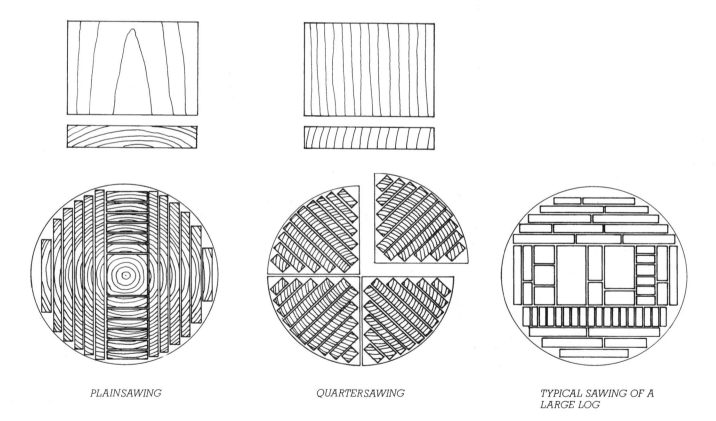

PLAINSAWING

QUARTERSAWING

TYPICAL SAWING OF A
LARGE LOG

FIGURE 4.9
Plainsawing produces boards with a broad grain figure, as seen in the end and top views above the plainsawed log. Quartersawing produces a vertical grain structure, which is seen on the face of the board as tightly spaced parallel summerwood lines. A large log of softwood is typically sawn to produce some large timbers, some plainsawed dimension lumber, and, in the horizontal row of small pieces seen just below the heavy timbers, some pieces of vertical-grain decking.

of the cell walls begins to evaporate. As the first of the bound water disappears, the wood starts to shrink, and the strength and stiffness of the wood begin to increase. The shrinkage, stiffness, and strength increase steadily as the moisture content decreases. Wood can be dried to any desired moisture content, but framing lumber is considered *seasoned* when its moisture content is 19 percent or less. For framing applications that require closer control of wood shrinkage, lumber seasoned to a 15 percent moisture content, labeled "MC 15," is available. It is of little use to season ordinary framing lumber to a moisture content below about 13 percent because wood is hygroscopic and will take on or give off moisture, swelling or shrinking as it does so, in order to stay in equilibrium with the moisture in the air.

Most lumber is seasoned at the sawmill, either by *air drying* in loose stacks for a period of months or, more commonly, by *kiln drying* under closely controlled conditions of temperature and humidity for a period of days (Figures 4.10 and 4.11). Seasoned lumber is stronger and stiffer than unseasoned *(green)* lumber, more dimensionally stable, and lighter and more economical to ship. Kiln drying is generally preferred to air drying because it is much faster and produces better quality lumber.

Wood does not shrink and swell uniformly with changes in moisture content. Moisture shrinkage along the length of the log *(longitudinal shrinkage)* is negligible for practical purposes. Shrinkage in the radial direction *(radial shrinkage)* is very large by comparison, and shrinkage around the circumference of the log *(tangential shrinkage)* is about half again greater than radial shrinkage (Figure 4.12). If an entire log is seasoned before sawing, it will shrink very little along its length, but it will grow noticeably smaller in diameter, and the difference between the tangential and radial shrinkage will cause it to *check,* that is, to split open all along its length (Figure 4.13).

These differences in shrinkage rates are so large that they cannot be ignored in building design. In constructing building frames of plainsawed lumber, a simple distinction is made between *parallel-to-grain shrinkage,* which is negligible, and *perpendicular-to-grain shrinkage,* which is considerable. The difference between radial and tangential shrinkage is not considered because the orientation of the annual rings in plainsawed lumber is random and unpredictable. As we will see in Chapter 9, wood building frames are carefully designed to equalize the amount of wood loaded perpendicular to grain from one side of the structure to the other in order to avoid the noticeable tilting of floors and tearing of wall finish materials that would otherwise occur.

The position in a log from which a piece of lumber is sawn determines in large part how it will distort as it dries. Figure 4.14 shows how the differences between tangential and radial shrinkage cause this to happen. These effects are pronounced and are readily predicted and observed in everyday practice.

FIGURE 4.10
For proper air drying, lumber is supported well off the ground. The stickers, which keep the boards separated for ventilation, are carefully placed above one another to avoid bending the lumber, and a watertight roof protects each stack from rain and snow. (*Courtesy of Forest Products Laboratory, Forest Service, USDA*)

FIGURE 4.11
Measuring moisture content in boards in a drying kiln. (*Courtesy of Western Wood Products Association*)

FIGURE 4.12
Shrinkage of a typical softwood with decreasing moisture content. Longitudinal shrinkage, not shown on this graph, is so small by comparison to tangential and radial shrinkage that it is of no consequence in wood buildings. (*Courtesy of Forest Products Laboratory, Forest Service, USDA*)

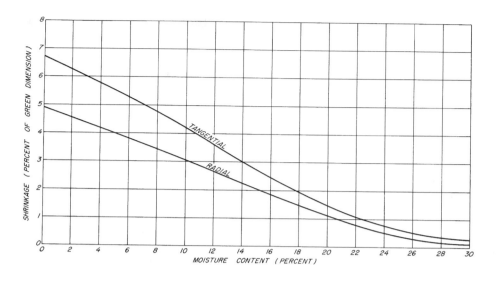

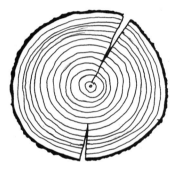

FIGURE 4.13
Because tangential shrinkage is so much greater than radial shrinkage, high internal stresses are created in a log as it dries, resulting inevitably in the formation of radial cracks called checks.

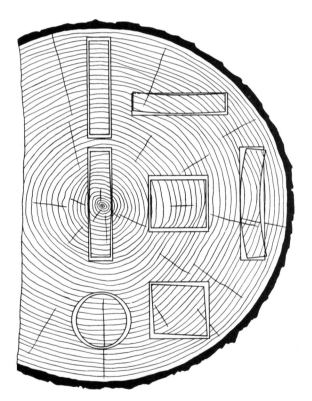

FIGURE 4.14
The difference between tangential and radial shrinkage also produces seasoning distortions in lumber. The nature of the distortion depends on the position the piece of lumber occupied in the tree. The distortions are the most pronounced in plainsawed lumber (upper right, extreme right, lower right). (*Courtesy of Forest Products Laboratory, Forest Service, USDA*)

Surfacing

Lumber is *surfaced* to make it smooth and more dimensionally precise. Rough (unsurfaced) lumber is often available commercially and is used for many purposes, but surfaced lumber is easier to work with because it is more square and uniform in dimension and less damaging to the hands of the carpenter. Surfacing is done by high-speed automatic machines whose rotating blades plane the surfaces of the piece and round the edges slightly. Most lumber is surfaced on all four sides *(S4S)*, but hardwoods are often surfaced on only two sides *(S2S)*, leaving the two edges to be finished by the craftsman.

Lumber is usually seasoned before it is surfaced, which allows the planing process to remove some of the distortions that occur during seasoning. However, for some framing lumber, this order of operations is reversed. The designation *S-DRY* in a lumber gradestamp indicates that the piece was surfaced (planed) when in a seasoned (dry) condition, and *S-GRN* indicates that it was planed when green.

Lumber Defects

Almost every piece of lumber contains one or more discontinuities in its structure caused by *growth characteristics* of the tree from which it came or manufacturing characteristics that were created at the mill (Figures 4.15 and 4.16). Among the most common growth characteristics are *knots,* which are places where branches joined the

trunk of the tree; *knotholes,* which are holes left by loose knots dropping out of the wood; *decay;* and *insect damage.* Knots and knotholes reduce the structural strength of a piece of lumber, make it more difficult to cut and shape, and are often considered detrimental to its appearance. Decay and insect damage that occurred during the life of the tree may or may not

affect the useful properties of the piece of lumber, depending on whether the organisms are still alive in the wood and the extent of the damage they have done.

Manufacturing characteristics arise largely from changes that take place during the seasoning process because of the differences in rates of shrinkage with varying orientations to

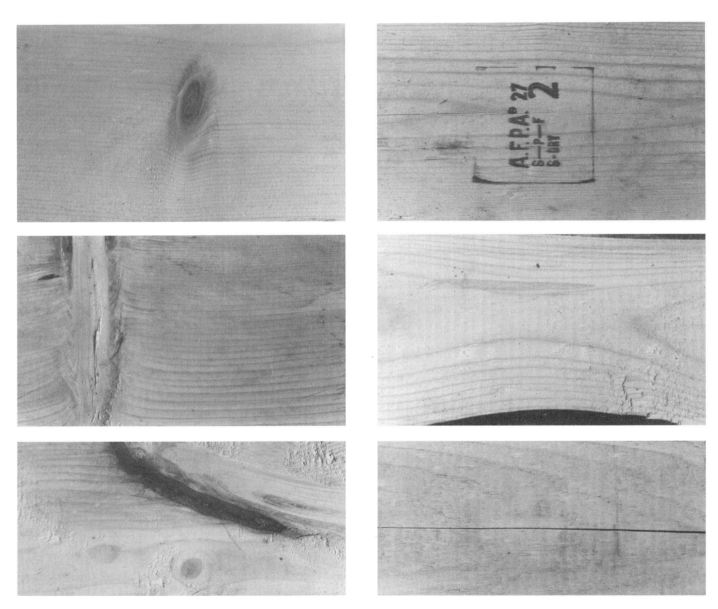

FIGURE 4.15
Surface features often observed in lumber include, in the left-hand column from top to bottom, a knot cut crosswise, a knot cut longitudinally, and a bark pocket. To the right are a gradestamp, wane on two edges of the same piece, and a small check. The gradestamp indicates that the piece was graded according to the rules of the American Forest Products Association, that it is #2 grade Spruce-Pine-Fir, and that it was surfaced after drying. The 27 is a code number for the mill that produced the lumber. (*Photos by Edward Allen*)

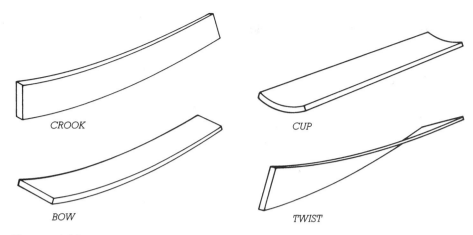

FIGURE 4.16
Four types of seasoning distortions in dimension lumber.

FIGURE 4.17
The effects of seasoning distortions can often be minimized through knowledgeable detailing practices. As an example, this wood baseboard, seen in cross section, has been formed with a relieved back, which is a broad, shallow groove that allows the piece to lie flat against the wall even if it cups (broken lines). The sloping bottom on the baseboard assures that it can be installed tightly against the floor despite the cup. The grain orientation in this piece is the worst possible with respect to cupping. If quartersawn lumber were not available, the next best choice would have been to mill the baseboard with the center of the tree toward the room rather than toward the wall.

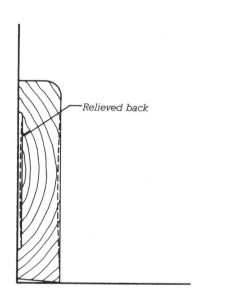

strength and stiffness, depending on its intended use, before it leaves the mill. Lumber is sold by species and grade; the higher the grade, the higher the price. Grading offers the architect and the engineer the opportunity to build as economically as possible by using only as high a grade as is required for a particular use. In a specific building, the main beams or columns may require a very high structural grade of lumber, while the remainder of the framing members will perform adequately in an intermediate, less expensive grade. For blocking, the lowest grade is perfectly adequate. For finish trim that will be coated with a clear finish, a high appearance grade is desirable; for painted trim, a lower grade will suffice.

Structural grading of lumber may be done either visually or by machine. In *visual grading,* trained inspectors examine each piece for ring density and for growth and manufacturing characteristics, then judge it and stamp it with a grade in accordance with industry-wide grading rules (Figure 4.18). In *machine grading,* an

the grain. *Splits* and *checks* are usually caused by shrinkage stresses. *Crooking, bowing, twisting,* and *cupping* all occur because of nonuniform shrinkage. Wane is an irregular rounding of edges or faces that is caused by sawing pieces too close to the perimeter of the log. Experienced carpenters judge the extent of these defects and distortions in each piece of lumber and decide accordingly where and how to use the piece in the building. Checks and shakes are of little consequence in framing lumber, but a joist or rafter with a crook in it is usually placed with the convex edge (the "crown") facing up, to allow the floor or roof loads to

straighten the piece. Carpenters straighten badly bowed joists or rafters by sawing or planing away the crown before they apply subflooring or sheathing over them. Badly twisted pieces are put aside to be cut up for blocking. The effects of cupping in flooring and interior baseboards and trim are usually minimized by using quartersawn stock and by shaping the pieces so as to reduce the likelihood of distortion (Figure 4.17).

Lumber Grading

Each piece of lumber is graded either for appearance or for structural

FIGURE 4.18
A grader marks the grade on a piece with a lumber crayon, preparatory to applying a gradestamp. (*Courtesy of Western Wood Products Association*)

DIMENSION LUMBER GRADES Table **2.1**

Product	Grades	WWPA Western Lumber Grading Rules Section Reference	Uses
Structural Light Framing (SLF) 2″ to 4″ thick 2″ to 4″ wide	SELECT STRUCTURAL NO.1 NO.2 NO.3	(42.10) (42.11) (42.12) (42.13)	Structural applications where highest design values are needed in light framing sizes.
Light Framing (LF) 2″ to 4″ thick 2″ to 4″ wide	CONSTRUCTION STANDARD UTILITY	(40.11) (40.12) (40.13)	Where high-strength values are not required, such as wall framing, plates, sills, cripples, blocking, etc.
Stud 2″ to 4″ thick 2″ and wider	STUD	(41.13)	An optional all-purpose grade designed primarily for stud uses, including bearing walls.
Structural Joists and Planks (SJ&P)	SELECT STRUCTURAL NO.1 NO.2 NO.3	(62.10) (62.11) (62.12) (62.13)	Intended to fit engineering applications for lumber 5″ and wider, such as joists, rafters, headers, beams, trusses, and general framing.

STRUCTURAL DECKING GRADES Table **2.2**

Product	Grades	WWPA Western Lumber Grading Rules Section Reference	Uses
Structural Decking 2″ to 4″ thick 4″ to 12″ wide	SELECTED DECKING	(55.11)	Used where the appearance of the best face is of primary importance.
	COMMERCIAL DECKING	(55.12)	Customarily used when appearance is not of primary importance.

TIMBER GRADES Table **2.3**

Product	Grades	WWPA Western Lumber Grading Rules Section Reference	End Uses
Beams and Stringers 5″ and thicker, width more than 2″ greater than thickness	DENSE SELECT STRUCTURAL* DENSE NO. 1* DENSE NO. 2* SELECT STRUCTURAL NO.1 NO.2	(53.00 & 170.00) (53.00 & 170.00) (53.00 & 170.00) (70.10) (70.11) (70.12)	Grades are designed for beam and stringer type uses when sizes larger than 4″ nominal thickness are required.
Post and Timbers 5″ × 5″ and larger, width not more than 2″ greater than thickness	DENSE SELECT STRUCTURAL* DENSE NO. 1* DENSE NO. 2* SELECT STRUCTURAL NO.1 NO.2	(53.00 & 170.00) (53.00 & 170.00) (53.00 & 170.00) (80.10) (80.11) (80.12)	Grades are designed for vertically loaded applications where sizes larger than 4″ nominal thickness are required.

*Douglas Fir or Douglas Fir-Larch only.

FIGURE 4.19
Standard structural grades for western softwood lumber. For each species of wood, the allowable structural stresses for each of these grades are tabulated in the structural engineering literature. (*Courtesy of Western Wood Products Association*)

automatic device assesses the structural properties of the wood and stamps a grade automatically on the piece. This assessment is made either by flexing each piece between rollers and measuring its resistance to bending or by scanning the wood electronically to determine its density. Appearance grading, naturally enough, is done visually. Figure 4.19 outlines a typical grading scheme for framing lumber, and Figure 4.20 outlines the appearance grades for nonstructural lumber. Light framing lumber for houses and other small buildings is usually ordered as "#2 and better" (a mixture of #1 and #2 grades) for floor joists and roof rafters and as "Stud" grade for wall framing.

The Structural Strength of Wood

The strength of a piece of wood depends chiefly on its species, its grade, and the direction in which the load acts with respect to the direction of grain of the piece. Wood is several times stronger parallel to grain than perpendicular to grain. With its usual assortment of defects, it is stronger in compression than in tension. *Allowable strengths* (structural stresses that include factors of safety) vary tremendously with species and grade. Allowable compressive strength parallel to grain, for example, varies from 325 to 1700 pounds per square inch (2.24 to 11.71 MPa) for commercially available grades and species of framing lumber, a difference of more than five times. Figure 4.21 compares the structural properties of an "average" framing lumber to those of the other common structural materials—brick masonry, steel, and concrete. Of the four materials, only wood and steel have useful tensile strength. Defect-

APPEARANCE LUMBER GRADES			Table **2.5**
Product	**Grades[1]**	**Equivalent Grades in Idaho White Pine**	**WWPA Grading Rules Section Number**
Highest Quality Appearance Grades			
Selects *(all species)*	B & BTR SELECT C SELECT D SELECT	SUPREME CHOICE QUALITY	10.11 10.12 10.13
Finish *(usually available only in Doug Fir and Hem-Fir)*	SUPERIOR PRIME E		10.51 10.52 10.53
Special Western Red Cedar Pattern Grades	CLEAR HEART A GRADE B GRADE		20.11 20.12 20.13
General Purpose Grades			
Common Boards (WWPA Rules) *(primarily in pines, spruces, and cedars)*	1 COMMON 2 COMMON 3 COMMON 4 COMMON 5 COMMON	COLONIAL STERLING STANDARD UTILITY INDUSTRIAL	30.11 30.12 30.13 30.14 30.15
Alternate Boards (WCLIB Rules) *(primarily in Doug Fir and Hem-Fir)*	SELECT MERCHANTABLE CONSTRUCTION STANDARD UTILITY ECONOMY		WCLIB[3] 118-a 118-b 118-c 118-d 118-e
Special Western Red Cedar Pattern[2] Grades	SELECT KNOTTY QUALITY KNOTTY		WCLIB[3] 111-e 111-f

[1]Refer to WWPA's *Vol. 2, Western Wood Species* book for full-color photography and to WWPA's *Natural Wood Siding* for complete information on siding grades, specification, and installation.

FIGURE 4.20
Nonstructural boards are graded according to appearance. (*Courtesy of Western Wood Products Association*)

Material	Allowable Tensile Strength	Allowable Compressive Strength	Density	Modulus of Elasticity
Wood (average)	700 psi (4.83 MPa)	1,100 psi (7.58 MPa)	30 pcf (480 kg/m³)	1,200,000 psi (8,275 MPa)
Brick masonry (average)	0	250 psi (1.72 MPa)	120 pcf (1,920 kg/m³)	1,200,000 psi (8,275 MPa)
Steel (ASTM A36)	22,000 psi (151.69 MPa)	22,000 psi (151.69 MPa)	490 pcf (7,850 kg/m³)	29,000,000 psi (200,000 MPa)
Concrete (average)	0	1,350 psi (9.31 MPa)	145 pcf (2,320 kg/m³)	3,150,000 psi (21,720 MPa)

FIGURE 4.21
Comparative physical properties of the four common structural materials: wood, masonry, steel, concrete.

free wood is comparable to steel on a strength-per-unit-weight basis, but with the ordinary run of defects, an average piece of lumber is somewhat inferior to steel by this yardstick.

When designing a wooden structure, the designer determines the maximum stresses that are likely to occur in each of the structural members and selects an appropriate species and grade of lumber for each. In a given locale, a limited number of species and grades are usually available in retail lumberyards, and it is from these that the selection is made. It is common practice to use a stronger but more expensive species (Douglas fir or Southern pine, for example) for highly stressed major members, and to use a weaker, less expensive species (such as Eastern hemlock) or species group (Hem–Fir, Spruce–Pine–Fir) for the remainder of the structure. Within each species, the designer selects grades based on published tables of allowable stresses. The higher the grade, the higher the allowable stress, but the lower the grade, the less costly the lumber.

There are many factors other than species and grade that influence the useful strength of wood. These include the length of time the wood will be subjected to its maximum load, the temperature and moisture conditions under which it serves, and the size and shape of the piece. Certain fire-retardant treatments also reduce the strength of wood slightly. All these factors are taken into account when engineering a building structure of wood.

Lumber Dimensions

Lumber sizes in the United States are given as *nominal dimensions* in inches. A piece nominally 1 by 2 inches in cross section is a 1 × 2 ("one by two"), a piece 2 by 10 inches is a 2 × 10, and so on. At the time a piece of lumber is sawn, it may approach these dimensions. Subsequent to sawing, however, seasoning and surfacing diminish its size substantially. By the time a kiln-dried 2 × 10 reaches the lumberyard, its actual dimensions are 1½ by 9¼ inches (38 × 235 mm). The relationship between nominal lumber dimensions (which are always written without inch marks) and *actual dimensions* (which are written *with* inch marks) is given in simplified form in Figure 4.22 and in complete form in Figure 4.23. Anyone who designs or constructs wooden buildings soon commits the simpler of these relationships to memory. Because of changing moisture content and manufacturing tolerances, however, it is never wise to assume that a piece of lumber will conform precisely to its intended dimension.

The experienced detailer of wooden buildings knows not to treat wood as if it had the dimensional accuracy of steel. The designer working with an existing wooden building will find a great deal of variation in lumber sizes. Wood members in hot, dry locations such as attics often will have shrunk to dimensions substantially below their original measurements. Members in older buildings may have been manufactured to full nominal dimensions or to earlier standards of actual dimensions such as 1⅝ inches or 1¾ inches (41 or 44 mm) for a nominal 2-inch member.

Pieces of lumber less than 2 inches in nominal thickness are called *boards*. Pieces ranging from 2 to 4 inches in nominal thickness are referred to collectively as *dimension lumber*. Pieces nominally 5 inches and more in thickness are termed *timbers*. Dimension lumber is usually supplied in 2-foot (610-mm) increments of length. The most commonly used lengths are 8, 10, 12, 14, and 16 feet (2.44, 4.05, 4.66, 4.27, and 4.88 m), but most retailers stock rafter material in lengths to 24 feet (7.32 m). Actual lengths are usually a fraction of an inch longer than nominal lengths.

Lumber in the United States is priced by the *board foot*. Board foot measurement is based on nominal dimensions, not actual dimensions. A board foot of lumber is defined as a

FIGURE 4.22
The relationship between nominal and actual dimensions for the most common sizes of kiln-dried lumber is given in this simplified chart, which is extracted from the complete chart in Figure 4.23.

Nominal Dimension	Actual Dimension
1"	¾" (19 mm)
2"	1½" (38 mm)
3"	2½" (64 mm)
4"	3½" (89 mm)
5"	4½" (114 mm)
6"	5½" (140 mm)
8"	7¼" (184 mm)
10"	9¼" (235 mm)
12"	11¼" (286 mm)
Over 12"	¾" less (19 mm less)

solid volume 12 square inches in nominal cross-sectional area and 1 foot long. A 1 × 12 or 2 × 6 that is 10 feet long contains 10 board feet. A 2 × 4 that is 10 feet long contains $[(2 \times 4)/12] \times 10 = 6.67$ board feet, and so on. Prices of dimension lumber and timbers in the United States are usually quoted on the basis of dollars per thousand board feet. In other parts of the world, lumber is sold by the cubic meter.

The architect and engineer specify lumber for a particular construction

STANDARD SIZES—FRAMING LUMBER — Table **2.4**
Nominal & Dressed (Based on *Western Lumber Grading Rules*)

Product	Description	Nominal Size: Thickness (inches)	Nominal Size: Width (inches)	Dressed Dimensions, Thicknesses & Widths (inches): Surfaced Dry	Dressed Dimensions, Thicknesses & Widths (inches): Surfaced Unseasoned	Length (feet)
DIMENSION	S4S	2 3 4	2 3 4 5 6 8 10 12 over 12	1½ 2½ 3½ 4½ 5½ 7¼ 9¼ 11¼ off ¾	1 9/16 2 9/16 3 9/16 4⅝ 5⅝ 7½ 9½ 11½ off ½	6′ and longer, generally shipped in multiples of 2′
				Thickness (unseasoned)	**Width (unseasoned)**	
TIMBERS	Rough or S4S (shipped unseasoned)	5 and larger		1/2 off nominal (S4S). See 3.20 of WWPA Grading Rules for Rough.		6′ and longer, generally shipped in multiples of 2′
		Thickness	**Width**	**Thickness (dry)**	**Width (dry)**	
DECKING	2″ (Single T&G)	2	5 6 8 10 12	1½	4 5 6¾ 8¾ 10¾	6′ and longer, generally shipped in multiples of 2′
	3″ and 4″ (Double T&G)	3 4	6	2½ 3½	5¼	

Abbreviations: FOHC—Free of Heart Center; S4S—Surfaced four sides; T&G—Tongued and grooved; Rough Full Sawn—Unsurfaced lumber cut to full specified size

FIGURE 4.23
A complete chart of nominal and actual dimensions for both framing lumber and finish lumber. (*Courtesy of Western Wood Products Association*)

use by designating its species, grade, seasoning, surfacing, nominal size, and chemical treatment, if any. When ordering lumber, the contractor must additionally give the required lengths of pieces and the required number of pieces of each length.

WOOD PRODUCTS

Much of the wood used in construction has been processed into laminated wood, structural composite wood, or wood panel products in order to overcome the various shortcomings of solid wood structural members.

Laminated Wood

Large structural members are often produced by joining many small strips of wood together with glue to form *laminated wood* (called *glulam* for short). There are three major reasons to laminate: size, shape, and quality. Any desired size of structural member can be laminated, up to the capacities of the hoisting and transportation machinery needed to deliver and erect it, without having to search for a tree of sufficient girth and height. Wood can be laminated into shapes that cannot be obtained in nature: curves, angles, varying cross sections (Figure 4.24). Quality can be specified and closely controlled in laminated members because defects can be cut out of the wood before laminating. Seasoning is carried out before the wood is laminated (largely eliminating the checks and distortions that characterize solid timbers), and the strongest, highest quality wood can be placed in the parts of the member that will be subjected to the highest structural stresses. The fabrication of laminated members obviously adds to the cost per board foot, but this is often overcome by the smaller size of the laminated member that can replace a solid timber of equal load-carrying

FIGURE 4.24
Glue laminating a U-shaped timber for a ship. Several smaller members have also been glued and clamped and are drying alongside the larger timber. (*Courtesy of Forest Products Laboratory, Forest Service, USDA*)

capacity. In many cases, solid timbers are simply not available at any price in the required size, shape, or quality.

Individual laminations are usually 1½ inches (38 mm) thick except in curved members with small bending radii where ¾-inch (19-mm) stock is used. End joints between individual pieces are either *finger jointed* or *scarf jointed*. These types of joints allow the glue to transmit tensile and compressive forces longitudinally from one piece to the next within a lamination (Figure 4.25). Adhesives are chosen according to the moisture conditions under which the member will serve. Any size member can be laminated, but standard depths range from 3 to 75 inches (76 to 1905 mm). Standard widths range from 2⅛ to 14¼ inches (54 to 362 mm).

Structural Composite Lumber

A number of wood producers have developed *structural composite lumber* products that are made up of ordinary plywood veneers. Unlike plywood, however, in these products the grains of all the veneers are oriented in the longitudinal direction of the piece of lumber to achieve maximum bending strength. One type of product, *laminated veneer lumber (LVL)*, uses the veneers in sheets and looks like a thick plywood with no crossbands (Figure 4.46). In another type, *parallel strand lumber (PSL)*, the veneers are sliced into narrow strands that are coated with adhesive, oriented longitudinally, pressed into a rectangular cross section, and cured under heat and pressure (Figure 4.26). LVL is generally pro-

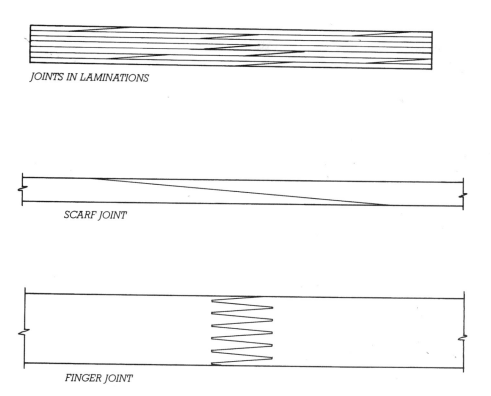

JOINTS IN LAMINATIONS

SCARF JOINT

FINGER JOINT

FIGURE 4.25
Joints within a lamination of a glue-laminated beam, seen in the upper drawing in a small-scale elevation view, must be scarf jointed or finger jointed to transmit the tensile and compressive forces from one piece of wood to the next. The individual pieces of wood are prepared for jointing by high-speed machines that mill the scarf or fingers with rotating cutters of the appropriate shape.

FIGURE 4.26
This parallel strand lumber, made up of strands of wood veneer, is largely free of defects, making it far stronger and stiffer than ordinary dimension lumber. (*Courtesy of Trus Joist MacMillan*)

FIGURE 4.27
Plywood is made of veneers selected to give the optimum combination of economy and performance for each application. This sheet of roof sheathing plywood is faced with a D veneer on the underside and a C veneer on the top side. These veneers, though unattractive, perform well structurally and are much less costly than the higher grades incorporated into plywoods made for uses where appearance is important. (*Courtesy of APA–The Engineered Wood Association*)

FIGURE 4.28
Five different wood panel products, from top to bottom: plywood, composite panel, waferboard, oriented strand board, and particleboard. (*Courtesy of APA–The Engineered Wood Association*)

duced in smaller sizes that approximate those of dimension lumber, whereas PSL is manufactured in a wide range of sizes, including large dimensions like those of glue-laminated timbers. Structural composite lumber products offer much the same benefits as laminated wood products, but they can be produced more economically through a higher degree of mechanization.

Wood Panel Products

Wood in panel form is advantageous for many building applications (Figure 4.27). The panel dimensions are usually 4 by 8 feet (1.22 × 2.44 m). Panels require less labor for installation than boards because fewer pieces must be handled, and wood panel products are fabricated in such a way that they minimize many of the problems of boards and dimension lumber. Panels are more nearly equal in strength in their two principal directions than solid wood, and shrinking, swelling, checking, and splitting are greatly reduced. Additionally, panel products make more efficient use of forest resources than solid wood products through less wasteful ways of reducing logs to building products and through utilization in some types of panels of material that would otherwise be thrown away—branches, undersized trees, and mill wastes.

Structural wood panel products fall into the following three general categories (Figure 4.28):

1. *Plywood panels,* which are made up of thin wood *veneers* glued together. The grain on the front and back face veneers runs in the same direction, while the grain in one or more interior crossbands runs in the perpendicular direction. There is always an odd number of layers in plywood, which equalizes the effects of moisture movement, but an interior layer may be made up of a single veneer or of two veneers with their grains running in the same direction.

2. *Composite panels,* which have two parallel face veneers bonded to a core of reconstituted wood fibers.

3. *Nonveneered panels,* which are of three different types:

a. *Oriented strand board* (OSB), which is made of long, strandlike wood particles compressed and glued into three to five layers. The strands are oriented in the same manner in each layer as the grains of the veneer layers in plywood. Because of the length and controlled orientation of the strands, oriented strand board is generally stronger and stiffer than the other two types of nonveneered panels.

b. *Waferboard,* which is composed of large, waferlike flakes of wood compressed and bonded into panels.

c. *Particleboard,* which is manufactured in several different classes, all of which are made up of smaller wood particles than oriented strand board or waferboard, compressed and bonded into panels.

APA–The Engineered Wood Association has established performance standards that allow considerable interchangeability among these various types of panels for many construction uses. The standards are based on the structural adequacy for a specified use, the dimensional stability under varying moisture conditions, and the durability of the adhesive bond that holds the panel together. Because it can be produced from small trees and even branches, oriented strand board is generally more economical than plywood and is gradually replacing it in many construction applications.

Plywood Production

Veneers for structural panels are *rotary sliced:* Logs are soaked in hot water to soften the wood, then each is rotated in a large lathe against a stationary knife that peels off a continuous strip of veneer, much as paper is unwound from a roll (Figures 4.29 and 4.30). The strip of veneer is clipped into sheets that pass through a drying kiln where in a few minutes their moisture content is reduced to roughly 5 percent. The sheets are then assembled into larger sheets, repaired as necessary with patches glued into the sheet to

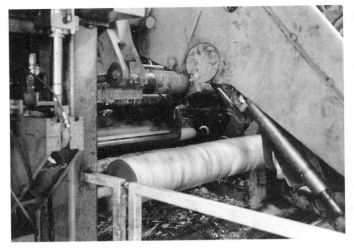

(a)

(b)

(c)

FIGURE 4.29

Plywood manufacture. (*a, b*) A 250-horsepower lathe spins a softwood log as a knife peels off a continuous sheet of veneer for plywood manufacture. (*c*) An automatic clipper removes unusable areas of veneer and trims the rest into sheets of the proper size for plywood panels. (*d*) The clipped sheets are fed into a continuous forced-air dryer, along whose 150-foot (45-m) path they will lose about half their weight in moisture. (*e*) Leaving the dryer, the sheets have a moisture content of about 5 percent. They are graded and sorted at this point in the process. (*f*) The higher grades of veneer are patched on this machine that punches out defects and replaces them with tightly fitted wood plugs. (*g*) In the layup line, automatic machinery applies glue to one side of each sheet of veneer and alternates the grain direction of the sheets to produce loose plywood panels. (*h*) After layup, the loose panels are prepressed with a force of 300 tons per panel to consolidate them for easier handling. (*i*) Following prepressing, panels are squeezed individually between platens heated to 300

(d)

degrees Fahrenheit (150°C) to cure the glue. (*j*) After trimming, sanding, or grooving as specified for each batch, the finished plywood panels are sorted into bins by grade, ready for shipment. (*Photos* b *and* i *courtesy of Georgia-Pacific; others courtesy of APA– The Engineered Wood Association*)

(e)

(f)

(g)

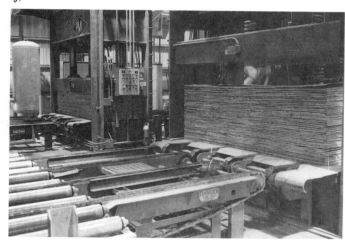

(h)

(i)

(j)

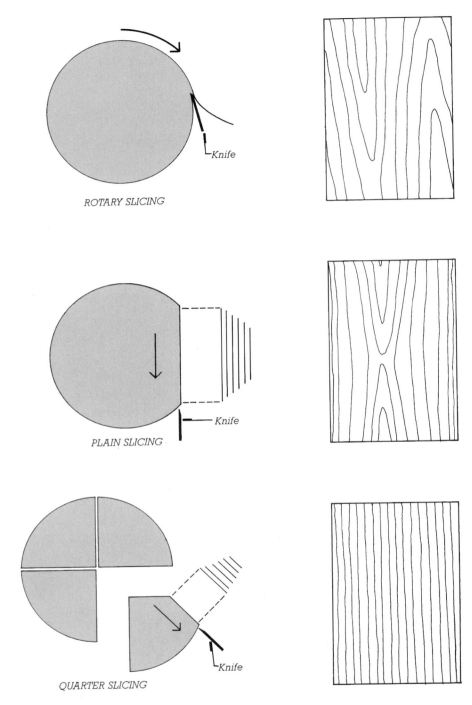

FIGURE 4.30
Veneers for structural plywood are rotary sliced, which is the most economical method. For better control of grain figure in face veneers of hardwood plywood, flitches are plain sliced or quarter sliced. The grain figure produced by rotary slicing, as seen in the detail to the right, is extremely broad and uneven. The finest figures are produced by quarter slicing, which results in a very close grain pattern with prominent rays.

fill open defects, and graded and sorted according to quality (Figure 4.31). A machine spreads glue onto the veneers as they are laid atop one another in the required sequence and grain orientations. The glued veneers are transformed into plywood in presses that apply elevated temperatures and pressures to create dense, flat panels. The panels are trimmed to size, sanded as required, graded, and gradestamped before shipping. Grade B veneers and higher are always sanded smooth, but panels intended for sheathing are always left unsanded because sanding slightly reduces the thickness, which diminishes the structural strength of the panel. Panels intended for subfloors and floor underlayment are lightly *touch sanded* to produce a more flat and uniform surface without seriously affecting their structural performance.

Veneers for hardwood plywoods intended for interior paneling and cabinetwork are usually sliced from square blocks of wood called *flitches* in a machine that moves the flitch vertically against a stationary knife (Figure 4.30). Flitch-sliced veneers are analogous to quartersawn lumber: They exhibit a much tighter and more interesting grain figure than rotary-sliced veneers. They can also be arranged on the plywood face in such a way as to produce symmetrical grain patterns.

Standard plywood panels are 4 by 8 feet (1220 × 2440 mm) in surface area and range in thickness from ¼ to 1 inch (6.4 to 25.4 mm). Longer panels are manufactured for siding and industrial use. Actual surface dimensions of the structural grades of plywood are slightly less than nominal. This permits the panels to be installed with small spaces between them to allow for moisture expansion. Composite panels and nonveneered panels are manufactured by analogous processes to the same set of standard sizes as plywood panels and to some larger sizes as well.

TABLE 1

VENEER GRADES

A Smooth, paintable. Not more than 18 neatly made repairs, boat, sled, or router type, and parallel to grain, permitted. Wood or synthetic repairs permitted. May be used for natural finish in less demanding applications.

B Solid surface. Shims, sled or router repairs, and tight knots to 1 inch across grain permitted. Wood or synthetic repairs permitted. Some minor splits permitted.

C Plugged — Improved C veneer with splits limited to 1/8-inch width and knotholes or other open defects limited to 1/4 x 1/2 inch. Wood or synthetic repairs permitted. Admits some broken grain.

C Tight knots to 1-1/2 inch. Knotholes to 1 inch across grain and some to 1-1/2 inch if total width of knots and knotholes is within specified limits. Synthetic or wood repairs. Discoloration and sanding defects that do not impair strength permitted. Limited splits allowed. Stitching permitted.

D Knots and knotholes to 2-1/2-inch width across grain and 1/2 inch larger within specified limits. Limited splits are permitted. Stitching permitted. Limited to Exposure 1 or Interior panels.

FIGURE 4.31
Veneer grades for softwood plywood. (*Courtesy of APA–The Engineered Wood Association*)

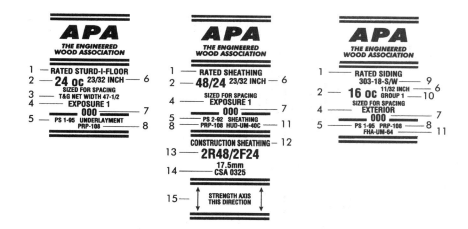

1 Panel grade
2 Span Rating
3 Tongue-and-groove
4 Exposure durability classification
5 Product Standard
6 Thickness
7 Mill number
8 APA's Performance Rated Panel Standard
9 Siding face grade
10 Species group number
11 HUD/FHA recognition
12 Panel grade, Canadian standard
13 Panel mark – Rating and end-use designation, Canadian standard.
14 Canadian performance-rated panel standard
15 Panel face orientation indicator

FIGURE 4.32
Typical gradestamps for structural wood panels. Gradestamps are found on the back of each panel. (*Courtesy of APA–The Engineered Wood Association*)

Specifying Structural Wood Panels

For structural uses such as subflooring and sheathing, wood panels may be specified either by thickness or by *span rating*. The span rating is determined by laboratory load testing and is given on the gradestamp on the back of the panel, as shown in Figures 4.32 and 4.33. The purpose of the span rating system is to permit the use of many different species of woods and types of panels to achieve the same structural objectives. A panel with a span rating of 32/16 may be used as roof sheathing over rafters spaced 32 inches (813 mm) apart or as subflooring over joists spaced 16 inches (406 mm) apart. The long dimension of the sheet must be placed perpendicular to the length of the supporting members. A 32/16 panel may be plywood, composite, or OSB, may be composed of any accepted wood species, and may be any of several thicknesses, so long as it passes the structural tests for a 32/16 rating.

The designer must also select from three *exposure durability classifications* for structural wood panels: *Exterior, Exposure 1,* and *Exposure 2.* Panels marked "Exterior" are suitable for use as siding or in other permanently exposed applications. "Exposure 1" panels have fully waterproof glue but do not have veneers of as high a quality as those of "Exterior" panels; they are suitable for structural sheathing and subflooring, which must often endure repeated wetting during construction. "Exposure 2" is suitable for panels that will be fully protected from weather and will be subjected to a minimum of wetting during construction. About 95 percent of structural panel products are classified "Exposure 1."

For panels intended as finish surfaces, the quality of the face veneers is of obvious concern and should be specified by the designer. For some types of work, fine flitch-sliced hardwood face veneers may be selected

TABLE 2

GUIDE TO APA PERFORMANCE RATED PANELS[a][b]
FOR APPLICATION RECOMMENDATIONS, SEE FOLLOWING PAGES.

APA RATED SHEATHING
Typical Trademark

Specially designed for subflooring and wall and roof sheathing. Also good for a broad range of other construction and industrial applications. Can be manufactured as plywood, as a composite, or as OSB. EXPOSURE DURABILITY CLASSIFICATIONS: Exterior, Exposure 1, Exposure 2. COMMON THICKNESSES: 5/16, 3/8, 7/16, 15/32, 1/2, 19/32, 5/8, 23/32, 3/4.

APA STRUCTURAL I
RATED SHEATHING[c]
Typical Trademark

Unsanded grade for use where shear and cross-panel strength properties are of maximum importance, such as panelized roofs and diaphragms. Can be manufactured as plywood, as a composite, or as OSB. EXPOSURE DURABILITY CLASSIFICATIONS: Exterior, Exposure 1. COMMON THICKNESSES: 5/16, 3/8, 7/16, 15/32, 1/2, 19/32, 5/8, 23/32, 3/4.

APA RATED STURD-I-FLOOR
Typical Trademark

Specially designed as combination subfloor-underlayment. Provides smooth surface for application of carpet and pad and possesses high concentrated and impact load resistance. Can be manufactured as plywood, as a composite, or as OSB. Available square edge or tongue-and-groove. EXPOSURE DURABILITY CLASSIFICATIONS: Exterior, Exposure 1, Exposure 2. COMMON THICKNESSES: 19/32, 5/8, 23/32, 3/4, 1, 1-1/8.

APA RATED SIDING
Typical Trademark

For exterior siding, fencing, etc. Can be manufactured as plywood, as a composite or as an overlaid OSB. Both panel and lap siding available. Special surface treatment such as V-groove, channel groove, deep groove (such as APA Texture 1-11), brushed, rough sawn and overlaid (MDO) with smooth- or texture-embossed face. Span Rating (stud spacing for siding qualified for APA Sturd-I-Wall applications) and face grade classification (for veneer-faced siding) indicated in trademark. EXPOSURE DURABILITY CLASSIFICATION: Exterior. COMMON THICKNESSES: 11/32, 3/8, 7/16, 15/32, 1/2, 19/32, 5/8.

(a) Specific grades, thicknesses and exposure durability classifications may be in limited supply in some areas. Check with your supplier before specifying.

(b) Specify Performance Rated Panels by thickness and Span Rating. Span Ratings are based on panel strength and stiffness. Since these properties are a function of panel composition and configuration as well as thickness, the same Span Rating may appear on panels of different thickness. Conversely, panels of the same thickness may be marked with different Span Ratings.

(c) All plies in Structural I plywood panels are special improved grades and panels marked PS 1 are limited to Group 1 species. Other panels marked Structural I Rated qualify through special performance testing. Structural II plywood panels are also provided for, but rarely manufactured. Application recommendations for Structural II plywood are identical to those for APA RATED SHEATHING plywood.

FIGURE 4.33
A guide to specifying structural panels. Plywood panels for use in paneling, furniture, and other applications where appearance is important are graded by the quality of their face veneers. (*Courtesy of APA–The Engineered Wood Association*)

rather than rotary-sliced softwood veneers, and the matching pattern of the veneers specified.

Other Wood Panel Products

Several types of nonstructural or semistructural panels of wood fiber are often used in construction. *Hardboard* is a thin, dense panel made of highly compressed wood fibers. It is available in general-purpose panels of standard dimension and in several thicknesses and surface finishes. *Cane fiberboard* is a thick, low-density panel with some thermal insulating value; it is used in wood construction chiefly as a nonstructural or semistructural wall sheathing. Panels made of recycled paper are low in cost and are useful for wall sheathing, carpet underlayment, and incorporation in certain proprietary types of roof decking and insulating assemblies.

CHEMICAL TREATMENT

Various chemical treatments have been developed to counteract two major weaknesses of wood: its combustibility and its susceptibility to attack by decay and insects. Fire-retardant treatment is accomplished by placing lumber in a pressure vessel and impregnating it with certain chemical salts that greatly reduce its combustibility. The cost of fire-retardant-treated wood is such that it is little used in single-family residential construction. Its major uses are roof sheathing in attached houses and nonstructural partitions and other interior components in buildings of fire-resistant construction.

Decay and insect resistance is very important in wood that is used in or very near to the ground, and in wood used for exposed outdoor structures such as marine docks, fences, decks, and porches. Decay-resistant treatment is accomplished by *pressure impregnation* with any of several types of preservatives. *Creosote*

is an oily derivative of coal that is widely used in engineering structures, but because of its odor, toxicity, and unpaintability, it is unsuitable for most purposes in building construction, and its use is banned in many local areas. *Pentachlorophenol* is also impregnated as an oil solution, and, as with other oily preservatives, wood treated with it cannot be painted. The most widely used preservatives in building construction are *waterborne salts*. The most common of these is chromated copper arsenate (CCA), which imparts a greenish color to the wood but permits subsequent painting or staining. A newer formulation is based on copper and quaternary ammonia and causes wood to weather to a warm brown color. Preservatives based on boron compounds are also available. Preservatives of any of these types can be brushed or sprayed onto wood, but long-lasting protection (30 years and more) can only be accomplished by pressure impregnation. Wood treated with waterborne salts is often sold without drying, which is appropriate for use in the ground, but for interior use, it must be kiln dried after treatment. Because of the poisonous nature of most types of wood preservatives, their use is often controlled by environmental regulations.

The heartwood of some species of wood is naturally resistant to decay and insects and can be used instead of preservative-treated wood. The most commonly used decay-resistant species are Redwood, Bald cypress, and Red and White cedars. The sapwood of these species is no more resistant to attack than that of any other tree, so "All-Heart" grade should be specified.

Wood–polymer composite planks intended for use in outdoor construction are a recent addition to the lumber market. A typical product of this type is made from wood fiber and recycled polyethylene. Its advantages are decay resistance and easy workability. Most such products are not as stiff as sawn lumber and require spe-

cial attention to structural considerations in design.

Most wood-attacking organisms need both air and moisture to live. Most can therefore be kept out of wood by constructing and maintaining a building so that its wood components are kept dry at all times. This includes keeping all wood well clear of the soil, ventilating attics and crawlspaces to remove moisture, using good construction detailing to keep wood dry, and fixing roof and plumbing leaks as soon as they occur.

WOOD FASTENERS

Fasteners have always been the weak link in wood construction. The interlocking timber connections of the past, laboriously mortised and pegged, were weak because much of the wood in a joint had to be removed to make the connection. In today's wood connections, it is usually impossible to insert enough nails, screws, or bolts in a connection to develop the full strength of the members being joined. Adhesives and toothed plates are often capable of achieving this strength, but their use is largely limited to the factory. Fortunately, most connections in wood structures depend primarily on the direct bearing of one member on another for their strength, and a variety of simple fasteners are adequate for the majority of purposes.

Nails

Nails are sharpened metal pins that are driven into wood with a hammer or a mechanical nail gun. *Common nails* and *finish nails* are the two types most frequently used. Common nails have flat heads and are used for most structural connections in light frame construction. Finish nails are virtually headless and are used to fasten finish woodwork, where they are less obtrusive than common nails (Figure 4.34).

In the United States, the size of a nail is measured in *pennies*. This

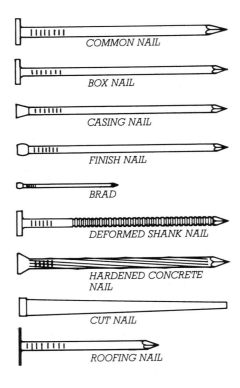

FIGURE 4.34
All nailed framing connections are made with common nails or their machine-driven equivalent. Box nails, which are made of lighter gauge wire, do not have as much holding power as common nails; they are used in construction for attaching wood shingles. Casing nails, finish nails, and brads are used for attaching finish components of a building. Their heads are set below the surface of the wood with a steel punch, and the holes are filled before painting. Deformed shank nails, which are very resistant to withdrawal from the wood, are used for such applications as attaching gypsum wallboard and floor underlayment, materials that cannot be allowed to work loose in service. Concrete nails can be driven short distances into masonry or concrete for attaching furring strips and sleepers. Cut nails, once widely used for framing connections, are now used mostly for attaching finish flooring because their square tips punch through the wood rather than wedge through, minimizing splitting of brittle woods. The large head on roofing nails is needed to apply sufficient holding power to the soft material of which asphalt shingles are made.

strange unit probably originated long ago as the price of a hundred nails of a given size and persists in use despite the effects of inflation on nail prices. Figure 4.35 shows the dimensions of the various sizes of common nails. Finish nails are the same length as common nails of the corresponding penny designation.

Nails are ordinarily furnished *bright,* meaning that they are made of plain, uncoated steel. Nails that will be exposed to the weather should be of a corrosion-resistant type, either *hot-dip galvanized,* aluminum, or stainless steel. (The zinc coating on *electrogalvanized* nails is very thin and is often damaged during driving.) Corrosion resistance is particularly important in nails for exterior siding, trim, and decks, which would be stained by rust leaching from bright nails.

The three methods of fastening with nails are shown in Figure 4.36. Each of these methods has its uses in building construction, as illustrated in Chapters 9 and 18. Nails are the favored means of fastening wood because they require no predrilling of holes under most conditions and they are extremely fast to install.

Wood Screws and Lag Screws

Screws are inserted into drilled holes and turned into place with a screwdriver or wrench (Figure 4.37). They are little used in ordinary wood light framing because they cost more and take much longer to install than nails, but they are often used in cabinetwork, in furniture, and for mounting hardware such as hinges. Screws form tighter, stronger connections than nails and can be backed out and reinserted if a component needs to be adjusted or remounted. Very large screws for heavy structural connections are called *lag screws.* They have square or hexagonal heads and are driven with a wrench rather than a screwdriver. Small *drywall screws,* which can be driven by a power screwdriver, without first drilling holes in most cases, are used to attach gypsum

FIGURE 4.35
Standard sizes of common nails, reproduced full size. The abbreviation "d" stands for "penny." The length of each nail is given below its size designation. The three sizes of nail most used in light frame construction, 16d, 10d, and 8d, are shaded.

FIGURE 4.36
Face nailing is the strongest of the three methods of nailing. End nailing is relatively weak and is useful primarily for holding framing members in alignment until gravity forces and applied sheathing make a stronger connection. Toe nailing is used in situations where access for end nailing is not available. Toe nails are surprisingly strong. Load tests show them to carry about five-sixths as much load as face nails of the same size.

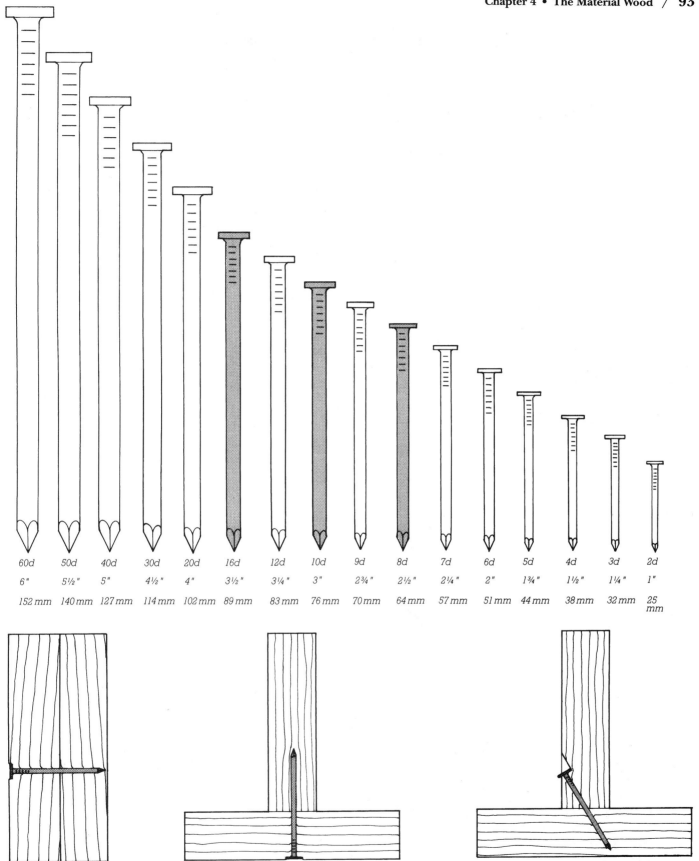

60d	50d	40d	30d	20d	16d	12d	10d	9d	8d	7d	6d	5d	4d	3d	2d
6"	5½"	5"	4½"	4"	3½"	3¼"	3"	2¾"	2½"	2¼"	2"	1¾"	1½"	1¼"	1"
152 mm	140 mm	127 mm	114 mm	102 mm	89 mm	83 mm	76 mm	70 mm	64 mm	57 mm	51 mm	44 mm	38 mm	32 mm	25 mm

FACE NAIL   *END NAIL*   *TOE NAIL*

board to wood framing members and can be at least as fast to install as nails. They are weak and extremely brittle, however, so they cannot be used for structural connections. Machine-driven screws of various kinds are also finding increasing use for attaching subflooring to joists in wood light frame construction; they result in a tighter floor construction that is less likely to squeak. *Deck screws* are now replacing nails as the preferred method of attaching outdoor decking boards to joists or girders because they are less likely to withdraw, and they hold the decking

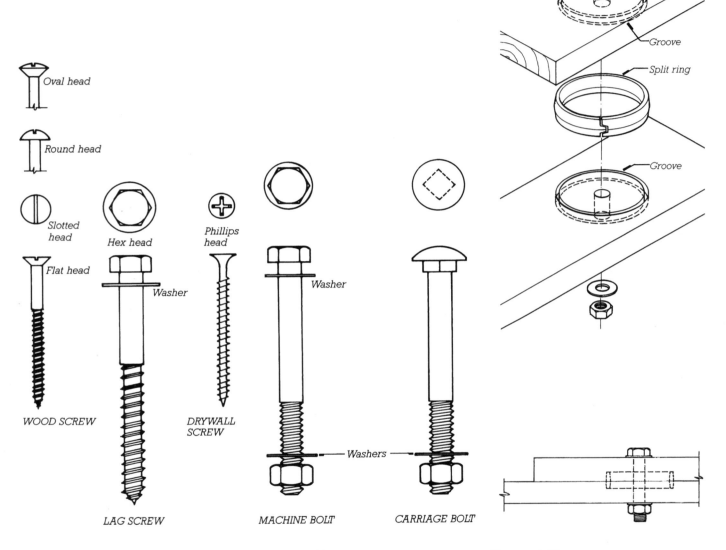

FIGURE 4.37
Flat-head screws are used without washers and are driven flush with the surface of the wood. Round-head screws are used with flat washers, and oval-head with countersunk washers. The drywall screw does not use a washer and is the only screw shown here that does not require a predrilled hole. Slotted, hex, and Phillips heads are all common.

FIGURE 4.38
Both machine bolts and carriage bolts are used in wood construction. Carriage bolts have a broad button head that needs no washer and a square shoulder under the head that is forced into the drilled hole in the wood to prevent the bolt from turning as the nut is tightened.

FIGURE 4.39
Split rings are high-capacity connectors used in heavily loaded joints of timber frames and trusses. After the center hole has been drilled through the two pieces, they are separated and the matching grooves are cut with a special rotary cutter driven by a power drill. The joint is then reassembled with the ring in place.

tighter to the frame, restraining it from warping and twisting.

Bolts

Bolts are used for major structural connections in heavy timber framing. Commonly used bolts range in diameter from ⅜ to 1 inch (9.53 to 25.4 mm), in any desired length. Flat steel disks called *washers* are inserted under heads and nuts of bolts to distribute the compressive force from the bolt across a greater area of wood (Figure 4.38).

Timber Connectors

Various types of timber connectors have been developed to increase the load-carrying capacity of bolts. The most widely used of these is the *split-ring connector* (Figure 4.39). The split ring is used in conjunction with a bolt and is inserted in matching circular grooves in the mating pieces of wood. Its function is to spread the load across a much greater area of wood than can be done with a bolt alone. The split permits the ring to adjust to wood shrinkage. Split rings are useful primarily in heavy timber construction.

Toothed Plates

Toothed plates (Figure 4.40) are used in factory-produced lightweight roof and floor trusses. They are inserted into the wood with hydraulic presses, pneumatic presses, or mechanical rollers and act as metal splice plates, each with a very large number of built-in nails (Figures 4.41 to 4.43). They are extremely effective connectors because their multiple, closely spaced points interlock tightly with the fibers of the wood.

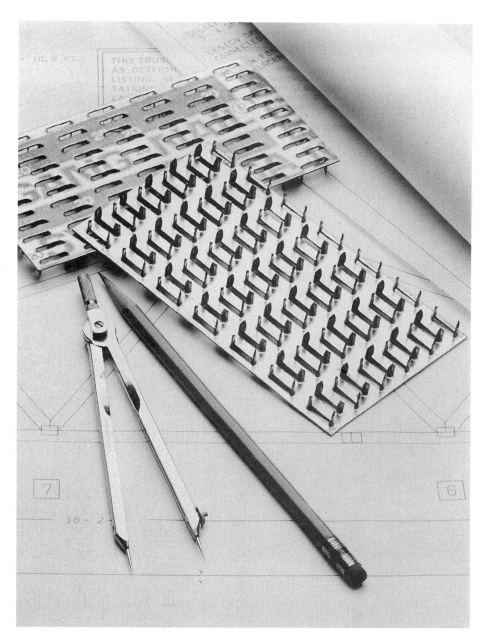

FIGURE 4.40
Manufacturers of toothed plate connectors also manufacture the machinery to install them and provide computer programs to aid local truss fabricators in designing and detailing trusses for specific buildings. The truss drawing in this photograph was generated by a plotter driven by such a program. The small rectangles on the drawing indicate the positions of toothed plate connectors at the joints of the truss. (*Courtesy of Gang-Nail System, Inc.*)

FIGURE 4.41
Factory workers align the wood members of a roof truss and position toothed plate connectors over the joints, tapping them with a hammer to embed them lightly and keep them in place. The roller marked "Gantry" then passes rapidly over the assembly table and presses the plates firmly into the wood. (*Courtesy of Wood Truss Council of America*)

FIGURE 4.42
The trusses are transported to the construction site on a special trailer. (*Courtesy of Wood Truss Council of America*)

FIGURE 4.43
Trusses can be designed and produced in almost any configuration. (*Courtesy of Wood Truss Council of America*)

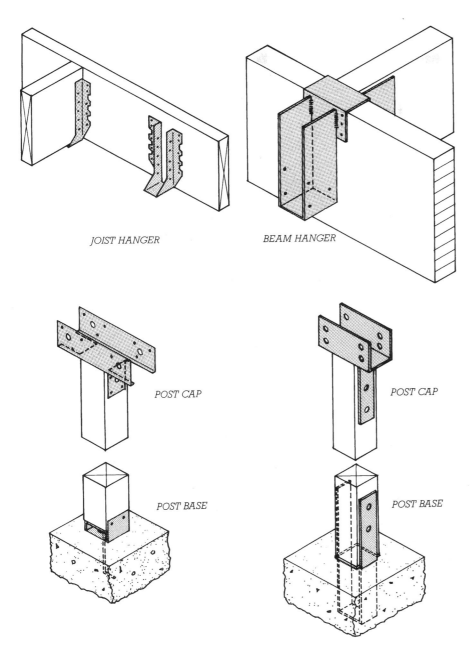

JOIST HANGER

BEAM HANGER

POST CAP

POST CAP

POST BASE

POST BASE

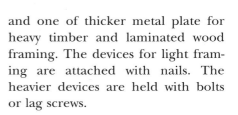

RAFTER ANCHOR

FRAMING ANCHORS

ANGLE

FIGURE 4.44
Joist hangers are used to make strong connections in floor framing wherever wood joists bear on one another at right angles. The heavier steel beam hangers are used primarily in heavy timber construction. Post bases serve the twofold function of preventing water from entering the end of the post and anchoring the post to the foundation. The bolts and lag screws used to connect the wood members to the heavier connectors are omitted from this drawing.

FIGURE 4.45
The sheet metal connectors shown in this diagram are less commonly used than those in Figure 4.44, but are invaluable in solving special framing problems and in reinforcing frames against wind uplift and earthquake forces.

Sheet Metal and Metal Plate Framing Devices

Dozens of ingenious sheet metal and metal plate devices are manufactured for strengthening common connections in wood framing. The most frequently used is the joist hanger, but all of the devices shown in Figures 4.44 and 4.45 find extensive use. There are two parallel series of this type of device, one made of sheet metal for use in light framing and one of thicker metal plate for heavy timber and laminated wood framing. The devices for light framing are attached with nails. The heavier devices are held with bolts or lag screws.

Machine-Driven Staples and Nails

The speed of wood frame construction can be increased significantly through the use of pneumatically operated *nail guns* and *staple guns*. The nails or staples are of special design and are prepackaged by the manufacturer in self-feeding strips, each containing a large number of fasteners. These are loaded into guns that are powered by high-pressure air fed through a hose from a small air compressor. They can be driven one by one as fast as the operator can pull the trigger. Several uses of machine-driven nails and staples are illustrated in Chapters 9, 10, 11, and 17.

Adhesives

Adhesives are widely employed in the factory production of plywood and panel products, laminated wood, wood structural components, and cabinetwork, but have relatively few uses on the construction job site. This disparity is explained by the need to clamp and hold adhesive joints under controlled environmental conditions until the adhesives have cured, which is easy to do in the shop but much more cumbersome and unreliable in the field. Mastic-type job site adhesives are widely used, however, to secure subflooring panels to wood framing. They are applied with a sealant gun and are clamped by simply nailing the panels to the framing. The nails usually serve as the primary structural connection, with the adhesive acting to eliminate squeaks in the floor.

WOOD-MANUFACTURED BUILDING COMPONENTS

Dimension lumber, structural panel products, mechanical fasteners, and adhesives are used in combination to manufacture a number of highly efficient structural components that offer certain advantages to the designer of wooden buildings.

Trusses

Trusses for both roof and floor construction are manufactured in small, highly efficient plants in every part of North America. Most are based on 2 × 4s and 2 × 6s joined with toothed plate connectors. The designer or builder needs to specify only the span, the roof pitch, and the desired overhang detail. The truss manufac-

FIGURE 4.46
The I-joist in the front of this photograph consists of two flanges of laminated veneer lumber (LVL) glued to a web of oriented strand board (OSB). The third member from the front is a beam made of LVL. The remainder of the beams and posts in this photograph are made of parallel strand lumber (PSL). (*Courtesy of Trus Joist MacMillan*)

turer either uses a preengineered design for the specified truss or employs a sophisticated computer program to engineer the truss and develop the necessary cutting patterns for its constituent parts. The manufacture of trusses is shown in Figures 4.41 to 4.43, and several uses of trusses are depicted in Chapter 9.

Roof trusses use less wood than a comparable frame of conventional rafters and ceiling joists. Like floor trusses, they span the entire width of the building in most applications, allowing the designer complete freedom to locate interior partitions anywhere they are needed. The chief disadvantages of roof trusses, as they are most commonly used, are that they make the attic space unusable and that they generally restrict the designer to the spatial monotony of a flat ceiling throughout the building. Truss shapes can be designed and manufactured to overcome both these limitations, however (Figure 4.43).

Wood I-Joists

Manufactured wood I-shaped members (I-joists) are useful for framing of both roofs and floors (Figure 4.46). The flanges of the members may be made from solid lumber or laminated veneer lumber. The webs consist of plywood or OSB. Like trusses, these components use wood more efficiently than conventional rafters and joists, and they can span farther between supports than dimension lumber. Further advantages include lighter weight than corresponding members made of dimension lumber, freedom from crooks and bows, and availability in lengths to 40 feet (12.2 m). Chapter 9 shows the use of I-joists in wood framing.

TYPES OF WOOD CONSTRUCTION

For residential construction, the wood light frame is the predominant system, accounting for over 95 percent of all new and remodeling work. The wood light frame is discussed in Chapter 9. Other wood systems such as the hand-hewn frames of centuries past, the stacked log structures of the early settlers, and the recent prefabricated panel systems are used as well. These systems are detailed in Chapters 20, 22, and 24.

C.S.I./C.S.C. MasterFormat Section Numbers for Materials

06100	ROUGH CARPENTRY
06105	Miscellaneous Carpentry
06120	Structural Panel
06130	Heavy Timber Construction
06150	Wood Decking
06160	Sheathing
06170	Prefabricated Structural Wood
06180	Glued-Laminated Construction
06200	FINISH CARPENTRY
06400	ARCHITECTURAL WOODWORK

Selected References

1. Hoadley, R. Bruce. *Understanding Wood.* Newtown, CT: Taunton Press, 1980.

Beautifully illustrated and produced, this volume vividly explains and demonstrates the properties of wood as a material of construction.

2. Western Wood Products Association. *Western Woods Use Book* (4th ed.). Portland, OR: Western Wood Products Association, 1996.

This looseleaf binder houses a complete reference library on the most common species of dimension lumber, timber, and finish lumber. It includes the National Design Specification for Wood Construction, the standard for engineering design of wood structures. Updated frequently.

3. Canadian Wood Council. *Wood Reference Handbook.* Ottawa: Canadian Wood Council, 1995.

A superbly illustrated encyclopedic reference on wood materials.

4. Forest Products Society. *Wood Handbook: Wood as an Engineering Material.* Madison, WI: Forest Products Society, 1999.

A comprehensive and scientific reference on wood and its structural capacity.

5. APA–The Engineered Wood Association. *Products and Applications.* Tacoma, WA: APA–The Engineered Wood Association, 1995–1999.

Updated frequently, this is a complete looseleaf guide to manufactured wood products, including plywood, oriented strand board, waferboard, structural insulated panels, and glue-laminated wood.

6. Callahan, Edward E. *Metal Plate Connected Wood Truss Handbook* (2nd ed.). Madison, WI: Wood Truss Council of America, 1997.

This is the standard reference on every phase of design, manufacturing, and installation of wood light trusses.

Key Terms and Concepts

bark
cambium
sapwood
heartwood
pit
cellulose
lignin
direction of grain
springwood (earlywood)
summerwood (latewood)
softwood
hardwood
tracheids
rays
fibers
vessels or pores
lumber
headsaw
sawyer
plainsawed
quartersawn
vertical grain
figure
free water
bound water
seasoning
air drying
kiln drying
green lumber

longitudinal, radial, tangential shrinkage
check
parallel-to-grain shrinkage
perpendicular-to-grain shrinkage
surfacing
S4S, S2S
S-DRY
S-GRN
growth characteristics
manufacturing characteristics
knot, knothole
decay
insect damage
split
crooking
bowing
twisting
cupping
wane
visual grading
machine grading
allowable strengths
nominal dimension
actual dimension
boards
dimension lumber

timbers
board foot
laminated wood (glulam)
finger joint, scarf joint
structural composite lumber
laminated veneer lumber (LVL)
parallel strand lumber (PSL)
structural wood panel
plywood panel
veneer
crossband
composite panel
nonveneered panel
oriented strand board (OSB)
waferboard
particleboard
rotary sliced
touch sanded
flitch
span rating
exposure durability classification
hardboard
cane fiber board
pressure impregnation

creosote
pentachlorophenol
waterborne salts
wood-polymer composite
nail
common nail
finish nail
penny
bright nail
hot-dip galvanized nail
electrogalvanized nail
screw
lag screw
bolt
washer
split ring connector
toothed plate
joist hanger
nail gun
staple gun
adhesive
truss
I-joist
sandwich panel
stressed-skin panel
manufactured housing
mobile home
sectional of modular home

REVIEW QUESTIONS

1. Discuss the changes in moisture content of the wood and their effects on a piece of dimension lumber, from the time the tree is cut, through its processing, until it has been in service in a building for an entire year.

2. Give the actual cross-sectional dimensions of the following pieces of kiln-dried lumber: 1 × 4, 2 × 4, 2 × 6, 2 × 8, 4 × 4, 4 × 12.

3. Why is wood laminated?

4. What is meant by a span rating of 32/16? What type of wood products are rated in this way?

5. For what reasons might you specify preservative-treated wood?

6. Which common species of wood have decay- and insect-resistant heartwood?

7. Why are nails the fasteners of choice in wood construction?

EXERCISES

1. Visit a nearby lumberyard. Examine and list the species, grades, and sizes of lumber carried in stock. For what uses are each of these intended? While at the yard, also look at the available range of fasteners.

2. Pick up a number of scraps of dimension lumber from a shop or construction site. Examine each to see where it was located in the log before sawing. Note any drying distortions in each piece: How well do these correspond to the distortions you would have predicted? Measure accurately the width and thickness of each scrap and compare your measurements to the specified actual dimensions for each.

3. Assemble samples of as many different species of wood as you can find. Learn how to tell the different species apart, by color, odor, grain figure, ray structure, relative hardness, and so on. What are the most common uses for each species?

4. Visit a construction site and list the various types of lumber and wood products being used. Look for a grade stamp on each and determine why the given grade is being used for each use. If possible, look at the architect's written specifications for the project and see how the lumber and wood products were specified.

MASONRY

5

Masonry is the simplest of building techniques: The mason stacks pieces of material (bricks, stones, concrete blocks, called collectively *masonry units*) atop one another to make walls. It is also the richest and most varied, however, with its endless selection of colors and textures. In addition, because the pieces of which it is made are small, masonry can take any shape, from a planar wall to a sinuous surface that defies the distinction of roof from wall.

Masonry is the material of earth, taken from the earth and comfortably at home in foundations, pavings, and walls that grow directly from the earth. With modern techniques of reinforcing, however, masonry can rise many stories from the earth, and in the form of arches and vaults, masonry can take wing and fly across space.

The most ancient of our building techniques, masonry remains labor intensive, requiring the patient skills of experienced and meticulous artisans to achieve a satisfactory result. It has kept pace with the times, however, and remains highly competitive technically and economically with other systems of structure and enclosure, the more so because one mason can produce in one operation a completely finished loadbearing wall, ready for use.

Masonry is durable. The designer can select masonry materials that are scarcely affected by water, air, or fire, ones with brilliant colors that will not fade, ones that will stand up to heavy wear and abuse, and make from them a building that will last for generations.

Masonry is a material of the small entrepreneur. One can set out to build a building of bricks with no more tools than a trowel, a shovel, a hammer, a measuring rule, a level, a square of scrap plywood, and a piece of string. Yet many masons can work together, aided by mechanized handling of materials, to put up projects as large as the human mind can conceive.

HISTORY

Masonry began spontaneously in the creation of low walls from stones or pieces of caked mud taken from dried puddles. *Mortar* was originally the mud smeared into the joints of the rising wall to impart stability and weathertightness. Where stone lay readily at hand, it was preferred to bricks; where stone was unavailable, *bricks* were made from local clays and silts. Changes came with the passing millennia: People learned to quarry, cut, and dress stone with increasing precision. Fires built against mud brick walls brought a knowledge of the advantages of burned brick, leading to the invention of the kiln. *Masons* learned the simple art of turning limestone into lime, and lime mortar gradually replaced mud in the joints of masonry.

By 4000 B.C., the peoples of Mesopotamia were building palaces

FIGURE 5.1
The Parthenon, constructed of marble, has stood on the Acropolis in Athens for more than 24 centuries. (*Photo by James Austin, Cambridge, England*)

and temples of stone and sun-dried brick. In the third millennium, the Egyptians erected the first of their stone temples and pyramids. In the last centuries prior to the birth of Christ, the Greeks perfected their temples of limestone and marble (Figure 5.1), and control of the Western world passed to the Romans, who made the first large-scale use of masonry arches and roof vaults in their basilicas, baths, palaces, and aqueducts. Medieval civilizations in both Europe and the Islamic world brought masonry vaulting to a very high plane of development. The Islamic craftsmen built magnificent palaces, markets, and mosques of brick and often faced them with brightly glazed clay tiles. The Europeans directed their efforts toward fortresses and cathedrals of stone, culminating in the pointed vaults and flying buttresses of the great Gothic churches (Figures 5.2 and 5.3). In Central America, South America, and Asia, other civilizations were carrying on a simultaneous evolution of building techniques in cut stone.

During the Industrial Revolution in Europe and North America, machines were developed that quarried and worked stone, molded bricks, and sped the transportation of these heavy materials to the building site. Sophisticated mathematics were applied for the first time to the analysis of the structure of masonry arches and to the art of stonecutting. Portland cement mortar came into widespread use, enabling the construction of masonry buildings of greater strength and durability.

In the late 19th century, masonry began to lose its primacy among the materials of construction. The very tall buildings of the central cities required frames of iron or steel to replace the thick masonry bearing walls that had limited the heights to which one could build. Reinforced concrete, poured rapidly and economically into simple forms made of wood, began to replace brick and

stone masonry in foundations and walls. The heavy masonry vault was supplanted by lighter floor and roof structures of steel and concrete.

The 19th-century invention of the hollow concrete block helped to avert the extinction of masonry as a craft. The concrete block was much cheaper than cut stone and required much less labor to lay than brick. It could be combined with brick or

stone facings to make lower cost walls that were still satisfactory in appearance. The brick cavity wall, an early-19th-century British invention, also contributed to the survival of masonry, for it produced a warmer, more watertight wall that was later to adapt easily to the introduction of thermal insulation when appropriate insulating materials became available in the middle of the 20th century.

FIGURE 5.2

Construction in ashlar limestone of the magnificent Gothic cathedral at Chartres, France, was begun in 1194 A.D. and was not finished until several centuries later. Seen here are the flying buttresses that resist the lateral thrusts of the stone roof vaulting. (*Photo by James Austin, Cambridge, England*)

FIGURE 5.3
The Gothic cathedrals were roofed with lofty vaults of stone blocks. The ambulatory roof at Bourges (built 1195–1275) evidences the skill of the medieval French masons in constructing vaulting to cover even a curving floor plan. (*Photo by James Austin, Cambridge, England*)

FIGURE 5.4
Despite the steady mechanization of construction operations in general, masonry construction in brick, concrete block, and stone is still based on simple tools and the highly skilled hands that use them. (*Courtesy of International Masonry Institute*)

Other 20th-century contributions to masonry construction include the development of techniques for steel-reinforced masonry, high-strength mortars, masonry units (both bricks and concrete blocks) that are higher in structural strength, and masonry units of many types that reduce the amount of labor required for masonry construction.

If this book had been written as recently as 125 years ago, it would have had to devote little space to materials of construction other than masonry and wood. Because other materials of construction were so late in developing, most of the great works of architecture in the world, and many of the best developed vernacular architectures, are built of masonry. We live amid a rich heritage of masonry buildings—there is scarcely a town in the world that is without a number of beautiful examples from which the serious student of masonry architecture can learn.

MORTAR

Mortar is as much a part of masonry as the masonry units themselves. Mortar serves to cushion the masonry units, giving them full bearing against one another despite their surface irregularities. Mortar seals between the units to keep water and wind from pene-

trating; it adheres the units to one another to bond them into a monolithic structural unit; and, inevitably, it is important to the appearance of the finished masonry wall.

The most characteristic type of mortar is made of portland cement, hydrated lime, an inert *aggregate* (sand), and water. The sand must be clean and must be screened to eliminate particles that are too coarse or too fine. The portland cement is the bonding agent in the mortar, but a mortar made only with portland cement is "harsh" and does not flow well on the *trowel* or under the brick, so *lime* is added to impart smoothness and workability. Lime is produced by burning limestone or seashells (calcium carbonate) in a kiln to drive off carbon dioxide and leave *quicklime* (calcium oxide). The quicklime is then slaked by allowing it to absorb as much water as it will hold, resulting in the formation of calcium hydroxide, called *slaked lime* or *hydrated lime*. The slaking process, which releases large quantities of heat, is usually carried out in the factory. The hydrated lime is subsequently dried, ground, and bagged for shipment. Until the late 19th and early 20th centuries, mortar was made without portland cement, and the lime itself was the bonding agent; it hardened by absorbing carbon dioxide from the air to become calcium carbonate, a very slow and uneven process.

Prepackaged *masonry cements* are also widely used for making mortar. Most are proprietary formulations that contain admixtures intended to contribute to the workability of the mortar. The formulations vary from one manufacturer to another. Two colors of masonry cement are commonly available: *light,* which cures to about the same light-gray color as ordinary concrete blocks, and *dark,* which cures to dark gray. Other colors are easily produced by the mason, either by adding pigments to the mortar at the time of mixing or by purchasing dry mortar mix that has been custom colored at the factory. Mortar mix can be obtained in shades ranging from pure white to pure black, including all the colors of the spectrum. Because mortar makes up a considerable fraction of the exposed surface area of a brick wall, typically about 20 percent, the color of the mortar is extremely important in the appearance of a brick wall and is almost as important in the appearance of stone or concrete masonry walls.

Four basic mortar types are defined, as summarized in Figure 5.5. Type N mortar is used for most purposes. Types M and S are suitable for higher strength structural walls and for severe weather exposures. Type O, the most economical, is used only in nonloadbearing interior work.

Mortar Type	Description	Construction Suitability	Minimum Average Compressive Strength at 28 days
M	High-strength mortar	Masonry subjected to high lateral or compressive loads or severe frost action; masonry below grade	2500 psi (17.25 MPa)
S	Medium high-strength mortar	Masonry requiring high flexural bond strength but subjected only to normal compressive loads	1800 psi (12.40 MPa)
N	Medium-strength mortar	General use above grade	750 psi (5.17 MPa)
O	Medium low-strength mortar	Nonloadbearing interior walls and partitions	30 psi (2.40 MPa)

FIGURE 5.5
Mortar types as defined by ASTM C270.

Portland cement mortar cures by hydration, not by drying: A complex set of chemical reactions takes up water and combines it with the constituents of the cement and lime to create a dense, strong, crystalline structure that binds the sand particles together. Mortar that has been mixed but not yet used can become too stiff for use, either by drying out or by commencing its hydration. If the mortar was mixed less than 90 minutes prior to its stiffening, it has merely dried and can safely be retempered with water to make it workable again. If the unused mortar is more than 2½ hours old, it must be discarded because it has already begun to hydrate and cannot be retempered without reducing its final strength.

CONCRETE MASONRY

Developed very late in the history of masonry, *concrete masonry units (CMUs)* are now the workhorses of masonry construction. In residential construction, they serve primarily in foundations, retaining walls, and fireplaces. CMUs are manufactured in three basic forms: bricks, larger hollow units that are commonly referred to as *concrete blocks*, and, less commonly, larger solid units. Considered crude and industrial looking by most people, residential concrete masonry is often covered with a veneer of brick, stone, stucco, or plaster, but split-face and other decorative units are available to make exposed concrete masonry more appealing to the eye (Figure 5.6).

Manufacture of Concrete Masonry Units

Concrete masonry units are manufactured by vibrating a stiff concrete mixture into metal molds, then immediately turning out the wet blocks or bricks onto a rack so that the mold can be reused, at the rate of a thousand or more units per hour. The racks of concrete masonry units

FIGURE 5.6 A combination of split-faced and regular CMUs are used in the walls and chimney of this residence. The masonry in the walls makes a visual reference to the chimney while adding color and texture to the walls. (*Photo by Rob Thallon*)

FIGURE 5.7
A forklift truck loads newly molded concrete masonry units into an autoclave for steam curing. (*Courtesy of Portland Cement Association, Skokie, Illinois*)

are cured at an accelerated rate by subjecting them to steam, either at atmospheric pressure or, for faster curing, at higher pressure (Figure 5.7). After steam curing, the units are dried to a specified moisture content and bundled on wooden pallets for shipping to the construction site.

Concrete masonry units are made in a variety of sizes and shapes (Figures 5.8 and 5.9) and in different densities of concrete, some of which use cinders,

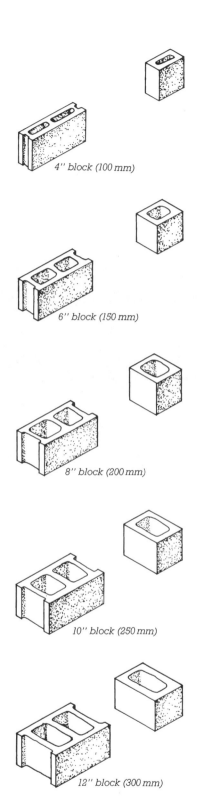

FIGURE 5.8
American standard concrete blocks and half blocks. Each full block is nominally 8 inches (200 mm) high and 16 inches (400 mm) long.

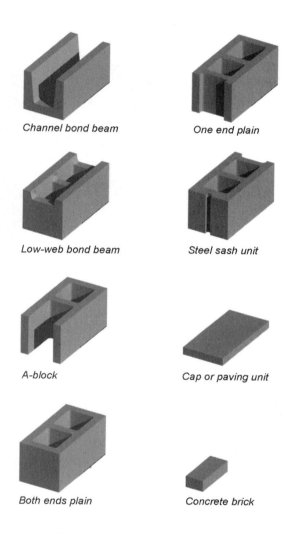

FIGURE 5.9
Other concrete masonry shapes. Bond beam units have space for horizontal reinforcing bars and grout and are used to tie a wall together horizontally. They are also used for reinforced block lintels. A-blocks are used to build walls with vertical reinforcing bars grouted into the cores in situations where there is insufficient space to lift the blocks over the tops of the projecting bars. Blocks with plain ends are used at outside corners where both the face and the end of the block will be exposed. Concrete bricks are interchangeable with modular clay bricks.

FIGURE 5.10
Concrete blocks and bricks can be cut very accurately with a water-cooled, diamond-bladed saw. For rougher sorts of cuts, a few skillful blows from the mason's hammer will suffice. (*Courtesy of Portland Cement Association, Skokie, Illinois*)

pumice, or expanded lightweight aggregates rather than crushed stone or gravel. Many colors and surface textures are available as standard items at masonry supply dealers, and special colors are relatively easy to produce by specifying a coloring agent and/or aggregate to be used in manufacturing of the units. When ordering specialty units, however, adequate lead time must be allowed for manufacturing and curing.

Hollow concrete block masonry is generally much more economical per unit of wall area or volume than brick or stone masonry. The blocks themselves are cheaper on a volumetric basis and are made into a wall much more quickly because of their larger size (a single standard concrete block occupies the same volume as 12 modular bricks). Concrete blocks can be produced in various degrees of strength, and because their hollow cores allow for the easy insertion of reinforcing steel and grout, they are widely used in masonry bearing wall construction. Where surface qualities not available in concrete masonry are

desired on a masonry wall, concrete blocks are often faced with brick or stone. Block walls also accept plaster, stucco, or tile work directly. Concrete masonry walls exposed to the weather, like any solid masonry walls, tend to leak in wind-driven rains. Except in dry climates, they should be painted on the outside with masonry paint or sealer.

Although innumerable special sizes, shapes, and patterns are available, American concrete blocks are standardized around an 8-inch (203.2-mm) cubic module. The most common block is nominally 8 × 8 × 16 inches (203 × 203 × 406 mm). The actual size of the block is 7⅝ × 7⅝ × 15⅝ inches (194 × 194 × 397 mm), which allows for a mortar joint ⅜ inch (9.5 mm) thick. This size block is designed to be lifted and laid conveniently with two hands (as compared to a brick, which is designed to be laid with one). Its double-cube proportions work well for bonding in straight walls, for corners, and for openings when combined with a single-cube half block. While concrete masonry units can be cut with a diamond-bladed power saw (Figure 5.10), it is more economical and produces better results if the designer lays out buildings of concrete masonry in dimensional units that correspond to the module of the block (Figure 5.11 and 5.12). Nominal 4-inch, 6-inch, and 12-inch block thicknesses (102 mm, 156 mm, and 308 mm) are also common, as is a solid concrete brick that is identical in size and proportion to a modular clay brick. A handy feature of the standard 8-inch block height is that it corresponds exactly to three courses of ordinary clay or concrete brickwork, making it easy to interweave blockwork and brickwork in composite walls.

Laying Concrete Blocks

Concrete block walls are laid in a step-by-step sequence, starting with the construction of the corners, which

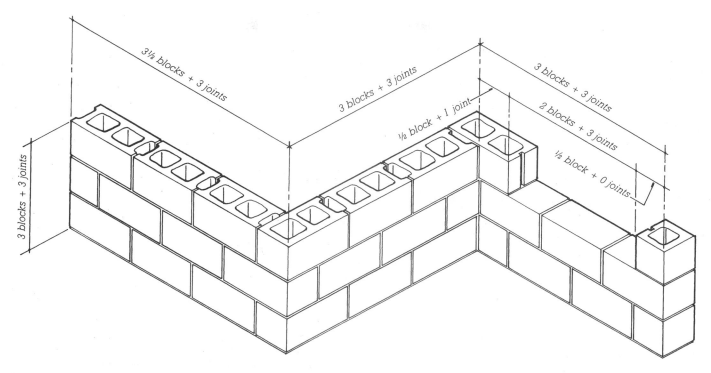

FIGURE 5.11
Concrete masonry buildings should be dimensioned to use uncut blocks except for special circumstances.

FIGURE 5.12
A small house by architects Clark and Menefee is cleanly detailed in ordinary concrete masonry units. All walls and openings are organized on an 8-inch module. A reinforced masonry retaining wall helps to tie the building visually into the site. (*Photo by Jim Rounsevell*)

are called *leads* (pronounced "leeds"). The leads establish the planes of the walls and the heights of the courses. The laying of leads is relatively labor intensive. A mason's rule or a *story pole* that is marked with the course heights is used to establish accurate course heights in the leads. The work is checked frequently with a spirit level to assure that surfaces are flat and plumb and courses are level. When the leads have been completed, a mason's line (a heavy string) is stretched between the leads, using an L-shaped *line block* at each end to

(a)

(b)

(c)

(d)

FIGURE 5.13
Laying a concrete masonry wall: (a) A bed of mortar is spread on the footing. (b) The first course of blocks for a lead is laid in the mortar. Mortar for the head joint is applied to the end of each block with the trowel before the block is laid. (c) The lead is built higher. Mortar is normally applied only to the face shells of the block and not to the webs. (d) As each new course is started on the lead, its height is meticulously checked with either a folding rule or, as shown here, a story pole marked with the height of each course.

locate the end of the line precisely at the top of each course of block.

The laying of the infill blocks between the leads is much faster and easier because the mason needs only to lay each block to the line to create a perfect wall. It follows that leads are expensive as compared to the wall surfaces between, so that where economy is important the designer should seek to minimize the number of corners in a masonry structure. The accompanying photographic sequence (Figure 5.13) illustrates the technique for laying concrete block walls.

(e)

(f)

(g)

(h)

(e, f) Each new course is also checked with a spirit level to be sure that it is level and plumb. Time expended in making sure the leads are accurate is amply repaid in the accuracy of the wall and the speed with which blocks can be laid between the leads. (g) The joints of the lead are tooled to a concave profile. (h) A soft brush removes mortar crumbs after tooling.

(i)

(j)

(k)

(l)

FIGURE 5.13 (*continued*)
(*i*) A mason's line is held taut between the leads on line blocks. (*j*) The courses of blocks between the leads are laid rapidly and are aligned only with the line; no story pole or spirit level is necessary. The mason has laid bed joint mortar and "buttered" the head joints for a number of blocks. (*k*) Each course of infill blocks is completed with a closer, which must be inserted between blocks that have already been laid. The head joints of the already laid blocks are buttered. (*l*) Both ends of the closer blocks are also buttered with mortar, and the block is lowered carefully into position. Some touching up of the head joint mortar is often necessary. (*All photos courtesy of Portland Cement Association, Skokie, Illinois*)

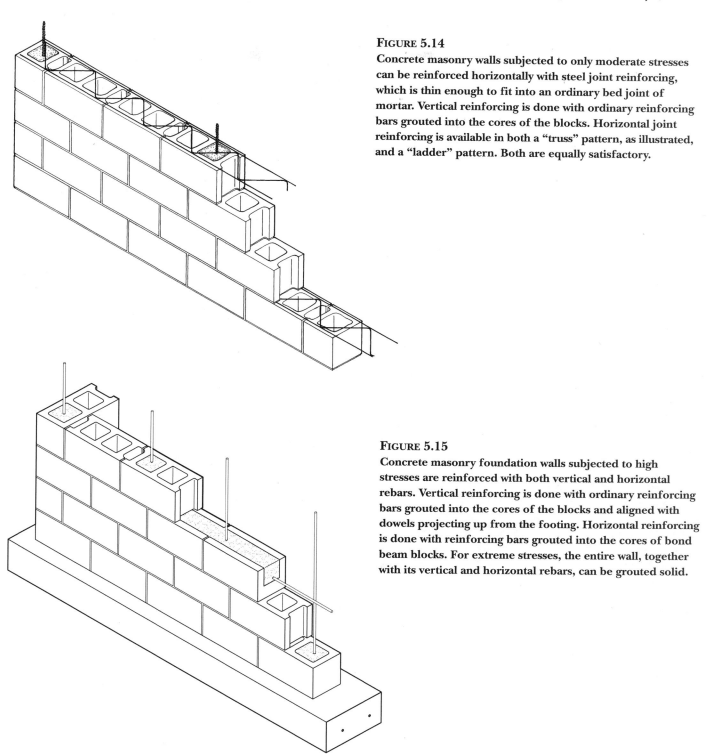

FIGURE 5.14
Concrete masonry walls subjected to only moderate stresses can be reinforced horizontally with steel joint reinforcing, which is thin enough to fit into an ordinary bed joint of mortar. Vertical reinforcing is done with ordinary reinforcing bars grouted into the cores of the blocks. Horizontal joint reinforcing is available in both a "truss" pattern, as illustrated, and a "ladder" pattern. Both are equally satisfactory.

FIGURE 5.15
Concrete masonry foundation walls subjected to high stresses are reinforced with both vertical and horizontal rebars. Vertical reinforcing is done with ordinary reinforcing bars grouted into the cores of the blocks and aligned with dowels projecting up from the footing. Horizontal reinforcing is done with reinforcing bars grouted into the cores of bond beam blocks. For extreme stresses, the entire wall, together with its vertical and horizontal rebars, can be grouted solid.

Concrete masonry is often reinforced with steel to increase its load-bearing capacity and its resistance to cracking. Horizontal reinforcing is usually inserted in the form of welded grids of small-diameter steel rods that are laid into the bed joints of mortar at the desired vertical intervals (Figure 5.14). In seismic zones where stronger horizontal reinforcing is required, bond beam blocks allow reinforcing bars to be placed in the horizontal direction (Figure 5.15). Vertical reinforcing bars are placed into the vertically aligned cores after the wall is complete. The cores are then filled with grout to unify steel and masonry into one continuous structural unit (Figure

5.16). In zones with severe seismic activity, all cells are usually grouted whether they contain bars or not, but it is also possible to fill only the cores that contain bars by holding the grout in the bond beams with a strip of metal mesh that bridges across the core openings beneath the bond beams. Lintels over openings in concrete block walls may be made of reinforced concrete or bond beam blocks with grouted horizontal reinforcing (Figure 5.17).

In recent years, *surface bonding* of concrete masonry walls has found application in buildings where the cost or availability of skilled labor is a problem. The blocks are laid without mortar, course upon course, to make a wall. Then a thin layer of a special cementitious compound containing short fibers of alkali-resistant glass is plastered on each side of the wall. This surface bonding compound, after it has cured, joins the blocks securely to one another both in tension and in compression. It serves also as a surface finish whose appearance resembles stucco.

Decorative Concrete Masonry Units

Concrete masonry units are easily and economically manufactured in an unending variety of surface patterns, textures, and colors intended for exposed use in exterior and interior walls. A few such units are diagrammed in Figure 5.18, and some of the resulting surface textures are depicted in Figures 5.19 and 5.20. Mold costs for producing special units are low when spread across the number of units required for medium to large housing developments, and many of the textured concrete masonry units that are now considered standard originated as special designs created by architects for particular buildings.

FIGURE 5.16
Grout is deposited in the cores of a reinforced concrete masonry wall using a grout pump and hose. (*Courtesy of Portland Cement Association, Skokie, Illinois*)

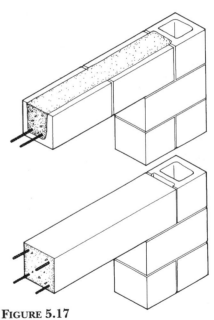

FIGURE 5.17
Lintels for openings in concrete masonry walls. At the top, a reinforced block lintel composed of bond beam units. At the bottom, a precast reinforced concrete lintel.

FIGURE 5.18
Some decorative concrete masonry units, representative of literally hundreds of designs currently in production. The scored-face unit, if the slot in the face is filled with mortar and tooled, produces a wall that looks as if it were made entirely of half blocks. The ribbed split-face unit is produced by casting "Siamese twin" blocks joined at the ribs, then shearing them apart.

Scored-face unit

Ribbed-face unit

Ribbed-face unit

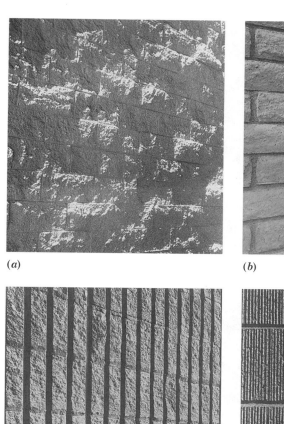
(a)

(b)

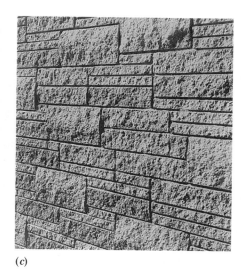
(c)

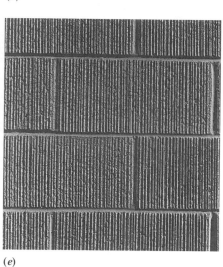
(d)

(e)

FIGURE 5.19
Some walls of decorative concrete masonry: (*a*) Split block. (*b*) Slump block, which is molded from relatively wet concrete and allowed to sag slightly after molding and before curing. (*c*) Split blocks of varying sizes laid in a random ashlar pattern. (*d*) Ribbed split-face blocks. (*e*) Striated blocks. (*Courtesy of National Concrete Masonry Association*)

FIGURE 5.20
Split-face blocks have been mingled with standard blocks in this foundation wall in order to complement a nearby rubble wall. (*Photo by Greg Thomson*)

BRICK MASONRY

Among the masonry materials, brick is special in two respects: fire resistance and size. As a product of fire, it is the most resistant to building fires of any masonry unit. Its size may account for much of the love that many people instinctively feel for brick: A traditional brick is shaped and dimensioned to fit the human hand. Hand-sized bricks are less likely to crack during drying or firing than larger bricks, and they are easy for the mason to manipulate. This small unit size makes brickwork very flexible in adapting to small-scale geometries and patterns and gives a pleasing scale and texture to a brick wall or floor.

One of the oldest masonry materials, brick has been used historically to make structural walls and vaults of immense proportions, and has been manipulated to make rich patterns and elaborate openings. Although still appreciated as a durable and beautiful building material, brick in modern residential construction is used primarily as a veneer for walls, floors, and fireplaces. This demotion in the status of brick has occurred primarily because a structural masonry wall can now be constructed much more efficiently with CMUs. The importance of thermal insulation has also contributed to the downfall of brick as a structural material because traditional structural brick walls are very difficult to insulate.

Molding of Bricks

Because of their weight and bulk, which make them expensive to ship over long distances, bricks are produced by a large number of relatively small, widely dispersed factories from a variety of local clays and shales. The raw material is dug from pits, crushed, ground, and screened to reduce it to a fine consistency. It is then tempered with water to produce a plastic clay ready for forming into bricks.

There are three major methods used today for forming bricks: the soft

FIGURE 5.21
A simple wooden mold produces seven water-struck bricks at a time. (*Photo by Edward Allen*)

mud process, the dry mud process, and the stiff mud process. The oldest is the *soft mud process,* in which a relatively moist clay (20 to 30 percent water) is pressed into simple rectangular molds, either by hand or with the aid of molding machines (Figure 5.21). To keep the sticky clay from adhering to the molds, the molds may be dipped in water immediately before being filled, producing bricks with a relatively smooth, dense surface that are known as *water-struck bricks.* If the wet mold is dusted with sand just before forming the brick, *sand-struck* or *sand-mold* bricks are produced, with a matte-textured surface.

The *dry press process* is used for clays that shrink excessively during drying. Clay mixed with a minimum of water (up to 10 percent) is pressed into steel molds by a machine working at a very high pressure.

The high-production *stiff mud process* is the one most widely used today. Clay containing 12 to 15 percent water is passed through a vacuum to remove any pockets of air, then extruded through a rectangular die (Figures 5.22 and 5.23). As the clay leaves the die, textures or thin mix-

tures of colored clays may be applied to its surface as desired. The rectangular column of moist clay is pushed by the pressure of extrusion across a cutting table, where automatic cutter wires slice it into bricks.

After molding by any of these three processes, the bricks are dried for 1 to 2 days in a low-temperature dryer kiln. They are then ready for transformation into their final form by a process known as *firing* or *burning.*

Firing of Bricks

Before the advent of modern kilns, bricks were most often fired by stacking them in a loose array called a *clamp,* covering the clamp with earth or clay, building a wood fire under the clamp, and maintaining the fire for a period of several days. After cooling, the clamp would be disassembled and the bricks sorted according to the degree of burning each had experienced. Bricks adjacent to the fire *(clinker bricks)* were often overburned and distorted, making them unattractive and therefore unsuitable for use in exposed brickwork. Bricks in a zone of the clamp near the fire would be fully burned but undistorted, suitable for exterior *facing bricks* with a high degree of resistance to weather. Bricks farther from the fire would be softer and would be set aside for use as backup bricks, while some bricks from around the perimeter of the clamp would not be burned sufficiently for any purpose and would be discarded. In the days before mechanized transportation, bricks for a building were often produced from clay obtained from the building site and were burned in clamps adjacent to the work.

Today, bricks are usually burned either in a *periodic kiln* or in a *continuous tunnel kiln.* The periodic kiln is a fixed structure that is loaded with bricks, fired, cooled, and unloaded (Figure 5.24). Bricks are passed continuously through a long tunnel kiln on special railcars to emerge at the far end fully burned. In either type of kiln, the first stages of burning are *water-smoking* and *dehydration,* which drive off the remaining water from the clay. The next stages are *oxidation* and *vitrification,* during which the temperature rises to 1800 to 2400 degrees Fahrenheit (1000 to 1300°C) and the clay is transformed into a ceramic material. This may be followed by a stage called *flashing,* in which the fire is regulated to create a reducing atmosphere in the kiln that develops color variations in the bricks. Finally, the bricks are cooled under controlled conditions to achieve the desired color and avoid thermal cracking. The cooled bricks are inspected, sorted, and packaged for shipment. The entire process of firing, monitored continuously to maintain product quality, takes from 40 to 150 hours. Considerable shrinkage takes place in the bricks during drying and firing; this must be taken into account when designing the molds for the brick. The higher the temperature, the greater the shrinkage and the darker the brick. Bricks are often used in a mixed range of colors, with the darker bricks inevitably being smaller than the lighter bricks. Even in bricks of uniform color, some size

FIGURE 5.22
A column of clay emerges from the die in the stiff mud process of molding bricks. (*Courtesy of Brick Institute of America*)

FIGURE 5.23
Rotating groups of parallel wires cut the column of clay into individual bricks, ready for drying and firing. (*Courtesy of Brick Institute of America*)

FIGURE 5.24
Three stages in the firing of water-struck bricks in a small factory: (*a*) Bricks stacked on a kiln car ready for firing. The open passages between the bricks allow the hot kiln gases to penetrate to the interior of the stack. The bed of the kiln car is made of a refractory material that is unaffected by the heat of the kiln. The rails on which the car runs are recessed in the floor. (*b*) The cars of bricks are rolled into the far end of this gas-fired periodic tunnel kiln. When a firing has been completed, the large door in the near end of the kiln is opened and the cars of bricks are rolled out on the rails that can be seen at the lower right of the picture. (*c*) After the fired bricks have been sorted, they are strapped into these "cubes" for shipping. (*Photos by Edward Allen*)

(*a*)

(*b*)

variation is to be expected, and bricks in general are subject to a certain amount of distortion from the firing process.

The color of a brick depends on the chemical composition of the clay or shale and the temperature and chemistry of the fire in the kiln. Higher temperatures, as noted in the previous paragraph, produce darker bricks. The iron that is prevalent in most clays turns red in an oxidizing fire and purple in a reducing fire. Other chemical elements interact in a similar way to the kiln atmosphere to make still other colors. For bright colors, the faces of the bricks can be glazed like pottery, during the normal firing or in an additional firing.

(*c*)

Standard Brick Sizes

Unit Name	Width	Length	Height
Modular	3½" or 3⅝" (90 mm)	7½" or 7⅝" (190 mm)	2¼" (57 mm)
Standard	3½" or 3⅝" (90 mm)	8" (200 mm)	2¼" (57 mm)
Engineer Modular	3½" or 3⅝" (90 mm)	7½" or 7⅝" (190 mm)	2¾" to 2¹³⁄₁₆" (70 mm)
Engineer Standard	3½" or 3⅝" (90 mm)	8" (200 mm)	2¾" (70 mm)
Closure Modular	3½" or 3⅝" (90 mm)	7½" or 7⅝" (190 mm)	3½" or 3⅝" (90 mm)
Closure Standard	3½" or 3⅝" (90 mm)	8" (200 mm)	3⅝" (90 mm)
Roman	3½" or 3⅝" (90 mm)	11½ or 11⅝" (290 mm)	1⅝" (40 mm)
Norman	3½" or 3⅝" (90 mm)	11½ or 11⅝" (290 mm)	2¼" (57 mm)
Engineer Norman	3½" or 3⅝" (90 mm)	11½ or 11⅝" (290 mm)	2¾" to 2¹³⁄₁₆" (70 mm)
Utility	3½" or 3⅝" (90 mm)	11½ or 11⅝" (290 mm)	3½" or 3⅝" (90 mm)
King Size	3" (75 mm)	9⅝" (240 mm)	2⅝" or 2¾" (70 mm)
Queen Size	3" (75 mm)	7⅝" or 8" (190 mm)	2¾" (70 mm)

FIGURE 5.25
Dimensions of bricks commonly used in North America, as established by the Brick Institute of America. This list gives an idea of the diversity of sizes and shapes available, and of the difficulty of generalizing about brick dimensions. Modular bricks are dimensioned so that three courses plus mortar joints add up to a vertical dimension of 8 inches (203 mm), and one brick length plus mortar joint has a horizontal dimension of 8 inches (203 mm). The alternative dimensions of each brick are calculated for ⅜-inch (9.5-mm) and ½-inch (12.7-mm) mortar joint thicknesses.

Brick Sizes

There is no truly standard brick. The nearest thing in the United States is the modular brick, dimensioned to construct walls in modules of 4 inches horizontally and 8 inches vertically, but the modular brick has not found ready acceptance in some parts of the country, and traditional sizes persist on a regional basis. Figure 5.25 shows the brick sizes that represent about 90 percent of all the bricks used in the United States. For most bricks in the normal range of sizes, three courses of bricks plus the accompanying three mortar joints add up to a height of 8 inches (203 mm). Length dimensions must be calculated specifically for the brick selected and must include the thicknesses of the mortar joints.

Bricks may be solid, *cored, hollow,* or *frogged* (Figure 5.26). By reducing the volume and thickness of the clay, cores and frogs permit more even drying and firing of bricks, reduce fuel costs for firing, reduce shipping costs, and create bricks that are lighter and easier to handle. Hollow bricks, which may contain up to 60 percent voids, are used primarily to enable the insertion and grouting of steel reinforcing bars in brickwork.

Choosing Bricks

We have already considered three important qualities that the designer must consider in choosing the bricks for a particular building: molding process, color, and size. There are also three *grades* of brick based on resistance to weathering and three *types* of facing bricks (bricks that will be exposed to view) based on the degree of uniformity in shape,

dimension, texture, and color from one brick to the next (Figure 5.27). The uses of the three grades are related to a map of weathering indices prepared from National Weather Service data on winter rainfall and freezing cycles (Figure 5.28). Grade SW is recommended for use in contact with the ground, or in situations where the brickwork is likely to be saturated with water, in any of the three regions of the map. Grade

FIGURE 5.26
From left to right: cored, hollow, and frogged bricks.

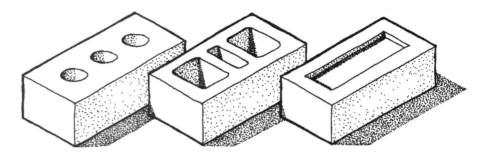

FIGURE 5.27
Grades and types of bricks, as defined by ASTM C62, C216, and C652.

Grades for Building and Facing Bricks	
Grade SW	Severe weathering
Grade MW	Moderate weathering
Grade NW	Negligible weathering

Types of Facing Bricks	
Type FBX	High degree of mechanical perfection, narrow color range, minimum size variation per unit
Type FBS	Wide range of color and greater size variation per unit
Tpe FBA	Nonuniformity in size, color, and texture per unit

FIGURE 5.28
Weathering regions of the United States, as determined by winter rainfall and freezing cycles. Grade SW brick is recommended for exterior use in the Severe Weathering region. (*Courtesy of Brick Institute of America*)

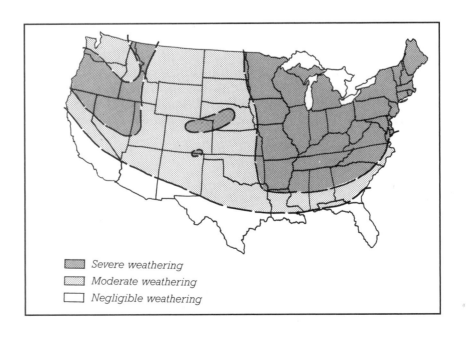

▨ Severe weathering
▧ Moderate weathering
☐ Negligible weathering

MW may be used above ground in any of the regions, but SW will provide greater durability. Grade NW is intended for use in sheltered or indoor locations. Bricks used for paving of walks, drives, and patios should be selected to minimize deterioration from freeze–thaw cycles.

The compressive strength of brickwork is of obvious importance in structural walls and piers and depends on the strengths of both the brick and the mortar. Typical allowable compressive stresses for unreinforced brick walls range from 75 to 400 pounds per square inch (0.52 to 2.76 MPa).

For brickwork exposed to very high temperatures, such as the lining of a fireplace or a furnace, *firebricks* are used. These are made from special clays *(fireclays)* that produce bricks with refractory qualities. Firebricks are laid in very thin joints of fireclay mortar.

Laying Bricks

Figure 5.29 shows a basic vocabulary of bricklaying. Bricks are laid in the various positions for visual reasons, structural reasons, or both. The simplest brick wall is a single *wythe* of *stretchers.* For walls two or more wythes thick, *headers* are used to bond the wythes together into a structural unit. *Rowlock* courses are often used for caps on garden walls and for sloping sills under windows, although such caps and sills are not durable in severe climates. Architects frequently employ *soldier* courses for visual emphasis in such locations as window lintels or tops of walls.

The problem of bonding multiple wythes of brick has been solved in many ways in different regions of the world, often resulting in surface patterns that are particularly pleasing to the eye. Figure 5.30 shows some popular bonding patterns for brickwork, two of which are designed to bond double-wythe structural walls. As a veneer on the exterior of buildings, the single wythe offers the designer little excuse to use anything but Running Bond. Inside a building, safely out of

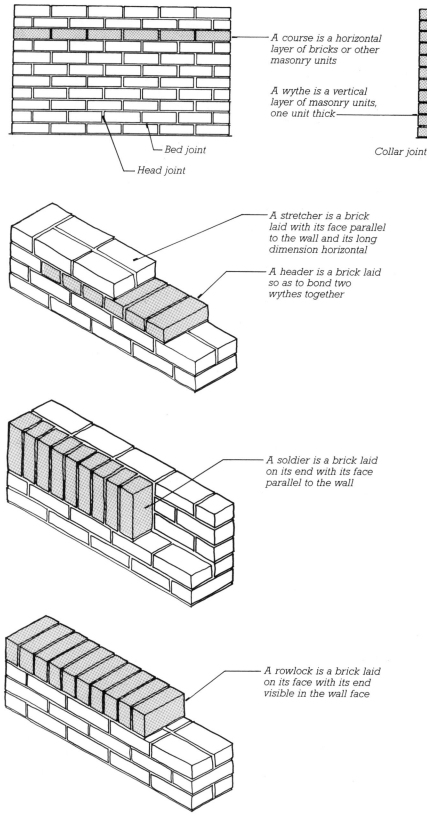

A course is a horizontal layer of bricks or other masonry units

A wythe is a vertical layer of masonry units, one unit thick

Collar joint

Bed joint

Head joint

A stretcher is a brick laid with its face parallel to the wall and its long dimension horizontal

A header is a brick laid so as to bond two wythes together

A soldier is a brick laid on its end with its face parallel to the wall

A rowlock is a brick laid on its face with its end visible in the wall face

FIGURE 5.29
Basic brickwork terminology.

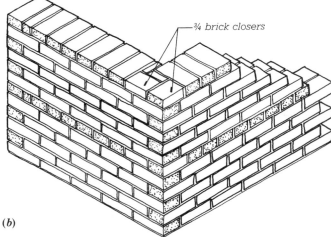

(a)

Running Bond consists entirely of stretchers

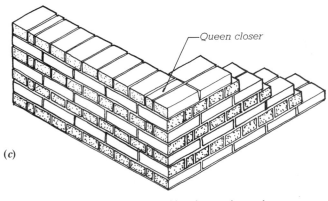

¾ brick closers

(b)

Common Bond (also known as American Bond) has a header course every sixth course. Notice how the head joints are aligned between the header and stretcher courses

Queen closer

(c)

English Bond alternates courses of headers and stretchers

FIGURE 5.30

Frequently used structural bonds for brick walls: (*a*) Running Bond, (*b*) Common Bond, and (*c*) English Bond. Partial closer bricks are necessary at the corners to make the header courses come out even while avoiding alignments of head joints in successive courses. The mason usually cuts the closers to length with a mason's hammer, but they are sometimes cut with a diamond saw. The Running Bond example shown in the photograph is from the late 18th century, with extremely thin joints, which require mortar made from very fine sand. Notice in the Common Bond wall (dating from the 1920s in this case) that the header course began to fall out of alignment with the stretcher courses, so the mason inserted a partial stretcher to make up the difference; such small variations in workmanship contribute to the visual appeal of brick walls. (*Photos by Edward Allen*)

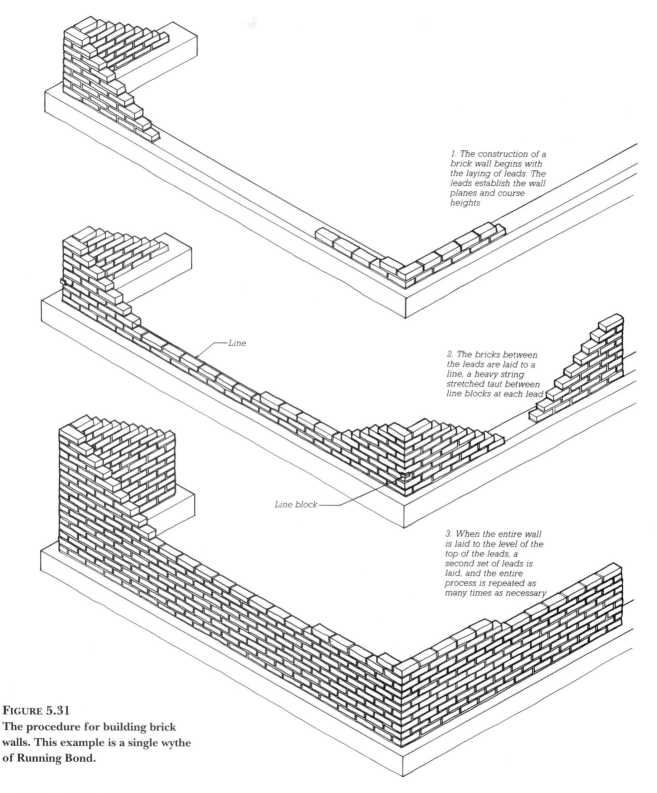

1. The construction of a brick wall begins with the laying of leads. The leads establish the wall planes and course heights

Line

2. The bricks between the leads are laid to a line, a heavy string stretched taut between line blocks at each lead

Line block

3. When the entire wall is laid to the level of the top of the leads, a second set of leads is laid, and the entire process is repeated as many times as necessary

FIGURE 5.31
The procedure for building brick walls. This example is a single wythe of Running Bond.

the weather and with no need for insulation, one may use solid brick walls in any desired bond. For fireplaces and other very small brick constructions, however, it is often difficult to create a long enough stretch of unbroken wall to justify the use of bonded brickwork.

The process of bricklaying is summarized in Figures 5.31 and 5.32. While conceptually simple, bricklaying requires both extreme care and considerable experience to produce a satisfactory result, especially where a number of bricklayers working side by side must produce identical work. Yet speed is essential to the economy of masonry construction. The work of

(a)

(b)

(c)

(d)

(e)

(f)

FIGURE 5.32

Laying a brick wall: (*a*) The first course of bricks for a lead is bedded in mortar, following a line marked on the foundation. (*b, c, d*) As each lead is built higher, the mason uses a spirit level to make sure that each course is level, straight, plumb, and in the same plane as the rest of the lead. A mason's rule or a story pole is also used to check the heights of the courses. (*e*) A finished lead. (*f*) A mason lays bricks to a line stretched between two leads. (*Courtesy of International Masonry Institute*)

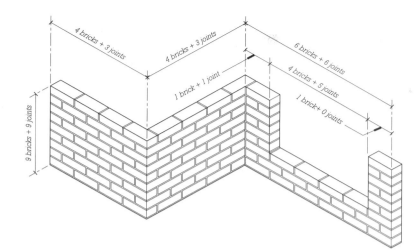

FIGURE 5.33
Dimensions for brick buildings are best worked out in advance by the designer based on the actual dimensions of the bricks and mortar joints to be used in the building. Bricks and mortar joints are carefully counted and converted to numerical dimensions for each portion of the wall.

a skilled mason is impressive both for its speed and for its quality. This level of expertise takes time and hard work to acquire, which is why the apprenticeship period for brickmasons is both long and demanding.

Bricks may be cut as needed, either with sharp, well-directed blows of the chisel-pointed end of a mason's hammer or, for greater accuracy and more intricate shapes, with a power saw that utilizes a water-cooled diamond blade (Figure 5.10). Cutting of bricks slows the process of bricklaying considerably, however, and ordinary brick walls should be dimensioned to minimize cutting (Figure 5.33).

Mortar joints can vary in thickness from about ¼ inch (6.5 mm) to more than ½ inch (13 mm). Thin joints work only when the bricks are identical to one another within very small tolerances and the mortar is made with fine sand. Very thick joints require a stiff mortar that is difficult to work with. Mortar joints are usually standardized at ⅜ inch (9.5 mm), which is easy for the mason and allows for considerable distortion and unevenness in the bricks. One-half-inch (12.7-mm) joints are also common.

The joints in brickwork are *tooled* an hour or two after laying as the mortar begins to harden, to give a neat appearance and to compact the mortar into a profile that meets the visual and weather-resistive requirements of the wall (Figures 5.34 and 5.35). Outdoors, the *vee joint* and *concave joint* shed water and resist freeze–thaw

(*a*)

(*b*)

(*c*)

FIGURE 5.34
(*a*) Tooling horizontal joints to a concave profile. (*b*) Tooling vertical joints to a concave profile. The excess mortar squeezed out of the joints by the tooling process will be swept off with a brush, leaving a finished wall. (*c*) Raking joints with a common nail held in a skate-wheel joint raker. The head of the nail digs out the mortar to a preset depth. (*Courtesy of Brick Institute of America*)

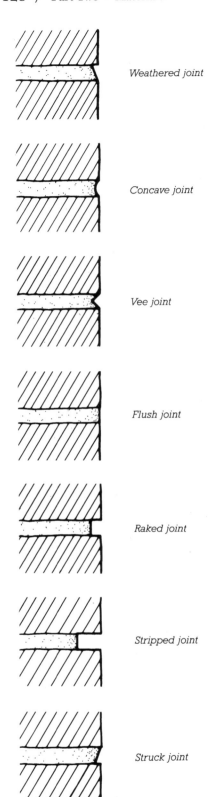

FIGURE 5.35
Joint tooling profiles for brickwork. The concave joint and vee joint are suitable for outdoor use in severe climates.

Weathered joint

Concave joint

Vee joint

Flush joint

Raked joint

Stripped joint

Struck joint

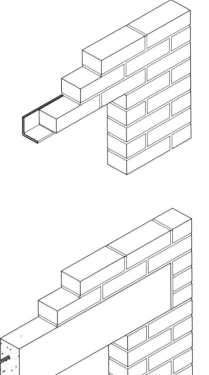

FIGURE 5.36
Two types of lintels for spanning openings in brick walls. The angle steel lintel (top) requires some trimming of the first courses of brick but is scarcely visible in the finished wall. The precast reinforced concrete lintel (bottom) is clearly visible. For short spans, cut stone lintels without reinforcing can be used in the same manner as the concrete lintel.

damage better than the others. Indoors, a *raked* or *stripped joint* can be used, if desired, to accentuate the pattern of bricks in the wall and deemphasize the mortar.

After joint tooling, the face of the brick wall is swept with a soft brush to remove the dry crumbs of mortar left by the tooling process. If the mason has worked cleanly, the wall is now finished, but most brick walls are later given a final cleaning by scrubbing with muriatic acid (HCl) and rinsing with water to remove mortar stains from the faces of the bricks. Because light-colored bricks can be stained by acids, they should be cleaned by other means.

Spanning Openings with Brick

Brick walls must be supported above openings for windows and doors. *Lintels* of reinforced concrete or steel

angles are equally satisfactory from a technical standpoint (Figure 5.36). The near invisibility of the steel lintel is a source of delight to some designers but dissatisfies those who prefer that a building express visually its means of support. Wood, which has been used historically for this purpose, is no longer used for lintels because of its tendency to burn, to decay, and to shrink and allow the masonry above to settle and crack.

The *corbel* is an ancient structural device of limited spanning capability, one that may be used for small openings in brick walls, for beam brackets, and for ornament (Figures 5.37 and 5.38). A good rule of thumb for designing corbels is that the projection of each course should not exceed half the course height; this results in a corbel angle of about 60 degrees to the horizontal and minimizes flexural stress in the bricks.

FIGURE 5.37
Corbelling has many uses in masonry construction. It is used in this example to span a door opening, and to create a bracket for support of a beam.

FIGURE 5.38
Ornamental corbelled brickwork in an 18th-century New England chimney. Step flashings of lead sheet waterproof the junction between the chimney and the wood shingles of the roof. (*Photo by Edward Allen*)

(a)

(b)

(c)

FIGURE 5.39

(a) Two rough brick arches under construction, each on its wooden centering. (b) The brick locations were marked on the centering in advance to be sure that no partial bricks or unusual mortar joint thicknesses will be required to close the arch. This was done by laying the centering on its side on the floor and placing bricks around it, adjusting their positions by trial and error to achieve a uniform spacing. Then the location of each brick was marked with pencil on the curved surface of the centering. (c) The brick arches whose construction is illustrated in the previous two photographs span a fireplace room that is roofed with a brick barrel vault. The firebox is lined with firebrick and the floor is finished with quarry tiles. (*Photos by Edward Allen*)

The brick *arch* is a structural form so widely used and so powerful, both structurally and symbolically, that entire books have been devoted to it. Given a *centering* of wood or steel (Figure 5.39), a mason can lay a brick arch very rapidly, although the *spandrel*, the area of flat wall that adjoins the arch, is slow to construct because of its numerous cut bricks. Some brick manufacturers are able to provide specially molded wedge-shaped bricks for an arch of any radius and shape. In an arch of *gauged brick*, each brick is rubbed to the required wedge shape on an abrasive stone, which is laborious and expensive. The *rough arch*, which depends on wedge-shaped mortar joints for its curvature, is therefore the type most commonly used in today's buildings (Figures 5.40 and 5.41).

FIGURE 5.40
A rough brick triple-rowlock arch spans a window opening. The recessing of the innermost rowlock course creates a shadow line that accentuates the arch. (*Photo by Edward Allen*)

FIGURE 5.41
A rough jack arch (also called a flat arch) in a wall of Flemish Bond brickwork. (*Photo by Edward Allen*)

STONE MASONRY

Stone masonry is the most expensive type of masonry used in residential construction and requires skilled and meticulous craftsmanship at the site. Because of this, the stonework of walls and fireplace facings for modern residences invariably consists of a veneer of stone over a structural CMU base (Figure 5.42). Stone paving is essentially a veneer over the ground (Figure 5.43).

Types of Building Stone

Building stone is obtained by taking rock from the earth and reducing it to the required shapes and sizes for construction. Three types of rock are commonly quarried to produce building stone:

• *Igneous rock,* which is rock that was deposited in a molten state

• *Sedimentary rock,* which is rock that was deposited by the action of water or wind

• *Metamorphic rock,* which was either igneous or sedimentary rock that has been transformed by heat and pressure into a different type of rock

Granite is the igneous rock most commonly quarried for construction in North America. It can be obtained in a range of colors that includes gray, black, pink, red, brown, buff, and green. Granite is nonporous, hard, strong, and durable, the most nearly permanent of building stones, suitable for use in contact with the ground or exposed to severe weathering. It is often used in its rough quarried state for garden walls or veneers, and its surface can be finished in any of a number of textures, including a mirrorlike polish applied to granite slabs and tiles used for countertops and fireplace facings. *Basalt* is another igneous rock often used in construction. Like granite, it is very dense and durable, but is usually found only in a dark-gray color and is seldom machined (Figure 5.42).

FIGURE 5.42
This tall stone retaining wall creates a level terrace for a house on the hill above it. The structure of the wall is made of reinforced CMUs that are covered with a basalt veneer. (*Landscape Architect: Ron Lovinger. Photo by Cynthia Girling*)

FIGURE 5.43
Granite paving laid without mortar in a bed of sand. Outdoor pavings of masonry may also be laid in mortar over concrete slabs. (*Photo by Edward Allen*)

Limestone and *sandstone* are the principal sedimentary rocks used in construction. Limestone is a porous stone that contains considerable groundwater (*quarry sap*) when quarried. While still saturated, most limestones are easy to work but are susceptible to frost damage. After seasoning in the air to evaporate the quarry sap, the stone becomes harder and is resistant to frost damage. Limestone colors range from almost white through gray and buff to iron oxide red. Sandstone has two familiar forms: *brownstone,* widely used in wall construction, and *bluestone,* a highly stratified, durable stone especially suitable for paving and wall copings. Neither limestone nor sandstone will accept a high polish.

Slate and *marble* are the major metamorphic stones utilized in construction. Slate was formed from clay and is a dense, hard stone with closely spaced planes of cleavage, along which it is easily split into sheets, making it useful for paving stones, roof shingles, and thin wall facings. It occurs in black, gray, purple, blue, green, and red. Marble is a recrystallized form of limestone. It is easily polished and cut into slabs and tiles. It occurs in white, black, and nearly every color, often with beautiful patterns of veining.

The stone industry is international in scope and operation. Architects and building owners select stone primarily on the basis of appearance, durability, and cost, often with little regard to national origin. As a result of these factors and the low cost of ocean freight relative to the value of stone, it is common to select a stone that is quarried in one country, cut and finished in another, and used for construction in a third. Because of the unique character of many domestic stones, both U.S. and Canadian quarries and mills ship millions of tons of stone to foreign countries each year, in addition to the even larger amount that is quarried, milled, and erected at home.

The construction industry uses stone in many different forms. *Fieldstone* is rough building stone that is obtained from riverbeds and rock-strewn fields. *Rubble stone* consists of irregular quarried fragments that have at least one good face to expose in a wall. *Dimension stone* is stone that has been quarried and cut into rectangular form; large slabs are often referred to as *cut stone*, and small rectangular blocks are called *ashlar*. The maximum sizes and minimum thicknesses of sheets of cut stone vary from one type of stone to another. *Flagstone* consists of thin slabs of stone that are used for flooring and paving, either rectangular or irregular in outline.

Stone Masonry

There are two simple distinctions that are useful in classifying patterns of stone masonry (Figures 5.44 and 5.45):

- *Rubble* masonry is composed of unsquared pieces of stone, whereas *ashlar* is made up of squared pieces.
- *Coursed* stone masonry has continuous horizontal joint lines, whereas *uncoursed* or *random* does not.

Rubble can take many forms, from rounded river-washed stones to broken pieces from a quarry. It may be either coursed or uncoursed.

Random Rubble

Coursed Rubble

Random Ashlar

Coursed Ashlar

FIGURE 5.44
Rubble and ashlar stone masonry, coursed and random.

FIGURE 5.45
Random granite rubble masonry (right) and random ashlar limestone (left). (*Photos by Edward Allen*)

FIGURE 5.46
In ages past, we entrusted our lives to the strength of stone and the craft of the stonemason. (*Photo by Rob Thallon*)

FIGURE 5.47
Rubble stonework at the base of a wall. The wall at the projecting bay in the background has yet to have the stonework laid. The omitted stone in the foreground allows a hose bibb to be hidden. (*Photo by Rob Thallon*)

FIGURE 5.48
A manufactured stone sample board is displayed at a masonry products supplier. Individual samples can be removed in order to examine their color and texture at the building site. (*Photo by Rob Thallon*)

Ashlar masonry may be coursed or uncoursed and may be made up of blocks that are the same size or several different sizes. The terms are obviously very general in their meaning, and the reader will find some variation in usage even among people experienced in the field of stone masonry.

Exceptional care is taken during construction to keep stonework clean: Nonstaining mortars are used, high standards of workmanship are enforced, and the work is kept covered as much as possible. Flashings in stone masonry must be of plastic or nonstaining metal. Stonework may be cleaned only with mild soap, water, and a soft brush.

From thousands of years of experience in building with stone, we have inherited a rich tradition of styles and techniques (Figure 5.46). This tradition is all the richer for its regional variations that are nurtured by the locally abundant stones: stern, gray granite and gray-streaked white Vermont marble in the northeastern

United States; red, brown, and blue sandstone in New York, Ohio, and Pennsylvania; buff and gray limestone from Indiana; gray-black basalt on the West Coast. Though the stone industry is now global in character, its greatest glories are often regional and local. Although stone masonry construction is less common now than it was a century ago, masons who specialize in stonework are typically dedicated to maintaining a high level of craft in their work (Figures 5.47 and 16.2).

Manufactured Stone

The beauty of stone masonry, in conjunction with the high cost of natural stone, has created a market for *manufactured stone*. Made in molds taken directly from real stone, manufactured stone is a thin material (approximately 1½ inches thick) made of lightweight concrete dyed to match the coloration of the authentic stone that it is meant to imitate (Figure 5.48). It is applied to the building as a veneer, more like tile than masonry, and its light weight even allows it to be applied from the top of the wall down to avoid spilling mortar on pieces that have already been laid. Manufactured stone is discussed in more detail in Chapter 11.

OTHER TYPES OF MASONRY UNITS

Bricks, stones, manufactured stones, and concrete blocks are the most commonly used types of masonry units, but other types of masonry units are available as well. *Glass blocks,* available in many textures and in clear, heat-absorbing, and reflective glass, are enjoying a second era of popularity after nearly disappearing from the market (Figure 5.49).

Autoclaved cellular concrete (ACC), though it has been manufactured and used in Europe for many years, has only recently begun to be made in North America (Figure 5.50). It is produced by mixing sand, lime, water, and a small amount of aluminum powder, and reacting these materials with steam to produce an aerated concrete that consists primarily of calcium silicate hydrates. ACC is available in precast panels for walls and roofs and in blocks that are laid in mortar like other concrete masonry units. Because of its entrapped gas bubbles, which are created by the reaction of the aluminum powder with the lime, the density of ACC is similar to that of wood, and it is easily sawed, drilled, and shaped. It has moderately good thermal insulating properties. It is not as strong as normal-density concrete, but it is sufficiently strong to serve as loadbearing walls, floors, and roofs in residential construction. Walls of ACC are too porous to be left exposed; they are usually stuccoed on the exterior and plastered on the interior.

FIGURE 5.49
Glass masonry units, commonly called glass blocks, are an appropriate material where conditions are moist or wet and daylight, privacy, and durability are all desired. Glass blocks are used effectively here to provide light and pattern to a bathroom and shower. (*Photo and design by James Givens*)

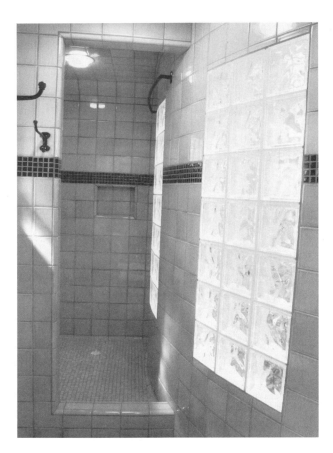

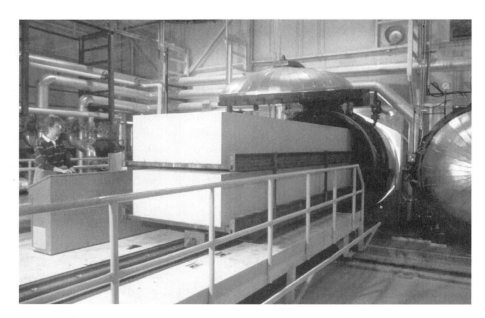

Figure 5.50
Two huge blocks of ACC material are removed from an autoclave, where they have been cured. These blocks will be wire cut into precisely dimensioned units 10 inches high x 25 inches long (250 x 635 mm) and 4 inches, 8 inches, or 10 inches thick (100, 200, or 250 mm). The solid units are laid in thin-set mortar. Two ACC factories have recently been completed in the American southeast. (*Courtesy of Hebel Building System*)

SPECIAL PROBLEMS OF MASONRY CONSTRUCTION

Mortar Joint Deterioration

Mortar joints are the weakest link in most masonry walls. Water running down a wall tends to accumulate in the joints, where cycles of freezing and thawing weather can gradually spall the mortar in an accelerating process of destruction that eventually creates water leaks and loosens the masonry units. To forestall this process as long as possible, a suitably weather-resistant mortar formulation must be used, and joints must be well filled and tightly compacted with a concave or vee tooling at the time the masonry is laid. Even with all these precautions, a masonry wall in a severe climate will show substantial joint deterioration after many years of weathering and may require *tuck pointing,* a process of raking and cutting out the defective mortar and replacing it with fresh mortar.

Moisture Resistance of Masonry

Most masonry materials, including mortar, are porous and can transmit water from the outside of the wall to the inside. Water can also enter through cracks between the masonry units and the mortar. To prevent water from entering a building through a masonry wall, the designer should begin by specifying appropriate types of masonry units, mortar, and joint tooling. The construction process should be supervised closely to be sure that all mortar joints are free of voids and that flashings are properly installed. Masonry walls should be protected against excessive wetting of the exterior surface of the wall insofar as practical through roof overhangs and proper roof drainage. Beyond these measures, consideration may also be given to coating the wall with stucco or water-repellent treatments. It is important that any exterior coating be highly permeable to water vapor to avoid blistering and rupture of the coating from outward vapor migration. Masonry primer/sealers and paints based on portland cement fill the pores of the wall without obstructing the outward passage of water vapor. Most other masonry paints are based on latex formulations and are also permeable to water vapor. Clear silicone sealers allow the natural color of the masonry to remain visible and are permeable, but they are not as durable as stucco coatings or masonry paints.

Cold and Hot Weather Construction

Mortar cannot be allowed to freeze before it has cured, or its strength and watertightness may be seriously damaged. In cold climates, special precautions are necessary if masonry work is carried out during the winter months. These include such measures as keeping masonry units and sand dry, protecting them from freezing temperatures prior to use, warming the mixing water (and sometimes the sand as well) to produce mortar at an optimum temperature for workability and curing, using a Type III (high early strength) cement to accelerate the curing of the mortar, and

mixing the mortar in smaller quantities in order that it not cool excessively before it is used. The masons' workstations should be protected from wind with temporary enclosures and heated if temperatures inside the enclosures do not remain above freezing. The finished masonry must be protected against freezing for at least 2 to 3 days after it is laid, and tops of walls should be protected from rain and snow. Chemical accelerators and so-called "antifreeze" admixtures are, in general, harmful to mortar and reinforcing steel and should not be used.

In hot weather, mortar may dry excessively before it cures. Some types of masonry units may have to be dampened before laying so that they do not absorb too much water from the mortar. It is also helpful to keep the masonry units and mortar ingredients, as well as the masons' workstations, in shade.

Efflorescence

Efflorescence is a fluffy powder, usually white, that sometimes appears on the surface of a wall of brick, stone, or concrete masonry (Figure 5.51). It consists of one or more water-soluble

salts that were originally present either in the masonry units or in the mortar. These were brought to the surface and deposited there by water that had seeped into the masonry, dissolved the salts, then migrated to the surface and evaporated. Efflorescence can usually be avoided by choosing masonry units that have been shown by laboratory testing not to contain water-soluble salts and by using clean ingredients in the mortar. Most types of efflorescence that form soon after the completion of construction are easily removed with water and a brush. Although efflorescence is likely to reappear after such a washing, it will diminish and finally disappear with time as the salt is gradually leached out of the wall. Efflorescence that forms for the first time after a period of years is an indication that water has only recently begun to enter the wall; it is best controlled by investigating and correcting the source of leakage.

Expansion and Contraction

Masonry walls expand and contract slightly in response to changes in both temperature and moisture content. Thermal movement is relatively easy

Figure 5.51
Efflorescence on a brick wall.
(***Photo by Edward Allen***)

to quantify (Figure 5.52). Moisture movement is more difficult: New clay masonry units tend to absorb water and expand slightly under moist conditions. New concrete masonry units usually shrink somewhat as they give off excess water following manufacture, which is the reason Type I units (moisture controlled) should be used where shrinkage must be minimized. Expansion and shrinkage in masonry materials are small as compared to moisture movement in wood or thermal movement in plastics or aluminum, but they must be taken into account in the design of the building. Large masonry buildings dissipate expansion and contraction by providing surface divider joints to avoid an excessive buildup of forces that could crack or spall (split off flakes of) the masonry, but residential-scale masonry structures generally use masonry for relatively small facades that do not require such detailing.

Material	in./in./°F	mm/mm/°C
Clay or shale brick masonry	0.0000036	0.0000065
Normal-weight concrete masonry	0.0000052	0.0000094
Lightweight concrete masonry	0.0000043	0.0000077
Granite	0.0000047	0.0000085
Limestone	0.0000044	0.0000079
Marble	0.0000073	0.0000131
Normal-weight concrete	0.0000055	0.0000099
Structural steel	0.0000055	0.0000117

FIGURE 5.52
Average coefficients of thermal expansion for some masonry materials.

C.S.I./C.S.C.
MasterFormat Section Numbers for Masonry

04060	**Masonry Mortar**
04070	**Masonry Grout**
04080	**Masonry Anchorage and Reinforcement**
04090	**Masonry Accessories**
04200	**MASONRY UNITS**
04210	**Clay Masonry Unit**
04220	**Concrete Masonry Unit**
04270	**Glass Masonry Unit**
04290	**Adobe Masonry Unit**
04400	**STONE**
04410	**Stone Materials**
04420	**Collected Stone**
04430	**Quarried Stone**
04800	**MASONRY ASSEMBLIES**
04810	**Unit Masonry Assemblies**
04820	**Reinforced Unit Masonry Assemblies**
04830	**Nonreinforced Unit Masonry Assemblies**
04850	**Stone Assemblies**

SELECTED REFERENCES

1. Beall, Christine. *Masonry Design and Detailing for Architects, Engineers, and Builders* (2nd ed.). New York: McGraw–Hill, 1987.

This 500-page book is the best general design reference on brick, stone, and concrete masonry.

2. Randall, Frank A., and Panarese, William C. *Concrete Masonry Handbook for Architects, Engineers, and Builders.* Skokie, IL: Portland Cement Association, 1991.

This is a clearly written, beautifully illustrated guide to every aspect of concrete masonry.

3. National Concrete Masonry Association. *Residential Technology, Volumes 1 and 2.* Herndon, VA: National Concrete Masonry Association, 2000.

A treasury of typical concrete masonry details is contained in this 212-page book.

4. Brick Institute of America. *BIA Technical Notes on Brick Construction.* McLean, VA: various dates.

This large ring binder contains the current set of technical notes with the latest information on every conceivable topic concerning bricks and brick masonry.

5. Brick Institute of America. *Principles of Brick Masonry.* Reston, VA: Brick Institute of America, updated frequently.

The 70 pages of this booklet present a complete curriculum in clay masonry construction for the student of building construction.

KEY TERMS AND CONCEPTS

masonry unit
mortar
brick
aggregate
trowel
lime
quicklime
slaked lime
hydrated lime
masonry cements
concrete masonry
   unit (CMU)
lead
story pole
line block
surface bonding
soft mud process
water-struck brick
sand-struck brick
sand-molded brick

dry press process
stiff mud process
clamp
clinker
face brick
backup brick
periodic kiln
tunnel kiln
water smoking
dehydration
oxidation
vitrification
flashing
cored brick
hollow brick
frogged brick
brick grades
brick types
firebrick

wythe
stretcher
header
rowlock
soldier
vee joint
concave joint
raked joint
lintel
corbel
centering
spandrel
gauged brick
rough arch
igneous rock
sedimentary rock
metamorphic rock
granite
basalt

limestone
sandstone
quarry sap
brownstone
bluestone
slate
marble
fieldstone
rubble
ashlar
dimension stone
cut stone
coursed stone
uncoursed stone
manufactured stone
glass block
autoclaved cellular
   concrete (ACC)
tuck pointing
efflorescence

REVIEW QUESTIONS

1. How many syllables are in the word "masonry"? (*Hint:* There cannot be more syllables in a word than there are vowels. Many people, even masons and building professionals, mispronounce this word.)

2. List the functions of mortar.

3. What are the ingredients of mortar? What is the function of each ingredient?

4. What are the advantages of concrete masonry units over other types of masonry units?

5. How may horizontal and vertical steel reinforcement be introduced into a CMU wall?

6. What are the molding processes used in manufacturing bricks? How do they differ from one another?

7. What is the function of a structural brick bond such as Common or Flemish Bond? Draw the three most popular brick bonds from memory.

8. What are the major types of stone used in construction? How do their properties differ?

9. In what ways are the laying of stone masonry different from the laying of bricks?

EXERCISES

1. Design a masonry gateway for one of the entrances to a college campus with which you are familiar. Choose whatever type of masonry you feel is appropriate and make the fullest use you can of the decorative and structural potentials of the material. Show as much detail of the masonry on your drawings as you can.

2. Visit a masonry supply company and view all the types of concrete masonry units that are available. What is the function of each?

3. Design a decorative CMU that can be used to build a richly textured wall.

4. What is the exact height of a brick wall that is 44 courses high, when three courses of brick plus their three mortar joints are 8 inches (203.2 mm) high?

5. Obtain sand, hydrated lime, several hundred bricks, and basic bricklaying tools from a masonry supply house. Arrange for a mason to help everyone in your class learn a bit of bricklaying technique. Use lime mortar (hydrated lime, sand, and water), which hardens so slowly that it can be retempered with water and used again and again for many weeks. Lay small walls in several different structural bonds. Experiment with rowlocks, soldiers, and corbels. Make simple wooden centering and construct an arch. Dismantle what you build at the end of each day, scrape the bricks clean, stack them neatly for reuse, and retemper the mortar with water, covering it with a sheet of plastic to keep it from drying out before it is used again.

6

CONCRETE

Concrete is the universal material of construction. The raw ingredients for its manufacture are readily available in every part of the globe, and concrete can be made into buildings with tools ranging from a primitive shovel to a computerized precasting plant. Concrete does not rot or burn; it is relatively low in cost; and it can be used for every building purpose, from lowly pavings to sturdy structural frames to handsome exterior claddings and interior finishes. However, it has no form of its own and no useful tensile strength. Before its limitless architectural potential can be realized, the designer must learn to combine concrete skillfully with steel to bring out the best characteristics of each material and to mold and shape it to forms appropriate to its qualities.

cement that he named *portland cement,* after English Portland limestone, whose durability as a building stone was legendary. His cement was soon in great demand, and the name portland remains in use to the present day.

Reinforced concrete, concrete that is combined with steel bars, was developed in the 1850s by several people simultaneously, including J. L. Lambot, who built several reinforced concrete boats in Paris in 1854, and an American, Thaddeus Hyatt, who made and tested a number of reinforced concrete beams. The combination of steel

HISTORY

The ancient Romans, while quarrying limestone for mortar, accidentally discovered a silica- and alumina-bearing mineral on the slopes of Mount Vesuvius that, when mixed with limestone and burned, produced a cement that exhibited the unique property of hardening underwater as well as in the air. This cement was also harder, stronger, much more adhesive, and cured much more quickly than the ordinary lime mortar to which they were accustomed. In time, this mortar not only became the preferred type for use in all their building projects but began also to alter the character of Roman construction. Masonry of stone or brick was used to build only the surface layers of piers, walls, and vaults, and the hollow interiors were filled entirely with large volumes of the new type of mortar (Figure 6.1). We now know that this mortar contained all the essential ingredients of modern portland cement and that the Romans were the inventors of concrete construction.

Knowledge of concrete construction was lost with the fall of the Roman Empire, not to be regained until the latter part of the 18th century, when a number of English inventors began experimenting with both natural and artificially produced cements. Joseph Aspdin, in 1824, patented an artificial

FIGURE 6.1

Hadrian's Villa, a large palace built near Rome between 125 and 135 A.D., used unreinforced concrete extensively for structures such as this dome. (*Photo by Edward Allen*)

142

and concrete, however, did not come into widespread use until a French gardener, Joseph Monier, obtained a patent for reinforced concrete flower pots in 1867 and went on to build concrete water tanks and bridges of the new material. By the end of the 19th century, engineering design methods had been developed for structures of reinforced concrete, and a number of major structures had been built.

For residential construction, concrete was readily adopted at an early date for its ability to replace the labor-intensive stone and mortar foundation with a less expensive, more regular, and equally durable alternative. As the use of the material became more widespread, experiments with its use for other parts of the house proliferated. In 1909, Thomas Edison produced a design to cast an entire concrete house in place (Figure 6.2). Although cast concrete houses never caught on, the use of concrete in residential construction has expanded beyond the foundation wall. The concrete slab on grade is the most significant

addition, providing basement floors, living space floors, garage floors, driveways, walkways, and terraces. Recently developed techniques that use insulated forms that are left in place to provide thermal insulation for walls have rapidly expanded the use of structural concrete for residential construction (Chapter 21). Early residential concrete work was rarely reinforced, but current practice typically combines reinforcing steel with the concrete for strength and stability.

CEMENT AND CONCRETE

Concrete is a rocklike material produced by mixing coarse and fine *aggregates*, portland cement, and water and allowing the mixture to harden. Coarse aggregate is normally gravel or crushed stone, and fine aggregate is sand. Portland cement, hereafter referred to simply as cement, is a fine gray powder. During the hardening of concrete, considerable heat (called the *heat of hydration*) is given off as the cement

combines chemically with water to form strong crystals that bind the aggregates together. During this *curing* process, especially as excess water evaporates from the concrete, concrete shrinks slightly.

In properly formulated concrete, the majority of the volume consists of coarse and fine aggregate, proportioned and graded so that the fine particles fill the spaces between the coarse ones (Figure 6.3). Each particle becomes completely coated with a paste of cement and water to join it fully to the surrounding particles.

Cement

Portland cement may be manufactured from any of a number of raw materials, provided that they are combined to yield the necessary amounts of lime, iron, silica, and alumina. Lime is commonly furnished by limestone, marble, marl, or seashells. Iron, silica, and alumina may be obtained in the form of clay or shale. The exact ingredients used depend on what is readily

FIGURE 6.2
A model of a cast concrete house designed under the direction of Thomas Edison. All of the parts of the house, including bathtubs, cupboards, fireplaces, and ornamental ceilings, were to be made from a single casting of concrete within a 500-piece cast iron formwork. When the forms were removed, only doors and windows needed to be added. A cluster of 11 houses was built in New Jersey using the techniques developed by Edison. (*From William A. Radford and Alfred S. Johnson,* Radford's Cyclopedia of Cement Construction, *Radford Architectural Co., Chicago, 1910*)

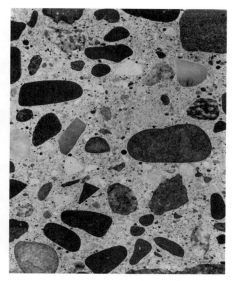

FIGURE 6.3
Photograph of a polished cross section of hardened concrete, showing the close packing of coarse and fine aggregates and the complete coating of every particle with cement paste. (*Courtesy of Portland Cement Association, Skokie, Illinois*)

available, and the recipe varies widely from one geographic region to another, often including slag or flue dust from iron furnaces, chalk, sand, ore washings, bauxite, and other minerals. The selected constituents are crushed, ground, proportioned, and blended, then conducted through a rotating kiln at temperatures of 2600 to 3000 degrees Fahrenheit (1400 to 1650°C) to produce *clinker* (Figure 6.4). After cooling, the clinker is pulverized (along with a small amount of gypsum to retard the curing process) to a powder finer than flour. This powder, portland cement, is either packaged in bags or shipped in bulk. In the United States, a standard bag of cement contains 1 cubic foot and weighs 94 pounds (43 kg).

There are eight different types of portland cement:

Type I   Normal

Type IA   Normal, air-entraining

Type II   Moderate resistance to sulfate attack

Type IIA Moderate resistance, air-entraining

Type III High early strength

Type IIIA High early strength, air-entraining

Type IV   Low heat of hydration

Type V   High resistance to sulfate attack

Type I cement is used for most purposes in construction. Types II and IV are used where the concrete will be in contact with water that has a high concentration of sulfates. Type III hardens more quickly than the other types and is employed in situations where a reduced curing period is desired as may be the case in cold weather, in the precasting of concrete structural elements, or when the construction schedule must be accelerated. Type IV is used in massive structures such as dams where the heat emitted by the curing concrete may accumulate and raise the temperature of the concrete to damaging levels. *Air-entraining* cements contain ingredients that cause microscopic air bubbles to form in the concrete during mixing (Figure 6.5). These bubbles, which usually comprise 2 to 8

percent of the volume of the finished concrete, give improved workability during placement of the concrete and, more important, greatly increase the resistance of the cured concrete to damage caused by repeated cycles of freezing and thawing. Air-entrained concrete is commonly used for paving and exposed architectural concrete in cold climates. With appropriate adjustments in the formulation of the mix, air-entrained concrete can achieve the same structural strength as normal concrete.

Aggregates and Water

Because aggregates make up roughly three-quarters of the volume of concrete, the structural strength of a concrete is heavily dependent on the quality of its aggregates. Aggregates for concrete must be strong, clean, resistant to freeze–thaw deterioration, chemically stable, and properly graded for size (Figure 6.6). An aggregate that is dusty or muddy will contaminate the cement paste with inert particles that weaken it, and an aggregate that contains any of a number of chemicals from sea salt to organic compounds can cause problems ranging from corrosion of reinforcing steel to retardation of the curing process and ultimate weakening of the concrete.

Size distribution of aggregate particles is important because a range of sizes must be included in each concrete mix to achieve close packing of the particles. A concrete aggregate is graded for size by passing a sample of it through a standard assortment of sieves with diminishing mesh spacings, then weighing the percentage of material that passes each sieve. This test makes it possible to compare the grading of an actual aggregate with the ideal grading for ng a sample of it through a standard assortment of sieves with diminishing mesh spacings, then weighing the percentage of material that passes each sieve. This test makes it possible to compare the grading of an actual

FIGURE 6.4
A rotary kiln manufacturing cement clinker. (*Courtesy of Portland Cement Association, Skokie, Illinois*)

aggregate with the ideal grading for a particular concrete mixture. Size of aggregate is also significant because the largest particle in a concrete mix must be small enough to pass easily between the most closely spaced rein-

forcing bars and to fit easily into the formwork. In general, the maximum aggregate size should not be greater than three-fourths of the clear spacing between bars or one-third the depth of a slab. For very thin slabs and toppings, a ⅜-inch (9-mm) maximum aggregate diameter is often specified. A ¾-inch or 1½ inch (19-mm or 38-mm) maximum is common for most slab and structural work, but aggregate diameters up to 6 inches (150 mm) are used in dams and other massive structures. Producers of concrete aggregates use screens to sort their product for size and can furnish aggregates graded to order.

Nonstructural lightweight concretes are made as insulating roof toppings in densities only one-fourth to one-sixth that of normal concrete. The aggregates in these concretes are usually expanded mica (*vermiculite*) or expanded volcanic glass (*perlite*), and the density of the concretes is further reduced by admixtures that entrain large amounts of air during mixing.

Mixing water for concrete must be free of harmful substances, especially organic material, clay, and salts such as chlorides and sulfates. Water that is suitable for drinking is generally suitable for concrete.

Admixtures

Ingredients other than cement, aggregates, and water are often added to concrete to alter its properties in various ways.

• *Air-entraining admixtures* may be put in the mix, if they are not already in the cement, to increase the workability of the wet concrete, to reduce freeze–thaw damage, or, in larger amounts, to create very lightweight nonstructural concretes with thermal insulating properties.

• *Water-reducing admixtures* allow a reduction in the amount of mixing water while retaining the same workability, which results in a higher strength concrete.

• *High-range water-reducing admixtures,* also known as *superplasticizers,* are organic compounds that transform a stiff concrete mix into a free-flowing liquid. They are used either to facilitate placement of concrete under difficult circumstances or to reduce the water content of a concrete mix in order to increase its strength.

• *Accelerating admixtures* cause the concrete to cure more rapidly, and *retarding admixtures* slow its curing to allow more time for working with the wet concrete.

• *Fly ash,* a fine powder that is a waste product from coal-fired power plants, increases concrete strength, decreases permeability, increases sulfate resistance, reduces temperature rise, reduces mixing water, and improves pumpability and workability of concrete.

• *Silica fume,* also known as *microsilica,* is a powder that is approximately 100 times finer than portland cement, consisting mostly of silicon dioxide. It is a byproduct of electronic chip manufacturing. When added to a concrete mix, it produces extremely high strength concrete that also has a very low permeability.

• *Blast furnace slag* is a byproduct of iron manufacture that can improve concrete workability, increase strength, reduce permeability, reduce temperature rise, and improve sulfate resistance.

• *Pozzolans* are various natural or artificial materials that react with the calcium hydroxide in wet concrete to form cementing compounds; they are used for purposes such as reducing the internal temperatures of curing concrete, reducing the reactivity of concrete with aggregates containing sulfates, or improving the workability of the concrete.

• *Workability agents* make the wet concrete easier to place in forms and finish by improving its plasticity. They include pozzolans and air-entraining admixtures, along with certain fly ashes and organic compounds.

• *Corrosion inhibitors* are used to reduce rusting of reinforcing steel in

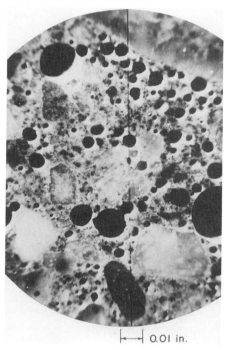

FIGURE 6.5
A photomicrograph of air-entrained concrete shows the bubbles of entrained air (0.01 inch equals 0.25 mm). (*Courtesy of Portland Cement Association, Skokie, Illinois*)

FIGURE 6.6
Taking a sample of coarse aggregate from a crusher yard for testing. (*Courtesy of Portland Cement Association, Skokie, Illinois*)

structures that are exposed to road deicing salts or other corrosion-causing chemicals.

• *Fibrous admixtures* are short fibers, usually of glass, steel, or polypropylene, that are added to a concrete mix to act as microreinforcing. Their most common use is to reduce plastic shrinkage cracking that sometimes occurs during curing of slabs.

• *Freeze protection admixtures* allow concrete to cure satisfactorily at temperatures as low as 20 degrees Fahrenheit (7°C).

• *Extended set-control admixtures* may be used to delay the curing reaction in concrete for any desired period, even several days. The *stabilizer* component, added at the time of initial mixing, defers the onset of curing indefinitely; the *activator* component, added when desired, reinitiates the curing process.

• *Coloring agents* are dyes and pigments used to alter and control the color of concrete for building components whose appearance is important.

The design of concrete mixtures is a science that can be described here only in its broad outlines. The starting point of any mix design is to establish the desired workability characteristics of the wet concrete, the desired physical properties of the cured concrete, and the acceptable cost of the concrete, keeping in mind there is no need to spend money to make concrete better than it needs to be for a given application. Concrete with ultimate compression strength as low as 2000 pounds per square inch (13.8 MPa) is satisfactory for some foundation elements. Concrete with ultimate compression strength of 22,000 pounds per square inch (150 MPa), produced with the aid of silica fume, fly ash, and superplasticizer admixtures, is currently being employed in the columns of some high-rise buildings, and higher strengths than this are certain to be developed in the near future. Acceptable workability is achievable at any of these strength levels.

Given a proper gradation of satisfactory aggregates, the strength of cured concrete is primarily dependent on the amount of cement in the mix and the *water–cement ratio*. While water is required as a reactant in the curing of concrete, much larger amounts of water must be added to a

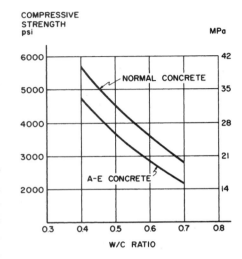

FIGURE 6.7

The effect of the water–cement ratio on the strength of concrete. A-E concrete is air-entrained concrete. (*Reprinted with permission of the Portland Cement Association from* Design and Control of Concrete Mixtures)

MAKING AND PLACING CONCRETE

Proportioning Concrete Mixes

The quality of cured concrete is measured by any of several criteria, depending on the end use of the concrete. For structural elements, compressive strength and stiffness are important. For paving and floor slabs, surface smoothness and abrasion resistance are also important. For exterior paving and concrete walls, a high degree of weather resistance is required. Watertightness is important in concrete walls. Regardless of the criterion to which one is working, however, the rules for making high-quality concrete are much the same: Use clean, sound ingredients; mix them in the correct proportions; handle the wet concrete properly to avoid segregating its ingredients; and cure the concrete carefully under controlled conditions.

FIGURE 6.8

Charging a transit-mix truck with measured quantities of cement, aggregates, admixtures, and water at a central batch plant. (*Courtesy of Portland Cement Association, Skokie, Illinois*)

concrete mix than are needed for the hydration of the cement, in order to give the wet concrete the necessary fluidity and plasticity for placing and finishing. The extra water eventually evaporates from the concrete, leaving microscopic voids that impair the strength and surface qualities of the concrete (Figure 6.7).

Absolute water–cement ratios by weight should be kept below 0.60 for most applications, meaning that the weight of the water in the mix should not be more than 60 percent of the weight of the portland cement. Higher water–cement ratios than this are often favored by concrete workers because they produce a fluid mixture that is easy to place in the forms, but the resulting concrete is likely to be deficient in strength and surface qualities. Low water–cement ratios make concrete that is dense and strong, but unless air-entraining or water-reducing admixtures are included in the mix, the concrete will not flow easily into the forms and will have large voids. It is important that concrete be formulated with the right quantity of water for each situation, enough to assure workability but not enough to

adversely affect the final properties of the material.

Most concrete in North America is proportioned at central batch plants, using up-to-date laboratory equipment and engineering knowledge to produce concrete with the specified properties. The concrete is *transit mixed* en route in a rotating drum on the back of a truck so that it is ready to pour by the time it reaches the job site (Figures 6.8 and 6.9). For very small jobs, concrete may be mixed at the job site, either in a small power-driven mixing drum or on a flat surface with shovels. For these small jobs, where the quality of the finished concrete generally does not need to be precisely controlled, proportioning is usually done by rule of thumb. Typically, the dry ingredients are measured volumetrically, using a shovel as a measuring device, in proportions such as one shovel of cement to two of sand to three of gravel, with enough water to make a wet concrete that is neither soupy nor stiff.

Each load of transit-mixed concrete is delivered with a certificate from the batch plant that lists its ingredients and their proportions. As

a further check on quality, a *slump test* may be performed at the time of pouring to determine if the desired degree of workability has been achieved without making the concrete too wet (Figures 6.10 and 6.11). When the strength of concrete is critical, such as in multistory concrete structures, standard test cylinders are also poured from each truckload. Within 48 hours of pouring, the cylinders are taken to a testing laboratory, cured for a specified period under standard conditions, and tested for compressive

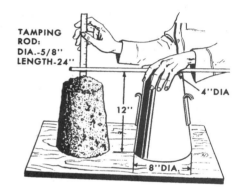

FIGURE 6.10
A slump test for concrete consistency. The hollow metal cone is filled with concrete and tamped with the rod according to a standard procedure. The cone is carefully lifted off, allowing the wet concrete to slump under its own weight. The slump in inches is measured in the manner shown. (*From the U.S. Department of Army, Concrete, Masonry, and Brickwork*)

FIGURE 6.9
A transit-mix truck discharges its concrete, which was mixed en route in the rotating drum, into a truck-mounted concrete pump, which forces it through a hose to the point in the building at which it is being poured. (*Photo by Rob Thallon*)

FIGURE 6.11
A slump test. (*Courtesy of Portland Cement Association, Skokie, Illinois*)

FIGURE 6.12
Inserting a standard concrete test cylinder into a structural testing machine, where it will be crushed to determine its strength. (*Photo courtesy of Portland Cement Association, Skokie, Illinois*)

strength (Figure 6.12). If the laboratory results are not up to the required standard, test cores are drilled from the actual members poured from the questionable batch of concrete. If the strength of these core samples is also deficient, the contractor will be required to cut out the defective concrete and replace it.

Handling and Placing Concrete

Wet concrete is not a liquid, but a slurry, an unstable mixture of solids and liquids. If wet concrete is vibrated excessively, dropped from very much of a height, or moved horizontally for any substantial distance in formwork, it is likely to *segregate*. The coarse aggregate works its way to the bottom of the form, and the water and cement paste rise to the top. The result is concrete of nonuniform and generally unsatisfactory properties. Segregation is prevented by depositing the concrete, fresh from the mixer, as close to its final position as possible. If concrete must be dropped a distance of more than 3 or 4 feet (a meter or so), it should be deposited through *dropchutes* that break the fall of the concrete. If concrete must be moved a large horizontal distance to reach inaccessible areas of the formwork, it should be pumped through hoses or conveyed in buckets or buggies, rather than pushed across or through the formwork (Figure 6.13).

Concrete must be compacted in the forms to eliminate trapped air and to fill completely around the reinforcing bars and into all the corners of the formwork. This may be done by repeatedly thrusting a rod, spade, or immersion-type vibrator into the concrete at closely spaced intervals throughout the formwork. Excessive agitation of the concrete must be avoided, however, or segregation will occur.

Curing Concrete

Because concrete cures by hydration and not by drying, it is essential that it be kept moist until its required strength is achieved. The curing reaction takes place over a very long period of time, but concrete is commonly designed to be used at the strength that it reaches after 28 days (4 weeks) of curing. If it is allowed to dry at any point during this time period, the strength of the cured concrete will be reduced, and its sur-

FIGURE 6.13
A worker directs concrete into footing formwork with the hose attached to a concrete pumper. The pumper is designed to move concrete uphill or over great distances, but it is becoming standard practice to use a pumper on most jobs regardless of terrain or scale. (*Photo courtesy of Portland Cement Association, Skokie, Illinois*)

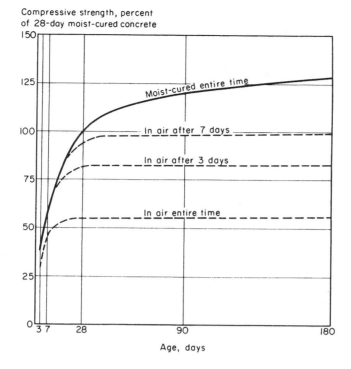

Compressive strength, percent
of 28-day moist-cured concrete

Moist-cured entire time

In air after 7 days

In air after 3 days

In air entire time

Age, days

FORMWORK

Because concrete is put in place as a shapeless slurry with no physical strength, it must be shaped and supported by *formwork* until it has cured sufficiently to support itself. Formwork is usually made of wood, metal, or plastic. It is constructed as a negative of the shape intended for the concrete. Formwork for a beam or slab serves as a temporary working surface during the construction process and as the temporary means of support for reinforcing bars. Formwork must be strong enough to support the considerable fluid pressure of wet concrete without excessive deflection. During curing, the formwork helps to retain the necessary water of hydration in the concrete. When curing is complete, the formwork must pull away cleanly from the concrete surfaces without damage either to the concrete or to the formwork, which is usually reused. This means that the formwork should have no reentrant corners that will trap or be trapped by the concrete. All formwork surfaces that are in contact with concrete must be coated with a *form-release compound,* which is an oil, wax, or plastic that prevents adhesion of the concrete to the form.

The quality of the concrete surfaces can be no better than the quality of the forms in which they are cast. Surface quality is not critical for footings or most basement walls, but for walls above grade, the requirements for surface quality and structural strength of formwork are rigorous. Top-grade wooden boards and plastic-overlaid plywoods are frequently used to achieve high-quality surfaces. The ties and temporary framing members that support the boards or plywood are closely spaced to avoid bulging of the forms under the extreme pressure of the wet concrete.

In a sense, formwork constitutes a temporary building that must be erected and demolished in order to produce a second, permanent building

face qualities will be adversely affected (Figure 6.14). Concrete elements cast in formwork are protected from dehydration on most of their surfaces by the formwork, but the top surfaces must be kept moist by repeatedly spraying or flooding with water, by covering with moisture-resistant sheets of paper or film, or by spraying on a *curing compound* that seals the surface of the concrete against loss of moisture. These measures are particularly important for concrete floor and paving slabs, whose large surface areas make them especially susceptible to drying. Premature drying is a particular danger when slabs are poured in hot or windy weather, which can cause a slab to crack even before it begins to cure. Temporary windbreaks may have to be erected, shade may have to be provided, and frequent fogging of the surface of the slab with a fine spray of water is required until the slab is hard enough to be covered or sprayed with a curing compound.

At low temperatures, the curing reaction in concrete proceeds at a much reduced rate. If concrete reaches subfreezing temperatures while curing, the curing reaction stops completely until the temperature of the concrete rises above the freezing mark. It is important that the concrete be protected from low temperatures and especially from freezing until it is fully cured. If freshly poured concrete is covered or insulated, its heat of hydration is often sufficient to maintain an adequate temperature in the concrete even at fairly low air temperatures. Under more severe winter conditions, the ingredients of the concrete may have to be heated before mixing, and both a temporary enclosure and a temporary source of heat may have to be provided.

In very hot weather, the hydration reaction is greatly accelerated, and concrete may begin curing before there is time to place and finish it. This tendency can be controlled by using cool ingredients and, under extreme conditions, by replacing some of the mixing water with an equal quantity of crushed ice, making sure that the ice has melted fully and the concrete has been thoroughly mixed before placing. Another method of cooling concrete is to bubble liquid nitrogen through the mixture at the batch plant.

of concrete. The cost of formwork is a major component of the overall cost of a concrete building frame.

REINFORCING

The Concept of Reinforcing

Concrete has no useful tensile strength (Figure 6.15) and was limited in its structural uses until the concept of steel *reinforcing* was developed. The compatibility of steel and concrete is a fortuitous accident. If the two materials had grossly different coefficients of thermal expansion, a reinforced concrete structure would tear itself apart during cycles of temperature variation. If they were chemically incompatible, the steel would corrode or the concrete would be degraded. If concrete did not adhere to steel, a very different and more expensive configuration of reinforcing would be necessary. Concrete and steel, however, change dimension at nearly the same rate in response to temperature changes; steel is protected from corrosion by the alkaline chemistry of concrete; and concrete bonds strongly to steel, providing a convenient means of adapting brittle concrete to structural elements that must resist tension, shear, and bending, as well as compression.

The basic theory of concrete reinforcing is extremely simple: Put the steel where tension occurs in a structural member and let the con-crete resist compression. This accounts fairly precisely for the location of most of the reinforcing steel that is used in a concrete structure. Steel is also used to resist cracking that might otherwise be caused by curing shrinkage and by thermal expansion and contraction in slabs and walls.

Steel Bars for Concrete Reinforcement

Reinforcing bars for concrete construction (commonly referred to as *"rebars"*) are made in mills by passing hot steel rods through a succession of rollers that press them into the desired shape. They are round in cross section, with surface ribs for better bonding to concrete (Figures 6.16 and 6.17). The bars are cut to a standard length [commonly 20 feet (6.1 m) for residential work], bundled, and shipped to local distributors.

Reinforcing bars are made in a limited number of standard diameters, as shown in Figures 6.18 and 6.19. Only the smallest three of these are commonly used in residential construction, with the #4 bar being employed more than 90 percent of the time. In the United States, bars are specified by a simple numbering system in which the number corresponds to the number of eighths of an inch (3.2 mm) of bar diameter. For example, a #4 reinforcing bar is ⅘ or ½ inch (12.7 mm) in diameter, and a #5 is ⅝ inch (15.9 mm). In Canada and most other countries outside the United States, a "hard" metric range of reinforcing bar sizes is standard (Figure 6.19).

In selecting reinforcing bars for a given basement wall or footing, the structural engineer knows from calculations the required cross-sectional area of the bars. This area may be achieved with a larger number of smaller bars, or a smaller number of larger bars, in any of a number of combinations. The final bar arrangement is selected based on the physical space available in the concrete member, the required cover dimensions (thickness of concrete between the bar and the surface of concrete), the clear spacing required between bars to allow passage of the concrete aggregate, and the sizes and numbers of bars that will be most convenient to install.

Reinforcing bars are available in grades 40, 50, and 60, corresponding to steel with yield strengths of 40,000, 50,000, and 60,000 pounds per square inch (275, 345, and 415 MPa). Grade 60 is the most readily available of the three, but grade 40 is almost always sufficient for residential work. There is little cost difference among the three grades, so engineers typically specify grade 40 and allow either of the two higher grades to be substituted if grade 40 is not available. The higher strength bars are useful where there is a restricted amount of space available for bars in a concrete member, and they are the most economical

Material	Allowable Tensile Strength	Allowable Compressive Strength	Density	Modulus of Elasticity
Wood (average)	700 psi (4.83 MPa)	1,100 psi (7.58 MPa)	30 pcf (480 kg/m³)	1,200,000 psi (8,275 MPa)
Brick masonry (average)	0	250 psi (1.72 MPa)	120 pcf (1,920 kg/m³)	1,200,000 psi (8,275 MPa)
Steel (ASTM A36)	22,000 psi (151.69 MPa)	22,000 psi (151.69 MPa)	490 pcf (7,850 kg/m³)	29,000,000 psi (200,000 MPa)
Concrete (average)	0	1,350 psi (9.31 MPa)	145 pcf (2,320 kg/m³)	3,150,000 psi (21,720 MPa)

FIGURE 6.15
Concrete, like masonry, has no useful tensile strength, but its compressive strength is considerable, and when combined with steel reinforcing, concrete can be used for every type of structure.

FIGURE 6.16
Glowing strands of steel are shaped into reinforcing bars as they snake their way through a rolling mill. (*Courtesy of Bethlehem Steel Company*)

FIGURE 6.17
The deformations rolled into the surface of a reinforcing bar help it to bond tightly to concrete. (*Photo by Edward Allen*)

ASTM Standard Reinforcing Bars

Bar Size		Nominal Dimensions					
		Diameter		Cross-Sectional Area		Weight (mass)	
American	Metric	in.	mm	in.²	mm²	lb/ft	kg/m
#3	#10	0.375	9.5	0.11	71	0.376	0.560
#4	#13	0.500	12.7	0.20	129	0.668	0.944
#5	#16	0.625	15.9	0.31	199	1.043	1.552
#6	#19	0.750	19.1	0.44	284	1.502	2.235
#7	#22	0.875	22.2	0.60	387	2.044	3.042
#8	#25	1.000	25.4	0.79	510	2.670	3.973
#9	#29	1.128	28.7	1.00	645	3.400	5.060
#10	#32	1.270	32.3	1.27	819	4.303	6.404
#11	#36	1.410	35.8	1.56	1006	5.313	7.907
#14	#43	1.693	43.0	2.25	1452	7.65	11.38
#18	#57	2.257	57.3	4.00	2581	13.6	20.24

FIGURE 6.18
American standard sizes of reinforcing bars. These sizes were originally established in conventional units of inches and square inches. More recently, "soft metric" designations have also been given to the bars without changing their sizes. Notice that the size designations of the bars in both systems of measurement correspond very closely to the rule-of-thumb values of ⅛ inch or 1 mm per bar size number.

FIGURE 6.19
These "hard metric" reinforcing bar sizes are used in most countries of the world.

Metric Reinforcing Bars

Size Designation	Nominal Mass (kg/m)	Nominal Dimensions	
		Diameter (mm)	Cross-Sectional Area (mm²)
10M	0.785	11.3	100
15M	1.570	16.0	200
20M	2.355	19.5	300
25M	3.925	25.2	500
30M	5.495	29.9	700
35M	7.850	35.7	1000
45M	11.775	43.7	1500
55M	19.625	56.4	2500

FIGURE 6.20
A rebar cutter/bender is demonstrated in a masonry supply yard. The rebar oriented parallel to the machine is being bent by the force applied to the lever arm. The rebar perpendicular to the machine has been cut by the action of the same lever arm. (*Photo by Lynne Clearfield*)

for vertical bars in columns where the bars share compressive loads with the concrete in which they are embedded.

Reinforcing steel is also produced in sheets or rolls of *welded wire fabric,* a grid of wires spaced 2 to 12 inches (50 to 300 mm) apart. The heaviest wires are just over ½ inch (12.7 mm) in diameter, and the lightest are ⅛ inch in diameter. The lighter styles of welded wire fabric resemble cattle fencing and are used to reinforce concrete slabs on grade. The most common size for residential slabs is 6 × 6–W1.4/1.4, which is a 6-inch grid of ⅛-inch-diameter wires. Many architects and engineers prefer a more widely spaced grid of individual #3 bars for slabs on grade, however, because the grid is stronger and workers are not so likely to step on the reinforcing during placement of the concrete, forcing it to the bottom of the slab where it cannot furnish adequate tensile resistance to the slab. The principal advantage of welded wire fabric over individual bars is economy of labor in placing the reinforcing, especially where a large number of

small bars can be replaced by a single sheet of material.

The small-diameter bars usually used for residential work are typically cut to length and bent as required at the site. A rebar cutter/bender is the simple tool that performs these tasks (Figure 6.20). Large-diameter bars for larger structures are usually precut and bent at a fabricating shop and shipped to the job site labeled to indicate their location in the building. Once in the forms, reinforcing bars are wired together to await pouring of the concrete. The wire has a temporary function only, which is to hold the reinforcement in position until the concrete has cured. Any transfer of load from one reinforcing bar to another in the completed building is done by the concrete. Where two bars must be spliced, they are overlapped a specified number of bar diameters, and forces are transferred from one to the other by the surrounding concrete.

The composite action of concrete and steel in reinforced concrete structural elements is such that the reinforcing steel is always loaded axially in tension or compression, and occasionally in shear, but never in bending. The bending stiffness of the reinforcing bars themselves is of no conse-

quence in imparting strength to the concrete.

Reinforcing a Simple Concrete Beam

In an ideal, simply supported beam under a uniform or distributed loading, such as a lintel over an opening in a masonry wall, compressive forces follow a set of archlike curves that create a maximum of compressive stress in the top of the beam at midspan, with progressively lower compressive stresses toward either end. A mirrored set of curves shows the lines of tensile force, with stresses again reaching a maximum at the middle of the span (Figure 6.21). In an ideally reinforced concrete beam, steel reinforcing bars would be bent to follow these lines of tension, and the bunching of the bars at midspan would serve to carry the higher stresses at that point. However, it is difficult to bend bars into these curves and to support the curved bars adequately in the formwork, so a simpler, rectilinear arrangement of reinforcing steel is substituted.

This is done with a set of bottom bars and stirrups. The *bottom bars* are placed near the bottom of the beam, leaving a specified amount of con-

Uniform Loading

FIGURE 6.21
The directions of force in a simply supported beam under a uniform loading. The solid lines represent compression, and the broken lines represent tension. Near the ends of the beam, the lines of tensile force move upward diagonally through the beam.

Diagonal forces are highest near the ends of the beam

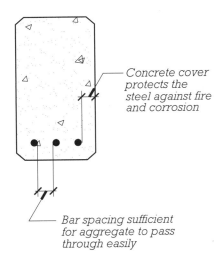

Concrete cover protects the steel against fire and corrosion

Bar spacing sufficient for aggregate to pass through easily

FIGURE 6.22
A cross section of a rectangular concrete beam, showing cover and bar spacing. The bars are located near the bottom of the beam where tension forces are greatest.

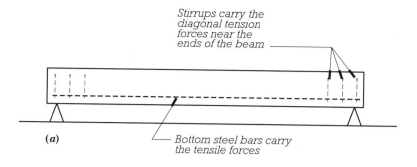

Stirrups carry the diagonal tension forces near the ends of the beam

(a)

Bottom steel bars carry the tensile forces

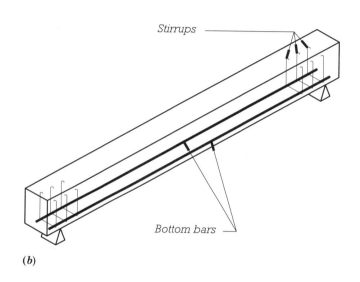

Stirrups

Bottom bars

(b)

FIGURE 6.23

(*a*) Steel reinforcing for a simply supported beam under a uniform loading. (*b*) A three-dimensional view of the same reinforcing.

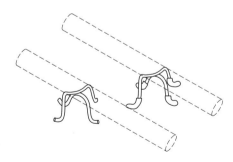

FIGURE 6.24
Chairs for supporting reinforcing bars in beams and slabs. The chair on the right has plastic-dipped feet. Chairs support only one or two bars each; wider supports, called bolsters, support several bars.

crete below and to the sides of the rods as *cover* (Figure 6.22). The concrete cover provides a full embedment for the reinforcing bars and protects them against fire and corrosion. The bars are most heavily stressed at the mid-point of the beam span, with progressively smaller amounts of stress toward each of the supports. The differences in stress are dissipated from the bars into the concrete by means of the adhesive *bond* forces between the concrete and the steel, aided by the ribs on the surface of the bars.

The bottom steel does the heavy tensile work in the beam, but lesser tensile forces remain in a diagonal orientation near the ends of the beam.

These are resisted by a series of *stirrups* (Figure 6.23). The stirrups furnish vertical tensile reinforcing to resist the cracking forces that run diagonally across them. Obviously, diagonal stirrups would work more efficiently, but economy of installation dictates that stirrups should be oriented vertically. In residential work, the stresses in short, lightly loaded beams such as lintels are often so minimal that stirrups are not required. The small tensile strength in the concrete itself is sufficient to resist the diagonal tensions.

When the simple beam of our example is formed, the bottom steel is supported at the correct cover height by steel *chairs* (Figure 6.24). These accessories remain in the concrete after pouring because, although their work is then finished, there is no way to get them out. In outdoor concrete work, the feet of the chairs and bolsters may rust where they come in contact with the face of the beam or slab, unless plastic or plastic-capped steel chairs are used. Where reinforced concrete is poured in direct contact with the soil, concrete bricks or small pieces of concrete are used instead of chairs to prevent rust from forming under the feet of the chairs and spreading up into the reinforcing bars.

Reinforcing a Simple Cantilever

In a simple cantilever such as a cantilevered concrete fireplace hearth (Figure 16.11), the tensile and compressive forces are at a maximum not in the middle of the span but rather at the end, where the cantilever springs from its support. The tensile forces are at the top of the concrete rather than at the bottom, so the reinforcing bars should be located near the top of the cantilevered member (Figure 6.25). It is important to anchor the bars into the mass of masonry or concrete from which the cantilever springs. This can be achieved, as in this example, by extending them back into the concrete of the supported portion of the hearth. A hearth or other cantilevered slab should have reinforcing bars distributed across its width. The same principles of bar placement and anchoring apply to a cantilevered beam with much narrower proportions.

Reinforcing a Concrete Retaining Wall

The reinforcing of a concrete retaining wall is much more complicated

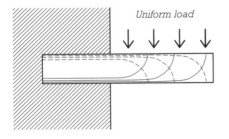

Uniform load

FIGURE 6.25
The directions of force in a cantilevered concrete member. The solid lines represent compression, and the broken lines represent tension. Reinforcing bars should be located near the lines of tension at the top of the member.

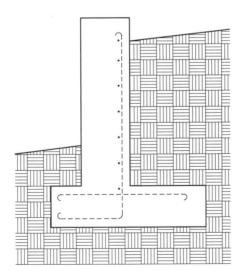

FIGURE 6.26
A cross section of a typical concrete retaining wall showing the location of the reinforcing bars.

than reinforcing either a simple beam or a cantilever. The forces that the wall must resist depend on the soil types under and uphill from the wall, and the estimation of these forces is a matter best attempted by an experienced engineer or architect. The placement of reinforcing bars in a retaining wall depends on whether the wall resists the lateral forces of the retained soil by acting as a cantilever from its footing, as a beam between its two ends, or both. If the wall acts as a vertical cantilever, the tensile forces in the wall, and therefore the main vertical bars, are on the uphill side, where they must be well anchored into the footing (Figure 6.26). Occasionally, a retaining wall is designed to span as a horizontal beam between its two ends. In this case, the tension in the wall is on the downhill side, and horizontal bars in this location resist the force of the retained soil. Because of the complexities of their design and the high incidence of failures of retaining walls, most jurisdictions require that retaining walls over 4 feet (1.2 m) tall be designed by an engineer or architect.

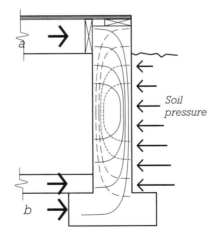

Soil pressure

FIGURE 6.27
A foundation wall acts like a vertically oriented beam. The load from the soil pressure is resisted by (*a*) the floor structure at the top of the wall and (*b*) the basement slab and footing at the base of the wall.

Reinforcing a Foundation Wall

Concrete foundation walls, whether for a crawlspace or a basement, are like retaining walls but with one important difference: Foundation walls are connected at their top to a floor structure. This connection to the floor braces the wall at the top, and the basement floor slab braces it at the bottom, enabling it to resist the pressure of the soil around the perimeter of the house (Figure 6.27). The wall acts essentially as a vertical beam between the basement floor slab and the main floor structure. In cases where the pressure of the backfilled soil is high, vertical reinforcing bars must be located near the inside force of the wall to provide tensile strength.

When a foundation wall is constructed on marginal soil that has little compressive strength, or on soil that is inconsistent from one end of the wall to the other, the wall must act as a horizontal beam to support the load of the house above. Because it cannot be predicted how the vertical forces along the wall will be distributed, the wall may be required to resist upward or downward bending forces at any point, so reinforcing bars are located at both the top and the bottom (Figure 6.28).

Reinforcing a Concrete Column

Site-cast concrete columns are seldom used in low residential construction, but occasionally occur in parking garages for multifamily structures, or in custom single-family houses. Columns contain two types of reinforcing: Vertical bars work with the concrete to share the compressive loads and to resist the tensile stresses that occur in columns when a building frame is subjected to wind or earthquake forces. *Ties* of smaller steel bars wrapped around the vertical bars help to prevent them from buckling under load: Inward buckling is prevented by the concrete core of the column, and outward buckling by the ties (Figure 6.29). The vertical bars may be arranged

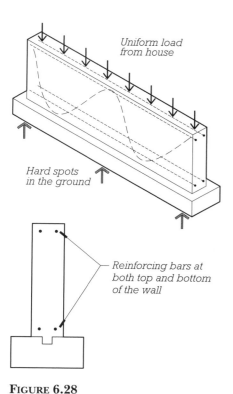

FIGURE 6.29
Reinforcing for concrete columns. To the left, a column with a rectangular arrangement of vertical bars and column ties. To the right, a circular arrangement of vertical bars with a column spiral. Either arrangement may be used in either a round or a square column.

FIGURE 6.28
Hard spots in the soil at the positions shown would cause tensile forces in a foundation wall to follow the curved line—illustrating the need to locate reinforcing bars at both the top and the bottom of a wall on inconsistent soil. It is wise to reinforce both the top and the bottom of the wall in most cases, whether the soil is considered stable or not.

either in a circle or in a rectangular pattern. The size and spacing of the ties are determined by the structural engineer. To minimize labor costs on the job site, the ties and vertical bars for each column are usually wired together in a horizontal position at ground level, and the finished *column cage* is later lifted into its final position.

ACI 301

Concrete structures are built under ACI 301, *Specifications for Structural Concrete for Buildings,* a publication of the American Concrete Institute. This is a comprehensive, detailed specification that covers every aspect of concrete work: formwork, accessories, reinforcement, concrete mixtures, handling and placing of concrete, and the use of concrete in exposed architectural surfaces. It leaves nothing to chance. It is a standard that is familiar to architects, engineers, contractors, and building inspectors, and furnishes the basis upon which everyone works in designing and constructing a concrete building.

C.S.I./C.S.C. MasterFormat Section Numbers for Concrete	
03100	**CONCRETE FORMS AND ACCESSORIES**
03110	Structural Cast-in-Place Form
03120	Architectural Cast-in-Place Form
03150	Concrete Accessories
03200	**CONCRETE REINFORCEMENT**
03210	Reinforcing Steel
03220	Welded Wire Fabric
03240	Fibrous Reinforcement
03300	**CAST-IN-PLACE CONCRETE**
03310	Structural Concrete
03330	Architectural Concrete
03350	Concrete Finishing
03390	Concrete Curing

SELECTED REFERENCES

1. Portland Cement Association. *Design and Control of Concrete Mixtures* (13th ed.). Skokie, IL: 1994.

The 15 chapters of this book summarize clearly and succinctly, with many explanatory photographs and tables, the state of current practice in making, placing, finishing, and curing concrete.

2. Concrete Reinforcing Steel Institute. *Manual of Standard Practice* (26th ed.). Schaumburg, IL: Concrete Reinforcing Steel Institute, 1997.

Specifications for reinforcing steel, welded wire fabric, bar supports, detailing, fabrication, and installation are standardized in this booklet.

3. American Concrete Institute. *ACI 301: Specifications for Structural Concrete for Buildings.* Farmington Hills, MI: ACI International, 1996.

This is the standard, detailed specification for every aspect of structural concrete.

KEY TERMS AND CONCEPTS

portland cement
reinforced concrete
concrete
aggregate
cement
heat of hydration
curing
clinker
air-entraining cement
lightweight aggregate
air-entraining admixture

water-reducing admixture
high-range water-
reducing admixture
superplasticizer
accelerating admixture
retarding admixture
fly ash
silica fume
blast furnace slag
pozzolan
workability agent

corrosion inhibitor
fibrous admixture
freeze protection admixture
extended set-control
admixture
stabilizer
activator
coloring agent
water–cement ratio
transit-mixed concrete
slump test

segregation
dropchute
curing compound
formwork
form-release compound
precast concrete
concrete reinforcing
reinforcing bar
rebar
welded wire fabric
bottom bars
cover

REVIEW QUESTIONS

1. What is the difference between cement and concrete?

2. List the conditions that must be met to make a satisfactory concrete mix.

3. List the precautions that should be taken to cure concrete properly. How do these change in very hot, very windy, and very cold weather?

4. What problems are likely to occur if concrete has too low a slump? Too high a slump? How can the slump be increased without increasing the water content of the concrete mixture?

5. Explain how steel reinforcing bars work in concrete.

6. Explain the role of stirrups in beams.

7. Explain the role of ties in columns.

EXERCISES

1. Design a simple concrete mixture. Mix it and pour some test cylinders for several water–cement ratios. Cure and test the cylinders. Plot a graph of concrete strength versus water–cement ratio.

2. Sketch from memory the pattern of reinforcing for a continuous concrete beam. Add notes to explain the function of each feature of the reinforcing.

3. Design, form, reinforce, and cast a small concrete beam, perhaps 6 to 12 feet (2 to 4 m) long. Get help from a professional, if necessary, in designing the beam.

4. Visit a construction site where concrete work is going on. Examine the forms, reinforcing, and concrete work. Observe how concrete is brought to the site, transported, placed, compacted, and finished.

PART THREE

WOOD LIGHT FRAME CONSTRUCTION

ROUGH SITEWORK

7

Before the actual construction of a house begins, several activities must be carried out on the site where the residence will be located. Excess vegetation and soil must be removed; the building's foundation must be laid out and excavated; water supply, sewer pipe, and utility cables must be installed and buried underground; and a drainage system for surface runoff must be put in place. All of these activities fall under the general description of rough sitework.

An important aspect of rough sitework is to organize the site as a temporary factory for the construction of the house. Although the construction phase generally lasts only 5 to 9 months, it will be a period of intense activity. While planning the efficient use of the site for construction, the experienced contractor also keeps in mind the long-term goal of sitework, which is to make an outdoor living environment that will complement the interior rooms of the house for many years to come.

This chapter will focus on the activities surrounding the short-term use of the site, primarily preparing the building foundation and locating utilities in the ground. Chapter 19, the final chapter in Part 3, will examine the conversion of the building property from a construction site into a landscaped residential environment.

DESIGN CONSIDERATIONS

Before the contractor starts any work on the site, the designer must complete a number of tasks:

1. Determine the exact location of the proposed building relative to the property lines. It is important to check the municipal *zoning ordinances,* which establish requirements such as *setbacks*—the distances required between buildings and property lines (Figure 7.1).

2. Locate the *utilities* that will support the site, such as power, water, gas, and telephone lines. Incorporate the position of the various pipes and lines, and their type (overhead or underground), into the design of the house.

3. For urban sites, locate the *municipal sewer line.* The designer must specify the floor elevations of the building in the design to assure the natural drainage of the house sewer into the municipal sewer.

4. For rural sites, obtain approval from the local board of health or land management agency for the installation of a *septic system* and secure a reliable source of *potable water* for the house.

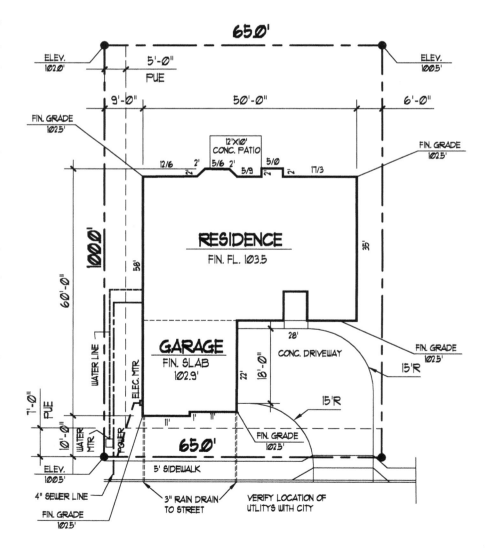

160

5. Design *storm drainage* in accordance with local ordinances.

6. If the ability of the soil to support the weight of a building is in doubt, a soil and/or foundation engineer should be consulted and his or her recommendations should be incorporated into the plans.

7. Investigate any other local ordinances, such as tree-cutting restrictions, wetlands regulations, historic district restrictions, and building code requirements, and make sure that these regulations have been complied with in the design of the house.

ORGANIZING THE SITE FOR CONSTRUCTION

The several-month period of actual construction is the most intensive use the site will ever experience. Heavy equipment of all types will move over the soil and and cut into it. Earth, gravel, and building materials will be deposited and stored on the surface of the site. Scores of workers will arrive daily in pickup trucks and vans that will park on or around the site.

Orchestrating all of this activity is one of the most important responsibilities of the contractor, and one that is crucial to the overall efficiency of the work. Access routes and storage locations need to be planned strategically so they are easy to reach, while avoiding locations where trenches will be dug and filled. Waste piles, temporary utilities, and parking areas

FIGURE 7.1
This architectural site plan shows the location of the proposed building on the property. The building official will check the drawing to assure that setback requirements are met. The contractor will use the drawing to help determine the location of the foundation. (*Courtesy of Dietz Design, Eugene, Oregon*)

for workers should be located in places that will not disrupt construction activity. On reasonably large, flat sites, this choreography is generally not difficult. On very small or steep sites, however, the management of all these materials and activities becomes much more difficult. Unforeseen circumstances will always call for adjustment, but a comprehensive plan prepared by the contractor for the use of the site throughout the course of construction will conserve much time and energy in the long run.

SITE PREPARATION

One of the first activities at the building site is a survey to establish the locations of the *property lines*. This is done by a *licensed surveyor,* who marks the corners of the property and notes relative elevations to be used for determining the floor levels of the house and establishing slopes of future site features such as driveways and drainage piping.

Before construction can begin, the contractor must clear major vegetation and debris from the portion of the site used for building. Any trees within the building footprint or very close to the building must be removed. Very small trees and brush should be cut near the ground and removed from the site. Medium-sized and larger trees should be cut above the ground so that excavators can later use the stumps as leverage when removing the roots. (The disposal of large stumps and roots can be a significant expense that is frequently overlooked.)

Once the property lines have been located and vegetation removed, the corners of the building are roughly situated and marked on property through a process known as *staking out.* Small stakes are driven to mark approximately the corners of the building's foundation. This is usually done either by the designer during the design process or by the

contractor just prior to construction.

The precise location of the building on the site is established by using the time-tested *batter board* system (Figure 7.2). Batter boards are constructed at least 4 feet outside the stakes at the corners of the foundation to allow room for excavation. Strings are stretched across the batter boards and, with the use of a plumb bob, are carefully poas bays, courts, and offsets (jogs in the foundation) are staked in or painted on the ground until after excavation.

When establishing the building's outline, it is important to make sure that the corners are square, forming a true 90-degree angle. A 3–4–5 triangle is often used to verify that the corners are square. Workers will measure out a multiple of 3 and 4 from the corner (say 6 and 8 feet) and will verify that the diagonal between the two points is a multiple of 5 (say 10 feet). Diagonal measurements between the outside corners of the building may also be supplied by the designer (Figure 7.3). Surveying instruments or other types of measuring equipment are also commonly used to establish the corners. On small sites, this work may be performed by a surveyor to assure that municipal setback requirements are recognized.

On sloping or irregular sites, measurements along the slope of the ground will not reflect the true horizontal distance between corners. In this situation, a *builder's level* is usually used in conjunction with batter boards (Figure 7.4). This instrument, through a magnifying telescope with crosshairs, is capable of precisely establishing vertical and horizontal planes. After one corner is located, the builder's level is set up accurately over this corner, and the angle between the second and third corners can be measured by rotating the instrument. Horizontal distances between corners are determined by establishing a horizontal plane with the level, measuring along this plane between corners, and projecting vertically to the ground with a

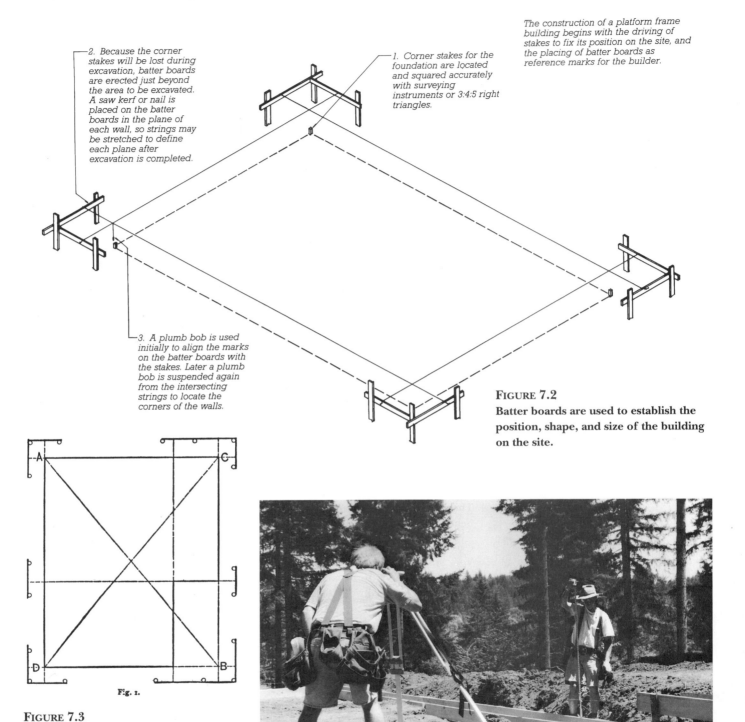

2. Because the corner stakes will be lost during excavation, batter boards are erected just beyond the area to be excavated. A saw kerf or nail is placed on the batter boards in the plane of each wall, so strings may be stretched to define each plane after excavation is completed.

1. Corner stakes for the foundation are located and squared accurately with surveying instruments or 3:4:5 right triangles.

The construction of a platform frame building begins with the driving of stakes to fix its position on the site, and the placing of batter boards as reference marks for the builder.

3. A plumb bob is used initially to align the marks on the batter boards with the stakes. Later a plumb bob is suspended again from the intersecting strings to locate the corners of the walls.

FIGURE 7.2
Batter boards are used to establish the position, shape, and size of the building on the site.

FIGURE 7.3
This illustration from 1905 shows a plan view of the same batter board system that is used today to lay out a foundation. Notice how extra boards are used to position a porch inset at the lower right corner of the drawing. Diagonal measurements *AB* and *CD* are used to assure that the building is square. (*From F. E. Kidder,* Building Construction and Superintendence, *William T. Comstock, New York, 1905*)

FIGURE 7.4
A common builder's level is used to establish level lines and to measure horizontal angles when staking a foundation. More sophisticated and expensive versions use a rotating laser beam to establish a horizontal plane from which levels can be measured. (*Photo by Rob Thallon*)

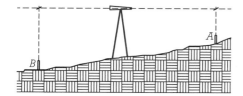

FIGURE 7.5

The horizontal distance between stake _A_ and point _B_ is measured along the horizontal line established by the builder's level. A plumb bob at point _B_ will locate the position on the ground for a stake at _B_. Measured along the ground, the distance from stake _A_ to stake _B_ would be greater than the horizontal distance between the two.

plumb bob from the measurement on the plane (Figure 7.5). Batter boards set back from these corners allow their position to be maintained throughout the excavation process.

EXCAVATION

After the site has been cleared and the approximate location of the corners of the building have been established, _excavation_ can begin. Depending on the soil type, the slope, the complexity of the building, and the availability of equipment, a number of machines, ranging from road grader to backhoe to bulldozer, may be used. (If the house is to be built on a rocky site, a blasting contractor might also be employed.) The excavator first removes from the surface all soil containing organic matter within the footprint of the house. This mixture of topsoil and plant material (grass and small roots) is stockpiled for future use as backfill.

Now the process of excavating to the correct depth begins. This depth, set by the designer, usually corresponds to the level at which the bottom of the building's footing will be located. (The footing is the concrete base that the building foundation will rest on.) The depth is set in relation to an existing _benchmark_ or datum established by the designer or the surveyor, such as a mark on the top of a street

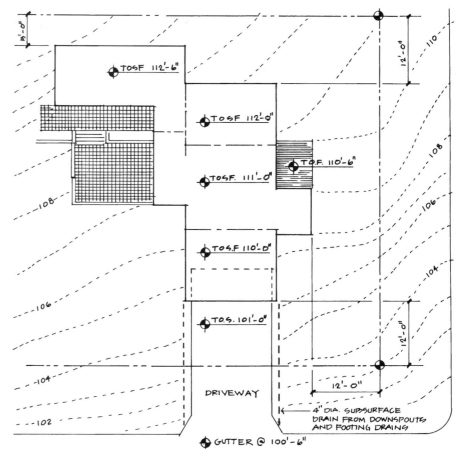

FIGURE 7.6

The elevations of the proposed finish floors are noted on the construction drawings. These elevations are referenced to a known elevation on the site known as a datum or benchmark—in this case, the gutter in front of the driveway. The contractor uses this information to establish the elevation of the excavation at the base of the footings.

FIGURE 7.7

Infrared laser levels such as this emit a highly concentrated beam of light across a level plane. Measurements taken from the laser beam establish grades for excavations, foundations, and floors. Laser levels are replacing optical levels as standard equipment on residential construction sites. Visible beam laser levels are also available for interior work. (_Photos © 2000 Topcon Corp._)

curb or a nail driven into a power pole or a special stake. The vertical distance between the benchmark and the base of the footing is indicated in the construction drawings (Figure 7.6).

Using a builder's level or a _laser level,_ the contractor and excavator

usually work as a team to measure the depth of the excavation. The level is firmly positioned beside the excavation, and the vertical distance between this instrument and the benchmark is measured using a measuring tape, a _surveyor's rod,_ or a _story pole_ (which is a

FIGURE 7.8
A rod with a sensor that emits a sound when struck by the laser beam allows depth measurements to be performed by one person. (*Photo by Rob Thallon*)

site-made calibrated stick) (Figure 7.8). Then the vertical distance between the benchmark and the base of the proposed footing is added to the first measurement to determine the total vertical distance between builder's level and base of excavation. If, for example, the benchmark is 50 inches below the level, and the bottom of the footing is designed to be 10 inches below the benchmark, the contractor would make a mark at 60 inches above the bottom of the surveyor's rod. The rod and level are used occasionally as excavation work proceeds to check the level of the bottom of the excavation, and the digging stops when the 60-inch mark is at the same horizontal plane as the level (Figure 7.9).

If the soil under the proposed house is not sufficiently strong to support the house, *compacted fill* may be required. The installation of compacted fill is directed by an engineer and involves the placement of 6- to 12-inch (150 to 300 mm) layers of soil or rock, called *lifts*. Each lift is compacted before the subsequent lift is placed. A variety of heavy rollers, vibrating rollers, and vibrating plates are used for this work (Figure 7.10). When soil is being compacted, it is important to monitor carefully the ratio of moisture to density of the material. This is crucial to the success of the operation as it relates the compaction work on the site to lab tests that determine the optimum moisture content for compaction of various soils (Figure 7.11).

It is common to encounter unforeseen circumstances during excavation work. Underground springs, buried boulders, and pockets of clay or other unstable soil are often uncovered. When this occurs, adjustments sometimes need to be made to the foundation or to the excavation. Often,

an engineer will be called in to examine the site and recommend the best method for proceeding with construction. Common solutions might include adding compacted gravel, doing additional compacting of the soil beneath the foundation, increasing the size of the footings, or increasing the amount of steel reinforcement in the foundation.

SITE UTILITIES

Temporary Utilities

For the convenience of the contractor, *temporary utilities* must be installed as early as possible. Electric power for construction equipment is required from the time footings are beginning to be made. The contractor usually sets up a temporary weathertight electrical panel, and the local utility company extends a permanent power line onto the property to supply the box with electricity (Figure 7.12). The temporary panel must be inspected by the electrical inspector before it can be connected, and the power company can often take weeks to schedule its work, so the experienced contractor starts this process well in advance of the need for temporary power on the site. If a stopgap solution is required, a gasoline-powered electric generator can be used to allow construction to proceed until electricity is available at the site.

Water is also needed early in the construction process, especially for concrete block foundations. Until a permanent water supply is installed on site, adequate quantities of water for construction can usually be trucked to the site in small barrels or borrowed from a neighbor.

Temporary toilet facilities for workers are required on the site from the beginning of construction. Usually, the contractor will rent a portable chemical toilet from a supply company, which delivers it to the site and cleans it weekly (Figure 7.13).

Until very recently, a temporary phone connection played an impor-

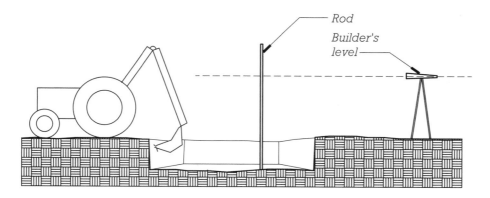

FIGURE 7.9
The rod measures the vertical distance from the bottom of the excavation to the horizontal plane established by the builder's level.

FIGURE 7.10
This machine is one of a variety designed to compact soil or gravel for the purpose of supporting construction. When used under a foundation, the material is compacted in layers or lifts, under the supervision of an engineer. (*Photo by Rob Thallon*)

FIGURE 7.11
A nuclear gauge can be used to test the strength of the soil. The density of the soil is measured by the machine with a probe inserted into a test hole. Soil strength is calculated by comparing the measured field density with the maximum theoretical dry density. (*Courtesy of Foundation Engineering, Eugene, Oregon*)

FIGURE 7.12
This temporary electrical panel with meter was set up and connected to a power supply before framing of the house began. Temporary lines and extension cords provide electricity for construction equipment from this source until the permanent power line is connected to the house. (*Photo by Rob Thallon*)

FIGURE 7.13
A chemically sanitized temporary toilet facility is a fixture on the site of new construction until the job is complete. (*Photo by Rob Thallon*)

tant role in site organization and overall construction coordination. With the recent introduction of cellular phones, hard-wired temporary site phones have all but disappeared.

Permanent Electrical Power

Residential electricity is supplied by the local utility from overhead or underground *high-voltage lines.* The high voltage is stepped down with a *transformer* to produce the 110/220 volt power used in the house (Figure 7.14). If the high-voltage lines are overhead, the transformer is located on a pole; if underground, the transformer is located on the ground. In either case, the *main service cable,* which connects the transformer to the new house, may be placed either overhead or underground. Overhead installations are less expensive and often less disruptive to a wooded site, while underground wires are aesthetically preferable and easier to maintain.

Electrical codes require overhead electric service wires to maintain a minimum clearance to the ground and from the limbs of surrounding trees. Underground cables are required to be located between 30 and 32 inches below the finished surface of the ground. At this depth, the cable will be well protected from everyday gardening activities. The trench may contain other utilities as well, so long as these are separated from the electrical cable by distances specified in the various codes. Electrical services are discussed further in Chapter 14.

Potable Water

In urban areas, residential water is generally drawn from a *municipal water supply.* The municipality collects the water from a river, from a rain-fed reservoir, or from wells, then filters and purifies it, and distributes it to customers through *water mains* buried in the street. To tap this source for a new house, the contractor digs a trench to the property line and a pipe is laid in it. The pipe is connected to

FIGURE 7.14
This transformer steps down high voltage from underground lines for use in the house. In this case, both high-voltage utility lines and 110/220-volt residential lines are underground. (*Photo by Rob Thallon*)

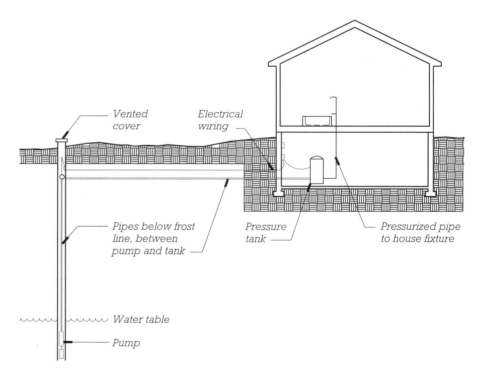

FIGURE 7.15
An electric pump at the base of a deep well or the top of a shallow well operates when the water pressure in the pressure tank drops to a specified level. Water pumped into the tank increases the tank pressure until an upper pressure limit is reached, which shuts off the pump. A pressurized cushion of air in the tank maintains water pressure in the pipes and prevents the pump from being activated every time a faucet is opened in the house.

the water main at one end and the house at the other, with a water meter somewhere along its length to measure water consumption. In cold climates, the water line must be buried below the frost line to keep it from freezing in winter, and in very cold climates the water meter is located inside the house with an electronic readout attached to the outside of the house.

In rural areas, residential water is often supplied by a well (Figure 7.15). The residential well is a six-inch hole dug by a hydraulic drilling rig to a depth at which an adequate supply of clean water is encountered. This may be as little as 15 feet in a sandy soil near a body of water or as much as several hundred feet on rocky sites. The well is *cased* with steel above the level of bedrock to prevent cave-ins and contamination. The well should be drilled at a location on the property where it can be easily serviced, yet it must also be placed far enough away from the septic system to prevent contamination. (Most codes require a minimum distance of 100 feet.) Banks usually require that a well produce 5 gallons per minute for financing to be approved. If the water volume of a newly drilled well is inadequate, a storage tank of 1500 gallons or more may solve the problem.

A *submersible electric pump* is lowered to the bottom of the well to pump water to the house. A *pressure tank* filled partly with air and partly with water is located in the line between well and house. This tank provides a pressurized cushion of air so that the pump is not activated each time a water tap in the house is opened. The air in the tank pressurizes the water in the pipes until the pressure in the tank reaches a preset lower limit. At that time, the electric pump is engaged automatically, water is forced into the tank, and the air cushion is repressurized by the rising volume of water.

Sewage Systems

The simplest way to dispose of sewage is to connect to a *municipal sewer* with a *building sewer.* The building sewer is typically a 3- or 4-inch plastic pipe, sloping away from the house at a minimum of ¼ inch per foot. It must be installed by a plumber or another licensed professional who specializes in that work. Usually, a city or town public works employee will connect the building sewer to the municipal sewer line.

In rural areas without a sewer, a septic system is installed on each site to digest the sewage and disperse the clear effluent into the subsoil. A septic system depends on good soil for proper operation, so the property owner contracts to have a *test pit* dug, and this pit is inspected by a government agency before a *septic permit* can be issued. From the pit, the inspector determines the soil type, the level of the water table, and the rate of water absorption based on a *percolation test.* Because a septic permit must be obtained before a building permit is issued, *septic approval* is usually a condition of the sale of rural residential property. A septic system has two main components—a *septic tank* and a *drain field* or leaching field (Figure 7.16). The septic tank is a large subterranean concrete or fiberglass con-

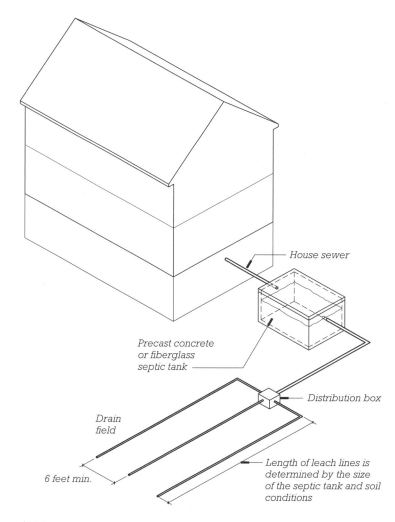

Labels:
House sewer
Precast concrete or fiberglass septic tank
Distribution box
Drain field
6 feet min.
Length of leach lines is determined by the size of the septic tank and soil conditions

FIGURE 7.16
A typical domestic septic system. Raw sewage flows by gravity from the house into the septic tank. After decomposition in the septic tank, a small amount of solid material falls as sludge to the bottom of the tank, and liquid effluent flows to the drain field where it leaches into the soil.

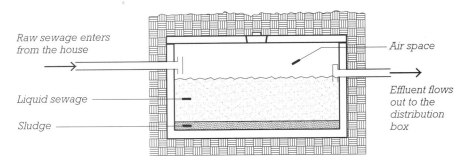

FIGURE 7.17
The septic tank is a large underground concrete or fiberglass container. Raw sewage flows in from one side, and partially digested liquid effluent exits from the other. A lid on the top provides access for occasional cleaning.

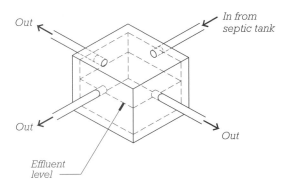

FIGURE 7.18
The distribution box receives liquid effluent from the septic tank. A number of holes at the same level below the supply line provide equal distribution of effluent to each of the leach lines.

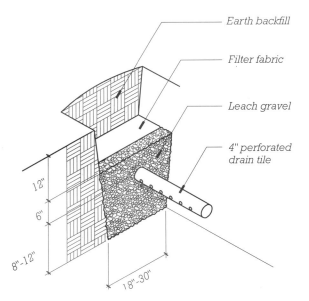

FIGURE 7.19
The perforated leach lines slope slightly and are laid in a bed of gravel to promote percolation of effluent into the soil. Some of the moisture moves downward to join the groundwater beneath the site, and some moves upward by capillary action to evaporate at the surface of the ground.

tainer that accepts the raw sewage from the house (Figure 7.17). Within the tank, anaerobic decomposition takes place—turning most of the sewage into liquid waste. The size of the tank depends on the number of bedrooms and plumbing fixtures in the house, but most houses have a 1000-gallon tank.

The building sewer runs from the house to the inlet of the septic tank. At the opposite end of the tank, and slightly lower than the inlet, is an outlet to which is connected a pipe leading to the drain field. The liquid effluent from the septic tank is carried through this pipe to a *distribution box* buried at the head of the drain field (Figure 7.18). This box is connected to a series of horizontal *leach lines* (Figure 7.19). The purpose of the distribution box is to equalize the flow of waste to each of the leach lines. The leach lines are perforated 4-inch (100-mm) pipes that are laid in a bed of gravel so that the *effluent* will percolate into the ground from them. The moisture from the effluent eventually drains down to the water table or moves upward through the soil by capillary action to evaporate into the atmosphere.

The locations on a site for the septic tank and drain field are restricted by the governmental authority granting the septic permit. Codes generally specify that the septic tank must be sufficiently deep in the ground so that waste from the building sewer can drain into it by gravity. It must also be a minimum prescribed distance from the house, usually 5 feet, and may not be situated beneath any surface where vehicles would normally travel, such as a road or driveway. The drain field must be in an area where the soil and slope have been approved as described in the septic permit, and no part of the drain field may be located beneath roads or driveways, where the soil would be compacted by vehicles. The system must also be a minimum distance from any stream or well on the property, and leach lines must lie lower than a specified distance above

FIGURE 7.20
A two-chambered dosing tank allows uphill pumping of effluent. The first chamber in the tank receives and digests raw sewage. Only liquid waste enters the second chamber, where a pump forces it uphill to the distribution box. An alarm system activated by a float warns the owner if the pump is not operating.

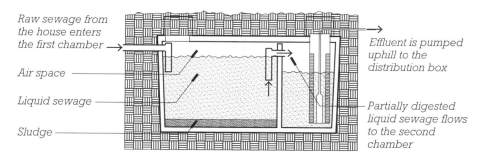

Raw sewage from the house enters the first chamber

Air space

Liquid sewage

Sludge

Effluent is pumped uphill to the distribution box

Partially digested liquid sewage flows to the second chamber

FIGURE 7.21
A new concrete dosing tank in the ground. Finish grading will cover the top of the tank with soil. (*Courtesy of Willamette Graystone, Eugene, Oregon*)

the water table. Practical considerations such as convenient vehicular access to the septic tank for periodic pumping and cleaning also play a role in locating the system on the site.

When there are problems with a site that prohibit the construction of a standard septic system, there may be remedies. A *sand filter system* provides percolation and evaporation of the effluent with the construction of a large engineered bed of sand and soil. Also, a standard drain field or sand filter system may actually be located above the elevation of the septic tank, provided a two-chambered *dosing tank* and a pump are used (Figures 7.20 and 7.21).

The septic system is generally designed and installed by a specialized subcontractor who is licensed for this work. Because it involves so much excavation, its installation is best scheduled during the dry part of the year.

Gas Lines

While almost all houses are connected to electrical power lines, not all houses are connected to gas lines. Of those that are, houses in incorpo-

rated areas generally use *natural gas*, which is delivered to the site in gaseous form through pipes, and those in rural areas use *propane*, which is delivered in liquid form by truck and pumped into an on-site storage tank. The location of the storage tank must conform with safety regulations and be conveniently positioned for access by the delivery truck.

Both natural gas and propane enter the house through a pipeline. When buried, this line may share a trench with other utilities, so long as the specified building code clearances are observed.

Telephone and Television Cable

In modern residential construction, telephone and communication cables are usually buried underground, although aboveground service is common in some areas. Each utility has "connector boxes" positioned along cables that run parallel to the public roadways. Individual residences tap into the utility by running a line from the house to the connector box. These lines may share a trench with other utilities, so long as specified building code clearances are observed.

SURFACE WATER DRAINAGE

Erosion Control

The extreme disruption of a site by the removal of vegetation and subsequent excavation can have a severe impact on downstream neighbors. An *erosion control plan* is therefore required of contractors by many municipalities before construction can begin. Regulatory agencies will consider several factors before approving an erosion control plan, including the season in which the proposed construction will take place and the contractor's method for removing storm water from the site during construction.

Storm Drainage System

The precipitation that falls on a site before the construction of a house filters into the ground until the ground is saturated, then runs off over the surface. When a house is built, however, the ground under it is no longer available to absorb *storm water,* even though the same amount of water still falls on the site. In addition, roofs, driveways, and other impermeable surfaces not only prevent absorption, but also redirect storm water to other locations on the property. This storm water must be collected, channeled, and disposed of in such a way that it does not cause long-term erosion of the land, especially in wet climates.

Water falling onto the roof of a house may be collected in gutters at the eave and channeled through downspouts to ground level. (The installation of gutters and downspouts is discussed in Chapter 10. The runoff flows into a *storm drain* (Figure 7.22). Modern residential storm drains are commonly made of 3- or 4-inch plastic pipe. This pipe is buried underground and sloped, ⅛-inch per foot minimum, to drain into a dry well or a storm sewer (discussed later). Typically, storm drain pipes are located at the perimeter of the house, where they collect runoff from the roof drainage system and can direct it by gravity to a catch basin or storm sewer.

Perimeter Drains

To prevent basement flooding or dampness in crawl spaces, a *perimeter drain* must be provided around the foundation. The perimeter drain is a 3- or 4-inch perforated plastic pipe laid in a porous bed of gravel at the base of the building's footing. The perforations in the pipe are positioned at the bottom so any water rising above this level will enter the pipe and flow by gravity away from the building (Figures 7.23 and 7.24). A *filter fabric* should be placed above the gravel to prevent small particles that might clog the pipes from entering the system. Perimeter drains are discussed further in Chapter 8.

Dry Well Versus Storm Sewer

Storm drains and perimeter drains need to dispose of the water they collect without causing erosion. In urban and suburban locations, the most common practice is to route the drains to a municipal *storm sewer* system, which channels the runoff to a nearby body of water.

In rural settings, or on very large tracts of land within municipalities, storm and perimeter drains are often

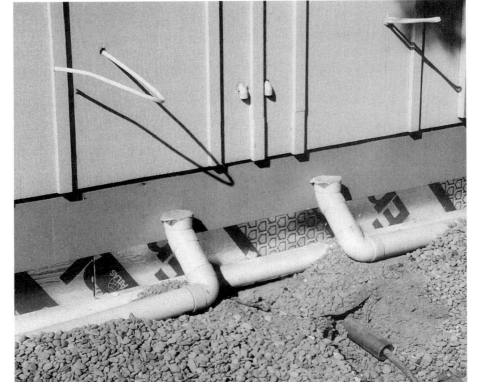

FIGURE 7.22
A residential storm drain is designed to carry runoff from downspouts to a storm sewer or dry well. The drain pipe should slope continuously at ⅛ inch per foot or more. The two storm drains shown here are at the party wall of a duplex and drain in opposite directions. The continuous pipe against the foundation is a perimeter drain with perforations. (*Photo by Rob Thallon*)

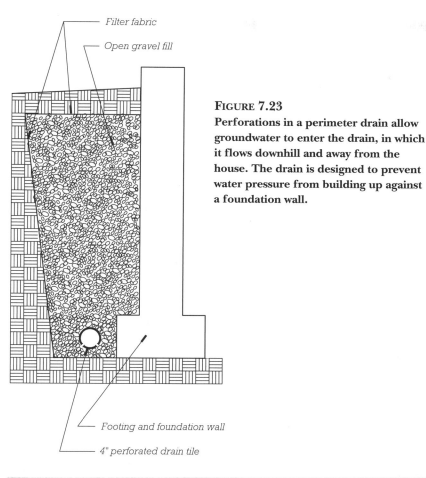

Filter fabric

Open gravel fill

FIGURE 7.23
Perforations in a perimeter drain allow groundwater to enter the drain, in which it flows downhill and away from the house. The drain is designed to prevent water pressure from building up against a foundation wall.

Footing and foundation wall

4" perforated drain tile

required to discharge into a *dry well,* which is essentially a large hole in the ground filled with rocks to prevent the sides from caving in. Runoff discharged into the dry well percolates into the ground. The design of dry wells is regulated by code, and their size and details are often specified by an engineer. In some cases, dry wells are not required, and runoff can be discharged to a *surface swale,* which is a holding area in a low part of the site from which runoff evaporates and seeps into the soil.

GRADING

The last phase of rough sitework involves *backfilling* and *rough grading.* Backfilling is the replacement around the foundation and into trenches of the soil and rock removed during excavation. Rough grading is the smoothing and leveling of the site to a level just below that of the final finish grade.

Backfilling

Once the foundation, the utilities, and the drainage pipes are in place, backfilling can begin. There is an advantage to backfilling as soon as possible after the other rough sitework has been completed, because it allows workers and equipment to have unimpeded access to the entire site for the duration of construction. However, when foundation walls are tall, backfilling against them should be delayed until the first floor framing is in place, because the framing is

FIGURE 7.24
Perimeter drainage is the key to a dry basement or crawlspace. Dampproofing of the foundation wall and the placement of gravel or a drainage mat against the wall above the drain will ensure that groundwater finds its way to the drain. (*Photo by Edward Allen*)

FIGURE 7.25
The floor structure at the top of this foundation wall braces it against the horizontal pressure from the backfill. Backfilling before the floor framing is in place can cause cracking in the foundation wall unless the wall is braced by other means. (*Photo by Edward Allen*)

FIGURE 7.26
Round-edged pea gravel will compact on its own and is often used under slabs. The round black object in the photo is a camera lens cap, about 2 inches in diameter. Gravel with larger stones up to 4 or 5 inches in diameter are often used for backfill where good drainage is required. (*Photo by Greg Thomson*)

needed to brace the top of the wall (Figure 7.25).

It is important to use the proper materials when backfilling. In most cases, an excavation may be backfilled with the same soil that was removed earlier. Backfilling with soil instead of gravel has two primary advantages: It is less expensive than backfilling with gravel, and it alleviates the need for disposing of the excavated soil. Soil that is used as backfill should always be free of organic material, which can attract termites and would eventually decompose and cause settlement. The primary disadvantage of backfilling with soil is that the soil must be compacted with a heavy vibrator or tamper to avoid settling, which can lead to structural, drainage, or visual problems.

Using gravel as backfill can solve the problem of compaction because some kinds of gravel are *self-compacting*, meaning that they have rounded edges and will naturally seek their lowest possible level, like marbles poured into a vessel (Figure 7.26). Gravel backfill also promotes

drainage around a foundation, which contributes to keeping basements and crawlspaces dry. The disadvantages of using gravel, as compared to soil, are that it is expensive to purchase and that it must be brought in from off site, then moved to the location on the property where it is to be used.

In some situations, especially in backfilling utility trenches under either roadways or floor slabs, settling can virtually be eliminated by backfilling with *controlled low-strength material (CLSM)*, which is made from portland cement and/or fly ash (a byproduct of coal-burning power plants), sand, and water. CLSM, sometimes called "flowable fill," is brought in concrete mixer trucks and poured into the excavation, where it compacts and levels itself, then hardens into soil-like material. The strength of CLSM is matched to the situation: For a utility trench, for example, CLSM is proportioned so that it is weak enough to be excavated easily by ordinary digging equipment when the pipe needs servicing, yet as strong as good-quality compacted backfill. CLSM has many

other uses in and around foundations. It is often used to pour *mud slabs*, which are weak concrete slabs used to create a level, dry base in an irregular, wet excavation. The mud slab serves as a working surface for the reinforcing and pouring of a foundation mat or basement floor slab. CLSM is also used to replace pockets of unstable soil that may be encountered beneath a substructure, and it may be used to create a stable volume of backfill around a basement wall.

Backfilling utility trenches is a fairly straightforward process. Once the pipe(s) and/or wire(s) are covered with a protective layer of backfill without rocks (preferably sand), the ditch merely needs to be filled to the top with more backfill. If several utilities are located in the same trench, the building codes require them to be separated at specified minimum distances (Figure 7.28). If soil is used as the backfill material, it will need to be compacted in lifts—especially if it passes under a driveway, terrace, or any other paved surface. Otherwise, it will settle, causing the paved surface to crack or distort.

FIGURE 7.27
A stone slinger is essentially a high-speed conveyor belt mounted to the rear of a gravel truck and can be aimed to place gravel or soil in locations that are otherwise difficult to access. In this case, gravel is being spread by the slinger while workers hand smooth it and check its grade. (*Photo by Rob Thallon*)

FIGURE 7.28
A utility trench may contain several different utilities, so long as they are separated at code-specified distances. The principle is to be able to dig up one line for service at a later date without disturbing the others. (*Photo by Rob Thallon*)

It is important to protect foundation walls or retaining walls from damage during the backfilling process. Foundations are likely to have waterproofing and/or rigid insulation applied to the exterior (Figure 7.29). Special care must be taken not to damage these or the foundation itself with backfilling equipment. *Protection board* is a layer of material that physically protects the foundation from damage by backfill.

Rough Grading

Rough grading is the smoothing of the surface of the site in preparation for the finish sitework that will take place later. The surface of the ground should be sloped away from the house to promote drainage. The level of rough grading is usually about 6 inches below the ultimate *finish grade* to allow for the addition of topsoil and various paved surfaces. Placing base gravel for the driveway is usually included as a part of rough grading, because it will provide a solid surface for delivery vehicles, which will help to compact it.

FIGURE 7.29
Gravel will be carefully placed (by hand) against the insulation installed on this foundation wall. The waterproofing layer beneath the insulation is protected by the insulation itself. The gravel protects the insulation from damage by soil backfill. *(Courtesy of National Concrete Masonry Association)*

C.S.I./C.S.C. MasterFormat Section Numbers for Rough Sitework	
01500	**TEMPORARY FACILITIES AND CONTROLS**
01510	**Temporary Utilities**
01520	**Construction Facilities**
01560	**Temporary Barriers and Enclosures**
02200	**SITE PREPARATION**
02210	**Subsurface Investigation**
02230	**Site Clearing**
02300	**EARTHWORK**
02310	**Grading**
02315	**Excavation and Fill**
02370	**Erosion and Sedimentation Control**
02500	**UTILITY SERVICES**
02510	**Water Distribution**
02520	**Well**
02530	**Sanitary Sewerage**
02540	**Septic Tank System**
02600	**DRAINAGE AND CONTAINMENT**
02620	**Subdrainage**
02630	**Storm Drainage**

SELECTED REFERENCES

1. Colley, B.C. *Practical Manual of Site Development*. New York: McGraw–Hill, 1986.

Written from a civil engineering perspective, the process of site development is explained in lay terms. Chapters on site analysis, earthwork, streets, sanitary sewers, storm drainage, and water supply are illustrated primarily with large-scale examples.

2. Landphair, Harlow C., and Klatt, Fred. *Landscape Architecture Construction.* Upper Saddle River, NJ: Prentice Hall, 1999.

Contains a detailed discussion of basic surveying, grading, storm water management, and erosion control as well as finish sitework topics discussed in Chapter 19 of this book.

KEY TERMS AND CONCEPTS

rough sitework
setback
site utilities
potable water
property line
registered land surveyor
staking out
batter board
builder's level
site excavation
benchmark
laser level
surveyor's rod
story pole
compacted fill
lift
temporary utilities
high-voltage line

transformer
main service cable
municipal water supply
water main
cased well
submersible pump
pressure tank
municipal sewer
building sewer
septic permit
test pit
percolation test
septic approval
septic tank
drain field
distribution box
leach line
effluent
sand filter system

dosing tank
natural gas
propane
erosion control plan
storm water
storm drain
perimeter drain
filter fabric
storm sewer
dry well
surface swale
backfill
self-compacting fill
controlled low-strength material (CLSM)
mud slab
protection board
rough grading
finish grade

REVIEW QUESTIONS

1. What actions does the contractor typically perform at a site before excavation proceeds?

2. What are the principal differences between the development of an urban site and a rural site?

3. Describe the utilities that are typically provided on a temporary basis at a residential construction site. Explain the need for these temporary utilities.

4. List and describe the extra components required for a septic system when it must be located at an elevation higher than the house.

5. Explain the difference between a storm drain and a perimeter drain.

EXERCISES

1. Find an undeveloped lot in your neighborhood and locate a hypothetical house on it. Draw a site plan that contains all information essential to a contractor for the development of the site. Locate the building(s), the utilities, the driveway and walk, and any other site features that will be constructed. Make an estimate about existing site elevations and use this information to provide floor levels, footing elevations, and other important information about vertical relationships.

2. Describe the development of the lot discussed in the preceding exercise to the point of backfilling against a newly poured foundation wall.

3. Document the process of site development for a local residential construction project. Photos taken from the same place(s) on subsequent days can illustrate the process clearly.

FOUNDATIONS

The function of a foundation is to transfer the structural loads from a building safely into the ground. Every building needs a foundation of some kind. A backyard tool shed will not be damaged by slight movements of its foundation and may need only wooden skids to spread its load across an area of the surface of the ground sufficient to support its weight. A house needs greater stability than a tool shed, so its foundation reaches through the unstable surface to underlying soil that is free of organic matter and unreachable by winter frost. The basic requirement of a residential foundation is to provide a permanent base for the building so that it does not settle in a way that might damage the structure or impair its function.

Foundations for light framing were originally made of stone or brick masonry. Today, in most cases, they are made of site-cast concrete or concrete masonry. Early concrete foundations were designed principally to support the building, and they were generally a great improvement over their predecessors. However, little attention was paid in the formative years of foundation design to drainage, waterproofing, or insulation—resulting in several generations' worth of damp crawlspaces, cold, wet basements, and cold floors. Now it is understood that a foundation wall of any material needs to be carefully dampproofed and drained to avoid flooding with groundwater and to prevent the buildup of water pressure in the surrounding soil that could cave it in (Figure 8.6). Likewise, concrete and concrete block foundations are highly conductive of heat and, when they enclose a habitable basement space, must be insulated to meet code requirements concerning energy conservation (Figures 8.25 and 8.32).

- Horizontal pressures of earth and water against basement walls
- During earthquakes, horizontal and vertical forces caused by the motion of the ground relative to the building

FOUNDATION SETTLEMENT

All foundations settle to some extent as the soil around and beneath them adjusts itself to these loads. Foundations on bedrock settle a negligible amount. Foundations on certain types of clay may settle to an alarming degree. Foundation settlement in most buildings is measured in millimeters or fractions of inches. If settlement occurs at roughly the same rate from one side of the building to the other (*uniform settlement*), no harm is likely to be done to the building. However, if large amounts of *differential settlement* occur, in which the various columns and loadbearing walls of the building settle by substantially different amounts, the frame of the building may become distorted, floors may slope, walls and glass may crack, and doors and windows may refuse to work properly (Figure 8.1).

FOUNDATION LOADS

A foundation supports a number of different kinds of loads:

- The *dead load* of the building, which is the sum of the weights of the frame; the floors, roofs, and walls; the electrical and mechanical equipment; and the foundation itself

- The *live load*, which is the sum of the weights of the people in the building; the furnishings and equipment they use; and snow, ice, and water on the roof

- *Wind loads*, which can apply lateral, downward, and uplift forces to a foundation

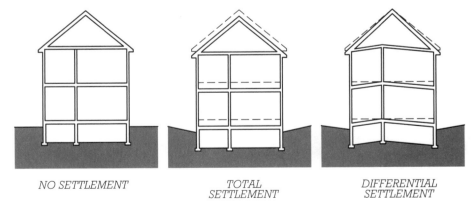

NO SETTLEMENT   TOTAL SETTLEMENT   DIFFERENTIAL SETTLEMENT

FIGURE 8.1
Uniform settlement is usually of little consequence in a building, but differential settlement can cause severe structural damage.

Accordingly, a primary objective in foundation design is to minimize differential settlement by loading the soil in such a way that equal settlement occurs under the various parts of the building. This is not difficult when all parts of the building rest on the same kind of soil. It can become a problem, however, when a building occupies a piece of ground that is underlain by two or more areas of different types of soil with very different loadbearing capacities such as original soil adjacent to a filled area. Most foundation failures are attributable to excessive differential settlement. Gross failure of a foundation, in which the soil fails completely to support the building, is extremely rare.

SOILS

Types of Soils

Soil is a general term referring to earth material that is particulate. The following distinctions may be helpful in acquiring an initial understanding of how soils are classified for engineering purposes:

• *Rock* is a continuous mass of solid mineral material, such as granite or limestone, that can only be removed by drilling and blasting. Rock is not completely monolithic, but is crossed by a system of joints that divide it into irregular blocks. Despite these joints, rock is generally the strongest and most stable material on which a building can be founded.

• *Gravel* consists of particles that can be lifted easily with thumb and forefinger. In the Unified Soil Classification System (Figure 8.2), gravels are classified visually as having more than half their particles larger than 0.25 inch (6.5 mm) in diameter.

• *Sand* is made up of individual particles that can be seen but are too small to be picked up individually. Sand particles range in size from about

			Group Symbols	Typical Names
Coarse-Grained Soils	Gravels	Clean Gravels	GW	Well-graded gravels, gravel–sand mixtures, little or no fines
			GP	Poorly graded gravels, gravel–sand mixtures, little or no fines
		Gravels with Fines	GM	Silty gravels, poorly graded gravel–sand–silt mixtures
			GC	Clayey gravels, poorly graded gravel–sand–clay mixtures
	Sands	Clean Sands	SW	Well-graded sands, gravelly sands, little or no fines
			SP	Poorly graded sands, gravelly sands, little or no fines
		Sands with Fines	SM	Silty sands, poorly graded sand–silt mixtures
			SC	Clayey sands, poorly graded sand–clay mixtures
Fine-Grained Soils	Silt and Clays	(Liquid limit greater than 50)	ML	Inorganic silts and very fine sands, rock flour, silty or clayey fine sands with plasticity
			CL	Inorganic clays of low to medium plasticity, gravelly clays, sandy clays, silty clays, lean clays
			OL	Organic silts and organic silty clays of low plasticity
		(Liquid limit less than 50)	MH	Inorganic silts, micaceous or diatomaceous fine sandy or silty soils, elastic silts
			CH	Inorganic clays of high plasticity, fat clays
			OH	Organic clays of medium to high plasticity
Highly Organic Soils			Pt	Peat and other highly organic soils

FIGURE 8.2

A soil classification chart based on the Unified Soil Classification System. The group symbols are a universal set of abbreviations for soil types, as used in code tables such as Figure 8.4.

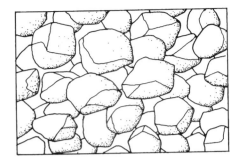

FIGURE 8.3
Silt particles (top) are approximately equidimensional granules, while clay particles (bottom) are platelike and generally much smaller than silt. Notice that the clay particles here are much smaller than the silt particles; a circular area of clay particles has been magnified to make their structure easier to see.

0.25 to 0.002 inch (6.5 to 0.06 mm). Sand and gravel are considered to be coarse-grained soils.

• *Silt* particles are approximately equidimensional and range in size from 0.002 to 0.00008 inch (0.06 to 0.002 mm). Because of their low surface-area-to-volume ratio, which approximates that of sand and gravel, the behavior of silt is controlled by the same mass forces that control the behavior of the coarse-grained soils.

• *Clay* particles are plate shaped rather than equidimensional (Figure 8.3) and smaller than silt particles, less than 0.00008 inch (0.002 mm). Clay particles, because of their smaller size and flatter shape, have a surface-area-to-volume ratio hundreds or thousands of times greater than

that of silt. The volume of the pores and the amount of water in the pores greatly influence the properties of a clay soil, which tends to be expansive (e.g., it expands and contracts with a change in moisture content).

• *Organic soils* such as peat and topsoil are not suitable for the support of building foundations. Because of their high content of organic matter, they are spongy and compress easily, and their properties can change over time due to changing water content or biological activity in the soil.

Properties of Soils

A building site is usually underlain by a number of superimposed layers (strata) of different soils. These strata were deposited one after another in very ancient times by volcanic action and by the action of water, wind, and ice. Most of the soils in these strata are mixtures of several different sizes and types of particles and bear such names as silty gravels, gravelly sands, clayey sands, silty clays, and so on (Figure 8.2). The distribution of particle sizes and types in a soil is important to know when designing a foundation because it is helpful in predicting the loadbearing capacity of the soil, its stability, and its drainage characteristics. Figure 8.4 gives some

typical ranges of loadbearing capacities for various types of soil. Soils for most regions in the United States have been classified and mapped by the Natural Resource Conservation Service and those in Canada by the Canadian Soil Information Service (CANSIS). Information on the soil type for a given building site can usually be found by contacting the local offices of these agencies.

The *stability* of a soil is its ability to retain its engineering properties under the varying conditions that may occur during the lifetime of the building. Rock, gravels, sands, and many silts tend to be stable soils. Many clays are dimensionally unstable under changing subsurface moisture conditions. Clay may swell considerably as it absorbs water and shrink as it dries. When wet clay is put under pressure, water can be squeezed out of it, with a corresponding reduction in volume. Taken together, these properties make clays that are subject to changes in water content the least stable and least predictable soils for supporting buildings.

The *drainage characteristics* of a soil are important in predicting how water will flow on and under building sites and around building substructures. Water passes readily through clean gravels and sands, slowly through very fine silts and sands, and almost

Presumptive Load-Bearing Values of Foundation Materials

Class of Material	Loadbearing Pressure (pounds per square foot)
Crystalline bedrock	12,000
Sedimentary and foliated rock	4,000
Sandy gravel and/or gravel (GW and GP)	3,000
Sand, silty sand, clayey sand, silty gravel and clayey gravel (SW, SP, SM, SC, GM and GC)	2,000
Clay, sandy clay, silty clay, clayey silt, silt and sandy silt (CI, ML, MH and CH)	1,500
For SI: 1 pound per square foot = 0.0479 kNIm2.	

FIGURE 8.4
Presumptive loadbearing values of various soil types as allowed by the 2000 International Residential Code.

not at all through many clays. The drainage characteristics of a soil can affect the foundation construction process by dictating which seasons are suitable for excavation and construction. Soil drainage can affect foundation design by requiring more or less gravel fill to promote proper foundation drainage.

DESIGNING FOUNDATIONS

Foundation design involves many decisions that depend on numerous variables at a range of scales. The most fundamental decision of whether to employ a slab, a crawlspace, or a basement foundation (Figure 8.5) depends on such things as soil type,

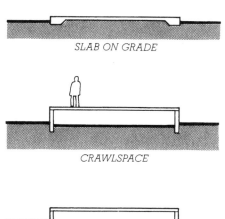

SLAB ON GRADE

CRAWLSPACE

BASEMENT

FIGURE 8.5

Three types of substructures using simple wall footings. The slab on grade is most economical under many circumstances, especially where the water table lies near the surface of the ground. A crawlspace is often used under a floor structure of wood or steel and gives much better access to underfloor piping and wiring than a slab on grade. A basement is often specified on a steep slope or in regions where the frost line requires deep footings.

site contours, spatial needs, and local construction costs. The choice of materials depends partially on aesthetic considerations, labor costs, and insulation methods. Designers must consider all of these factors together before determining the final form and shape of the foundation. The typical basement foundation shown in Figure 8.6 illustrates the general areas that require consideration.

Residential foundation design is usually relatively simple from an engineering standpoint because foundation loads are low as compared to those of large, multistory buildings. The uncertainties that do exist in foundation design at the scale of a house can usually be reduced with reasonable economy by adopting a large factor of safety in calculating the bearing capacity of the soil. Unless the designer has reason to suspect poor soil conditions, residential foundations are usually designed using rule-of-thumb *allowable soil stresses* and standardized proportions for the footings (the base of the foundation) (Figure 8.4). The designer then examines the actual soil when the excavations have been made. If it is not of the quality that was expected, the footings can be hastily redesigned using a revised estimate of the *bearing capacity* of the soil before construction continues. If unexpected groundwater is encountered, better drainage provisions may have to be provided around the foundation or the depth of the foundation may have to be decreased.

The design of the foundation is influenced as much by the building above as it is by the soil type below. Tables found in the code specify footing size by whether the foundation will support a one-, two-, or three-story house (Figure 8.7). The overall size of the foundation is coordinated with the dimensions of the framing (often laid out on a 2-foot module to use standard lumber sizes efficiently). Knowing that foundation construc-

tion is expensive relative to wood framing, the designer will simplify the foundation as much as possible. For example, small projections in the floor plan may be framed with wood over a simple, straight foundation wall.

Both crawlspace and basement walls may be designed using tables in the code. The tables take into account the wall thickness, wall height, wall reinforcement, backfill height, and soil type (Figure 8.8). A licensed architect or engineer must be employed to design walls that exceed the minima prescribed in the tables. The typical residential concrete or concrete masonry foundation wall is 8 inches (200 mm) thick and is reinforced with #4 reinforcing steel bar (rebar) at 48 inches (1.2 m) on center in both directions. Walls on extremely stable soil may be thinner (6 inches) and require no rebar. Taller-than-normal walls on normal soil generally are thicker and have more rebar or rebar of larger diameter.

Knowing that most residential foundations are made of either site-cast concrete or concrete block, the residential designer must often choose between the two. Both are very similar in terms of durability, strength, and appearance, so some designers will leave their options open until after the bidding process. The principal differences are as follows:

• A concrete foundation can be made to any dimension, whereas block is practically limited to the module of a masonry unit (8 inches) in both horizontal and vertical dimensions.

• A concrete foundation uses formwork, whereas a block foundation does not. The lack of formwork can be a significant advantage in terms of time and cost, especially where the configuration of the foundation is complicated.

• A concrete foundation has a smooth finish surface, whereas block has a built-in pattern. A concrete masonry foundation can have an attractive, stonelike texture if split-face block is used.

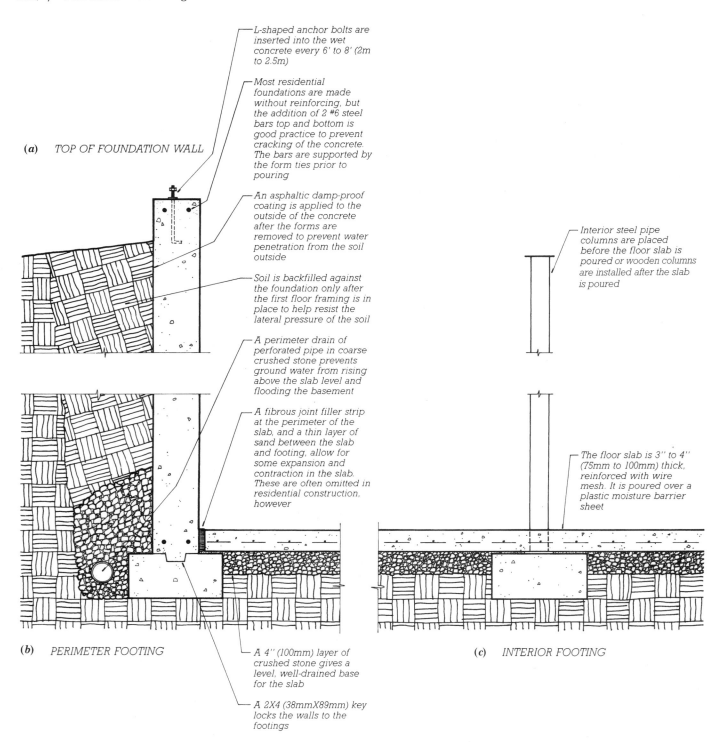

(a) TOP OF FOUNDATION WALL

L-shaped anchor bolts are inserted into the wet concrete every 6' to 8' (2m to 2.5m)

Most residential foundations are made without reinforcing, but the addition of 2 #6 steel bars top and bottom is good practice to prevent cracking of the concrete. The bars are supported by the form ties prior to pouring

An asphaltic damp-proof coating is applied to the outside of the concrete after the forms are removed to prevent water penetration from the soil outside

Soil is backfilled against the foundation only after the first floor framing is in place to help resist the lateral pressure of the soil

A perimeter drain of perforated pipe in coarse crushed stone prevents ground water from rising above the slab level and flooding the basement

A fibrous joint filler strip at the perimeter of the slab, and a thin layer of sand between the slab and footing, allow for some expansion and contraction in the slab. These are often omitted in residential construction, however

Interior steel pipe columns are placed before the floor slab is poured or wooden columns are installed after the slab is poured

The floor slab is 3" to 4" (75mm to 100mm) thick, reinforced with wire mesh. It is poured over a plastic moisture barrier sheet

(b) PERIMETER FOOTING

(c) INTERIOR FOOTING

A 4" (100mm) layer of crushed stone gives a level, well-drained base for the slab

A 2X4 (38mmX89mm) key locks the walls to the footings

FIGURE 8.6
Typical details for a poured concrete basement foundation. A block foundation would be the same except that block would be substituted for concrete at the wall. A crawlspace foundation has shorter walls and does not require a slab.

Minimum Width of Concrete or Masonry Footings (inches)

	Loadbearing Value of Soil (psf)					
	1,500	2,000	2,500	3,000	3,500	4,000
Conventional light-frame construction						
1-story	16	12	10	8	7	6
2-story	19	15	12	10	8	7
3-story	22	17	14	11	10	9

FIGURE 8.7
This table from the International Residential Code relates building weight (assumptions derived from construction type and number of stories) to soil type.

Reinforced Concrete and Masonry Foundation Walls

		Minimum Vertical Reinforcement Size and Spacing for 8-inch Nominal Wall Thickness		
		Soil classes		
Maximum Wall Height (feet)	Maximum Unbalanced Backfill Height (feet)	GW, GP, SW and SP soils	GM, GC, SM, SM-SC and ML soils	SC, MH, ML-CL and inorganic CL soils
6	5	#4 at 48" o.c.	#4 at 48" o.c.	#4 at 48" o.c.
	6	#4 at 48" o.c.	#4 at 40" o.c.	#5 at 48" o.c.
7	4	#4 at 48" o.c.	#4 at 48" o.c.	#4 at 48" o.c.
	5	#4 at 48" o.c.	#4 at 48" o.c.	#4 at 40" o.c.
	6	#4 at 48" o.c.	#5 at 48" o.c.	#5 at 40" o.c.
	7	#4 at 40" o.c.	#5 at 40" o.c.	#6 at 48" o.c.
8	5	#4 at 48" o.c.	#4 at 48" o.c.	#4 at 40" o.c.
	6	#4 at 48" o.c.	#5 at 48" o.c.	#5 at 40" o.c.
	7	#5 at 48" o.c.	#6 at 48" o.c.	#6 at 40" o.c.
	8	#5 at 40" o.c.	#6 at 40" o.c.	#6 at 24" o.c.
9	5	#4 at 48" o.c.	#4 at 48" o.c.	#5 at 48" o.c.
	6	#4 at 48" o.c.	#5 at 48" o.c.	#6 at 48" o.c.
	7	#5 at 48" o.c.	#6 at 48" o.c.	#6 at 32" o.c.
	8	#5 at 40" o.c.	#6 at 32" o.c.	#6 at 24" o.c.
	9	#6 at 40" o.c.	#6 at 24" o.c.	#6 at 16" o.c.

For SI: 1 inch = 25.4 mm, 1 foot = 304.8 mm.

FIGURE 8.8
In this International Residential Code table, basement wall strength is related to the lateral load on the wall as determined by the height of the backfill and the soil type.

FOOTINGS

A *footing* is a wide element at the base of a foundation that is designed to distribute building loads over the surface of the soil. A *column footing* is a square (or circular) block of concrete, with or without steel reinforcing, that accepts the concentrated load placed on it from above by a building column and spreads this load across an area of soil large enough that the allowable bearing stress of the soil is not

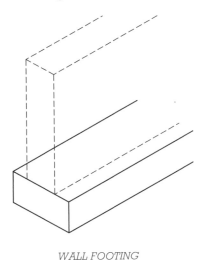

WALL FOOTING

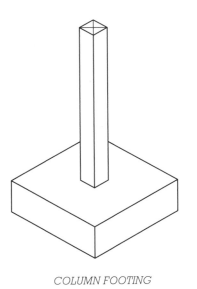

COLUMN FOOTING

FIGURE 8.9
A column footing and wall (strip) footing of concrete. The steel reinforcing bars have been omitted for clarity.

exceeded. A *wall footing* or *strip footing* is a continuous strip of concrete that serves the same function for a load-bearing wall (Figure 8.9).

To minimize settlement, footings are placed on *undisturbed soil*. The last inches of soil in the excavation for a footing are often removed with hand tools to avoid the loosening of the bearing layer of soil that would be caused by digging machinery. Also, to eliminate *heaving*, an upward thrusting of the foundation that results from the expansion of soil when it freezes, footings are constructed with their base below the *frost line*, the depth to which soil might be expected to freeze during a severe winter. The frost line depths for all regions of North America are specified in the code.

When soil is very poor or irregular, adequate bearing for footings may be achieved in a number of ways. Some buildings are built on *engineered fill*, which is earth that has been deposited in thin layers across an entire building site at a controlled moisture content and compacted in accordance with detailed procedures that assure a known degree of long-term stability. *Compacted gravel fill* also may be distributed under an entire building or more locally under each footing. Alternatively, controlled low-strength material (CLSM), a self-leveling mix of fly ash, cement, and sand, also called "flowable fill" or "controlled-density fill," may be used to stabilize questionable soils. This material is pumped into overexcavated trenches where it hardens to a consistency such that it can be excavated by a backhoe but not by a shovel.

Even when all reasonable efforts have been made to assure that the soil will provide an adequate bearing surface, it is prudent in many cases to place rebar in the footing to minimize the effects of potential differential settlement. The codes do not require rebar in most cases, but when it is required, the standard is to have two #4 bars running longitudinally in each footing. If the footing needs to extend farther than twice the width

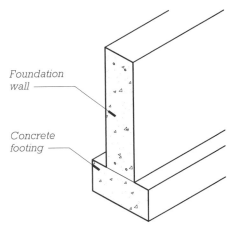

Foundation wall

Concrete footing

FIGURE 8.10
A simple footing and foundation wall. In most cases, the wall is centered on the footing.

of the wall it supports to achieve the necessary bearing area, additional reinforcing in both directions is generally required.

Footings appear in many forms in different foundation systems. In climates with little or no ground frost, a concrete slab on grade with thickened edges is the least expensive foundation and floor system and is applicable to one- and two-story buildings of any type of construction. For floors raised above the ground, either over a crawlspace or a basement, concrete strip footings support concrete or masonry foundation walls (Figure 8.10). When building on slopes, it is often necessary to step the footings to maintain the required depth of footing at all points around the building (Figure 8.11). If soil conditions, steep slopes, or earthquake precautions require it, column footings may be linked together with reinforced concrete tie beams to avoid possible differential slippage between footings.

Constructing the Footings

Footings are laid out at the site by the foundation subcontractor, using the same batter boards used by the excavator (Figure 7.2). The elevation at the top of the footing is determined from the same benchmark used for

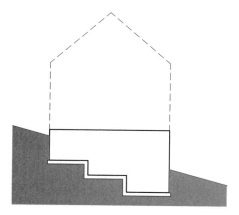

STEPPED FOOTING

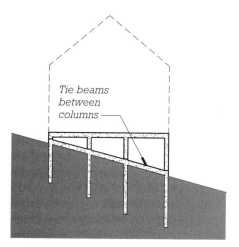

Tie beams between columns

COLUMN FOUNDATION

FIGURE 8.11
Foundations on sloping sites, viewed in a cross section through the building. The broken line indicates the outline of the superstructure. Wall footings are stepped to provide a level bearing surface between footing and soil and to maintain the necessary distance between the bottom of the footing and the surface of the ground. Separate column foundations are often connected with reinforced concrete tie beams to reduce differential movement between the columns.

FIGURE 8.12
Forms for footings on a sloped site showing steps in the footing to accommodate the change in grade. The forms have not yet been staked to the ground. (*Photo by Rob Thallon*)

FIGURE 8.13
Concrete will be poured to the top of the form boards, completely filling the space between them and encasing the rebar to make a strong footing. This example happens to be formed on engineered gravel fill. (*Photo by Rob Thallon*)

excavation. The sides of the footings are formed with boards (usually 2 × 8s) staked to the ground with wood or metal stakes. With the forms in place, the horizontal reinforcing steel, if any, is roughly positioned between them. The two sides of the forms are then tied with scraps of wood nailed across the top, and the rebar is supported with wire tied to these ties (Figure 8.13).

At this stage, the building inspector must verify that the loadbearing soil is adequate, the footings have been properly sized, and the rebar has been positioned correctly. Following this inspection, the concrete can be poured into the forms to create the footings. Depending on the site, the concrete can flow directly from the truck chute, it can be moved with wheelbarrows, or it can be pumped using a concrete pumper. After pouring, the top of the footing is straight-edged. No further finishing operations are required (Figure 8.14). The footing is left to cure for at least a day before the wall forms are erected.

To connect the footing mechanically to the foundation wall that will later be built on top of it, there are two basic strategies. The less expensive is to form a groove, called a *key*, in the top of the footing with strips of wood that are temporarily embedded in the wet concrete. When more

FIGURE 8.14
These footings have been poured and the forms stripped. The projecting rebar dowels will tie the foundation wall to the footing. A key in the footing can also be used to tie a concrete foundation wall and footing. (*Photo by Rob Thallon*)

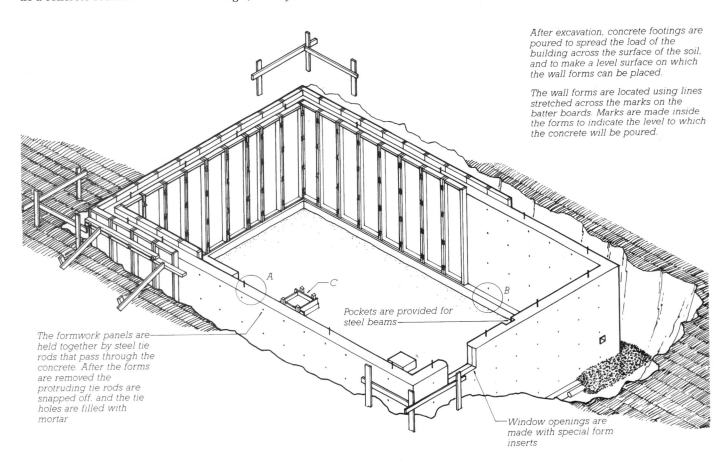

After excavation, concrete footings are poured to spread the load of the building across the surface of the soil, and to make a level surface on which the wall forms can be placed.

The wall forms are located using lines stretched across the marks on the batter boards. Marks are made inside the forms to indicate the level to which the concrete will be poured.

Pockets are provided for steel beams

The formwork panels are held together by steel tie rods that pass through the concrete. After the forms are removed the protruding tie rods are snapped off, and the tie holes are filled with mortar

Window openings are made with special form inserts

FIGURE 8.15
A typical excavation and basement foundation. A shorter crawlspace foundation is constructed in the same fashion. The letters A, B, and C indicate portions of the foundation that are detailed in Figure 8.6.

strength is required in the connection, vertically projecting *dowels* of steel reinforcing bars are installed in the footing just after it has been straightedged. These tie the footing structurally to the foundation wall. If vertical reinforcing bars are used to reinforce the wall, they are overlapped with these projecting bars and tied to them with wire before the formwork for the wall is completed.

CONCRETE FOUNDATION WALLS

Concrete foundation walls are cast in place on top of the poured concrete strip footings by the same subcontractor who constructs the footings. Typically, the concrete foundation follows the perimeter of the building and the

garage, but in a large or complex house, foundation walls may also be located under some interior walls. Concrete foundation walls are very strong and durable, and the wet concrete with which they are made is fluid, so they can be formed and cast into any size or shape.

Casting Concrete Foundation Walls

The first step in building the foundation walls is to assemble the temporary wall forms into which the concrete will be poured (Figures 8.16 and 8.17). These forms are reusable panels of plywood or metal that are placed on top of the completed footing. Wall forms may be custom built of lumber and plywood for each job, but it is more common for standard

prefabricated formwork panels to be employed. The panels for one side of the form are coated with a *form-release compound*, set on the footing, aligned carefully, and braced. At this time, preformed plastic crawlspace vents, if any, are attached to the forms, and other openings in the wall such as windows or access doors are blocked out with site-built or special manufactured forms attached to the formwork panels.

If the foundation walls are to be reinforced, the steel reinforcing is installed next, with the bars wired to one another at the intersections. The vertical bars overlap corresponding dowels projecting from the footing, when they exist. At wall corners, L-shaped horizontal bars are installed to maintain full structural continuity between the two walls.

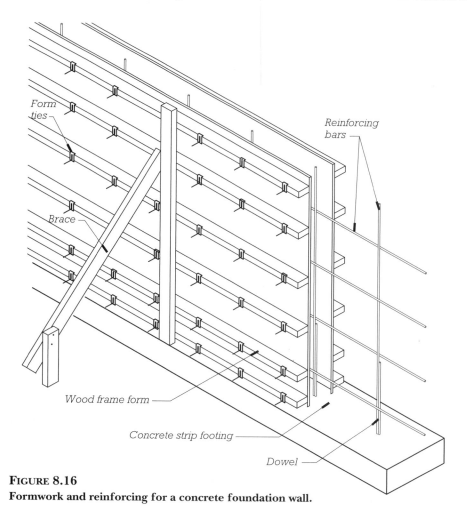

FIGURE 8.16
Formwork and reinforcing for a concrete foundation wall.

FIGURE 8.17
Formwork and reinforcing for a residential foundation wall. Vertical rebar projecting from the footing has been wired to bars that extend nearly to the top of the foundation wall. Horizontal rebar has been wired in place, and forms on one side of the wall have been placed. There is a line on the forms that marks the level to which the concrete will be poured.
(Photo by Rob Thallon)

The amount of rebar and its location are determined by soil conditions, by the height and length of the foundation wall, and by the amount of backfill to be placed against the wall. In ideal conditions, there may be no need for any reinforcement in either footing or foundation wall. In extreme conditions, reinforcement may be called for in both footing and wall. Codes specify reinforcement requirements for simple buildings (Figure 8.8), but for complex buildings or situations not covered in code tables, an engineer is consulted to specify the size and location of reinforcing steel.

When the first side of the forms is in place and the reinforcing has been installed, the formwork for the second side of the wall is erected. These second-side forms need to be exactly parallel to the forms on the first side and need to be tied to them. The *form ties* are a critical part of the formwork because they must hold the formwork together under the pressure of the wet, heavy concrete. For short walls, the ties may be simple metal straps nailed to the base of the forms and wood straps at the top. However, for tall walls, in which the pressure of wet concrete is considerably greater, ties made of specially shaped steel rods are inserted through holes provided in the formwork panels and secured to the outside faces of the form by devices supplied with the form ties. Both ties and fasteners vary in detail from one manufacturer to another (Figure 8.18). The ties pass straight through the concrete wall from one side to another and will remain in the wall after it is poured.

With reinforcing steel, blocked-out openings, and ties in place, the forms are checked to be sure that they are straight, plumb, correctly aligned, and adequately braced. A surveyor's transit is used to establish the exact height to which the concrete will be poured, and this height is marked all around the inside of the forms. After an inspection by the building official, pouring may then proceed (Figure 8.19).

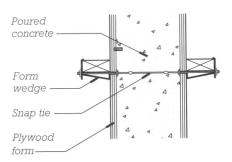

FIGURE 8.18
Cross section of a proprietary concrete form tie.

Poured concrete

Form wedge

Snap tie

Plywood form

FIGURE 8.19
The formwork has been completed and braced. Metal spacers nailed to the top of the formwork maintain proper form width. (*Photo by Rob Thallon*)

FIGURE 8.20
The forms are being stripped from these freshly poured concrete foundation walls. The ties have not yet been twisted off. Note the crawlspace vents cast into the top of the distant wall. (*Photo by Rob Thallon*)

Concrete is then transported to the top of the wall, and workers standing on top of the forms deposit the concrete in the forms, agitating it to eliminate air pockets with a shovel or stick for short walls or a mechanical vibrator for tall walls. When the form has been filled and compacted up to the level that is marked inside the formwork, hand floats are used to smooth and level the top of the wall. *Anchor bolts* and *hold-downs* (when required) are set into the wet concrete, projecting up in anticipation of the framing that will be bolted to them. The top of the form is then covered with a plastic sheet or canvas, and the wall is left to cure.

After a day or two of curing, the bracing is taken down, the connectors are removed from the ends of the form ties, and the formwork is stripped from the wall (Figure 8.20). Tall walls are left bristling both inside and out with projecting ends of form ties. These are twisted off with heavy pliers, and the holes that they leave in the surfaces of the wall are carefully filled with grout. If required, major defects in the wall surface caused by flaws in the form-

work or inadequate vibration of the concrete can be repaired at this time. The wall is now complete, ready for the addition of a *mudsill* and the beginning of framing. The foundation may be waterproofed and insulated (as discussed later in this chapter) at this stage, but backfilling against tall walls is delayed until the floor framing is in place.

Other methods of forming and casting concrete foundation walls are used as well. Metal forms that flare at the base are used for short walls on flat sites to facilitate pouring of footing and wall simultaneously. Insulated concrete forms (ICFs) that are left in

place to add thermal insulation to a concrete wall are rapidly gaining popularity and are discussed in Chapter 21. Finally, precast foundation walls, manufactured in large sections before being transported to the site, are gaining acceptance in some areas.

CONCRETE MASONRY FOUNDATION WALLS

The concrete masonry (concrete block) foundation wall, like a poured concrete wall, rests on the completed footing. Block walls are somewhat simpler to construct than concrete walls, however, because there are no forms to build. The work is usually performed by a masonry subcontractor who may or may not have constructed the footings. For most work, standard 8×8×16-inch block is used with half blocks as needed at corners and openings and bond blocks for horizontal rebar (Figure 8.21). For extra strength, 12-inch-deep blocks are sometimes required. Vents, windows, and other openings are made by omitting blocks and creating a bond beam lintel across the opening (Figure 5.17). If the foundation is designed to an 8-inch (200-mm) block module, very few blocks will have to be cut.

Constructing a Concrete Masonry Foundation Wall

To start the work, the outside corners of the wall are marked on the footing,

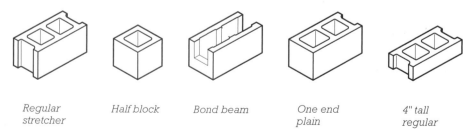

Regular stretcher   Half block   Bond beam   One end plain   4" tall regular

FIGURE 8.21
Residential block foundation walls are usually built with just a few simple block types. Decorative blocks such as split-face block require the same types but, because they are more expensive and more difficult to waterproof, are only used above grade.

FIGURE 8.22
A typical block foundation wall under construction. The block lying on its side next to the worker's hand will have a screen adhered to its inside face to make a crawlspace vent. (*Photo by Rob Thallon*)

FIGURE 8.23
The first course of block has been laid on the footing, and vertical rebar dowels can be seen projecting through the cells of the block. (*Photo by Rob Thallon*)

FIGURE 8.24
Rebar laid in a bond beam. The corner block has been cut with a masonry saw to make it a bond beam block, typically because it is too expensive for masonry supply stores to stock right- and left-hand corner bond beam blocks. (*Photo by Rob Thallon*)

and strings are stretched between them to define the outer face of the future wall. The concrete blocks are then laid in mortar using a running bond (Figure 8.22). Like poured concrete foundation walls, block walls may be built with or without reinforcing. Unreinforced block walls cannot carry such high stresses as reinforced walls and are not suitable for areas with high seismic risk. If the wall is to be reinforced, horizontal rebar is laid in bond beam blocks as the wall is being constructed, and vertical rebar is inserted from the top, after the wall has been completed, into the cells containing the rebar dowels that project up from the footing (Figures 8.23 and 8.24).

After the masonry is complete and the rebar is all in place, the wall is inspected by the building inspector to assure proper reinforcing. It is then grouted to unite the block and rebar into a single structural unit. At a minimum, cells that contain vertical bars are grouted. For a stronger wall, all cells are grouted. Grout is generally delivered in a concrete truck and pumped into the top of the wall with a concrete pump or a smaller grout pump. Anchor bolts and hold-downs are set into the wet grout at the top course(s) of block, which is usually a bond beam. The day after the anchor bolts are grouted in place, the mudsill can be bolted to the wall, and the framing can begin.

CONCRETE SLAB FOUNDATIONS

A concrete slab on grade is a level surface of concrete for the ground floor of a building. The concrete is supported directly by the soil and is thickened at its perimeter to provide a footing under the exterior walls (Figure 8.25b). Concrete slabs are especially prevalent in warm southern climates where footings can be shallow and where their inherent thermal mass can help to keep the house cool. Because the foundation and subfloor can be constructed in one relatively simple operation, slab-on-grade construction is inexpensive. In addition, a beautiful (and inexpensive) finished floor can be made with the addition of color and sealer to the concrete surface.

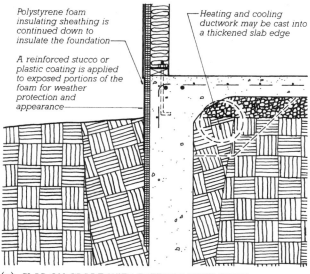

Polystyrene foam insulating sheathing is continued down to insulate the foundation

A reinforced stucco or plastic coating is applied to exposed portions of the foam for weather protection and appearance

Heating and cooling ductwork may be cast into a thickened slab edge

(a) *SLAB ON GRADE WITH EXTERIOR INSULATION*

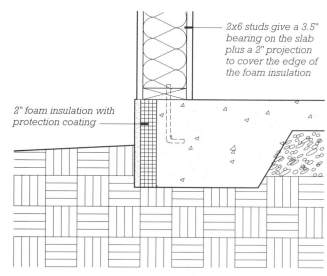

2x6 studs give a 3.5" bearing on the slab plus a 2" projection to cover the edge of the foam insulation

2" foam insulation with protection coating

(b) *SLAB ON GRADE WITH EXTERIOR INSULATION*

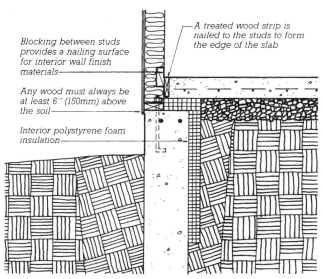

Blocking between studs provides a nailing surface for interior wall finish materials

Any wood must always be at least 6" (150mm) above the soil

Interior polystyrene foam insulation

A treated wood strip is nailed to the studs to form the edge of the slab

(c) *SLAB ON GRADE WITH INTERIOR INSULATION*

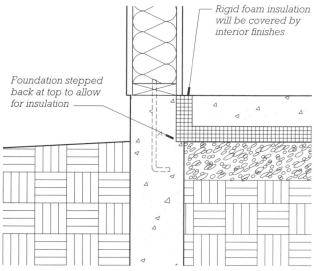

Rigid foam insulation will be covered by interior finishes

Foundation stepped back at top to allow for insulation

(d) *SLAB ON GRADE WITH INTERIOR INSULATION*

FIGURE 8.25
Some typical concrete slab-on-grade details with thermal insulation. The exterior insulation at (a) and (b) may be used with a deep footing (a) or a turned-down footing (b) and may be installed when the slab is poured or at a later date. The interior insulation at (c) and (d) must be used with a deep footing and is installed before the slab is poured.

The thickened edges at the perimeter of the slab are sized as if they were footings for a perimeter foundation wall. The slab can also be thickened at specific interior locations to provide footings for loadbearing columns or walls (Figure 8.26). In cold climates or on sloped sites, where the footings may have to be so deep as to make the thickened edge footing impractical, poured concrete or concrete block stem walls provide the foundation, and the slab is poured independently of these walls (Figure 8.25c, d). The thickness of the interior nonloadbearing slab usually ranges from 3 to 4 inches (75 to 100 mm) for a residential floor and is often increased to around 6 inches (180 mm) for a garage floor.

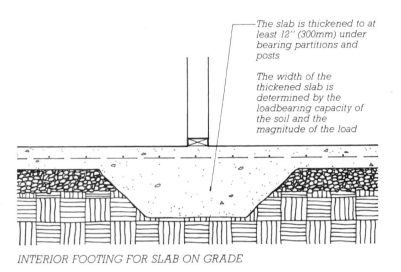

INTERIOR FOOTING FOR SLAB ON GRADE

The slab is thickened to at least 12" (300mm) under bearing partitions and posts

The width of the thickened slab is determined by the loadbearing capacity of the soil and the magnitude of the load

FIGURE 8.26
A typical interior footing to support a loadbearing column or wall.

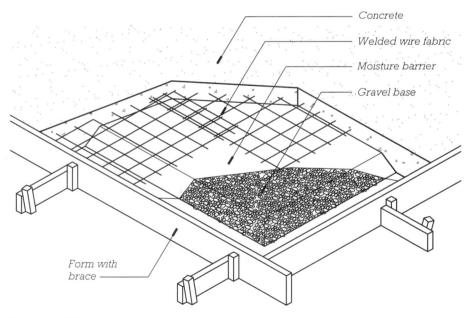

Concrete

Welded wire fabric

Moisture barrier

Gravel base

Form with brace

FIGURE 8.27
The construction of a concrete slab on grade. Notice how the wire reinforcing fabric overlaps where two sheets of fabric join.

Forming and Preparing a Slab on Grade

To prepare for the placement of a slab on grade, topsoil is scraped away to expose the subsoil beneath. A 4-inch-deep (100-mm) layer of crushed stone is compacted over the subsoil as a drainage layer to keep water away from the slab. A moisture barrier (usually a heavy sheet of polyethylene plastic) is laid over the crushed stone. The thickened parts of the slab that act as footings bear directly on the soil, without crushed stone beneath. A simple edge form is constructed of wood or metal around the perimeter of the area to be poured and is coated with a form-release compound to prevent the concrete from sticking. The top edge of the form is leveled carefully, and metal anchor straps that will later tie the mudsill to the foundation are fastened to it.

A reinforcing mesh of *welded wire fabric*, cut to a size just a bit smaller than the dimensions of the slab, is laid over the moisture barrier (Figure 8.27). The fabric most commonly used for lightly loaded slabs, such as those in houses, is 6 × 6–W1.4/W1.4, which has a wire spacing of 6 inches (150 mm) in each direction and a wire diameter of 0.135 inch (3.43 mm). (For slabs in garages, a fabric made of heavier wires or a grid of reinforcing bars is often used instead.) The reinforcing helps protect the slab against cracking that might be caused by concrete shrinkage, temperature stresses, concentrated loads, frost heaves, or settlement of the ground beneath. *Fibrous admixtures* are finding increasing acceptance as a means of controlling the plastic shrinkage cracking that often takes place during curing of a slab. Fibrous reinforcing, however, is not strong enough to replace wire fabric for general crack control in slabs.

Concrete is highly conductive of heat, so slab-on-grade foundations usually must be insulated at their perimeter to meet code requirements for energy conservation (Figure 8.25). In all climates, *perimeter insulation* helps to maintain comfortable indoor temperatures—keeping the slab cool in warm climates and keeping it warm in cold climates. In moderate and cold climates, there may be a layer of rigid insulation installed under the entire slab to isolate the mass of the concrete from the cool soil below (Figure 8.28).

One disadvantage of slabs is the difficulty of running utilities (plumbing, heating, and electrical systems) within them. Once these systems have been cast into or under the concrete, it is virtually impossible to repair or remodel them. Therefore, designers usually find a way to keep heating ducts and wiring out of the slab, but there is no way to eliminate under-floor plumbing (Figure 12.11). Drain lines are set by the plumber in the crushed stone below the slab and, after they have been tested by the plumber and inspected (by the plumbing inspector), are covered with more crushed stone to isolate them from the concrete. All pipes that pass through the slab must also be isolated from

the concrete (usually with fiberglass wrap) to prevent abrasion from thermal expansion and contraction.

Control joints must be provided at intervals in a slab on grade. A control joint is a straight, intentional crack that is formed with a fiberboard strip or tooled into the surface of the slab before the concrete has hardened. Alternatively, a control joint may be created by sawing a shallow groove into the top of the slab after it has hardened. The function of a control joint is to provide a place where the forces that cause cracking can be relieved without disfiguring the slab. The reinforcing mesh is discontinued at each control joint as a further inducement for cracking to occur in this location.

Pouring and Finishing the Slab on Grade

Pouring of the slab commences with the placing of concrete into the form-

work (Figure 8.29a). This may be done directly from the chute of a transit-mix truck or with wheelbarrows, concrete buggies, or a concrete pump and hoses. The method selected will depend on the scale of the job and the accessibility of the slab to the truck delivering the concrete. The concrete is spread by hand with shovels or rakes until the form is full, and the same tools are used to agitate the concrete slightly, especially around the edges, to eliminate air pockets. Next, using hand-held hooks, the concrete masons reach into the wet concrete and raise the welded wire fabric to the midheight of the slab, so that it will be able to resist tensile forces caused by forces acting either upward or downward.

The first operation in finishing the slab is to *strike off* or *straightedge* the concrete by drawing a stiff plank of wood or metal across the top edges of the formwork to achieve a level surface (Figure 8.29b). This is done with an end-to-end sawing motion that avoids tearing the projecting pieces of coarse aggregate from the surface of the wet concrete. A bulge of concrete is maintained in front of the straightedge as it progresses across the slab, so that, when a low point is encountered, concrete from the bulge will flow in to fill it. When straightedging has been completed, the top of the slab is level but rather rough.

The slab is next *floated* to produce a smoother surface (Figure 8.29c, d). The masons wait until the watery sheen has evaporated from the surface, then smooth the concrete with a flat tool called a float. The working surfaces of floats are made of wood or of metal with a slightly rough surface. As the float is drawn across the surface, its friction vibrates the concrete gently and brings cement paste to the surface where it is smoothed. If too much floating is done, however, an excess of paste and free water rises to the surface to form puddles, and it is almost impossible to get a good finish. Experience on the part of the mason is essential to floating, as it is

FIGURE 8.28
This slab on grade with a turned-down footing is ready to pour. Rigid insulation has been added under the slab, and the perimeter will be insulated after the slab is poured. Underfloor plumbing is wrapped with insulation to eliminate contact with the concrete. Slab reinforcement in this case is with rebar rather than the more typical welded wire fabric. (*Photo by Rob Thallon*)

to all slab finishing operations, in order to know just when to begin each operation and just when to stop.

For a completely smooth, dense surface, the slab must also be troweled. This is done either by hand with a smooth rectangular steel trowel (Figure 8.29*e*) or with a rotary power trowel. Troweling is done several hours after floating, when the slab is becoming quite firm. If the concrete mason cannot reach all areas of the slab from around the edges, kneeboards are placed on the surface of the concrete to distribute the mason's weight without making indentations. Any marks left by the kneeboards are removed by the trowel as the mason works backward across the surface from one edge to the other.

When the finishing operations have been completed, the slab should be cured under damp conditions for at least a week; otherwise, its surface may crack or become dusty from premature drying. Damp curing may be accomplished by covering the slab with an absorbent material, such as sawdust, earth, sand, straw, or burlap, and maintaining the material in a damp condition for the required length of time. Alternatively, an impervious sheet of plastic or waterproof paper may be drawn over the slab soon after troweling to prevent the escape of moisture from the concrete (Figure 8.29f). The same effect can be obtained by spraying the concrete surface with one or more applications of a liquid curing compound, which forms an invisible moisture barrier membrane over the slab.

(a)

(b)

(c)

(d)

(e)

(f)

FIGURE 8.29

Constructing and finishing a concrete slab on grade: (*a*) Distributing the concrete in the forms is hard work and must be done rapidly, so a large crew is generally on hand even if a concrete pumper is used to place the concrete in the forms. Workers must take care not to penetrate the moisture barrier during this part of the process. (*b*) Striking off the surface of a concrete slab on grade just after pouring, using a motorized straightedging device. The motor vibrates the straightedge end to end to work the wet concrete into a level surface. This operation is often performed by hand. (*c*) A bull float can be used for preliminary smoothing of the surface immediately after straightedging. (*d*) Machine floating brings cement paste to the surface and produces a plane surface. Floating can also be done by hand. (*e*) Steel troweling several hours after floating produces a dense, hard, smooth surface. (*f*) One method for damp curing a slab is to cover it with polyethylene sheeting to retain moisture inside the concrete. (*Photo* a *by Rob Thallon; photos* b, c, d, e, *and* f *courtesy of Portland Cement Association, Skokie, Illinois*)

WOOD FOUNDATIONS

The preservative-treated wood foundation (Figure 8.31), first developed in the 1970s, is now a popular alternative to the standard concrete or concrete block foundation. Called the *permanent wood foundation (PWF),* this system accounts for about 5 percent of all foundations in the United States and about 20 percent of foundations in Canada. It is easily insulated in the same manner as the frame of the house it supports and can be constructed in any weather by the same crew of carpenters who will frame the building. A further advantage of permanent wood foundations is that they allow for easy installation of electrical wiring, plumbing, and interior finish materials in the basement.

Permanent wood foundations do not rely on a concrete footing for support, but rest instead on a leveled bed of crushed stone that also provides

FIGURE 8.30
A recently poured concrete slab with form boards still in place. Anchor straps embedded in the concrete will tie the wood framing to the slab foundation. (*Photo by Rob Thallon*)

FIGURE 8.31
Erecting a preservative-treated wood foundation. One worker applies a bead of sealant to the edge of a panel of preservative-treated wood components, as another prepares to push the next panel into position against the sealant. The panels rest on a horizontal preservative-treated plank, which, in turn, rests on a drainage layer of crushed stone. A major advantage of a wood foundation is that it can be insulated in the same way as the superstructure of the building. (*Courtesy of APA–The Engineered Wood Association*)

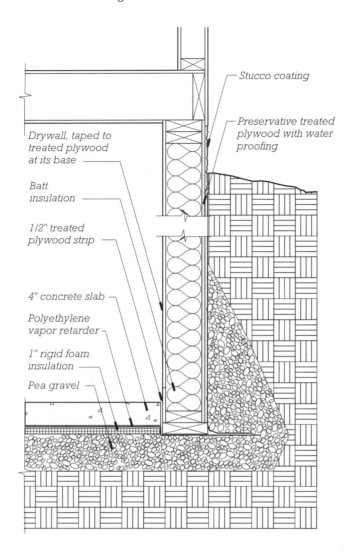

FIGURE 8.32
A permanent wood foundation is essentially a below-grade preservative-treated stud wall that is engineered to withstand lateral pressure from backfill. A slab resists soil pressure at the base of the wall, and the floor holds the wall at its top. Gravel fill and waterproofing keep moisture away from the wall. Insulation is located in the stud cavities just as it is in an above-grade wall.

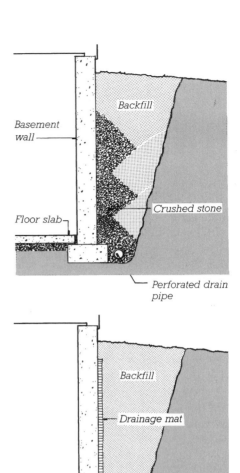

FIGURE 8.33
Two methods of relieving water pressure around a building substructure. The gravel drain (top) is the standard method, but it is hard to do well because of the difficulty of depositing the stone and backfill soil in neatly separated alternating layers. The drainage mat or drainage panel (bottom) is much easier and often more economical to install. It is a manufactured component that may be made of a loose mat of inert fibers, a plastic egg-crate structure, or some other very open, porous material. It is faced on the outside with a filter fabric that prevents fine soil particles from entering and clogging the drainage passages in the mat or panel. Water approaching the wall falls through the porous material to the drain pipe at the footing.

drainage (Figure 8.32). The size and spacing of studs used for wood foundations depend on the height and type of backfill. Resistance to the lateral thrust of backfill is provided at the base of the wall by a 4-inch (100-mm) concrete slab, poured before backfilling occurs. Because of the need for resistance to sliding at the base of the wall, wood foundations are most practical for basement foundations, where the floor slab prevents the walls from sliding inward.

DRAINAGE AND WATERPROOFING

Before any effort is made to keep groundwater from penetrating the foundation, every attempt should be made to control surface runoff. The ground should slope away from the foundation whenever possible to promote drainage away from the building. Roof drainage water should be contained in gutters, downspouts, and

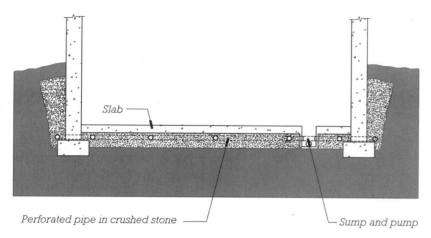

FIGURE 8.34
For a high degree of security against substructure flooding, drainage both around and under the basement is required.

storm drains and channeled away from the building. These principles are discussed in more detail in Chapter 7.

Once surface runoff has been controlled, an excellent way to keep a basement or crawlspace dry is to drain it by surrounding it with a thick layer of clean gravel or crushed stone. Large stone with no fine particles is best for this purpose because it leaves large, open passages between the stones through which water can flow freely. Water passing through the soil toward the building cannot reach the foundation wall without first falling to the bottom of the gravel layer, from where it can be drawn off by perimeter drains (also called footing drains) made of perforated pipes (Figure 8.33). It is good practice to place filter fabric around the entire volume of gravel to keep out small soil particles that might eventually clog the open passages between stones. The perforated pipes are at least 4 inches (100 mm) in diameter and serve to maintain an open channel in the crushed stone bed through which water can flow. The pipes drain by gravity either "to daylight" below the building on a sloping site, to a sump pit that is periodically pumped dry, or to a dry well.

An excellent alternative to gravel is *drainage matting* that is placed against the outside of the foundation wall. The matting contains an integral filter fabric and maintains an airspace next to the wall through which water falls to the perimeter drain at its base.

A properly installed drainage system will prevent the buildup of potentially destructive water pressure against basement walls and under slabs (Figure 8.34). However, many houses also have some type of membrane waterproofing, which serves to keep soil humidity from permeating the basement walls. Residential foundations in relatively dry regions are often coated on the outside with an inexpensive asphalt *dampproofing*, which is not a reliable barrier to the passage of water, but will prevent most moisture from coming through the wall. More reliable *waterproofing* systems include natural clay that swells when moist to become impervious, portland cement plasters with water-repellent admixtures, rubber or plastic membranes, and bitumen-modified urethanes. These systems are variously applied with a brush, a trowel, a roller, a sprayer, or mechanically in sheets by the general contractor or a specialized subcontractor.

FIGURE 8.35
A CMU basement foundation wall. The lower portion of the wall has been coated with dampproofing, and a perimeter drain has been installed at the base. Rigid insulation will be placed against the dampproofing and gravel will be placed (by hand) against the insulation. The gravel will provide a drainage channel to the perimeter drain and will protect the insulation and dampproofing from damage by soil backfill. (*Photo by Rob Thallon*)

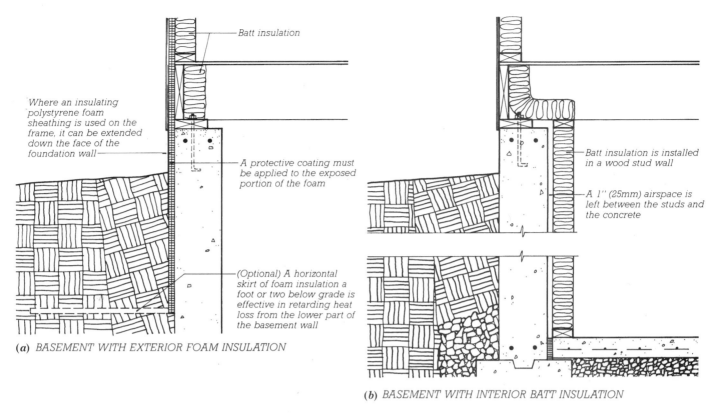

Where an insulating polystyrene foam sheathing is used on the frame, it can be extended down the face of the foundation wall

Batt insulation

A protective coating must be applied to the exposed portion of the foam

(Optional) A horizontal skirt of foam insulation a foot or two below grade is effective in retarding heat loss from the lower part of the basement wall

(*a*) BASEMENT WITH EXTERIOR FOAM INSULATION

Batt insulation is installed in a wood stud wall

A 1'' (25mm) airspace is left between the studs and the concrete

(*b*) BASEMENT WITH INTERIOR BATT INSULATION

Cantilevered joists allow use of foam on the basement portion of the house only

(*c*) BASEMENT WITH EXTERIOR FOAM INSULATION

Batt insulation is stapled to the header joist and extended down the concrete wall and 2' (600mm) onto the floor of the crawlspace

A plastic moisture barrier sheet keeps the crawlspace and insulation dry

(*d*) CRAWLSPACE WITH INTERIOR BATT INSULATION

FIGURE 8.36

Most building codes require thermal insulation of the foundation. Here are three different ways of adding insulation to a poured concrete or concrete masonry basement, and one way of insulating a crawlspace foundation. The crawlspace may alternatively be insulated on the outside with panels of plastic foam. The interior batt insulation shown in (*b*) is commonly used but raises unanswered questions about how to avoid possible problems arising from moisture accumulating between the insulation and the wall.

FOUNDATION INSULATION

Because foundation walls are generally located at the perimeter of the building, there is usually a need to insulate them in order to complete the thermal envelope of the house. Basement walls must be insulated to protect living spaces or utilities located in the basement (Figure 8.36a–c). Crawl space foundations are also sometimes insulated (Figure 8.36d). Slab foundations in all but the warmest climates must be insulated because they are continuous with the slab floor of the house.

When foundation insulation is below grade at the exterior of the foundation, a rigid insulation such as extruded polystyrene foam that does not deteriorate in the soil is best. The installation of this rigid insulation is often done by the general contractor (rather than the insulation subcontractor) because it is easy to apply and must be in place before backfilling can occur. At the interior of foundation walls, either rigid or conventional batt-type insulation may be applied later in the construction process by the insulation subcontractor. Recent problems with insect infestations in rigid foam insulation that runs from below grade to framing have brought about code restrictions regarding this practice. This problem and the general topic of insulation are discussed in Chapter 15.

A recently developed alternative for insulating foundation walls is the insulating concrete form (ICF) that serves both to form the concrete and to remain in place permanently as thermal insulation. This material is discussed in Chapter 21.

Basement Insulation

Basement insulation may be applied either inside or outside the foundation wall. Outside the wall, rigid insulation boards, typically 2 to 4 inches thick (50 to 100 mm), are applied with adhesive or mechanical fasteners and held ultimately by the pressure of the soil (Figures 8.36a,b). Above grade, this rigid insulation must be protected from mechanical abrasion and ultraviolet degradation. Depending on the situation, flashing or stucco is usually used for this protection (Figure 8.36a). Proprietary products are available that combine insulation board with a drainage mat in a single assembly. Alternatively, either batt insulation or rigid insulation boards may be installed later in the construction process at the interior of the foundation wall (Figure 8.36c).

Crawlspace Insulation

When the crawlspace is used as a return or supply plenum for a forced-air heating system (Figure 13.8), or when it is unventilated and sealed (Figure 15.29), the perimeter foundation walls must be thermally insulated. In these situations, the crawlspace walls are typically insulated on the inside with batt insulation that is stapled to the mudsills, hangs down the walls, and runs out 2 to 4 feet onto the floor of the crawlspace (Figure 8.36d).

Slab-on-Grade Insulation

In all but the warmest of climates, slabs are required by code to be insulated at their perimeter where they are in contact with outside air. Because some portion of the insulation is always in contact with the soil, rigid insulation board is used (Figure 8.37).

FIGURE 8.37
This new slab has 2-inch rigid insulation board installed at the perimeter. Metal flashing protects the top portion of the insulation from abrasion. The worker is finishing a porch slab poured against the completed insulation. (*Photo by Rob Thallon*)

Shallow Frost-Protected Foundations

Codes in some cold climates permit polystyrene foam insulation boards to be used in a special configuration that allows the construction of footings above the normal frost line in the soil, resulting in lower excavation costs. Continuous layers of insulation board are placed around the perimeter of the building in such a way that heat flowing into the soil in winter from the interior of the building keeps the soil beneath the footings at a temperature above freezing (Figure 8.38).

BACKFILLING

After the basement walls have been waterproofed or dampproofed, the insulating boards have been applied, and the drainage features installed, the area around the foundation in an open excavation is backfilled to restore the level of the ground. Backfilling should not commence, however, until the foundation walls are braced to resist the considerable pressure of the backfill with internal supporting constructions such as walls and floors (Figure 7.25). The backfilling operation involves placing soil back against the outside of the basement walls and compacting it there in layers, taking care not to damage drainage or waterproofing components or to place excessive pressure against the walls. An open, fast-draining soil such as sand or gravel is preferred for backfilling

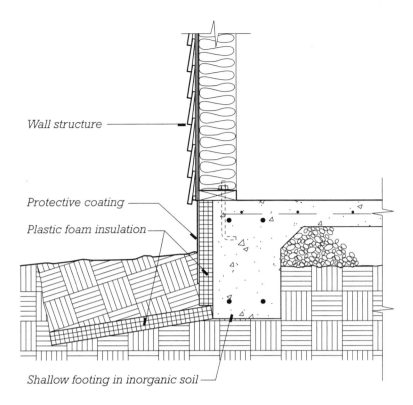

Wall structure

Protective coating

Plastic foam insulation

Shallow footing in inorganic soil

FIGURE 8.38
A typical detail for a shallow frost-protected footing.

because it allows the perimeter drainage system around the basement to do its work. Compaction must be sufficient to minimize subsequent settling of the backfilled area.

FOUNDATION DESIGN AND THE BUILDING CODES

Building codes contain numerous provisions relating to the design and construction of foundations, which

we have discussed throughout this chapter. The codes define which soil types are considered satisfactory for bearing the weight of buildings and set forth maximum assumed loadbearing values for each soil type (Figure 8.4). They establish minimum dimensions for footings and foundation walls and require engineering design of retaining walls. Through these provisions, the codes attempt to ensure that every building will rest upon secure foundations and a dry substructure.

C.S.I./C.S.C. MasterFormat Section Numbers for Foundations	
02450	Foundations and Loadbearing Elements
02480	Foundation Wall
06140	Treated Wood Foundation
07100	DAMPPROOFING AND WATERPROOFING

SELECTED REFERENCES

1. Ambrose, James E. *Simplified Design of Building Foundations* (2nd ed.). New York: John Wiley & Sons, 1988.

After an initial summary of soil properties, this small book covers simplified foundation computation procedures for both shallow and deep foundations.

2. Liu, Chen, and Evett, Jack B. *Soils and Foundations*. Englewood Cliffs, NJ: Prentice Hall, 1981.

This is a fairly detailed discussion of the engineering properties of soils, subsurface exploration techniques, soil mechanics, and shallow and deep foundations, but is well suited to the beginner.

KEY TERMS AND CONCEPTS

dead load
live load
wind load
uniform settlement
differential settlement
soil
rock
gravel
sand
silt
clay
expansive soil
organic soil
soil stability
soil drainage characteristics
allowable soil stress
bearing capacity of soil

footing
column footing
wall footing
strip footing
undisturbed soil
frost line
heaving
engineered fill
compacted gravel fill
controlled low-strength material (CLSM)
concrete foundation wall
form-release compound
form tie
anchor bolt
hold-down
mudsill
concrete masonry foundation wall

slab on grade
welded wire fabric
fibrous admixture
control joint
strike off
straightedge
float
trowel
kneeboard
permanent wood foundation
drainage matting
dampproofing
waterproofing
perimeter insulation
shallow frost-protected foundation

REVIEW QUESTIONS

1. What is the nature of the most common type of foundation failure?

2. Explain in detail the differences among fine sand, silt, and clay, especially as they relate to the foundations of buildings.

3. What are some ways to achieve adequate bearing for footings on poor soil?

4. Starting with a poured concrete strip footing, describe the steps in the typical process of forming and casting a concrete foundation wall.

5. Name some advantages and disadvantages of permanent wood foundations.

6. Describe several precautions that should be taken to keep a basement dry.

7. What is the difference between dampproofing and waterproofing?

8. List a number of ways to insulate a basement wall.

9. What can go wrong during backfilling around a basement? What be done to minimize the pitfalls of backfilling?

EXERCISES

1. Obtain the foundation drawings for a nearby residential project. What sorts of soils are found beneath the site? How deep is the water table? Before looking at the drawings, decide what type of foundation should be used in this situation (keeping in mind the relative weight of the building). Now look at the foundation drawings. What type is actually used? Why? Is the design derived from tables in the building code or was an engineer involved?

2. What type of foundation and substructure is normally used for houses in your area? Why?

FRAMING

HISTORY

Wood light frame construction was the first uniquely American building system. It was developed in the first half of the 19th century when builders recognized that the closely spaced vertical members used to infill the walls of a heavy timber building frame were themselves sufficiently strong that the heavy posts of the frame could be eliminated. Its development was accelerated by two technological breakthroughs of the period. The advent of the water-powered sawmill lowered the cost of boards and small wood framing members to an affordable level for the first time in history. Additionally, machine-made nails were introduced, which were remarkably inexpensive compared to the hand-forged nails that preceded them.

The earliest version of wood light framing, the *balloon frame*, was a building system framed solely with slender, closely spaced wooden members: joists for the floors, studs for the walls, rafters for the sloping roofs. Heavy posts and beams were completely eliminated and, with them, the difficult, expensive, mortise-and-tenon joinery they required. There was no structural member in a frame that could not be handled easily by a single carpenter, and each of the hundreds of joints was made with lightning rapidity using two or three nails. The impact of this new building system was revolutionary. In 1865, G. E. Woodward wrote, "A man and a boy can now attain the same results, with ease, that twenty men could on an old-fashioned frame."

The balloon frame (Figure 9.2) used full-length studs that ran continuously for two stories from foundation to roof. In time, these were recognized as being too long to erect efficiently, and the tall, hollow spaces between studs acted as multiple chimneys in the event of a fire, spreading the blaze rapidly to the upper floors

unless they were closed off with wood or brick firestops at each floor line. Several modified versions of the balloon frame were subsequently developed in an attempt to overcome these difficulties, and the most recent of these, the *platform frame,* is now the universal standard.

THE PLATFORM FRAME

While complex in its details, the platform frame is very simple in concept. A floor platform is built. Loadbearing walls are erected upon it. A second floor platform is built upon these walls, and a second set of walls upon

FIGURE 9.1
Carpenters apply plywood roof sheathing to a platform-framed apartment building. The ground floor is a concrete slab on grade. The edge of the wooden platform of the upper floor is clearly visible between the stud walls of the ground and upper floors. Most of the diagonal bracing is temporary, but permanent let-in diagonal braces occur between the two openings at the lower left and immediately above in the rear building. The openings have been framed incorrectly, without supporting studs for the headers. (*Courtesy of Southern Forest Products Association*)

this platform. The attic and roof are then built upon the second set of walls. There are, of course, many variations: A concrete slab that lies directly on the ground is sometimes substituted for the ground-floor platform; a building may be one or three stories tall instead of two; and several types of roofs that do not incorporate attics are frequently built. The essentials remain, however: A floor platform is completed at each level, and the walls bear upon the platform rather than directly upon the walls of the story below.

The advantages of the platform frame over the balloon frame are that it uses short, easily handled lengths of lumber for the wall framing; its vertical hollow spaces are automatically firestopped at each floor; and its platforms function as convenient working surfaces for the carpenters who build the frame. The major disadvantage of the platform frame is that each platform constitutes a thick layer of wood whose grain runs horizontally. This leads inevitably to a relatively large amount of vertical shrinkage in the frame as the excess moisture dries from the wood, which can lead to distress in the exterior and interior finish surfaces.

A conventional platform frame is made entirely of nominal 2-inch members, which are actually 1½ inches (38 mm) in thickness. They can be ordered in a variety of lengths and are available in 2-foot (600-mm) increments. At the building site, the carpenters measure and saw the members to the exact lengths required for the frame. All connections are made with nails, using either face nailing, end nailing, or toe nailing, as required by the characteristics of each joint (Figures 4.36 and 9.8). Nails are driven either by hammer or by handheld pneumatic nailing machines (nail guns). In either case, the connection is quickly made because the nails are installed without drilling holes or otherwise preparing the joint.

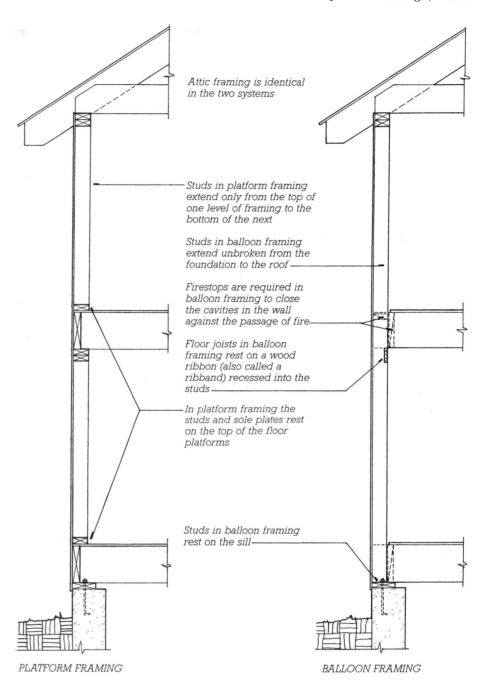

Attic framing is identical in the two systems

Studs in platform framing extend only from the top of one level of framing to the bottom of the next

Studs in balloon framing extend unbroken from the foundation to the roof

Firestops are required in balloon framing to close the cavities in the wall against the passage of fire

Floor joists in balloon framing rest on a wood ribbon (also called a ribband) recessed into the studs

In platform framing the studs and sole plates rest on the top of the floor platforms

Studs in balloon framing rest on the sill

PLATFORM FRAMING

BALLOON FRAMING

FIGURE 9.2

Comparative framing details for platform framing (left) and balloon framing (right). Platform framing is easier to erect but settles considerably as the wood dries and shrinks. If nominal 12-inch (300-mm) joists are used to frame the floors in these examples, the total amount of loadbearing cross-grain wood between the foundation and the attic joists is 33 inches (838 mm) for the platform frame, and only 4½ inches (114 mm) for the balloon frame.

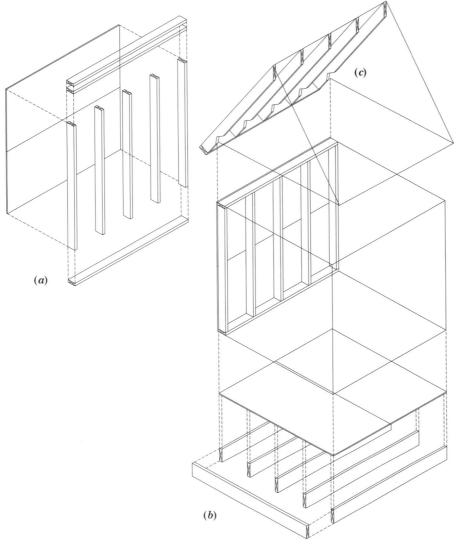

FIGURE 9.3
The basic components of platform frame construction. (*a*) Floors are made of repetitive joists, (*b*) walls are made of repetitive studs, and (*c*) roofs are made of repetitive rafters. Each of these repeated elements is fastened to cross members at the ends and covered with sheathing to form a structural plane.

Each plane of structure (floor platform, wall frame, etc.) is basically constructed in the same way. Pieces of framing lumber are aligned parallel to one another with equal spacing between. Crosspieces are then placed at right angles to the parallel framing lumber at either end. The crosspieces are nailed to the parallel members to maintain their spacing and flatness. The assembled framing pieces are then covered with *sheathing*, a facing layer of boards or panels that join and

stabilize the pieces into a single structural unit, ready for the application of finish materials inside and out (Figure 9.3).

Sheathing is a key component of platform framing. The end nails that connect the crosspieces to the parallel framing members have little holding power against the considerable forces exerted on the building by wind, but the sheathing connects the frame into a single, strong unit from foundation to roof. The rectilinear geom-

etry of the parallel framing members has no useful resistance to wracking by lateral forces such as wind, but rigid sheathing panels or diagonal sheathing boards brace the building effectively against these forces. Sheathing also furnishes a surface to which shingles, siding, and flooring are nailed for finish surfaces. In buildings without sheathing, or with sheathing materials that are too weak to tie and brace the frame, diagonal bracing must be applied to the walls to impart lateral stability.

Openings are required in all of these planes of structure: for windows and doors in the walls; for stairs and chimneys in the floors; and for chimneys, skylights, and dormers in the roofs. In each case, framing members are added around the opening to restore stiffness to the plane of structure and to hold the parallel members in position. Floors, walls, and roofs each have unique structural requirements for openings in their structural plane, and each has specific framing members for this purpose, which will be discussed in the following sections.

BUILDING THE FRAME

Planning the Frame

While it is true that an experienced carpenter can frame a simple building from the most minimal of drawings, a platform frame of wood for a custom-designed building should be planned as carefully as a frame of steel or concrete in a larger building. The designer should determine an efficient layout and the appropriate sizes for joists and rafters, and communicate this information to the carpenters by means of framing plans (Figures 9.8 and 9.33). For most purposes, member sizes can be determined using the standardized structural tables in the building codes and in other references such as Reference 3 at the end of this chapter. Detailed section drawings, similar to those seen throughout this chapter, are also

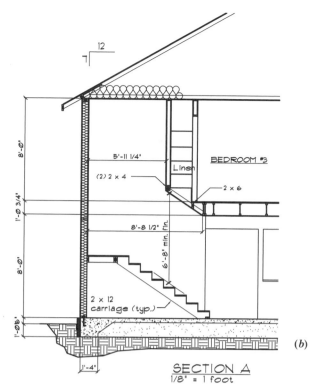

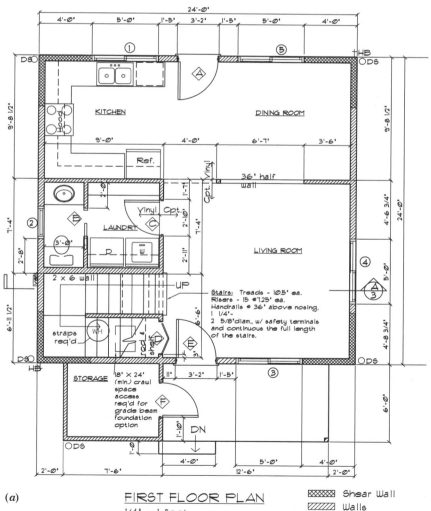

FIGURE 9.4

The information for framing a house is all contained in the construction drawings. For the simplest buildings, a floor plan and a section are adequate. (*a*) The floor plan shows the location of the walls and the location and size of window and door openings. (*b*) The section shows vertical dimensions and roof pitch. Spacing of joists, studs, and rafters are indicated on the section. More complicated buildings require floor framing plans, roof framing plans, and numerous details. (*Courtesy of Rainbow Valley Design and Construction, Eugene, Oregon*)

prepared for the major connections in the building. The architectural floor plans serve to indicate the locations and dimensions of all the walls, partitions, and openings (Figure 9.4), and the exterior elevations are drawings that show the outside faces of the building, with vertical framing dimensions indicated as required. For most buildings, sections are also drawn that cut completely through the building, showing the dimensional relationships of the various floor levels and roof planes, as well as the slopes of the roof surfaces.

Erecting the Frame

The erection of a typical platform frame (referred to as *rough carpentry* in architects' specifications) can best be understood by following the sequential isometric diagrams that begin with Figure 9.6. Notice the basic simplicity of the building process: A platform is built, walls are assembled horizontally on the platform and tilted up into place, and another platform or a roof is built on top of the walls. Most of the work is accomplished without the use of ladders or scaffolding, and temporary bracing is needed only to support the walls until the next level of framing is installed and sheathed.

The details of a platform frame are not left to chance. While there are countless local and personal variations in framing details and techniques, the sizes, spacings, and connections of the members in a platform frame are closely regulated by building codes, down to the sizes and spacings of the studs, the size and number of nails for each connection (Figure 9.7), and the thicknesses and nailing of the sheathing panels.

Attaching the Frame to the Foundation

A *sill*, often called a *mudsill*, which functions as a base for the wood framing, is bolted (or strapped) to the foundation of the building. Preferably, the sill should be made of preservative-treated or naturally decay-resistant wood. A single sill, as shown in Figure 9.9, is all that is required by most codes, but in better quality work, the sill is often doubled for greater stiff-

ness. The top of the foundation is usually somewhat uneven, so at low spots the sill must be shimmed with wood shingle wedges to be able to transfer loads from the frame to the foundation. A compressible, fibrous sill sealer should be inserted between the sill and the foundation to reduce air infiltration through the gap (Figure 9.9). The normal foundation bolts are sufficient to hold most buildings on their foundations, but

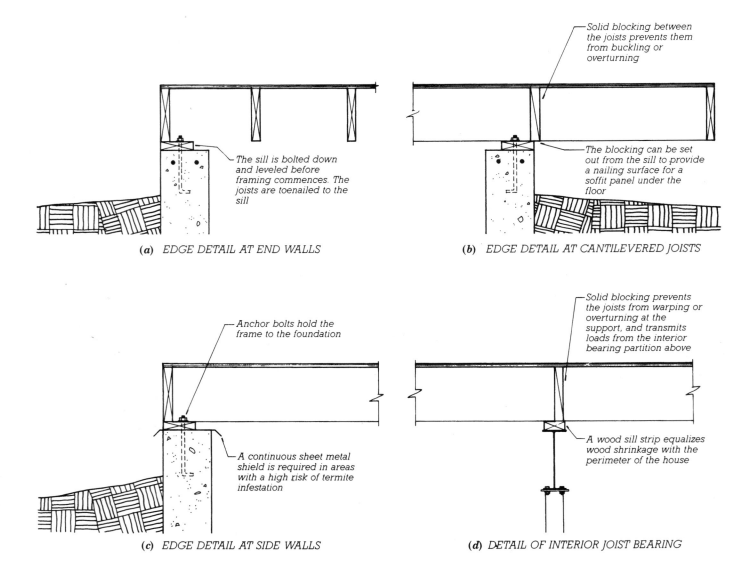

(a) EDGE DETAIL AT END WALLS

The sill is bolted down and leveled before framing commences. The joists are toenailed to the sill

(b) EDGE DETAIL AT CANTILEVERED JOISTS

Solid blocking between the joists prevents them from buckling or overturning

The blocking can be set out from the sill to provide a nailing surface for a soffit panel under the floor

(c) EDGE DETAIL AT SIDE WALLS

Anchor bolts hold the frame to the foundation

A continuous sheet metal shield is required in areas with a high risk of termite infestation

(d) DETAIL OF INTERIOR JOIST BEARING

Solid blocking prevents the joists from warping or overturning at the support, and transmits loads from the interior bearing partition above

A wood sill strip equalizes wood shrinkage with the perimeter of the house

FIGURE 9.5
Ground-floor framing details, keyed to the lettered circles in Figure 9.7. A fibrous sill sealer material should be installed between the sill and the top of the foundation to reduce air leakage, but the sealer material is not shown on these diagrams. Using accepted architectural drafting conventions, continuous pieces of lumber are drawn with an X inside, and intermittent blocking with a single slash.

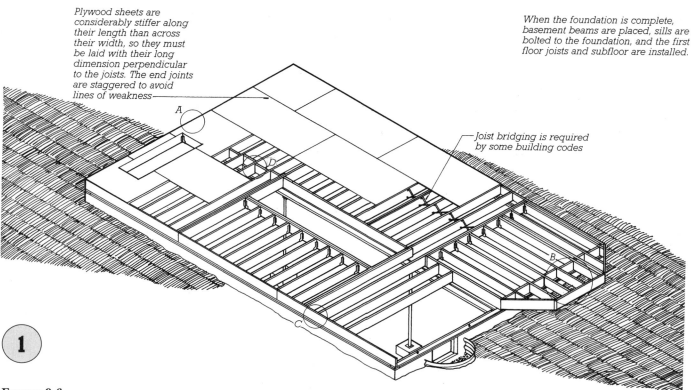

Plywood sheets are considerably stiffer along their length than across their width, so they must be laid with their long dimension perpendicular to the joists. The end joints are staggered to avoid lines of weakness

When the foundation is complete, basement beams are placed, sills are bolted to the foundation, and the first floor joists and subfloor are installed.

Joist bridging is required by some building codes

①

FIGURE 9.6
Step 1 in erecting a typical platform frame building: the ground-floor platform. Compare this drawing with the framing plan in Figure 9.4. Notice that the direction of the joists must be changed to construct the cantilevered bay on the end of the building. A cantilevered bay on a long side of the building could be framed by merely extending the floor joists over the foundation.

Connection	Common Nail Size
Stud to sole plate	4-8d toe nails or 2-16d end nails
Stud to top plate	2-16d end nails
Double studs	2-16d end nails or toe nails
Corner studs	10d face nails 12" apart
Sole plate to joist or blocking	16d face nails 24" apart
Double top plate	16d face nails 16" apart
Lap joints in top plate	10d face nails 16" apart
Rafter to top plate	2-10d face nails
Rafter to ridge board	3-8d toe nails
Jack rafter to hip rafter	2-16d end nails
Floor joists to sill or beam	3-8d toe nails
Ceiling joists to top plate	3-16d toe nails
Ceiling joist lap joint	3-10d face nails
Ceiling joist to rafter	3-10d face nails
Collar tie to rafter	3-10d face nails
Bridging to joists	2-8d face nails each end
Left-in diagonal bracing	2-8d face nails each stud or plate
Tail joists to headers	Use joist hangers
Headers to trimmers	Use joist hangers
Ledger to header	3-16d face nails per tail joist

FIGURE 9.7
Platform framing members are fastened according to this nailing schedule, which framing carpenters know by memory, and which is incorporated into most building codes.

frames in areas subject to high winds or earthquakes may require more elaborate attachments (Figure 9.31).

Floor Framing and Bridging

In a floor structure, the parallel pieces are the *floor joists,* and the crosspieces at the ends of the joists are called *headers, rim joists,* or *band joists,* depending on local custom. The sheathing on a floor is known as the *subfloor.*

Houses are typically wider than can be spanned by a single floor joist. As an approximation, 2 × 6 joists will span up to 8 feet (2.4 m), 2 × 8 joists up to 11 feet (3.4 m), 2 × 10 joists up to 14 feet (4.3 m), and 2 × 12 joists up to 17 feet (5.2 m). It is therefore extremely common for at least one structural *girder* or structural wall to

be located at the interior of the building to break the joist span into two or more sections. To support the first-floor framing, there is usually a girder that runs the length of the building and is itself supported at its ends by the foundation wall and at the center of the house by columns in the basement or crawlspace (Figure 9.5D). At the level of the second (or third) floor, the joists are usually supported near the center of the building by structural walls (Figure 9.26).

Floor framing is usually laid out in such a way that the ends of uncut subfloor panels will fall directly over joists; otherwise, many panels will have to be cut, wasting both materials and time (Figure 9.7). The standard joist spacings are 16 or 24 inches o.c. (406 or 610 mm o.c.; "o.c." stands for

"on center," meaning that the spacing is measured from center to center of the joists). Occasionally, a joist spacing of 19.2 inches (486 mm) is used. Any of these spacings automatically provides a joist at every panel end.

Subflooring should be glued to the joists to prevent squeaking and increase floor stiffness (Figure 9.14). Plywood and OSB panels must be laid with the grain of their face layers perpendicular to the direction of the joists, because these panels are considerably stiffer in this orientation. Sheathing and subflooring panels are normally manufactured ⅛ inch (3 mm) short in each face dimension so that they may be spaced slightly apart at all of their edges. This spacing prevents the floor from buckling when rain dampens the wood panels during construction and causes them to expand.

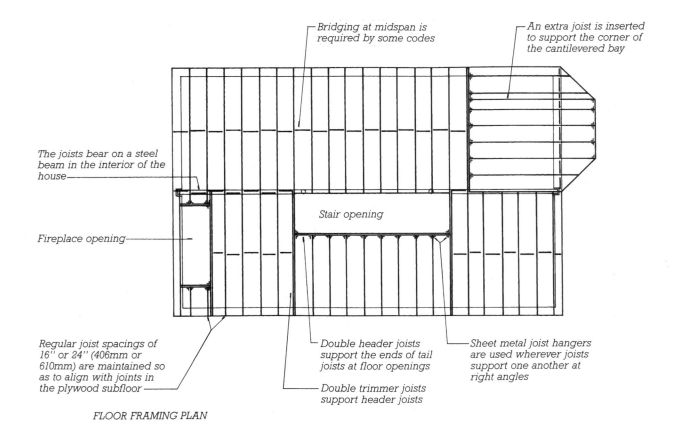

Bridging at midspan is required by some codes

An extra joist is inserted to support the corner of the cantilevered bay

The joists bear on a steel beam in the interior of the house

Stair opening

Fireplace opening

Regular joist spacings of 16" or 24" (406mm or 610mm) are maintained so as to align with joints in the plywood subfloor

Double header joists support the ends of tail joists at floor openings

Double trimmer joists support header joists

Sheet metal joist hangers are used wherever joists support one another at right angles

FLOOR FRAMING PLAN

FIGURE 9.8
A framing plan for the ground-floor platform of the building shown in Figure 9.7. Where joists are cantilevered to create an overhanging bay, the cantilever distance should not exceed one-quarter of the total length of the joist.

FIGURE 9.9
Carpenters apply a preservative-treated wood sill to a site-cast concrete foundation. Fluffy glass fiber sill sealer has been placed on the top of the concrete wall, and the sill has been drilled to fit over the projecting anchor bolts. Before each section of sill is bolted tightly, it is leveled as necessary with wood shingle shims between the concrete and the wood. As the anchor bolts are tightened, the sill sealer squeezes down to a negligible thickness. A length of completed sill is visible at the upper right. The basement windows were clamped into reusable steel form inserts and placed in the formwork before the concrete was cast, which causes the window to become integral with the basement wall. After the concrete was cast and the formwork was stripped, the steel inserts were removed, leaving a neatly formed concrete frame around each window. A pocket in the top of the wall for a steel beam can be seen at the upper left. (*Photo by Edward Allen*)

FIGURE 9.10
Installing floor joists. Blocking will be inserted between the joists over the two interior beams to prevent overturning of the joists. (*Photo by Edward Allen*)

FIGURE 9.11
Various types of manufactured joists and floor trusses are often used instead of dimension lumber. This I-joist has laminated veneer lumber (LVL) flanges and a plywood web. I-joists are manufactured in very long pieces and tend to be straighter, stronger, stiffer, and lighter in weight than sawn joists. See also Figure 4.46. (*Courtesy of Trus Joist MacMillan*)

FIGURE 9.12
These floor trusses (shown here being set up for a demonstration house in a parking lot) are made of sawn lumber members joined by toothed plate connectors. The oriented strand board (OSB) web at each end of the truss allows workers to shorten the truss with a saw if necessary. Trusses are deeper than sawn joists or I-joists, but can span farther between supports and offer large passages for ductwork and pipes. (*Courtesy of Wood Truss Council of America*)

FIGURE 9.13
Applying OSB subflooring. (*Courtesy of APA–The Engineered Wood Association*)

Bridging, which is the crossbracing or solid blocking between joists at midspan, is a traditional feature of floor framing (Figure 9.15). Its function is to hold the joists straight and to help them share concentrated loads. Although many building codes no longer require it, bridging should be used on better quality buildings because it makes a noticeable difference in the rigidity of a floor.

Openings in the floor for stairs and chimneys are framed with *headers* and *trimmers* (Figure 9.4). The headers and trimmers are usually made of the same dimension lumber as the floor joists and are doubled at large openings to support the higher loads placed on them by the presence of the opening.

Where ends of joists butt into supporting headers, such as around stair openings and at changes of joist direction for projecting bays, end nails and toe nails cannot carry the full weight of the joists, so sheet metal *joist hangers* must be used. Each provides a secure pocket for the end of the joist and punched holes into which a number of special short nails are driven to make a safe connection.

FIGURE 9.14
For stiffness and squeak resistance, subflooring should be glued to the joists. The adhesive is a thick mastic that is squeezed from a sealant gun. (*Courtesy of APA–The Engineered Wood Association*)

FIGURE 9.15
Bridging between joists may be solid blocks of joist lumber, diagonal wood crossbridging, or, as seen here, diagonal steel crossbridging. During manufacture, the thick steel strip was folded across its width into a V-shaped section to make it stiff. Steel crossbridging requires only one nail per piece and no cutting, so it is the fastest to install. (*Courtesy of American Plywood Association*)

Manufactured *wood I-joists* and *floor trusses* are increasingly used in place of sawn joists because they can span farther between supports and they tend to be straighter (Figures 9.11 and 9.12).

Wall Framing, Sheathing, and Bracing

In a wall structure, the vertical framing member is called a *stud* and is generally made of 2 × 4 or 2 × 6 lumber. The crosspiece at the bottom of the wall is called the *sole plate*, and the crosspiece at the top (which is doubled for strength if the wall bears a load from above) is called the *top plate*. Wall framing, like floor framing, is laid out in such a way that a framing member, in this case a stud, occurs under each (vertical) joint between sheathing panels. The lead carpenter initiates wall framing by marking the stud locations on the two pieces of lumber destined to be the top plate and sole plate of each wall (Figure 9.20). Other carpenters fol-

low behind to cut the studs and headers and assemble the walls in a horizontal position on the subfloor. As each wall frame is completed, the carpenters tilt it up and nail it into position, bracing it temporarily as needed (Figures 9.21 to 9.24).

The sheathing, which is made of rigid panels (usually OSB), acts as permanent bracing for the walls and is applied as soon as possible after the wall is framed. Many builders apply sheathing to each wall while it still lies flat on the floor platform. Some types of sheathing panels such as insulating plastic foam are intended only as thermal insulation and have no structural value. Where these are used, let-in *diagonal bracing* is recessed into the outer face of the studs of each wall before it is erected (Figure 9.25). The let-in brace is customarily a wood 1 × 4, although steel braces are sometimes substituted. Walls braced with structural sheathing (or let-in bracing) are called *shear walls*. They are typically, though not necessarily, located only

at the exterior of the building, where they support upper floors and the roof, provide a nailing surface for exterior finish materials, and brace the building against wind and other lateral forces. Heavily loaded shear walls may require special anchoring and are sometimes made with thicker-than-normal sheathing or sheathing on both sides of the wall.

At the interior of the building, walls are not typically braced, and they may or may not support loads from above. Interior walls that support loads from upper floors or from the roof are called *bearing walls*, and those that carry no loads are referred to as *non-earing walls* or *partition walls* (Figures 9.16 and 9.18).

Corners and intersections between bearing and partition walls must provide sufficient nailing surfaces for the edge of each plane of exterior and interior finish materials. This requires a minimum of three studs at each intersection, unless special metal clips are used to reduce the number to two (Figure 9.19).

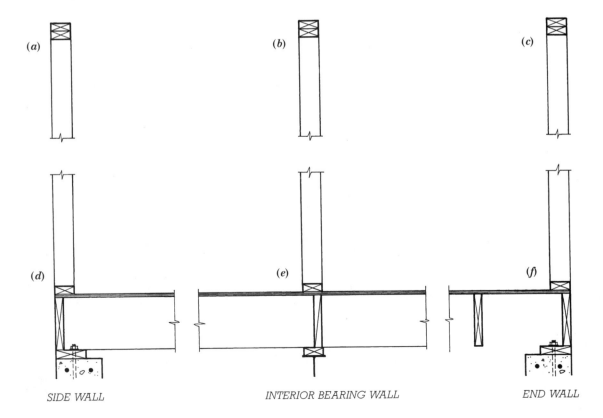

FIGURE 9.16
Typical ground-floor wall framing details, keyed by letter to Figure 9.17.

(a)   (b)   (c)

(d)   (e)   (f)

SIDE WALL   INTERIOR BEARING WALL   END WALL

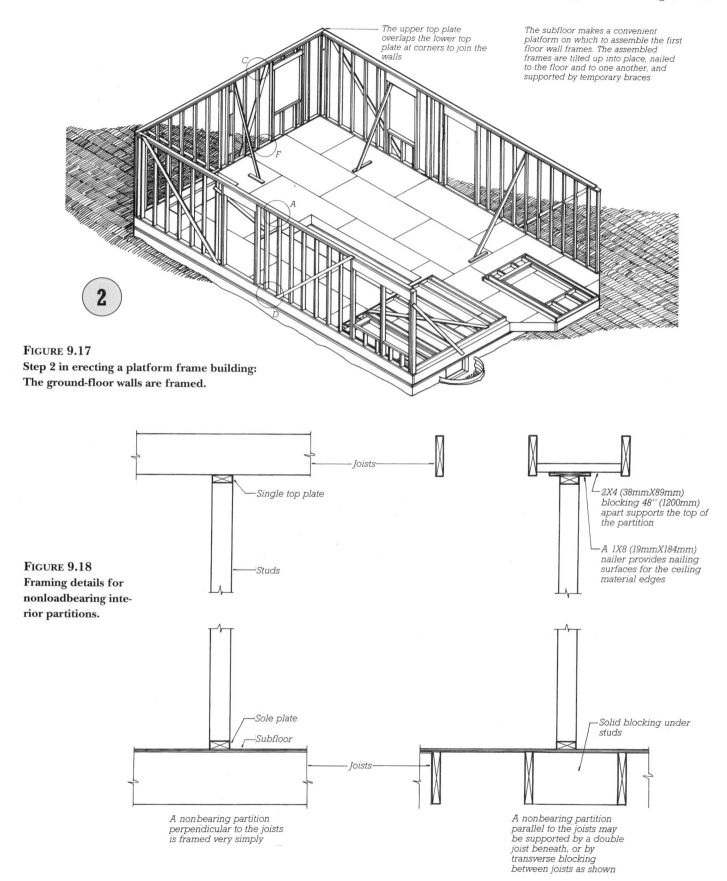

The upper top plate overlaps the lower top plate at corners to join the walls

The subfloor makes a convenient platform on which to assemble the first floor wall frames. The assembled frames are tilted up into place, nailed to the floor and to one another, and supported by temporary braces.

C

F

A

D

2

FIGURE 9.17
Step 2 in erecting a platform frame building:
The ground-floor walls are framed.

Joists

Single top plate

Studs

FIGURE 9.18
Framing details for nonloadbearing interior partitions.

2X4 (38mmX89mm) blocking 48″ (1200mm) apart supports the top of the partition

A 1X8 (19mmX184mm) nailer provides nailing surfaces for the ceiling material edges

Sole plate

Subfloor

Joists

Solid blocking under studs

Joists

A nonbearing partition perpendicular to the joists is framed very simply

A nonbearing partition parallel to the joists may be supported by a double joist beneath, or by transverse blocking between joists as shown

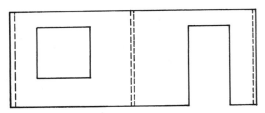

1. This is the layout of a typical exterior wall. It meets two other exterior walls at the corners, and a partition in the middle. It has two rough openings, one for a window and one for a door.

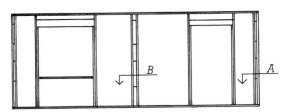

2. The framer begins by marking all the stud and opening locations on the sole plate and top plate. The "special" studs are cut and assembled first: two corner posts, a partition intersection, and full-length studs and supporting studs for the headers over the openings.

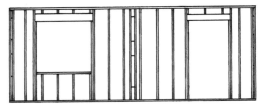

3. The wall is next filled with studs on a regular 16" (400mm) or 24" (600mm) spacing, to provide support for edges of sheathing panels.

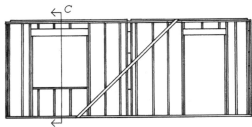

4. Diagonal bracing, usually 1X4 (19mmX89mm), is let into the face of the frame if the building will not have rigid sheathing. The second top plate may be added before the wall is tilted up, or after.

FIGURE 9.19
Procedure and details for wall framing.

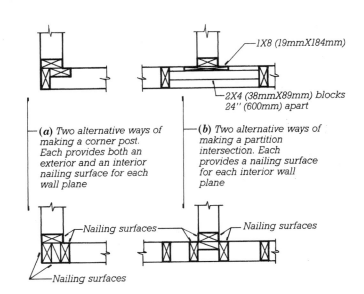

1X8 (19mmX184mm)

2X4 (38mmX89mm) blocks 24" (600mm) apart

(a) Two alternative ways of making a corner post. Each provides both an exterior and an interior nailing surface for each wall plane

(b) Two alternative ways of making a partition intersection. Each provides a nailing surface for each interior wall plane

Nailing surfaces

Nailing surfaces

Nailing surfaces

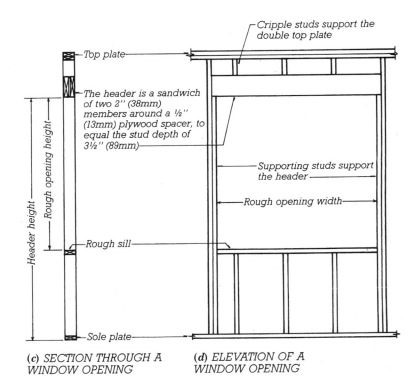

Cripple studs support the double top plate

Top plate

The header is a sandwich of two 2" (38mm) members around a ½" (13mm) plywood spacer, to equal the stud depth of 3½" (89mm)

Supporting studs support the header

Rough opening width

Rough sill

Header height

Rough opening height

Sole plate

(c) SECTION THROUGH A WINDOW OPENING

(d) ELEVATION OF A WINDOW OPENING

FIGURE **9.20**
The lead carpenter aligns the top plate and sole plate side by side on the floor platform and marks each stud location on both of them simultaneously. This is the first step in constructing a wall frame. (*Photo by Edward Allen*)

FIGURE **9.21**
Assembling studs to a plate, using a pneumatic nail gun. The triple studs are for a partition intersection. (*Courtesy of Senco Products, Inc.*)

FIGURE 9.22
Tilting an interior partition into position. The gap in the upper top plate will receive the projecting end of the upper top plate from another partition that intersects at this point. (*Courtesy of APA–The Engineered Wood Association*)

FIGURE 9.23
Fastening a wall to the floor platform. The horizontal blocks between the studs will receive horizontal lines of nails used to attach vertical wood siding. (*Courtesy of Senco Products, Inc.*)

FIGURE 9.24
Ground-floor wall framing is supported by temporary diagonal bracing. After the upper-floor framing is in place and wall bracing or sheathing is complete, the frame becomes completely self-bracing and the temporary bracing is removed. The outer walls of this building are framed with 2 × 6s (38 × 140 mm) to allow for a greater thickness of thermal insulation, whereas the interior partitions are made of 2 × 4s (38 × 89 mm). (*Photo by Joseph Iano*)

Long studs for tall walls often must be larger than a 2 × 4 in order to resist wind forces. Because long pieces of sawn dimension lumber tend not to be perfectly straight, studs manufactured of laminated veneer lumber, parallel strand lumber, or finger-jointed lumber are sometimes used instead. Some building codes require blocking to be installed at the midheight of the studs in wall frames that are taller than 8 feet (2.4 m). The purpose of this blocking is to stop off the cavities between studs to restrict the spread of fire.

Openings in walls for windows and doors require a structural header at their upper edge to support loads from floors and/or a roof above. *Window headers* and *door headers* must be sized in accordance with building code criteria. Typically, a header consists of two nominal 2-inch members standing on edge, separated by a plywood spacer that serves to make the header as thick as the depth of the wall studs (Figure 9.19). Headers for long spans and/or heavy loads are often made of laminated veneer lumber or parallel strand lumber, either of which is stronger and stiffer than dimension lumber because of an absence of knots. At the bottom of a window opening, a *rough sill* serves to keep the studs below the window in alignment.

Wall Framing for Increased Thermal Insulation

The 2 × 4 (38 × 89 mm) has been the standard wall stud since light framing was invented. In recent years, however, pressures for heating fuel conservation have led to building codes that often require more thermal insulation than can be inserted in the cavities of a wall framed with 2 × 4s. Some designers and builders have simply adopted the 2 × 6 (38 × 140 mm) as the standard stud, usually at a spacing of 24 inches (610 mm). Others have stayed with the 2 × 4 stud but covered the wall with

FIGURE 9.25
Applying a panel of insulating foam sheathing. Because this type of sheathing is too weak to brace the frame, diagonal bracing is inserted into the outside faces of the studs at the corners of the building. Steel bracing, nailed at each stud, is used in this frame and is visible just to the right of the carpenter's leg. (*Courtesy of Celotex Corporation*)

insulating plastic foam sheathing, either inside or out, thus reaching an insulation value about the same as that of a 2 × 6 wall. Many use both 2 × 6 studs and insulating sheathing to achieve even better thermal performance. Some designers in very cold climates provide still more space for thermal insulation with exterior walls that are framed in two separate layers or with vertical truss studs made up of two ordinary studs joined at intervals by plywood plates. Some of these constructions are illustrated in Figures 15.3 and 15.4.

Stair Framing

Interior stairways are typically framed using 2 × 12 lumber that is notched for the rise and run of the stair. This structural part of the stair is called a *stringer* or *stair jack*, and there are usually three of them, one on each side of the stair and one in the middle (Figure 9.28). Side stringers adjacent to framed walls are normally spaced away from the wall with a 2 × 4 in order to allow easy application of finish wall materials without notching for the stairs.

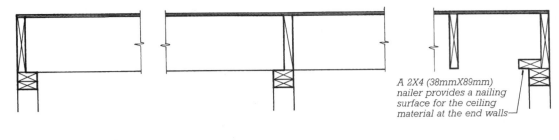

(a) SIDE WALL   (b) INTERIOR BEARING WALL   (c) END WALL

A 2X4 (38mmX89mm) nailer provides a nailing surface for the ceiling material at the end walls

FIGURE 9.26
Details of the second-floor platform, keyed to the letters on Figure 9.27. The extra piece of lumber on top of the top plate in detail *C*, End Wall, is continuous blocking whose function is to provide a nailing surface for the edge of the finish ceiling material, which is usually either gypsum board or veneer plaster base.

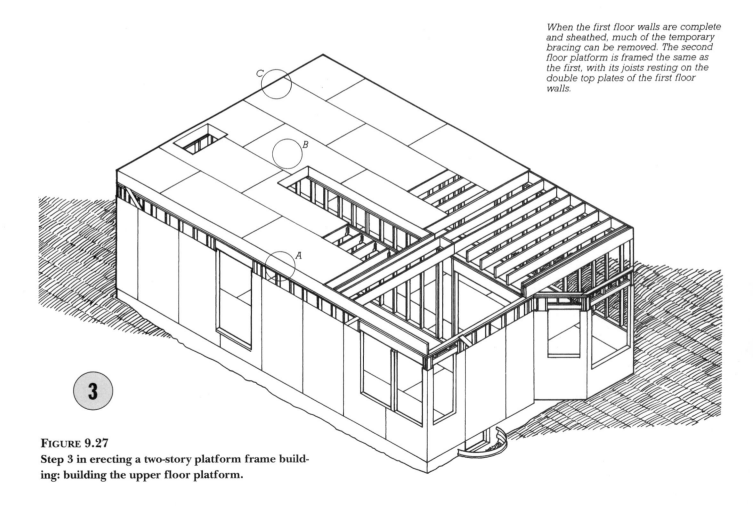

When the first floor walls are complete and sheathed, much of the temporary bracing can be removed. The second floor platform is framed the same as the first, with its joists resting on the double top plates of the first floor walls.

3

FIGURE 9.27
Step 3 in erecting a two-story platform frame building: building the upper floor platform.

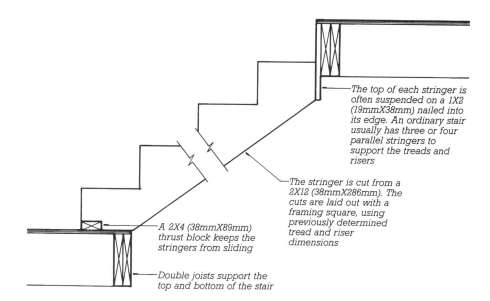

The top of each stringer is often suspended on a 1X2 (19mmX38mm) nailed into its edge. An ordinary stair usually has three or four parallel stringers to support the treads and risers

The stringer is cut from a 2X12 (38mmX286mm). The cuts are laid out with a framing square, using previously determined tread and riser dimensions

A 2X4 (38mmX89mm) thrust block keeps the stringers from sliding

Double joists support the top and bottom of the stair

FIGURE 9.28
Interior stairways are usually framed as soon as the upper floor platform is completed. This gives the carpenters easy up-and-down access during the remainder of the work. Temporary treads of joist scrap or plywood are nailed to the stringers. These will be replaced by finish treads after the wear and tear of construction are finished.

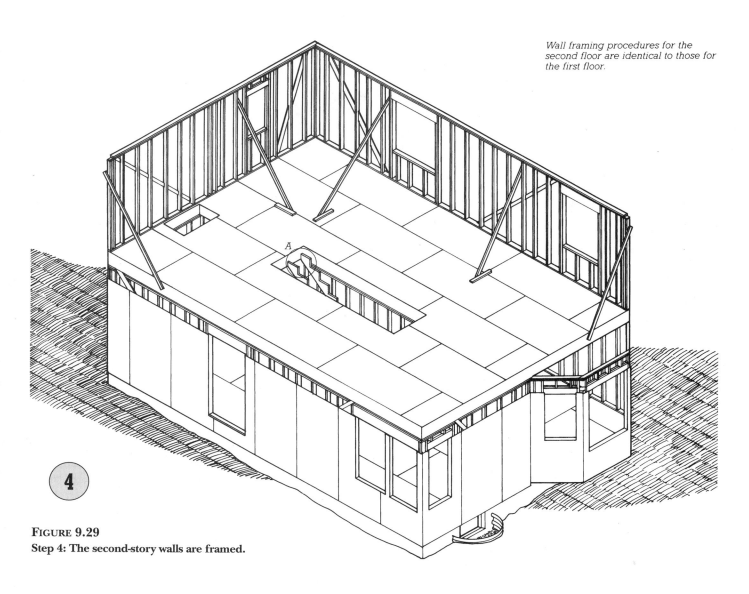

Wall framing procedures for the second floor are identical to those for the first floor.

4

FIGURE 9.29
Step 4: The second-story walls are framed.

FIGURE 9.30
Installing upper floor subflooring. The grade of the plywood panels used in this building is C-C Plugged, in which all surface voids are filled and the panel is lightly sanded to allow carpeting to be installed directly over the subfloor without additional underlayment. The long edges of the panels have interlocking tongue-and-groove joints to prevent excessive deflection of the edge of a panel under a heavy concentrated load such as a standing person or the leg of a piano. (*Courtesy of APA–The Engineered Wood Association*)

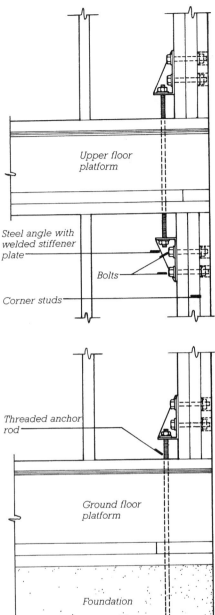

FIGURE 9.31
Platform frame buildings in areas subject to high winds or earthquakes must sometimes be reinforced against uplift at the corners. Hold-downs of the type shown here tie the entire height of the building securely to the foundation. To compensate for wood shrinkage, the nuts should be retightened after the first heating season, which can mean that access holes must be provided through the interior wall surfaces.

Roof Framing

The generic roof shapes for wood light frame buildings are shown in Figure 9.32. These are often combined to make roofs that are suited to covering more complex plan shapes and building volumes.

In a sloping roof, the parallel members are called *rafters,* and these connect at their base to the wall top plates and at the peak to a *ridge board* or *ridge beam.* In order to make a strong connection with the top plate, the rafters are notched with a *bird's mouth.* This allows the rafters to be toe nailed to the wall top plate where blocking is also required to prevent their rotation. At the ridge, the rafters are either toe nailed or end nailed, depending on the thickness of the ridge itself.

For structural stability, roofs with a ridge must be securely connected with nails at the base of their rafters to *ceiling joists* to make what is, in effect, a series of triangular trusses. In some cases, the designer may wish to eliminate the ceiling joists, in order to expose the sloping underside of the roof as the finished ceiling surface. In those cases, a beam or bearing wall must be inserted at the ridge, unless a system of exposed ties is designed to replace the ceiling joists (Figure 9.32).

The edge of the roof is where there is least conformity of residential framing details. Some roofs have large overhangs, while others have none at all. At the *eave,* the horizontal roof edge, there may be an enclosed *soffit* or exposed rafter tails. At the *rake,* the sloped portion of the roof edge, concealed *lookouts* may support the overhang, or it may be supported with exposed *brackets,* or there may be no rake overhang (Figure 9.39).

Once the roof framing is in place, it is sheathed with plywood or OSB (Figure 9.46). If the underside of this sheathing is to be visible at the eave or rake in the finished building, then a higher grade of sheathing that is rated for weather exposure or solid boards with a tongue-and-groove edge detail must be used.

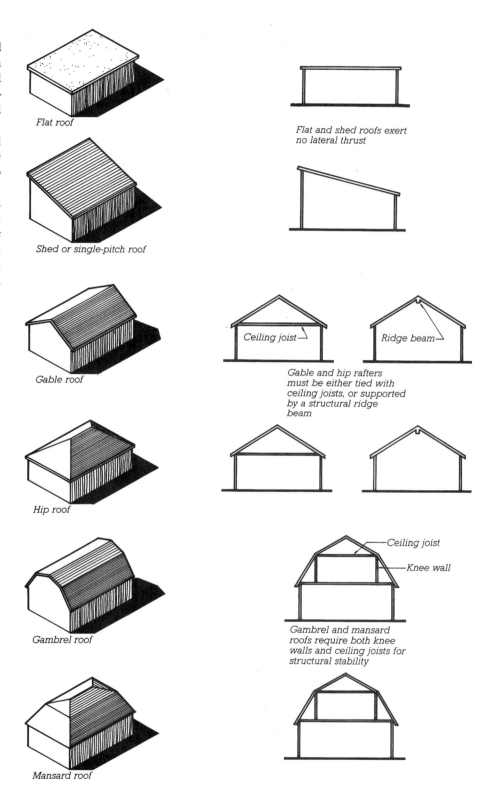

Flat roof

Shed or single-pitch roof

Gable roof

Hip roof

Gambrel roof

Mansard roof

Flat and shed roofs exert no lateral thrust

Ceiling joist

Ridge beam

Gable and hip rafters must be either tied with ceiling joists, or supported by a structural ridge beam

Ceiling joist

Knee wall

Gambrel and mansard roofs require both knee walls and ceiling joists for structural stability

FIGURE 9.32
Basic roof shapes for wood light frame buildings.

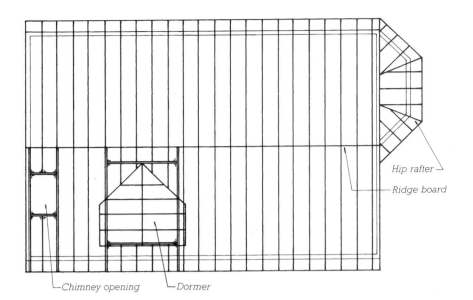

Hip rafter

Ridge board

Chimney opening

Dormer

FIGURE 9.33
A roof framing plan for the building
illustrated in Figure 9.34. The dormer
and chimney openings are framed with
doubled header and trimmer rafters.
The dormer is then built as a separate
structure that is nailed to the slope of
the main roof.

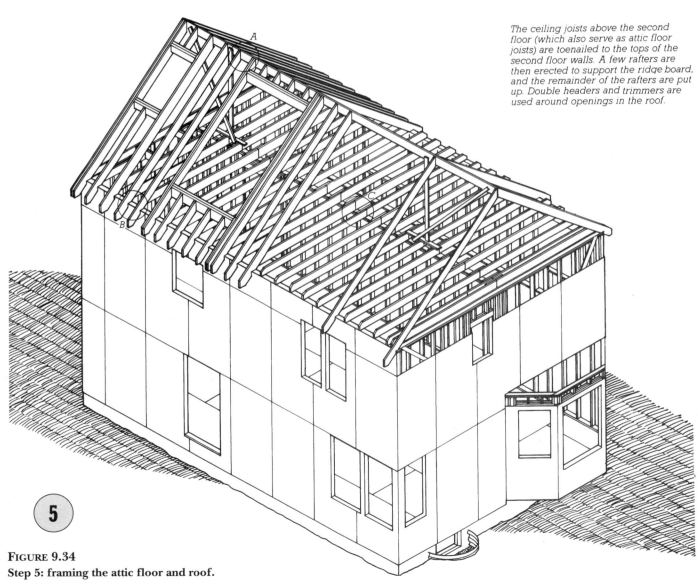

*The ceiling joists above the second
floor (which also serve as attic floor
joists) are toenailed to the tops of the
second floor walls. A few rafters are
then erected to support the ridge board,
and the remainder of the rafters are put
up. Double headers and trimmers are
used around openings in the roof.*

5

FIGURE 9.34
Step 5: framing the attic floor and roof.

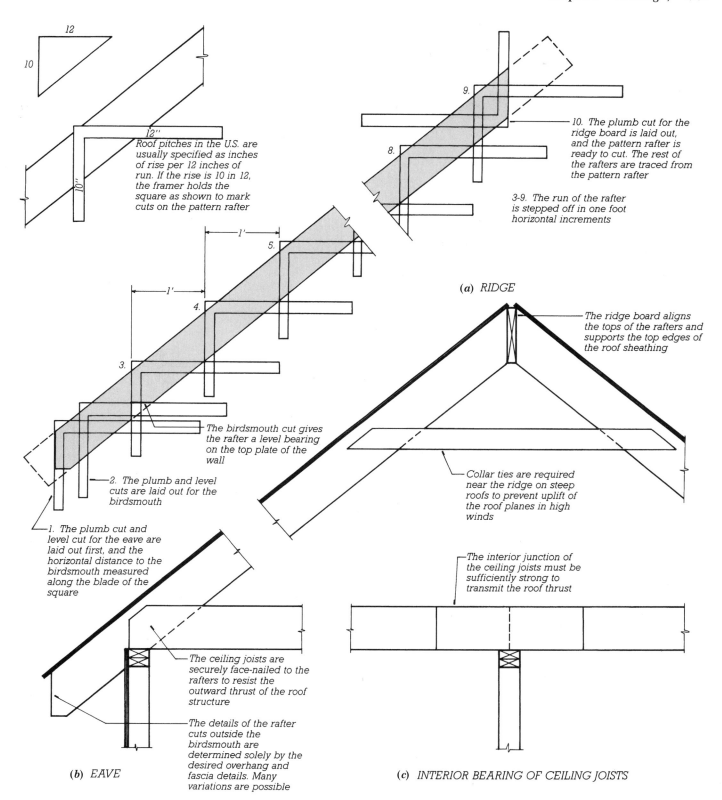

12

10

12"

10"

Roof pitches in the U.S. are usually specified as inches of rise per 12 inches of run. If the rise is 10 in 12, the framer holds the square as shown to mark cuts on the pattern rafter

9.

8.

10. The plumb cut for the ridge board is laid out, and the pattern rafter is ready to cut. The rest of the rafters are traced from the pattern rafter

3-9. The run of the rafter is stepped off in one foot horizontal increments

(*a*) *RIDGE*

1'

5.

1'

4.

3.

The birdsmouth cut gives the rafter a level bearing on the top plate of the wall

2. The plumb and level cuts are laid out for the birdsmouth

1. The plumb cut and level cut for the eave are laid out first, and the horizontal distance to the birdsmouth measured along the blade of the square

The ridge board aligns the tops of the rafters and supports the top edges of the roof sheathing

Collar ties are required near the ridge on steep roofs to prevent uplift of the roof planes in high winds

The interior junction of the ceiling joists must be sufficiently strong to transmit the roof thrust

The ceiling joists are securely face-nailed to the rafters to resist the outward thrust of the roof structure

The details of the rafter cuts outside the birdsmouth are determined solely by the desired overhang and fascia details. Many variations are possible

(*b*) *EAVE*

(*c*) *INTERIOR BEARING OF CEILING JOISTS*

FIGURE 9.35
Roof framing: The lettered details are keyed to Figure 9.34, facing page. The remainder of the page shows how a framing square is used to lay out a pattern rafter, reading from the first step at the lower end of the rafter to the last step at the top.

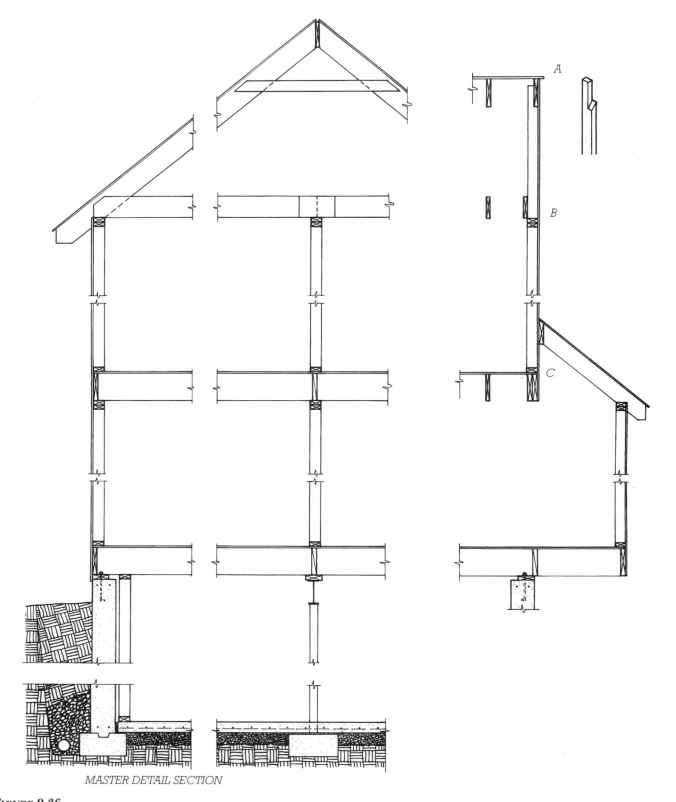

MASTER DETAIL SECTION

FIGURE 9.36

A summary of the major details for the structure shown in Figure 9.37, aligned in relationship to one another. The lettered details are keyed to Figure 9.37. The gable end studs are cut as shown in detail *A* and face nailed to the end rafter.

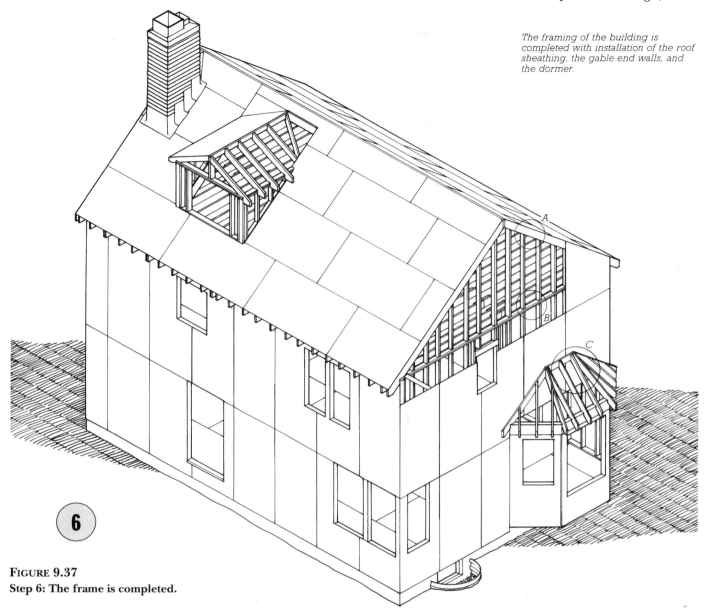

The framing of the building is completed with installation of the roof sheathing, the gable end walls, and the dormer.

⑥

FIGURE 9.37
Step 6: The frame is completed.

While a college-graduate architect or engineer would find it difficult to lay out the rafters for a sloping roof using trigonometry, a carpenter, without resorting to mathematics, has little problem making the layout if the *pitch* (slope) is specified as a ratio of *rise* to *run*. Rise is the vertical dimension and run is the horizontal. In the United States, pitch is usually given on the architect's drawings as inches of rise per foot (12 inches) of run. An old-time carpenter uses these two figures on the two edges of a framing square to lay out the rafter as shown in Figures 9.35 and 9.40. The actual length of the rafter is never figured, nor does it need to be, because the square is used to make all the measurements as horizontal and vertical distances. Today, many carpenters prefer to do rafter layout with the aid of tables that give actual rafter lengths for various pitches and horizontal distances; these tables are stamped on the framing square itself or printed in pocket-sized booklets.

Hips and *valleys* introduce another level of trigonometric complexity in rafter layout, but the experienced carpenter has little difficulty even here. Again, he or she can use published tables for hip and valley rafters or do the layout the traditional way, as illustrated in Figure 9.38. The head carpenter lays out only one rafter of each type by these procedures. This then becomes the *pattern rafter* from which other carpenters trace and cut the remainder of the rafters (Figure 9.41). Because they are so slender and have a tendency to buckle, trusses must be securely braced as soon as they are installed, following truss industry recommendations. Failure to install bracing has resulted in many disastrous collapses and even loss of lives.

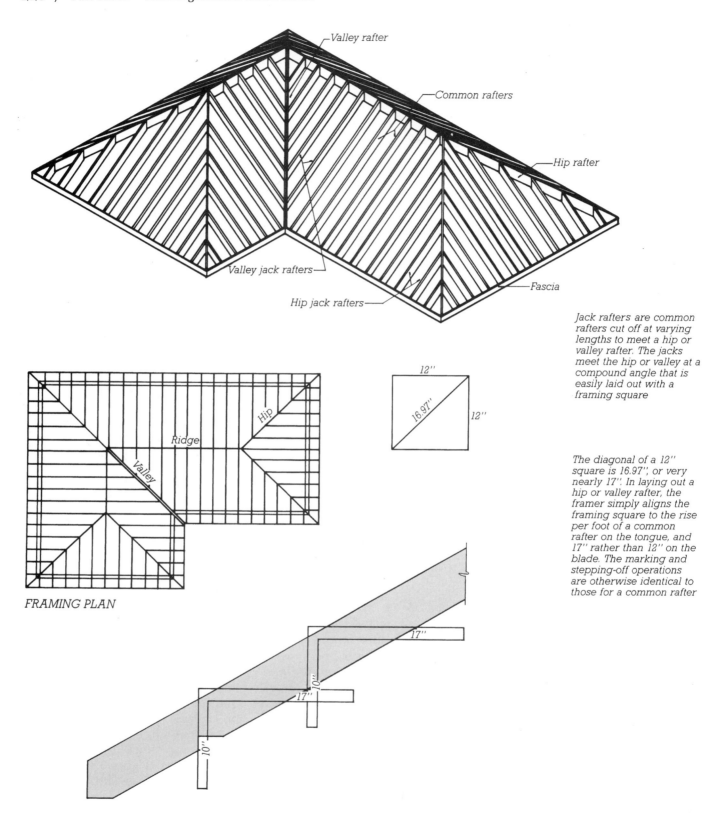

Valley rafter

Common rafters

Hip rafter

Valley jack rafters

Hip jack rafters

Fascia

Jack rafters are common rafters cut off at varying lengths to meet a hip or valley rafter. The jacks meet the hip or valley at a compound angle that is easily laid out with a framing square

12''

16.97''

12''

The diagonal of a 12'' square is 16.97'', or very nearly 17''. In laying out a hip or valley rafter, the framer simply aligns the framing square to the rise per foot of a common rafter on the tongue, and 17'' rather than 12'' on the blade. The marking and stepping-off operations are otherwise identical to those for a common rafter

FRAMING PLAN

Ridge

Hip

Valley

17''

17''

10''

10''

FIGURE 9.38
Framing for a hip roof. The difficult geometric problem of laying out the diagonal hip rafter is solved easily by using the framing square in the manner shown.

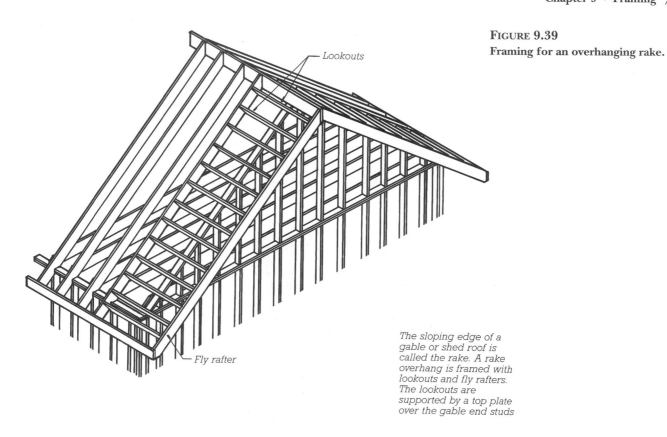

FIGURE 9.39
Framing for an overhanging rake.

Lookouts

Fly rafter

The sloping edge of a gable or shed roof is called the rake. A rake overhang is framed with lookouts and fly rafters. The lookouts are supported by a top plate over the gable end studs

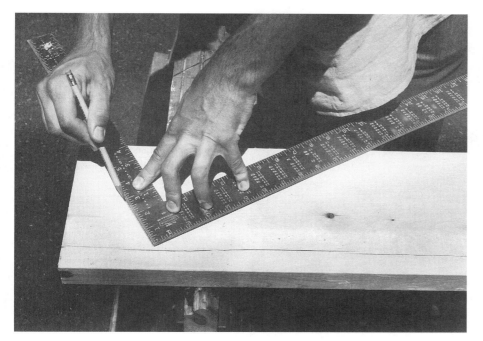

FIGURE 9.40
A framing square being used to mark rafter cuts. The run of the roof, 12 inches, is aligned with the edge of the rafter on the blade of a square, and the rise, 7 inches in this case, is aligned on the tongue of the square. A pencil line along the tongue will be perfectly vertical when the rafter is installed in the roof, and one along the blade will be horizontal. True horizontal and vertical distances can be measured on the blade and tongue, respectively. (*Photo by Edward Allen*)

FIGURE 9.41
Tracing a pattern rafter to mark cuts for the rest of the rafters. The corner of the building behind the carpenters has let-in corner braces on both floors, and most of the rafters are already installed. (*Courtesy of Southern Forest Products Association*)

FIGURE 9.42
I-joists may be used as rafter material instead of solid lumber. (*Courtesy of Trus Joist MacMillan*)

Prefabricated Framing Assemblies

Prefabricated *roof trusses* (and, to a lesser extent, *floor trusses*) have found widespread use in platform frame buildings because of their speed of erection, economy of material usage, and long spans. Most are light enough to be lifted and installed by two carpenters (Figures 9.43 and 9.44).

Manufactured wall panels have been adopted more slowly than roof and floor trusses, except by large builders who mass-market hundreds or thousands of houses per year. For the smaller builder, wall framing can be done on site with the same amount of material as manufactured panels and with little or no additional overall expenditure of labor, especially when the building requires walls of varying heights and shapes.

FIGURE 9.43
Roof framing with prefabricated trusses. Sheet metal clips and nails anchor the trusses to the top plate. These particular trusses have a raised heel, which allows more depth of insulation above the exterior wall than does a typical truss. (*Photo by Rob Thallon*)

FIGURE 9.44
Prefabricated trusses are delivered to the building site in bundles. (*Courtesy of Gang-Nail Systems, Inc.*)

FIGURE 9.45
Trusses are typically lifted to the roof by a boom mounted on the delivery truck. This is one of a series of attic trusses, which will frame a habitable space under the roof. (*Photo by Rob Thallon*)

FIGURE 9.46
Applying plywood roof sheathing to a half-hipped roof. Blocking between rafters at the wall line has been drilled with large holes for attic ventilation. The joint in the sheathing above the walls occurs because the sheathing outside of the walls will be exposed from the underside and must be rated for exposure to the weather while the rest of the sheathing does not. The line of horizontal blocking between studs is to support the edges of plywood siding panels applied in a horizontal orientation. (*Courtesy of APA–The Engineered Wood Association*)

FIGURE 9.47
A house frame sheathed with OSB panels. (*Courtesy of APA–The Engineered Wood Association*)

FIGURE 9.48
This proprietary fire wall consists of light-gauge metal framing, noncombustible insulation, and gypsum board. The metal framing is attached to the wood structure on either side with special clips (not shown here) that break off if the wood structure burns through and collapses, leaving the wall supported by the undamaged clips on the adjacent building. (*Courtesy of Gold Bond Building Products, Charlotte, North Carolina*)

FIGURE 9.49
This townhouse has collapsed completely as the result of an intense fire, but the houses on either side, protected by fire walls of the type illustrated in Figure 9.48, are essentially undamaged. (*Courtesy of Gold Bond Building Products, Charlotte, North Carolina*)

WOOD FRAMING AND THE BUILDING CODES

Wood platform framing is probably regulated more closely by codes than any other construction type. In general, code sections that relate to framing address the issues of durability of the building and the safety of its occupants.

To prevent decay of the frame, minimum clearances from the ground are specified and preservative-treated or naturally resistant wood is required at vulnerable locations. Protection of the frame from the weather is also required by means of approved roofing materials and siding in conjunction with a weather barrier. In regions that are prone to termite damage, chemical or physical barriers to prevent these pests from gaining access to the framing are mandatory.

There are innumerable provisions in the code to assure the strength of the building under normal use. Tables specify allowable spans for girders, joists, and rafters; the stabilization of these members with blocking, bracing, and sheathing is also prescribed. There are explicit specifications for fasteners, rules about the size and location of notches and holes for utilities, and detailed drawings that show exactly how certain parts of the structure must be assembled (Figure 9.50).

Many sections of the code are written to minimize damage to the structure due to natural catastrophe. In the United States, the country is divided into seismic zones that are ranked according to the probable intensity of earthquakes. In the most severe zones, the building frame must be braced and anchored to the foundation with measures that far exceed the bracing and anchorage measures required for wind (Figure 9.51). The code also identifies regions of high wind risk in which provisions such as increased anchorage between the roof and the walls of a house are incorporated into regulations in order to decrease property loss and personal injury. These provisions have been significantly extended in recent years due to the severe damage inflicted in Florida by Hurricane Andrew in 1992.

The codes devote considerable attention to minimizing the loss of life and property due to the ravages of fire. With regard to framing, code provisions attempt to stop the rapid spread of fire in a structure once it has started. Thus, fire blocking is required to stop the passage of fire through any framed cavities that pass from one level to another in a structure. Blocking is required, for example, at the stud spaces around a stair landing to retard the passage of flames from the space below the landing. Other sections of the code strive to prevent fire from igniting the framing in the first place. Framing must be covered by drywall or plaster in especially vulnerable locations and must be separated by minimum clearances from flues and fireboxes.

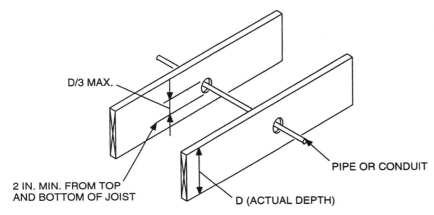

FIGURE 9.50
One of numerous explicit guidelines for framing residential buildings. (*Reprinted with permission from International Code Council, Inc.*)

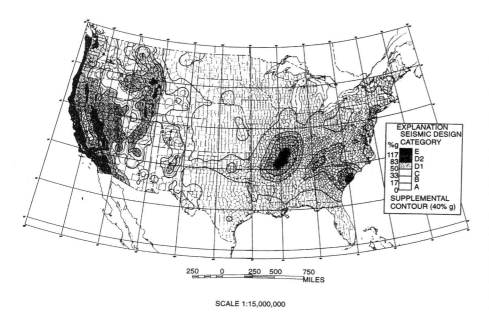

SCALE 1:15,000,000

FIGURE 9.51
Seismic design categories in the contiguous continental United States. (*Reprinted with permission from the International Residential Code for One- and Two-Family Dwellings, 2000*)

THE UNIQUENESS OF WOOD LIGHT FRAME CONSTRUCTION

Wood light framing is popular because it is an extremely flexible and economical way of constructing small buildings. Its flexibility stems from the ease with which carpenters with ordinary tools can create buildings of astonishing complexity in a variety of geometries. Its economy can be attributed in part to the relatively unprocessed nature of the materials from which it is made, and in part to mass-market competition among the suppliers of components and materials as well as local competition among small builders.

Platform framing is the one truly complete and open system of construction that we have. It incorporates structure, enclosure, thermal insulation, mechanical installations, and finishes into a single constructional concept. Thousands of products are made to fit it: dozens of competing brands of windows and doors; a multitude of interior and exterior finish materials; and scores of electrical, plumbing, and heating products. For better or worse, platform framing can be dressed up to look like a building of wood or of masonry in any architectural style from any era of history. Architects have failed to exhaust its formal possibilities and engineers have failed to invent a new environmental control system that it cannot assimilate.

At the other end of the scale, platform frame construction has led to the creation of countless horrors around the perimeters of our cities. To create the minimum-cost, detached single-family house, one can begin by reducing the number of exterior wall corners to four, because extra corners cost extra dollars for framing and finishing. Then the roof slope can be made as shallow as asphalt shingles will permit, to minimize roof area and to allow roofers to work without scaffolding. Roof overhangs can also be eliminated to save material and labor. Windows can be made as small as building codes will allow, because a square foot of window costs more than a square foot of wall. The thinnest, cheapest finish materials can be used inside and out, even if they will begin to look shabby after a short period of time, and all ceilings can be made flat and as low as is legally permitted, to save materials and labor. The result is an uninteresting, squat, dark residence that weathers poorly; has little meaningful relation to its site; and contributes to a lifeless, uninspiring neighborhood. The same principles, with roughly the same result, can be applied to the construction of apartments. In most cases, things only get worse if one tries to dress up this minimal box building with inexpensive shutters, gingerbread, window boxes, and cheap imitations of fancy materials. The single most devastating fault of Wood light frame construction is not its combustibility or its tendency to decay, but its ability to be reduced to the irreducibly minimal — scarcely human — building system. Yet one can look to the best examples of the Carpenter Gothic, Queen Anne, Shingle Style, and Craftsman Style buildings of the late 19th and early 20th centuries, or the Bay Region and Modern styles of our own time, to realize that wood light framing gives to the designer the freedom to make a finely crafted building that nurtures life and elevates the spirit.

C.S.I./C.S.C.
MasterFormat Section Numbers for Framing

06100	**ROUGH CARPENTRY**
06105	**Miscellaneous Carpentry**
06120	**Structural Panel**
06150	**Wood Decking**
06160	**Sheathing**
06170	**Prefabricated Structural Wood**

SELECTED REFERENCES

1. Thallon, Rob. *Graphic Guide to Frame Construction.* Newtown, CT: Taunton Press, 1991.

A complete visual handbook of wood frame construction for architects, builders, students and owner/builders. "An invaluable reference for experienced designers and builders; an essential aid for beginners," according to the NAHB.

2. International Code Council, Inc. *2000 International Residential Code for One- and Two-Family Dwellings.*

The definitive legal guide for platform frame residential construction. Updated regularly.

3. American Forest and Paper Association. *Span Tables for Joists and Rafters.* Washington, DC: American Forest and Paper Association.

This is the standard reference for the structural design of platform frame apartments and single-family residences. Updated frequently.

KEY TERMS AND CONCEPTS

balloon frame	header	window header	rake
platform frame	trimmer	door header	lookout
sheathing	joist hanger	rough sill	bracket
rough carpentry	I-joist	stringer	pitch
sill	floor truss	stair jack	rise
mudsill	stud	rafter	run
floor joist	sole plate	ridge board	hip
rim joist	top plate	ridge beam	valley
band joist	diagonal bracing	bird's mouth	pattern rafter
subfloor	shear wall	ceiling joist	roof truss
girder	bearing wall	eave	fire blocking
bridging	partition wall	soffit	

REVIEW QUESTIONS

1. Draw a series of very simple section drawings to illustrate the procedure for erecting a platform frame building, starting with the foundation and continuing with the ground floor, the ground-floor walls, the second floor, the second-floor walls, and the roof. Do not show details of connections but simply represent each plane of framing as a heavy line in your section drawing.

2. Draw from memory the standard detail sections for a two-story platform frame dwelling. (*Hint:* The easiest way to draw a detail section is to draw the pieces in the order in which they are put in place during construction. If your simple drawings from the preceding question are correct, and if you follow this procedure, you will not find this question so difficult.)

3. What are the differences between balloon framing and platform framing? What are the advantages and disadvantages of each? Why has platform framing become the method of choice?

4. Why is firestopping not usually required in platform framing?

5. Why is a steel beam or glue-laminated wood beam preferred to a solid wood beam at the foundation level?

6. How is a platform frame building braced against wind and earthquake forces?

7. Light framing of wood is highly combustible. In what different ways does a typical building code take this fact into account?

EXERCISES

1. Visit a building site where a wood platform frame is being constructed. Compare the details that you see on the site with the ones shown in this chapter. Ask the carpenters why they do things the way they do. Where their details differ from the ones illustrated, make up your own mind about which is better, and why.

2. Develop floor framing and roof framing plans for a building you are designing. Estimate the approximate sizes of the joists and rafters using the rules of thumb on page 209.

3. Make thumbnail sketches of 20 or more different ways of covering an L-shaped building with combinations of sloping roofs. Start with the simple ones (a single shed, two intersecting sheds, two intersecting gables) and work into the more elaborate ones. Note how the varying roof heights of some schemes could provide room for a partial second-story loft or for high spaces with clerestory windows. How many ways do you think there are of covering an L-shaped building with sloping roofs?

4. Look around you as you travel through areas with wood frame buildings, especially older areas, and see how many ways designers and framers have roofed simple buildings in the past. Build up a collection of sketches of ingenious combinations of sloping roof forms.

5. Build a scale model of a platform frame from basswood or pine, reproducing accurately all its details, as a means of becoming thoroughly familiar with them. Better yet, build a small frame building for someone at full scale, perhaps a toolshed, playhouse, or garage.

10

FINISHING THE ROOF

A building's roof is its first line of defense against the weather. The roof protects the interior of the building from rain, snow, and sun. The roof helps to insulate the building from extremes of heat and cold and to control the accompanying problems with condensation of water vapor. Like any front-line defender, it must itself take the brunt of the attack. A roof is subject to the most intense solar radiation of any part of a building. At midday, the sun broils a roof with radiated heat and ultraviolet light. On clear nights, a roof radiates heat to the blackness of space and becomes colder than the surrounding air. From noon to midnight of the same day, it is possible for the surface temperature of a roof to vary from near boiling to below freezing. Rain falling on a roof causes abrasion on impact and then washes over the surface, causing further erosion. In cold climates, snow and ice cover a roof after winter storms, and cycles of freezing and thawing gnaw at the materials of the roof. A roof is vital to the sheltering function of a building, yet it is singularly vulnerable to the destructive forces of nature.

The *roofing* is generally applied to a house as soon as its framing is complete so as to protect the framing from the weather. A dry working environment is also required for plumbing, heating, and electrical subcontractors, who will be scheduled to begin work inside the structure as soon as the roofing is in place. The roofing is generally installed by a roofing subcontractor.

PREPARATION FOR ROOFING

Historically, roofs were surfaced with a single roofing material. Thatched cottages were roofed with thatch alone, and barns might be covered only with hand-split shakes. Today, roofs often include such complexities as dormers, skylights, and chimney penetrations, which require underlayment and flashing as well as the primary roofing material.

Underlayment

Before any roofing material is installed, the roof sheathing is generally covered with a layer (or two) of 15-lb or 30-lb building felt. Called *underlayment,* this layer serves to protect the building from precipitation before the roofing is applied. It also provides a permanent second layer of defense to back up the roofing. This flexible layer can be installed rather quickly, and when it is in place, the building is said to be "dried in" (Figure 10.1). In very cold climates, a special underlayment called an *ice-and-water shield* is required at the eaves in order to pre-

FIGURE 10.1
The roof framing has been completed and underlayment applied to the roof of this residence even before the walls are sheathed. The temporary boards nailed to the steeper part of the roof allow workers to walk safely on the inclined surface. (*Photo by Rob Thallon*)

240

FIGURE 10.2
Code requires an ice-and-water shield in very cold climates. The shield must extend 24 inches (610 mm) over the insulated portion of the building.

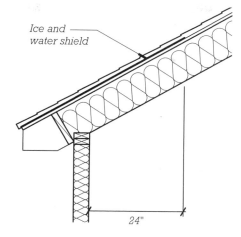

Ice and water shield

24"

FIGURE 10.3
Typical flashing locations: (*a*) rake, (*b*) valley, (*c*) chimney and other roof penetrations, (*d*) sloped edge of roof at wall, (*e*) level edge of roof at wall, (*f*) eave. Code roof flashing requirements vary according to the type of roofing.

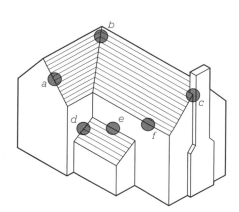

vent damage to the structure as a result of ice dams (Figures 10.2 and 10.13). Ice dams are discussed in Chapter 15.

Flashing

Roof flashing comprises metal sheet materials that are used to protect against water leakage at the junctions and edges of roof surfaces. The earliest flashings were made of lead or copper. Today, galvanized sheet steel and aluminum are the principal flashing materials, although enamel-coated sheet steel, copper, and stainless steel may also be used (Figure 10.3).

Some roof flashing materials, such as edge flashing, are readily available in standardized forms that work in a variety of situations. However, many roof flashings must be custom made to fit a particular roof slope, a particular roofing material, or a special roof form. Standard flashings are made in large factories and purchased in bulk by roofing contractors, while custom flashings are fabricated in local sheet metal shops (Figure 10.4). The roofing subcontractor usually confers with the sheet metal subcontractor to determine the size and configuration of custom flashing, while the designer specifies the material of which it is made and may also recommend a specific shape.

FIGURE 10.4
Sheet metal shops such as this stay busy manufacturing custom flashings and gutters for both residential and commercial projects. Most of the work takes place on waist-high tables where sheet metal is measured, cut, formed, and soldered.
(*Photo by Rob Thallon*)

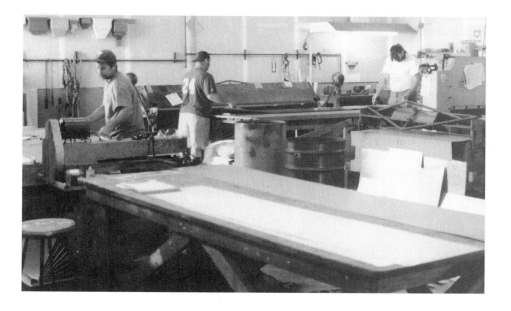

FIGURE 10.5
A steep roof can be made waterproof with any of a variety of materials. This thatched roof is being constructed by fastening bundles of reeds to the roof structure in overlapping layers, in such a way that only the butts of the reeds are left exposed to the weather. (*Courtesy of Warwick Cottage Enterprises*)

ROOF SLOPE

Roofs are classified according to their steepness or *slope*. The slope of a roof is the ratio of the (vertical) rise to the (horizontal) run of the roof (Figure 9.34). In the United States, the ratio is given in inches of rise per 12-inch run. A 6:12 roof, for example, rises 6 inches for every 12 inches of horizontal distance. In Canada, slope is expressed as a ratio of 1:x so that a Canadian roof that slopes 1:2 is the equivalent of a 6:12 roof in the United States.

Roofing materials can be organized conveniently into two groups: those that work on *steep roofs* and those that work on *low-slope roofs*, roofs that are nearly flat. The distinction is important: A steep roof drains itself quickly of water, giving wind and gravity little opportunity to push or pull water through the roofing material. Therefore, steep roofs can be covered with roofing materials that are fabricated and applied in small, overlapping units—shingles of wood, slate, or artificial composition; tiles of fired clay or concrete; or even tightly wrapped bundles of reeds, leaves, or grasses (Figure 10.5). There are several advantages to these materials: Many of them are inexpensive. The small, individual units are easy to handle and install. Repair of localized damage to the roof is easy. The effects of thermal expansion and contraction, and of movements in the structure that supports the roof, are minimized by the ability of the small roofing units to move with respect to one another. Water vapor vents itself easily from the interior of the building through the loose joints in the roofing material. Finally, a steep roof of well-chosen materials skillfully installed can be a delight to the eye.

Low-slope roofs have none of these advantages. Water drains relatively slowly from the surfaces, and small errors in design or construction

Material	Sheathing	Lowest Allowed Slope	Requirements Below 4:12 Slope
Asphalt shingles	Solid	2:12	Double underlayment
Wood shingles	Solid or spaced	3:12	
Wood shakes	Solid or spaced	3:12	
Clay tiles	Solid or spaced	2½:12	Double underlayment
Concrete tiles	Solid or spaced	2½:12	Double underlayment
Slate	Solid	4:12	
Metal panel	Solid or spaced	3:12	
Roll roofing	Solid	1:12	
Membrane	Solid	¼:12	

FIGURE 10.6
Common residential roofing materials with required slopes, sheathing, and underlayment. The table is derived from Chapter 9 of the 2000 International Residential Code.

can cause them to trap puddles of standing water. Slight structural movements can tear the membrane that keeps the water out of the building. Water vapor pressure from within the building can blister and rupture the membrane. Low-slope roofs also have advantages: A low-slope roof can cover a building of any horizontal dimension, whereas a steep roof becomes uneconomically tall when used on a very broad building. And, when appropriately detailed, low-slope roofs can serve as balconies, decks, patios, and even landscaped parks.

Steep Roofs

A roof with a pitch of 4:12 or greater is referred to as a *steep roof* by the National Roofing Contractors Association. There are numerous roof coverings in this category, and many can be used on roofs with a slope lower than 4:12 by increasing the number of layers of underlayment (Figure 10.6). As the roof slope increases, the performance of roof coverings generally improves, but at slopes steeper than about 7:12, workers cannot move easily without slipping, and roofing contractors increase their prices to allow for the installation of *roof jacks* or other safety devices (Figure 10.7).

Sheathing for Steep Roofs

Most steep roofs are constructed with *solid sheathing* made of OSB or plywood panels similar to wall and subfloor sheathing. The sheathing supports the underlayment over which the roofing material is laid. *Sheathing clips* may be used to maintain the alignment of panel edges between rafters, permitting thinner sheathing to be used. At eaves or rakes that are exposed from the underside, panel sheathing must be rated for exposure to the weather or tongue-and-groove boards may be used instead (Figure 10.8).

Figure 10.7
This very steep roof is being reroofed with cedar shingles. Steel roof jacks nailed to the structure hold horizontal boards that allow workers access to the roof surface. When the job is complete, the jacks are removed, but the nails that hold them remain in the structure. Roof jacks are generally required on roofs with 8 in 12 or steeper pitch. (*Photo by Rob Thallon*)

Figure 10.8
This roof is being sheathed with solid sheathing because it will be roofed with asphalt shingles, the most common roofing material in North America. The eaves, because they will be exposed from below, must be sheathed with a weather-resistant material — in this case, tongue-and-groove boards. (*Photo by Rob Thallon*)

FIGURE 10.9
The roof on this new residence is being sheathed with spaced boards because it will be roofed with cedar shingles, which have the strength to span between sheathing boards and last longer when allowed to breathe from beneath. (*Photo by Donald Corner*)

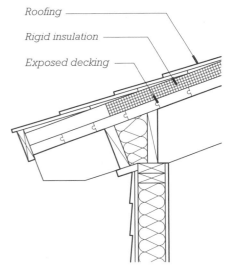

FIGURE 10.10
Section of a roof with sheathing exposed to an interior room. Ventilation of this roof is not required because the rigid insulation does not allow air to be trapped in the assembly. Some roofing manufacturers will not warranty their products on such a roof because roof surface temperatures are not moderated by ventilation.

Many steep roof coverings have structural integrity of their own and thus can be supported intermittently on *spaced sheathing* (also called open or skip sheathing) made of 1 × 4 or 1 × 6 boards nailed across the rafters with open spaces between (Figure 10.9). Materials such as wood shingles endure longer when they are installed over spaced sheathing, which allows them to dry from the underside between storms.

where the underside of the sheathing is to be left exposed as a finish ceiling, a vapor retarder and rigid insulation panels should be applied above the sheathing, just below the roofing (Figure 10.10). A layer of plywood or OSB is then nailed over the insulation panels as a nail base for fastening the shingles or sheet metal. There are also special composite insulation panels that include an integral nail base layer on top.

Insulation and Vapor Retarder for Steep Roofs

The *thermal insulation* and *vapor retarder* in most steep roofs are installed below the roof sheathing. Typical details of this practice are shown in Chapter 15. In places

Steep Roof Materials

Common roof coverings for steep roofs fall into four general categories: shingles, tiles, architectural sheet metal, and roll roofing. *Shingles* are small, lightweight units applied in overlapping layers with staggered vertical shingle, tile, and sheet metal roofs are available and range in price from economical to expensive.

Each type of material must be laid on a roof deck that slopes sufficiently to assure leakproof performance. The manufacturer specifies minimum slopes for each material. Slopes greater than the minimum should be used in locations where water is likely to be driven up the roof surface by strong storm winds.

Asphalt Shingles

Asphalt shingles are the most widely used roof covering in North America and are employed on approximately 90 percent of single-family houses. They are inexpensive to buy, quick and easy to install, moderately fire resistant, and have an expected lifetime of 15 to 25 years, depending on their exact composition. Asphalt shingles are die cut from heavy sheets of asphalt-impregnated felt. Most felts contain glass fibers for strength and stability, but some have an older composition that consists primarily of cellulose. The sheets are faced with mineral granules that act as a wearing layer and decorative finish. The most common type of asphalt shingle is 12 inches by 36 inches (305 mm by 914 mm) in size. (A metric shingle 337 mm × 1000 mm is also widely marketed.) Each shingle is slotted twice in the center and has a half-slot at each end to produce a roof that looks as though it were made of smaller shingles (Figure 10.11). Many other shingle styles are also available, including hexagonal patterns and thicker shingles that are laminated from several layers of material.

Asphalt shingles are flexible, which allows them to be used in unique ways. For example, asphalt shingles can bend across a valley or a ridge—eliminating the need for flashing at these locations. Because asphalt shingles have no stiffness, they must be applied over solid sheathing that has been covered with underlayment (Figures 10.12 to 10.14).

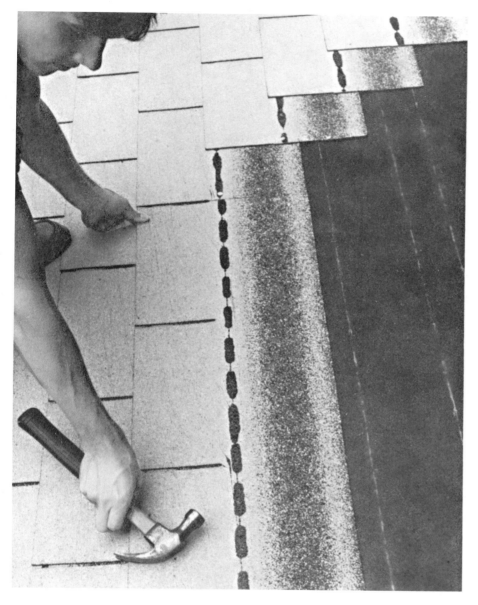

FIGURE 10.11
Installing asphalt shingles. To give a finer visual scale to the roof, the slots make each shingle appear as if it were three smaller shingles when the roof is finished. Many different patterns of asphalt shingles are available, including ones that do not have slots. The dark strips running the length of the shingles are adhesive to seal the lower edges of the shingles against uplift from wind. (*Photo by Edward Allen*)

joints. The overlap between courses is such that the entire surface of the roof is covered by a minimum of two layers of roofing. *Roof tiles* are small but heavy units of concrete or fired clay. They overlap and interlock so that the roof surface is covered by only one layer. *Sheet metal* roofing is made of very lightweight metal panels that overlap slightly or are seamed, so that the roof is covered by one layer of roofing. *Roll roofing* is a flexible asphalt-impregnated felt that comes in rolls 36 inches (915 mm) wide that may be applied to provide single or double coverage. A wide variety of

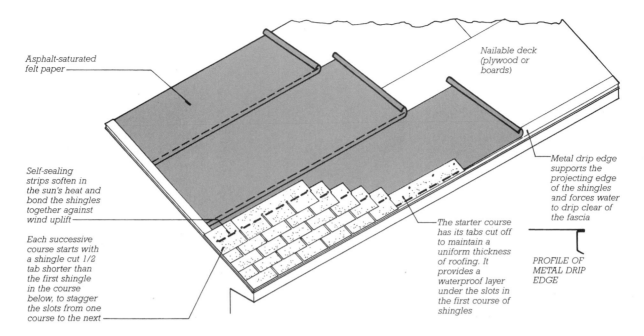

Asphalt-saturated felt paper

Nailable deck (plywood or boards)

Self-sealing strips soften in the sun's heat and bond the shingles together against wind uplift

Each successive course starts with a shingle cut 1/2 tab shorter than the first shingle in the course below, to stagger the slots from one course to the next

The starter course has its tabs cut off to maintain a uniform thickness of roofing. It provides a waterproof layer under the slots in the first course of shingles

Metal drip edge supports the projecting edge of the shingles and forces water to drip clear of the fascia

PROFILE OF METAL DRIP EDGE

FIGURE 10.12
Starting an asphalt shingle roof. Most building codes require the installation of an ice-and-water barrier beneath the shingles along the eave in regions with cold winters; its function is to prevent the entry of standing water that might be created by ice dams. The most effective form of barrier is a 3-foot-wide (900-mm) strip of modified bitumen sheet that replaces the lowest course of asphalt-saturated felt paper. The bitumen self-seals around the shanks of the roofing nails as they are driven through it.

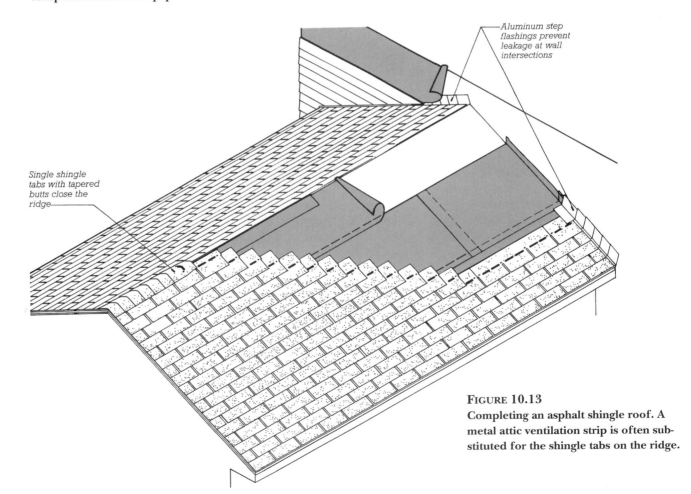

Aluminum step flashings prevent leakage at wall intersections

Single shingle tabs with tapered butts close the ridge

FIGURE 10.13
Completing an asphalt shingle roof. A metal attic ventilation strip is often substituted for the shingle tabs on the ridge.

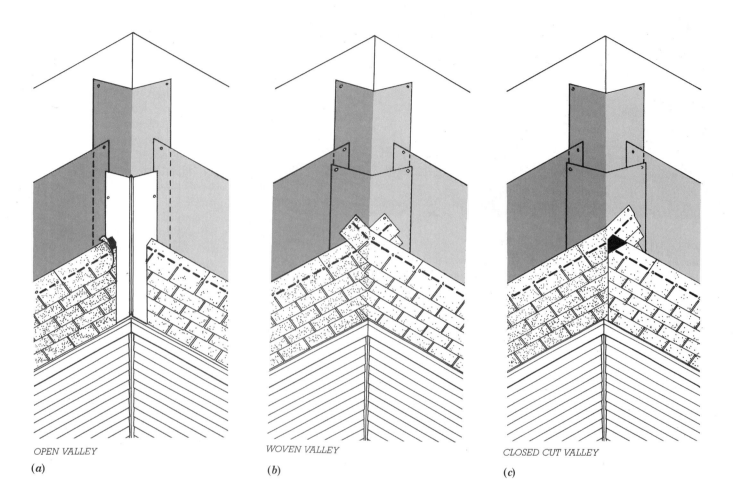

OPEN VALLEY

(*a*)

WOVEN VALLEY

(*b*)

CLOSED CUT VALLEY

(*c*)

FIGURE 10.14
Three alternative methods of making a valley in an asphalt shingle roof. (*a*) The open valley uses a sheet metal flashing; the ridge in the middle of the flashing helps prevent water that is coming off one slope from washing up under the shingles on the opposite slope. The woven valley (*b*) and cut valley (*c*) are favorites of roofing contractors because they require no sheet metal. The solid black areas on shingles in the open and closed cut valleys indicate areas to which asphaltic roofing cement is applied to adhere shingles to each other.

FIGURE 10.15
Shingles and shakes, although made from the same material, are very different in appearance and durability. Shingles, seen here on the walls, are sawn on both sides so that both faces are relatively smooth, and all shingles taper uniformly from bottom to top. Shakes, seen here on the roof, are split from a short section of log called a bolt and are typically sawn diagonally after splitting to create the taper. The sawn face of a shake is its underside. The split upper side of a shake appears rough and irregular as compared to the sawn face of a shingle. (*Photo by Edward Allen*)

Wood Shingles and Shakes

Wood shingles are thin, tapered slabs of wood sawn from short pieces of tree trunk, with the grain of the wood running approximately parallel to the length of the shingle. *Shakes* are split from the wood, rather than sawn, and are thicker, with a much more irregular face texture than wood shingles (Figure 10.15). Most wood shingles and shakes in North America are made from red cedar, white cedar, or redwood, because of the natural decay resistance of these woods (Figure 10.16). Wood shingles and shakes are moderately expensive to purchase and install. They are not resistant to fire unless they have been pressure treated with fire-retardant chemicals.

Both wood shingle and shake roofs can be applied over spaced sheathing (Figure 10.17). Wood shakes naturally breathe better than shingles because their uneven texture tends to allow more air circulation between

FIGURE 10.16
A house is both roofed and sided with red cedar shingles to feature its sculptural qualities. Note that the shadow lines created by the shingles are horizontal. (*Architect: William Isley. Photo by Paul Harper. Courtesy of Red Cedar Shingle and Handsplit Share Bureau*)

FIGURE 10.17
Installing red cedar shakes on spaced sheathing. Small corrosion-resistant nails are driven near each edge at the mid-height of the shingle. Each succeeding course covers the joints and nails in the course below. The spaces between the sheathing boards allow the undersides of the shingles to dry out between storms. (*Photo by Rob Thallon*)

FIGURE 10.18
Shake application over a new roof deck using air-driven, heavy-duty staplers for greater speed. The strips of asphalt-saturated felt have all been placed in advance with their lower edges unfastened. Each course of shakes is laid out, slipped up under its felt strip, then quickly fastened by roofers walking across the roof and inserting staples as fast as they can pull the trigger. (*Courtesy of Senco Products, Inc.*)

layers. However, the looseness of fit also increases the opportunity for wind-driven rain and snow to penetrate the roofing. To counteract this shortcoming, most codes require that shakes be applied with an 18-inch-wide strip of 30-lb. asphalt-impregnated *interlayment* woven between each pair of courses (Figure 10.18).

Tile Roofs

Tile roofs in general are heavy, durable, highly resistant to fire, and have a relatively high first cost. Their weight (up to 10 psf, which is five times the weight of asphalt shingles) can require a stronger roof structure.

Clay tiles have been used on roofs for thousands of years. It has been said that the tapered barrel tiles traditional to the Mediterranean region (similar to the mission tiles in Figure 10.19) were originally formed on the thighs of the tilemakers. Many other patterns of clay tiles are now available, both glazed and unglazed. Clay tiles are most commonly used in the southwestern United States, where

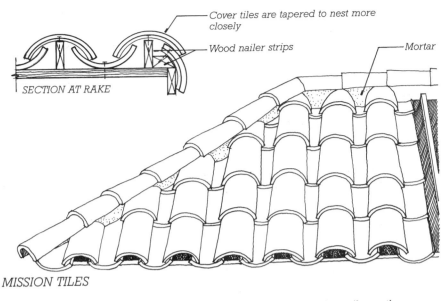

Cover tiles are tapered to nest more closely

Wood nailer strips

Mortar

SECTION AT RAKE

MISSION TILES

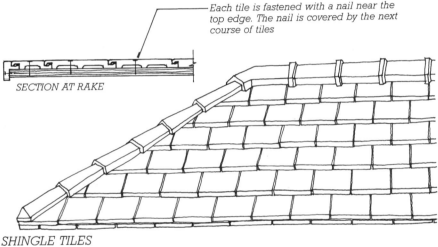

Each tile is fastened with a nail near the top edge. The nail is covered by the next course of tiles

SECTION AT RAKE

SHINGLE TILES

FIGURE 10.19
Two styles of clay tile roofs. The mission tile has very ancient origins.

they are related historically and climatically to a Spanish heritage.

Concrete tiles are generally less expensive than clay tiles and are available in many similar patterns. Made of high-density concrete, these tiles are also more consistent in shape and dimension than clay tiles because they are not subject to the distortions that are often caused by high-temperature firing. The relative precision of their manufacture allows concrete tiles to lie flatter on the roof with less need for packing of gaps between tiles with mortar (Figure 10.19). Concrete tiles are coated with a resin to make them waterproof and are typically installed over 30-lb underlayment and preservative-treated horizontal nailing battens (Figure 10.20).

Slate tiles form a fire-resistant, long-lasting roof that is suitable for buildings of the finest materials. It is relatively costly but is among the most durable of all roofing materials. Slate for roofing is delivered to the site split, trimmed to size, and punched or drilled for nailing (Figure 10.21). Both in their shape and by the method by which they are laid on the roof, slate tiles resemble wood shingles or shakes (Figure 10.22), although the natural stone material is similar to clay or concrete tile.

FIGURE 10.20
A roof set up for the installation of concrete tiles. Plywood sheathing has been covered with underlayment, and preservative-treated nailing battens have been attached. Furring strips oriented with the slope of the roof hold the battens above the level of the underlayment so that the flow of water will not be restricted. The battens are spaced at intervals that correspond with the length of roofing tiles. The tiles will be held in place by means of a lip at their top edge, which will hook over the battens. Some tiles will also be mechanically fastened to the battens with screws or nails that pass through predrilled holes in the tiles. (*Photo by Rob Thallon*)

FIGURE 10.21
Splitting slate for roofing. The thin slates in the background will next be trimmed square and to dimension, after which nail holes will be punched in them. (*Photo by Flournoy. Courtesy of Buckingham-Virginia Slate Corporation*)

FIGURE 10.22
A slate roof during installation. (*Courtesy of Buckingham-Virginia Slate Corporation*)

Sheet Metal Roofing

Sheets of lead and copper have been used for roofing since ancient times. Both metals form self-protecting oxide layers that are very beautiful and last for many decades. Today's metal roofing materials include enamel-coated galvanized steel, copper, lead-coated copper, stainless steel, terne, and terne-coated stainless steel. All are relatively high in first cost, but they can be expected to last for many decades. They are installed in small sheets using ingenious systems of joining and fastening to maintain watertightness at the seams. The modern *standing-seam* roof is made with long sheets of metal crimped together on the roof (Figures 10.23 and 10.24). The seams create a strong visual pattern that can be manipulated by the designer to emphasize the qualities of the roof shape.

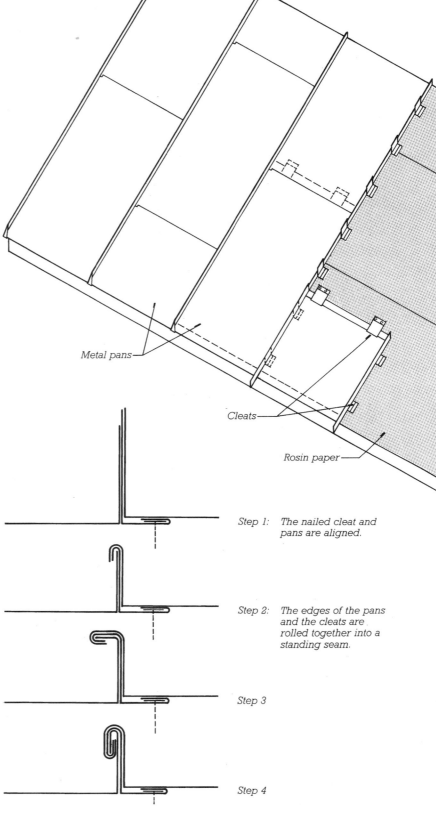

Metal pans

Cleats

Rosin paper

FIGURE 10.23
An automatic roll seamer, moving under its own power, locks standing seams in a copper roof. A cleat is just visible at the lower right. (*Courtesy of Copper Development Association, Inc.*)

Step 1: The nailed cleat and pans are aligned.

Step 2: The edges of the pans and the cleats are rolled together into a standing seam.

Step 3

Step 4

FIGURE 10.24
Installing an architectural standing-seam metal roof.

FIGURE 10.25
A metal panel roof such as the one on this residence is long lasting and can be recycled when it is replaced. Many people expect a metal roof to be noisy in a rainstorm, but thermal insulation generally provides acoustical insulation as well so that the sound of rain on the roof does not reach the interior of the house.
(*Photo by Rob Thallon*)

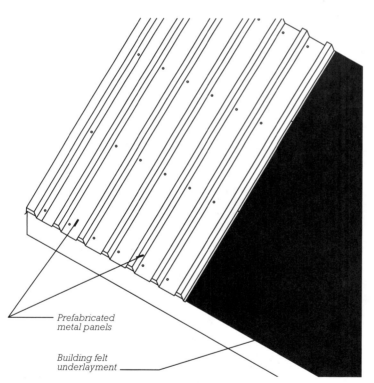

Prefabricated
metal panels

Building felt
underlayment

FIGURE 10.26
Installing a prefabricated metal panel roof.

A new generation of metal roofing, *panel roofing*, has gained popularity in recent years (Figure 10.25). Like the low-cost corrugated metal roofing used on agricultural buildings, panel roofing is preformed and requires no specialized equipment for installation at the site. The panels are made of long sheets of galvanized or aluminized steel, usually coated with long-lasting polymeric coatings in various colors. The most common panels are 2 feet (610 mm) wide and come in a variety of profiles with ridges running lengthwise. Panels are precut to length (from eave to ridge) at the factory and are fastened to the roof with exposed screws sealed with neoprene washers (Figure 10.26). Narrower panels approximately 12 inches wide are also available. These panels are produced with raised edge seams that include a means of concealed attachment to the roof deck and a snap-together interlocking mechanism

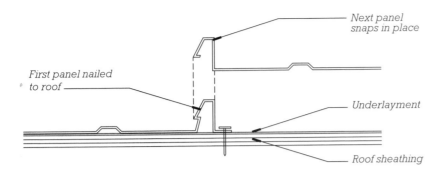

First panel nailed to roof

Next panel snaps in place

Underlayment

Roof sheathing

FIGURE 10.27
A concealed-fastener snap-lock metal panel roofing system. During installation, the leading edge of the metal panel is nailed to the roof sheathing. The trailing edge of the next panel snaps onto the leading edge of the previously installed panel, covering the fastener.

FIGURE 10.28
A galvanic series for metals used in buildings. Each metal is corroded by all those that follow it in the list. The wider the separation between two metals on the list, the more severe the corrosion is likely to be. Some families of alloys, such as stainless steels, are difficult to place with certainty because some alloys within the family behave differently from others. To be certain, the designer should consult with the manufacturer of any metal product before installing it in contact with dissimilar metals in an outdoor environment.

Aluminum
Zinc and galvanized steel
Chromium
Steel
Stainless steel
Cadmium
Nickel
Tin
Lead
Brass
Bronze
Copper

(Figure 10.27). Architectural panel roofs are considerably less costly than traditional forms of sheet metal roofing. Most types (as well as standing-seam types) can be used on slopes below 4:12 (Figure 10.6).

The same metal should be used for every component of a sheet metal roof, including the fasteners and flashings. If this is impossible, metals of similar galvanic activity should be used. This is because when strongly dissimilar metals touch in the presence of rainwater, which is generally acidic, the galvanic action will cause rapid corrosion (Figure 10.28).

Roll Roofing

The same sheet material from which asphalt shingles are cut is also manufactured in rolls 3 feet (900 mm) wide as asphalt *roll roofing*. Roll roofing is very inexpensive and is used primarily on storage sheds and very low cost residential buildings. It can be manipulated by skilled designers, however, to make an attractive roof (Figure 10.29). Its chief drawbacks are that thermal expansion of the roofing or drying shrinkage of the wood deck can cause unsightly ridges to form in the roofing and that thermal contraction can tear it.

FIGURE 10.29
Roll roofing is generally reserved for agricultural and utility buildings, but it is used on residences as well. This house has two colors of roll roofing: one at the edge, the other in the field of the roof. (*Photo by Rob Thallon*)

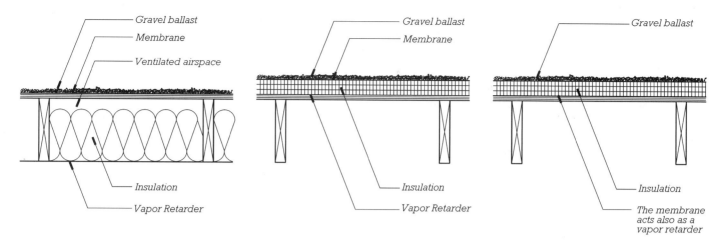

FIGURE 10.30
Low-slope roofs with thermal insulation in three different positions, shown here with a wood joisted roof deck. At left, insulation below the deck, with a vapor retarder on the warm side of the insulation. In the center, insulation between the deck and the membrane, with a vapor retarder on the warm side of the insulation. At right, a protected membrane roof, in which the insulation is above the membrane.

LOW-SLOPE ROOFS

A roof with a pitch lower than 4:12 is referred to as a *low-slope roof*. The distinction is made at this pitch because most of the roofing materials suitable for steep roofs do not perform well below the 4:12 pitch. At the steeper end of this range (4:12 down to 2:12), several of the roofing materials used for steep roofs, such as asphalt shingles or sheet metal panels, can be employed, provided that extra precautions are taken to compensate for the low pitch (Figure 10.6). Roll roofing can be used on slopes as low as 1:12. Below this, none of the materials used on steep roofs will perform adequately, and an entirely different approach must be taken.

To prevent the penetration of water at such low slopes, roofing materials need to be entirely continuous and are referred to as *membranes*. A membrane is an impervious sheet of material that keeps water out of the building. A membrane roof is a complex, highly interactive assembly made up of several components. The *deck* is the structural surface that supports the roof. *Thermal insulation* is installed to slow the passage of heat into and out of the building. A *vapor retarder* is essential in colder climates to prevent moisture from accumulating within the insulation. *Drainage* components remove the water that runs off the membrane. Around the membrane's edges and wherever it is penetrated by pipes, vents, expansion joints, electrical conduits, or roof hatches, special flashings and details must be designed and installed to prevent water penetration.

The Roof Deck

Roof sheathing for a low-slope roof is referred to as the *roof deck*. To provide drainage, the roof must slope toward drainage points at an angle sufficient to drain reliably and overcome any structural deflections. A slope of at least ¼ inch per foot of run (1:50) is recommended. To create this slope in a low-slope roof (often inaccurately referred to as a flat roof), two basic methods may be employed. In one, the joists or beams that support the deck may be sloped. This approach creates a sloped ceiling surface below if the joists are not tapered. Alterna-tively, the slope may be created over a dead-level deck with a system of tapered rigid insulation boards or a tapered fill of lightweight insulating concrete.

If a roof is insufficiently sloped, puddles of water will stand for extended periods of time in the low spots that are inevitably present, leading to premature deterioration of the roofing materials in those areas. The roof membrane must be laid over a smooth surface. A wood deck that is to receive a roof membrane should have gaps or knotholes filled with a nonshrinking paste filler.

Thermal Insulation

Thermal insulation for a low-slope roof may be installed in any of three positions: below the structural deck, between the deck and the membrane, or above the membrane (Figure 10.30).

Insulation Below the Deck
Below the deck, fibrous batt insulation is installed between framing members above a vapor retarder made of polyethylene sheeting. A ventilated airspace should be pro-vided between the insulation and the

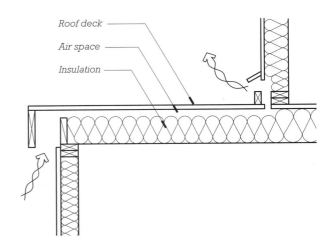

FIGURE 10.31

Low-slope roofs with insulation below the deck must be vented. Venting the joist cavities from the eave is usually adequate, but venting from two opposite sides provides better airflow. The drawing illustrates ventilation air entering through an eave or outside edge and exiting through a parapet or wall at the opposite side of the roof.

deck to dissipate stray water vapor. Vents should be located at opposite sides of the roof, at eaves, parapets, or walls (Figure 10.31). This organization of roofing, insulation, vapor retarder, and ventilation space is the same basic arrangement found in most steep roofs. Insulation in this position in low-slope roofs is relatively economical and trouble-free, but it leaves the membrane exposed to the full range of outdoor temperature fluctuations.

Insulation Between the Deck and the Membrane

The traditional position for low-slope roof insulation is between the deck and the membrane. Insulation in this position must be in the form of low-density rigid panels in order to support the membrane. The insulation protects the deck from temperature extremes and is itself protected from weather by the membrane. The roof membrane in this type of installation is subjected to extreme temperature variations, and any water or water vapor that may accumulate in the insulation will become trapped beneath the membrane, which can cause the insulation and roof deck to decay. In addition, vapor pressure from the trapped moisture can cause the membrane to blister and eventually to rupture. (See Chapter 15 for an explanation of insulation and vapor problems.)

Two precautions are advisable in cold climates for insulation that is located between the deck and the membrane. A vapor retarder should be installed below the insulation, and the insulation should be ventilated to allow the escape of any moisture that may accumulate there. Ventilation is accomplished by installing *topside vents*, one per thousand square feet (100 m²) or so, that allow water vapor to escape upward through the membrane. Topside vents are most effective with a loose-laid membrane, which allows trapped moisture to work its way toward the vents from any part of the insulating layer.

Insulation Above the Membrane: The Protected Membrane Roof

Insulation above the roof membrane, known as a *protected membrane roof (PMR)*, is a relatively new concept (Figure 10.32). It offers two major advantages. First, the membrane is protected from extremes of heat and cold. Second, the membrane is on the warm side of the insulation, where it is immune to vapor blistering problems. Because the insulation itself is exposed to water when placed above the membrane, the insulating material must be one that retains its insulating value when wet and does not decay or disintegrate. Extruded polystyrene foam board is the one material that has all these qualities.

The insulating board is either laid loose or embedded in a coat of hot asphalt to adhere it to the membrane below. It is held down and protected from sunlight (which disintegrates polystyrene) by a layer of ballast. The ballast may consist of crushed stone, a thin layer of concrete that has been factory laminated to the upper surface of the insulating board, or interlocking concrete blocks (Figure 10.33). Critics of the protected membrane roof system originally predicted that the membrane would disintegrate quickly because of its continual exposure to dampness trapped under and around the insulating boards. However, experience of more than 20 years has shown that the membrane ages little when protected from sunlight and temperature extremes by the insulation boards, despite the presence of moisture.

Vapor Retarders for Low-Slope Roofs

When the insulation is above the membrane, the membrane itself also serves as the vapor retarder. In all other low-slope roof constructions, a separate vapor retarder is advisable except in warm, humid climates, where wintertime condensation is not a problem and summertime air conditioning can cause water vapor to migrate inward through the roof. In every case, the potential migration of moisture through the insulation should be considered in relation to the membrane, which always acts as a vapor retarder (no matter what its position relative to the insulation).

FIGURE 10.32
A cutaway detail of a proprietary type of protected membrane roof shows, from bottom to top, the roof deck, the membrane, polystyrene foam insulation, a polymeric fabric that separates the ballast from the insulation, and the ballast. (*Courtesy of Dow Chemical Company*)

FIGURE 10.33
Roofers embed stone ballast in hot asphalt to hold down and protect the panels of rigid insulation in a protected membrane roof. The area of the membrane behind the wheelbarrow has not yet received its insulation. (*Courtesy of Celotex Corporation*)

The Low-Slope Roof Membrane

The membranes used for low-slope roofing fall into three general categories: the built-up roof membrane, the single-ply roof membrane, and the fluid-applied roof membrane.

The Built-Up Roof Membrane

A *built-up roof membrane (BUR)* is assembled in place from multiple layers of asphalt-impregnated felt bedded in bitumen (Figures 10.34 and 10.35). The felt fibers may be cellulose, glass, or synthetic. The felt is saturated with asphalt at the factory and delivered to

the site in rolls. The bitumen is usually asphalt derived from the distillation of petroleum, but for protected membrane or very low-slope roofs, coal-tar pitch is used instead, because of its greater resistance to standing water. Both asphalt and coal-tar pitch are applied hot in order to merge

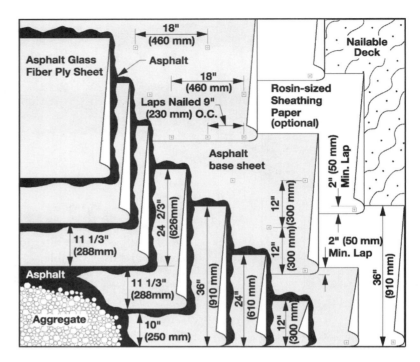

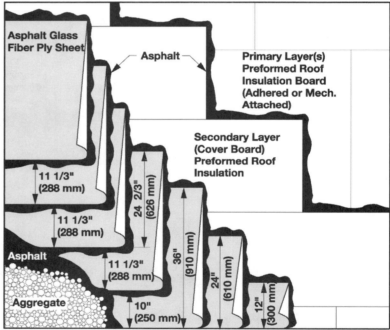

FIGURE 10.34
Two typical built-up roof constructions, as seen from above. The top diagram is a cut-away view of a built-up roof over a plywood roof deck. The membrane is made from plies of felt overlapped in such a way that it is never less than four plies thick. A non-permeable base sheet isolates the membrane from the deck, which will move slightly due to changes in loading and moisture content.

The insulation for this roof is below the deck between the structural joists. The bottom diagram shows how rigid insulation boards are attached to a plywood roof deck to provide a firm, smooth base for application of the membrane. The insulation acts in the same manner as the base sheet in the top diagram to isolate the membrane from movement in the roof deck. A three-ply membrane is shown. In cold climates, a vapor retarder should be installed between the layers of insulation. (*Courtesy of National Roofing Contractors Association.*)

FIGURE 10.35
Overlapping layers of roofing felt are hot mopped with asphalt to create a four-ply membrane. (*Courtesy of Manville Corporation*)

with the saturant bitumens in the felt and form a single-piece membrane. The felt is laminated in overlapping layers (plies) to form a membrane that is two to four plies thick. The more plies used, the more durable the roof. To protect the membrane from sunlight and physical wear, a layer of aggregate (crushed stone or other mineral granules) is embedded in the surface (Figure 10.33).

Cold-applied mastics may be used in lieu of hot bitumen in built-up roof membranes. A roofing mastic is compounded of asphalt and other substances to bond to felts or to synthetic fabric reinforcing mats at ordinary ambient temperatures. The mastic may be sprayed or brushed on and hardens by the evaporation of solvents.

Single-Ply Roof Membranes

Single-ply membranes are a diverse and rapidly growing group of sheet materials that are applied to the roof in a single layer. As compared to built-up membranes, they require less on-site labor, and they are usually more elastic and therefore less prone to cracking and tearing. They are affixed to the roof by any of several means: with adhesives, by the weight of ballast, by fasteners concealed in the seams between sheets, or, if they are sufficiently flexible, with ingenious mechanical fasteners that do not penetrate the membrane.

The materials currently used for single-ply membranes fall into two general groups, thermoplastic and thermosetting. *Thermoplastic* materials may be softened by applying heat. Sheets of thermoplastic membrane may be joined at seams by heat welding or solvent welding. *Thermosetting* materials cannot be softened by heat. They must be joined at seams by adhesives or pressure-sensitive tapes.

The most common thermoplastic membrane is made of polyvinyl chloride (PVC), a relatively low cost compound commonly known as vinyl. PVC sheet for roofing is 0.045 to 0.060 inch thick (1.14 to 1.5 mm). It may be laid loose, mechanically attached, adhered, or used as a protected membrane. Other thermoplastic membranes with increased flexibility, cohesion, toughness, resistance to ultraviolet deterioration, and fire resistance are also available.

The most widely used thermosetting material for single-ply roof membranes is EPDM (ethylene propylene diene monomer). Relatively low in cost, it is a synthetic rubber manufactured in sheets from 0.030 to 0.060 inch (0.75 to 1.5 mm) in thickness. It may be laid loose, adhered, mechanically fastened, or used in a protected membrane roof. Other thermosetting materials are also available.

Fluid-Applied Membranes

Fluid-applied membranes are used primarily for domes, shells, and other complex shapes that are difficult to roof by conventional means. Such shapes are often too flat on top for shingles, but too steep on the sides for built-up roof membranes, and, if doubly curved, are difficult to fit with single-ply membranes. Fluid-applied membranes are applied in liquid form with a roller or spray gun, usually in several coats, and cure to form a rubbery membrane. Materials applied by this method include neoprene (with a weathering coat of chlorosulfonated polyethylene), silicone, polyurethane, butyl rubber, and asphalt emulsion.

Traffic Decks

Often, a terrace is desired on the upper floor of a house over a low-slope roof covering a portion of the floor below. In this case, the membrane is generally applied directly to the sheathing, and the insulation is in the structure below (Figure 10.36). Because people and furniture can be expected to be moving about on the terrace, it is necessary to protect the membrane from abrasion. One common approach for small terraces is to use concrete pavers

FIGURE 10.36
A heat-sealed, single-ply roof at the low corner of a small roof terrace. The single-ply roofing has been applied directly to plywood sheathing and heat sealed at the joints. A copper drain in the floor leads to a scupper and downspout. The hole in the wall is an overflow drain that will allow water to escape in case the main drain gets clogged. Siding will lap over the edges of the roofing, and concrete pavers set on building felt will protect the roofing against abrasion. (*Photo by Rob Thallon*)

that are set on a thin bed of sand or on a 30-lb felt slip sheet that acts as a cushion between pavers and membrane. The pavers must be retained at all edges, so this approach works best when the terrace is bounded by a solid railing with a scupper for drainage (Figure 10.37). Another common detail for this situation employs a thin lightweight concrete slab over the membrane. The structure below must be very stiff to prevent cracking in the slab, especially if the terrace is large. Wood duckboards, essentially a thin wooden deck on sleepers, may also be used to protect terrace membranes.

ROOF EDGE DETAILS

The outer edge of a steep roof, where it meets the wall at the eave and rake, requires special attention on the part of the building's designer. Several objectives must be kept in mind. The eaves and rakes should be designed so that the siding can be easily attached. The edges of the roofing material should be positioned and supported in such a way that water flowing over them will drip free of the trim and siding below. The eaves must be ventilated to allow for the free circula-

tion of air beneath the roof sheathing (Figure 10.38). Finally, provision must be made to attach gutters or to provide another means to drain rainwater and snowmelt from the roof without damaging the structure below.

Details at the edges of low-slope roofs have similar requirements. Where overhangs occur, details should be similar to eaves for steep roofs. Parapets require details that ensure that the roof membrane is turned up onto the wall a sufficient distance to contain rainwater and snowmelt and that siding has sufficient clearance

FIGURE 10.37
A completed roof terrace similar to the one under construction in Figure 10.36. (*Photo by Rob Thallon*)

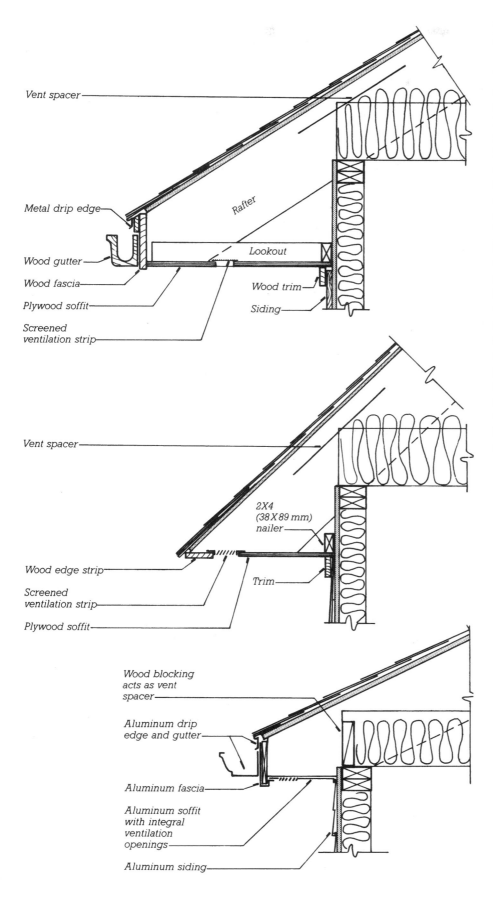

Figure 10.38
Three ways, from among many, of finishing the eaves of a Wood light frame building. The top detail has a wood fascia and a wood gutter. The gutter is spaced away from the fascia on wood blocks to help prevent decay of the gutter and fascia. The width of the overhang may be varied by the designer, and a metal or plastic gutter may be substituted for the wood one. The sloping line at the edge of the ceiling insulation indicates a vent spacer as shown in Figure 15.25. The middle detail has no fascia or gutter; it works best for a steep roof with a sufficient overhang to drain water well away from the walls below. The bottom detail is finished entirely in aluminum. It shows wood blocking as an alternative to vent spacers for maintaining free ventilation through the attic. Sometimes designers entirely eliminate eave overhangs and gutters from their buildings for the sake of a "clean" appearance, but this is ill advised because the water from the roof washes over the walls, which leads to staining, leaking, decay, and premature deterioration of the windows, doors, and siding.

from wet surfaces. Where neither eave nor parapet are present at the edge of a low-slope roof, venting is usually provided through the wall or fascia. Some typical details of low-slope roofs are presented in Figures 10.40 to 10.42.

All are shown with built-up roof membranes, but details for single-ply membranes are similar in principle.

The construction of eaves, rakes, and parapets is done by the framing contractor and is generally completed before the roofing is installed, though the trim is usually left off until the time when the siding is installed. If weather permits, the eaves and rakes are often painted before the roofing is installed.

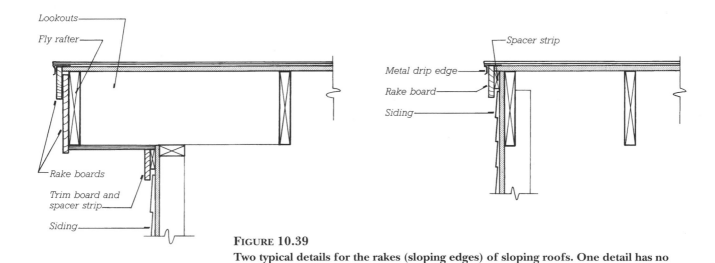

FIGURE 10.39
Two typical details for the rakes (sloping edges) of sloping roofs. One detail has no rake overhang, and the other has an overhang supported on lookouts.

FIGURE 10.40
A roof edge for a conventional built-up roof. The membrane consists of four plies of felt bedded in asphalt with a gravel ballast. The base flashing is composed of two additional plies of felt that seal the edge of the membrane and reinforce it where it bends over the curb. The curb directs water toward interior drains or scuppers rather than allowing it to spill over the edge.

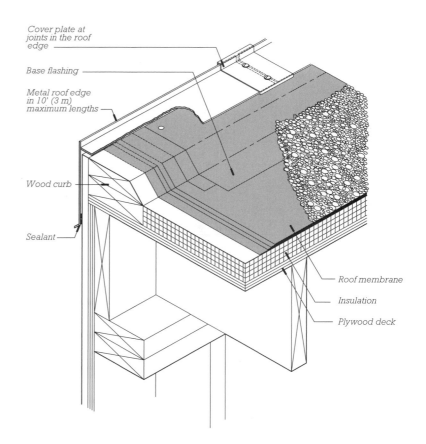

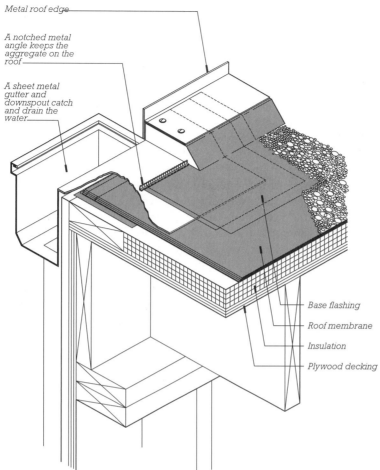

Metal roof edge

A notched metal angle keeps the aggregate on the roof

A sheet metal gutter and downspout catch and drain the water

Base flashing

Roof membrane

Insulation

Plywood decking

FIGURE 10.41
Detail of a scupper. The curb is discontinued to allow water to spill off the roof into a gutter and downspout. Additional layers of felt, called stripping, seal around the sheet metal components. Large roofs use interior drains rather than scuppers as their primary means of drainage.

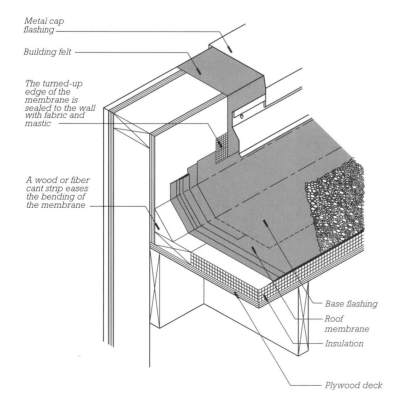

Metal cap flashing

Building felt

The turned-up edge of the membrane is sealed to the wall with fabric and mastic

A wood or fiber cant strip eases the bending of the membrane

Base flashing

Roof membrane

Insulation

Plywood deck

FIGURE 10.42
A conventional parapet design.

ROOF DRAINAGE

Gutters with *downspouts* (downspouts are also called leaders) are usually installed along the eaves of a sloping roof to remove rainwater and snowmelt without allowing them to wet the walls or cause splashing or erosion on the ground below (Figure 10.43). By far, the most common gutter system in use today is the *continuous metal gutter*, which is formed at the site in uninterrupted lengths of up to 40 feet or more (Figure 10.44). Usually made of steel or aluminum with a baked-enamel finish, continuous gutters come in a variety of shapes and virtually eliminate joints, which are the most common source of gutter failure.

Gutters are usually spiked to the eave of the building (Figure 10.45) and are gently sloped toward the points at which vertical downspouts drain away the collected water. The spacing and flow capacities of downspouts are determined by formulas based on local rainfall history. At the bottom of each downspout, a means must be provided for getting the water away from the building, in order to prevent soil erosion and basement flooding. A system of underground storm drain pipes can collect the water from all the downspouts and conduct it to a storm sewer, a dry well, or a drainage ditch (Figure 7.22). In locations where storm drainage is not required, *splash blocks* at the base of each downspout minimize erosion and direct runoff away from the building.

Gutters made of plastic are generally more durable than metal gutters because they will return to their original shape after being deformed. Plastic gutters are more expensive than metal, however, and require many more fittings because they are not as strong or as stiff, so they require more time to install. Nevertheless, they remain popular in some geographical areas and with owner/builders. Wooden gutters, although antiquated, expensive, and impermanent, are still used in some regions for aesthetic reasons. Custom-made gutters that are recessed into the surface of the roof are popular with architects because they are virtually

FIGURE 10.43
Gutters collect rainwater runoff at the level eave of a roof, and downspouts conduct this runoff down to the ground, where it is discharged. In this case, the downspout discharges onto a lower roof over which the water will run until it reaches another gutter at the eave. Common gutters and downspouts are made with a wide range of standard parts that allows them to control the runoff from virtually any roof. (*Photo by Rob Thallon*)

FIGURE 10.44
Most gutters are now made with a trailer-mounted gutter machine. A roll of sheet metal is fed into one end of the machine, which forms it into continuous lengths of gutter. Workers need only install end caps and downspout fittings to the gutter before it is mounted on the house. Most such machines are owned and operated by sheet metal shops. (*Photo by Rob Thallon*)

invisible (Figure 10.46), but they are expensive and difficult to construct sufficiently well to prevent water from entering the building if the gutters should become clogged with debris.

Some codes allow rainwater gutters to be omitted entirely if the overhang of the roof is sufficient, thus avoiding the problems of clogging and ice buildup commonly associated with gutters. Typical minimum overhangs for gutterless buildings are 1 foot (300 mm) for single-story buildings, and 2 feet (600 mm) for two-story buildings. To prevent soil erosion and mud spatter from dripping water, the drip line at ground level below the gutterless eave must be protected with a bed of stone (Figure 10.47).

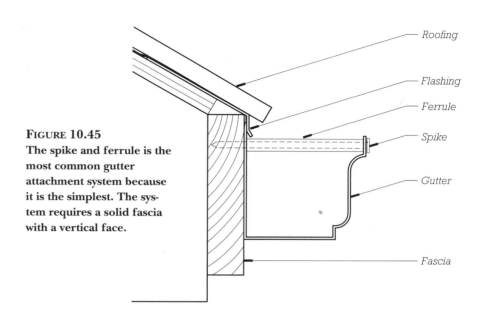

FIGURE 10.45
The spike and ferrule is the most common gutter attachment system because it is the simplest. The system requires a solid fascia with a vertical face.

Roofing

Flashing

Ferrule

Spike

Gutter

Fascia

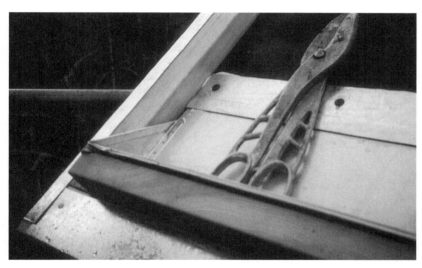

FIGURE 10.46
This custom gutter system will be hidden behind a flashing at the base of a standing seam roof. (*Photo by Rob Thallon*)

FIGURE 10.47
There are ways other than downspouts to control storm water runoff from a roof. Runoff from the left side of this roof will be conducted to the ground over the surface of a chain. Runoff from the right side will drip directly from the roof into a stone splash bed. (*Photo by Rob Thallon*)

ROOFING AND THE BUILDING CODES

Manufacturing standards and installation procedures for roofing materials are specified by all the model building codes. Building codes also regulate the type of roofing that may be used on a building, based on a required level of resistance to flame spread and fire penetration as measured by ASTM (American Society for Testing and Materials) procedure E108. Roofing materials are grouped into the following four classes:

• Class A roof coverings are effective against severe fire exposure. They include slate, concrete tiles, clay tiles, most asphalt shingles, most built-up and single-ply roofs, and other materials certified as Class A by approved testing agencies. They may be used on any building in any type of construction.

• Class B roof coverings are effective against moderate fire exposure and include many built-up and single-ply roofs, sheet metal roofings, and some composition shingles.

• Class C roof coverings are effective against light fire exposure. They include fire-retardant-treated wood shingles and shakes.

• Nonclassified roof coverings such as untreated wood shingles may be used on most residential construction and on some agricultural, accessory, and storage buildings.

ASTM E108 applies to whole roof assemblies, including deck, insulation, membrane or shingles, and ballast, if any. Therefore, the classifications given in the preceding list should be taken only as a general guide. It is difficult to summarize with precision the classification of any particular type of shingle or membrane without knowing the other components of the assembly. The required class of roofing for a particular building may also be affected by an urban fire zone in which the building is located and by the proximity of the building to its neighbors.

C.S.I./C.S.C.
MasterFormat Section Numbers for Finishing the Roof

07300	SHINGLES, ROOF TILES, AND ROOF COVERINGS
07310	Shingle
07320	Roof Tile
07400	ROOFING AND SIDING PANELS
07410	Metal Roof and Wall Panel
07500	MEMBRANE ROOFING
07510	Built-up Bituminous Roofing
07520	Cold-Applied Bituminous Roofing
07530	Elastomeric Membrane Roofing
07540	Thermoplastic Membrane Roofing
07550	Modified Bituminous Membrane Roofing
07560	Fluid-Applied Roofing
07580	Roll Roofing
07600	FLASHING AND SHEET METAL
07620	Sheet Metal Flashing and Trim
07650	Flexible Flashing
07700	ROOF SPECIALTIES AND ACCESSORIES
08600	SKYLIGHTS
08610	Roof Window
08620	Unit Skylight

SELECTED REFERENCES

1. National Roofing Contractors Association. *Roofing and Waterproofing Manual.* Rosemont IL: National Roofing Contractors Association. Updated frequently.

Covers all types of roofs, both low and high slope. It's the roofing industry bible. Updated frequently.

2. Patterson, Stephen, and Madan, Mehta. *Roofing Design and Practice.* Saddle River, NJ: Prentice Hall, 2001.

This is a general text, well-illustrated and readable, that covers all kinds of roofing.

KEY TERMS AND CONCEPTS

roofing
underlayment
ice-and-water shield
roof flashing
roof slope
steep roof
roof jack
solid sheathing
sheathing clip
spaced sheathing
thermal insulation
vapor retarder
shingle
asphalt shingle
wood shingle

shake
interlayment
tile
clay tile
concrete tile
slate tile
sheet metal roofing
standing seam
panel roofing
roll roofing
low-slope roof
membrane
roof deck
insulation below the deck

insulation between the deck and membrane
insulation above the membrane
protected membrane roof
built-up roof (BUR)
single-ply membrane
thermoplastic material
thermosetting material
fluid-applied roof membrane
traffic deck
gutter
downspout
continuous metal gutter
splash block
Class A, B, C roofing

REVIEW QUESTIONS

1. What are the major differences between a low-slope roof and a steep roof? What are the advantages and disadvantages of each type?

2. Discuss the three positions in which thermal insulation may be installed in a low-slope roof and the advantages and disadvantages of each.

3. Explain in precise terms the function of a vapor retarder in a low-slope roof.

4. Compare a built-up roof membrane to a single-ply roof membrane.

5. What is the difference between cedar shingles and cedar shakes?

6. What are some reasons for and against adding gutters to the roof of a house?

What precautions should be considered in a case where gutters are not used?

7. What are the advantages and disadvantages of the various types of gutter systems?

8. At what point in exterior finishing operations can interior finishing operations begin?

EXERCISES

1. Sketch the eave detail of a steep roof system of your choice. Now sketch the rake detail of the same system and figure out how the eave and rake connect at the corner. You may have to make a scale model to understand this completely.

2. Find a low-slope roof system being installed and take notes on the process until the roof is completed. Ask questions of the roofers, the architect, or your instructor about anything you don't understand.

3. Examine a number of existing roofs around your neighborhood, looking for evidence of problems such as cracking, blistering, and leaking. Photograph the problem area and explain to your class the reasons for each problem that you discover.

11

FINISHING THE EXTERIOR WALLS

269

The exterior walls of a building are similar to the roof in that they form a barrier between the inside and the outside. They protect the interior rooms from sun, rain, snow, and wind, and they insulate from the heat and cold of the outdoors to help maintain a comfortable indoor temperature. Because they are vertical, the walls are not as severely attacked by weather as the roof, but because walls are punctuated with numerous door and window openings, they must be carefully designed and constructed if they are not to leak.

Though manufacturers have spent countless resources developing improved windows, doors, sidings, and sealants, walls function best and last longest when they are protected from rain and sun by roof overhangs. Wall elements that are not protected by overhangs will deteriorate much more quickly than those that are so protected.

Though roof overhangs protect the walls from rain, snow, and sun, all of which come from above, walls need protection from below as well. Rain dripping from a roof or falling from the sky can splash when it hits the ground, soaking the base of the wall. Snow can accumulate in tall drifts that pile up against walls. The splashing problem (called splashback) can be relieved by the addition of gutters and downspouts, which are discussed in Chapter 10. Both splashing and snow problems may be addressed by protecting the base of the wall with materials that do not deteriorate rapidly when they come into contact with moisture (Figure 11.2).

There is a logical sequence for finishing exterior walls that involves layering materials, one over another, to make an envelope that is as weathertight as possible. Starting at the sheathing of the framed walls, a membrane weather barrier is added to protect the framing from moisture that might make its way through the outer protective layers. Next, the windows and doors are installed, providing edges to which the siding will be fitted. The siding and its accompanying trim are finally added, and exterior coats of paint or stain finish the job. The membrane weather barrier and windows and doors are customarily installed by the framing crew or the general contractor. The siding may be installed by the same people or by a specialized siding subcontractor. The painting is almost always done by a painting subcontractor.

FIGURE 11.1
Architect Bernard Maybeck understood the principles of protecting walls from the weather. The R. H. Mathewson House, built in 1915 in Berkeley, California, has large roof overhangs, and the exposed wooden parts of the house are kept well above the ground. (*Photo by Rob Thallon*)

FIGURE 11.2
This house, built in the year 2000, employs roof overhangs over each window and door and has water-resistant cement board siding at the base. Carefully planned roof overhangs can control sunlight as well as protect from the weather. (*Photo by Rob Thallon*)

THE WEATHER BARRIER MEMBRANE

Once the frame of the house has been completed, a *weather barrier* membrane is applied to the wall sheathing. This membrane has two primary functions: It serves as an air barrier, reducing the leakage of air between indoors and outdoors. It also protects the structural shell of the building from water during construction and,

in the long term, acts as a backup waterproofing layer beneath the siding. Tests have shown that even the most professionally applied siding cannot completely eliminate the penetration of rainwater or snowmelt that is driven by wind or pulled through by capillary action or gravity. The weather barrier intercepts this moisture before it reaches the sheathing, while allowing water vapor to pass freely so that it does not accumulate in the wall.

Prior to the 1980s, the weather barrier membrane usually consisted of 15-lb *asphalt-saturated felt* (commonly known as "tarpaper"). This material comes in 3-foot-wide (610 mm) rolls and is stapled to the sheathing in successive overlapping courses, starting at the base of the wall. Felt is still used by many builders and provides an excellent barrier (Figure 11.42). However, a universal concern with heating fuel efficiency has led to the development of airtight, vapor-per-

meable *housewraps*. These are manufactured strips up to 12 feet wide that can wrap an entire story of a house in one continuous band of material, thus minimizing air leakage through seams (Figure 11.3). To reduce air leakage most effectively, seams should be minimized. Such seams as cannot be avoided must be cemented together or sealed with tape, and the edges around windows and doors should be carefully sealed. (Other types of weather barrier membranes are discussed in Chapter 15.)

Whichever membrane is used, it should lap over the flashings at the tops of windows and doors. It should also lap over itself at horizontal seams so that water running down its face will always be directed away from the building. Using tape or sealant, the barrier should be sealed to window casings or nailing flanges (Figure 11.4).

FIGURE 11.3
Housewrap completely covers the walls of this new house in preparation for the application of siding. Lightweight wraps such as this have almost entirely replaced felt building paper for use as a weather barrier.
(*Photo by Rob Thallon*)

FIGURE 11.4
The strip of weather barrier around this window is first stapled to the framing around the rough opening. Next, as the window is installed, it is caulked to the strip. When the complete weather barrier is applied later, it will be taped to the strip around the window, making the barrier continuous. Flashing will be added at the top of the window to divert rainwater. (*Photo by Edward Allen*)

WINDOWS

The word "window" is thought to have originated in an old English expression that means "wind eye" or "wind hole." The earliest windows in buildings were open holes through which smoke could escape and fresh air could enter. Control devices of various kinds were soon added to the holes: Hanging skins, mats, or fabric were used to regulate airflow. Shutters were added to provide shading and to keep out burglars. Translucent membranes of oiled paper or cloth and, eventually, glass were developed to admit light while preventing the entry of air, water, and snow. When a translucent membrane was eventually mounted in a moving sash, light and air could be controlled independently of one another. With the addition of woven insect screens, windows permitted air movement while keeping out mosquitoes and flies. Further improvements followed one upon another as the centuries passed. A typical window today is an intricate, sophisticated mechanism with many layers of controls: curtains, shade or blind, sash, glazings, insulating airspace, low-emissivity coatings, insect screen, weatherstripping, and perhaps a storm sash or shutters.

Windows and doors are very special components of walls. Windows allow for simultaneous control of the passage of light, air, and visual images through walls. Doors permit people, goods, and, in some cases, vehicles to pass through walls. Windows and doors, in addition to these important functions, play a large role in establishing the character and personality of a building, much as our eyes, nose, and mouth play a large role in our personal appearance. At the same time, windows and doors are the most complex, expensive, and potentially troublesome parts of a wall. The experienced designer exercises great care and wisdom in selecting them and detailing their installation to achieve satisfactory results.

The size, proportion, and location of windows can have a significant impact not only on the character of a house, but also on its energy performance. Even the most energy-efficient window allows 4 to 10 times as much heat to pass through as does a reasonably well insulated wall, so the amount of window area in an exterior wall should be judiciously monitored by the designer. In fact, building codes set prescriptive limits for the percentage of the gross area of exterior wall that can be made of glass. The International Residential Code, for example, specifies a maximum of 15 percent *glazed* (glass) area for normally insulated houses. Windows can also contribute to heating a house when oriented to take advantage of passive solar heat gain. Codes typically boost the maximum allowable gross area of glass in exterior walls when 50 percent or more of the glass is located in a south-facing wall. Cooling energy savings can also be realized by locating operable windows in positions to take advantage of natural cross ventilation.

The basic components of a window (Figure 11.5) are the *sash*, which holds the glass; the *jamb*, which is the frame around the sash that fits into the rough opening in the wall; and the *casing*, which surrounds the jamb and covers the gap between the window and the framing, both inside and out. At the base of the window, the sash rests on an exterior *sill*, which slopes away from the building to carry away rainwater. When two sashes are adjacent, the structural piece between them, whether horizontal or vertical, is called a *mullion*. Divisions within a sash are made with slender pieces called *muntins*.

A *prime window* is a window that is permanently installed in a building. A *storm window* is a removable auxiliary unit that is added seasonally to a prime window to improve its thermal performance. A *combination window,* which is an alternative to a storm win-

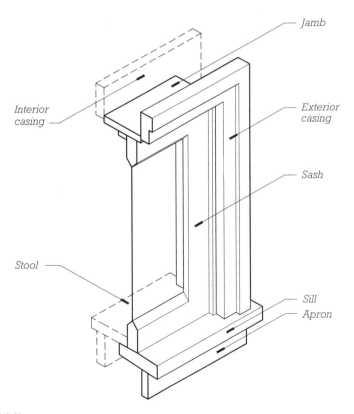

FIGURE 11.5
Window nomenclature follows a tradition that has developed over hundreds of years. The jamb and sill are known collectively as the frame.

dow, is an auxiliary unit that incorporates both glass and insect screening; a portion of the glass is mounted in a sash that can be opened in summer to allow ventilation through the screening. A combination window is normally left in place year-round. Some windows are designed and manufactured specifically as *replacement windows* that install easily in the openings left by deteriorated windows that have been removed from older buildings.

Types of Windows

Figure 11.6 illustrates in diagrammatic form the window types most commonly used in residential buildings. Fixed windows are the least expensive and the least likely to leak air or water because they have no operable com-

ponents. *Single-hung* and *double-hung* windows have one or two moving sashes that slide up and down in tracks in the frame of the window. In older windows, the sashes were held in position by cords and counterweights, but today's double-hung windows usually rely on a system of springs to counterbalance the weight of the sashes. A *sliding window* is essentially a single-hung window on its side and shares with single-hung and double-hung windows the advantage that tracks in the frame hold the sash securely along two opposite sides. This inherently stable construction allows single-hung, double-hung, and sliding windows to be designed in an almost unlimited range of sizes and proportions. It also allows the sashes to be more lightly built than those in *pro-*

jected windows, a category that includes principally *casement windows* and *awning windows*. All projected windows have sashes that rotate outward from their frames and therefore must have enough structural stiffness to resist wind loads while also supporting their own weight at only two corners.

With the exception of the rare triple-hung window, no window with sashes that slide can be opened to more than half of its total area. By contrast, projected windows can be opened to their full area. Casement windows are helpful in catching passing breezes and inducing ventilation through the building; they are generally narrow (30 inches, or 762 mm maximum) in width but can be joined to one another and to sashes of fixed glass to fill wider openings (Figure 11.7). Awning windows can be broad but are not usually very tall. Awning windows have the advantages of protecting an open window from water during a rainstorm and of lending themselves to a building-block approach to the design of window walls (Figure 11.8). *Tilt/turn windows* have elaborate but concealed hardware that allows each window to be operated either as an in-swinging casement or as a hopper. Windows in roofs are specially constructed and flashed for watertightness and may be either fixed (*skylights*) or openable (*roof windows*).

Projected windows are usually provided with synthetic rubber (elastomeric) *weatherstripping* that seals by compression around all the edges of the sash when it is closed. Single-hung, double-hung, and sliding windows generally must rely on brush-type weatherstripping because it does not exert as much friction as elastomeric materials against a sliding sash. Brush-type materials do not seal as tightly as compression weatherstripping, and they are also subject to more wear than elastomeric weatherstripping over the life of the window. As a result, projected windows are generally somewhat more resistant to air leakage than windows that slide in their frames.

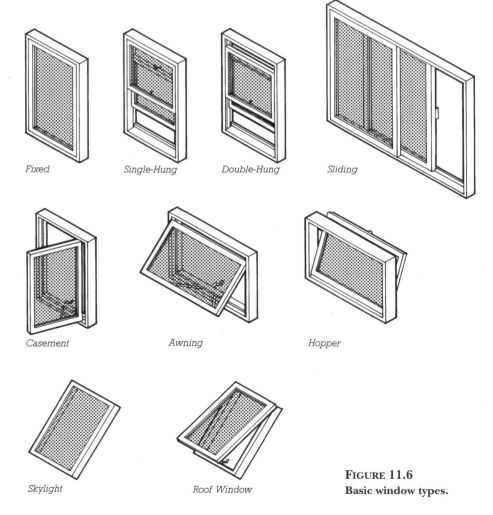

Fixed *Single-Hung* *Double-Hung* *Sliding*

Casement *Awning* *Hopper*

Skylight *Roof Window*

FIGURE 11.6
Basic window types.

FIGURE 11.7
Windows can be "ganged" at the factory or at the site. Ganging of windows consists of fastening standard windows together into a single unit. It is most efficient to gang at the factory, but if the assembly is very large, collections of sashes ganged at the factory must be further joined at the site. (*Photo by Rob Thallon*)

FIGURE 11.8
Awning and fixed windows in coordinated sizes offer the architect the possibility of creating patterned walls of glass.
(*Courtesy of Marvin Windows and Doors*)

Insect screens may be mounted only on the interior side of the sash in casement and awning windows. To open the window with the screen in place, most manufacturers use a crank mechanism. Screens are mounted on the outside in other window types.

Glass tends to attract and hold dust and dirt on both inside and outside surfaces and must be washed at intervals if it is to remain transparent and attractive. Inside surfaces are usually easy for window washers to reach. Outside surfaces are often hard to reach, requiring ladders or scaffolding. Accordingly, most operable windows are designed to allow people to wash the outside surface of glass while standing inside the building. Casement and awning windows are usually hinged in such a way that there is sufficient space between the hinged edge of the sash and the frame when the window is open to allow access to the outer surface of glass for ease of cleaning. Double-hung and sliding windows are often designed to allow each sash to be rotated out of its track for cleaning (Figure 11.9).

Window Construction

Windows were formerly made on the construction site by highly skilled carpenters, but nearly all are produced now in factories (Figure 11.10). Factory-produced windows cost less and, more important, are of better quality. Windows need to be made to a very high standard of precision if they are to operate easily and maintain a high degree of weathertightness over a period of many years.

The traditional material for window frames and sashes is wood, although aluminum, plastics, and combinations of these three materials have also come into widespread use. Wood is a fairly good thermal insulator and, if free of knots, is easily worked into sash. However, in service, wood shrinks and swells with changing moisture content and requires repainting every few years. When dampened by weather, leakage, or condensate, wood windows are subject to decay. Knot-free wood is becoming increasingly rare and expensive, for which reason wood products made of short lengths finger-jointed together, oriented wood strands, or veneers are now used more frequently in window construction than solid lumber. These substitute materials, while functionally very satisfactory, are not attractive to the eye, so they are usually covered (clad) with wood veneer on the inside and an exterior *cladding* of plastic or aluminum (Figures 11.11 and 11.12). Clad windows currently account for about three-quarters of the market for wood windows.

Vinyl windows, which are capturing a continuously increasing proportion of the residential window market, never need painting and are fairly good thermal insulators. The most common material for vinyl window frames is

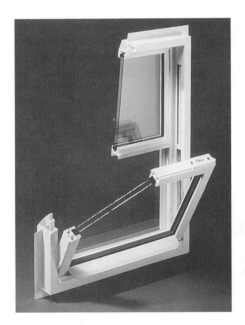

FIGURE 11.9
For ease of washing the exterior surfaces of the glass, the sashes of this plastic window can be unlocked from the frame and tilted inward. (*Courtesy of Vinyl Building Products, Inc.*)

FIGURE 11.10
Most new windows and doors in North America are made in very large factories such as this. Assembly line production and economy of scale result in high-quality products at prices with which it is difficult for small companies to compete. (*Photo courtesy of Andersen Windows, Inc.*)

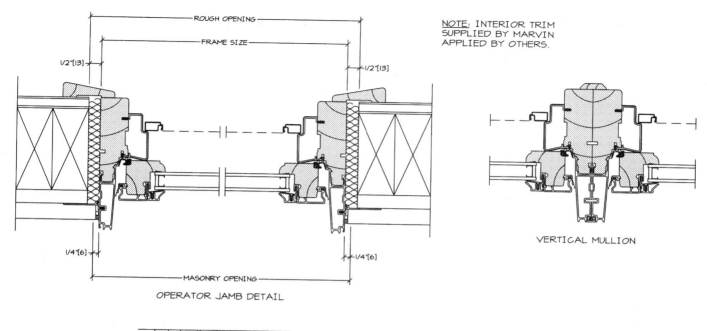

NOTE: INTERIOR TRIM
SUPPLIED BY MARVIN
APPLIED BY OTHERS.

ROUGH OPENING

FRAME SIZE

1/2"[13]

1/2"[13]

1/4"[6]

1/4"[6]

MASONRY OPENING

OPERATOR JAMB DETAIL

VERTICAL MULLION

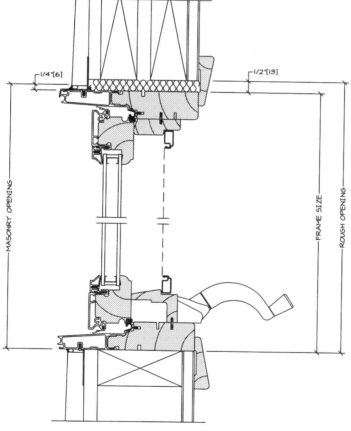

1/4"[6]

1/2"[13]

MASONRY OPENING

FRAME SIZE

ROUGH OPENING

HEAD JAMB & SILL DETAIL

FIGURE 11.11
Manufacturer's catalog details for an aluminum-clad, wood-framed casement window
with double glazing and an interior insect screen. (*Courtesy of Marvin Windows and
Doors*)

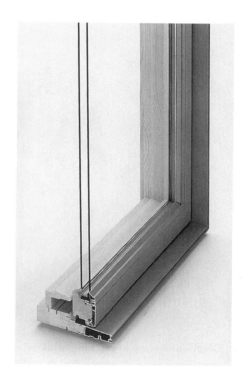

FIGURE 11.12
Cutaway sample of an aluminum-clad,
wood-framed window. (*Courtesy of Marvin
Windows and Doors*)

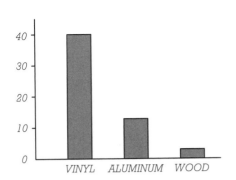

FIGURE 11.13
A comparison of the coefficients of thermal expansion for vinyl, aluminum, and wood. Vinyl expands about 10 times as much as wood so allowances for differential movement must be made when joining these two materials in a building. Units in the graph are in./in./°F $\times$ 10^{-6}.

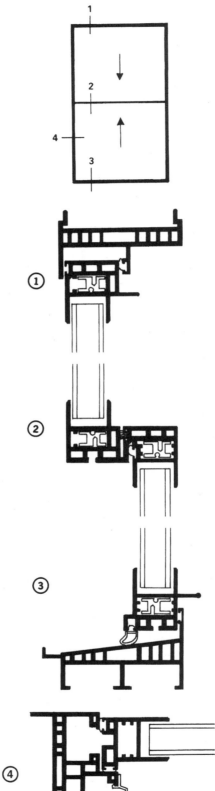

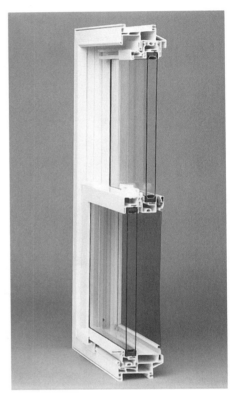

FIGURE 11.15
Cutaway sample of a plastic double-hung window with double glazing and an external half-screen. (*Courtesy of Vinyl Building Products, Inc.*)

FIGURE 11.14
Details of a vinyl single-hung residential window. The small inset drawing at the top of the illustration shows an elevation view of the window with numbers that are keyed to the detail sections below. (*American Architectural Manufacturers Association Window Selection Guide—1988. Courtesy of AAMA*)

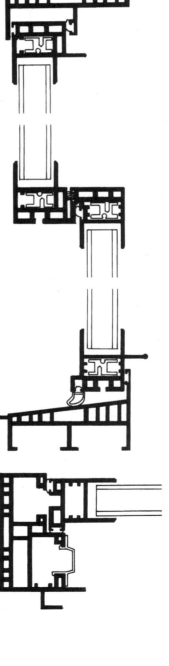

polyvinyl chloride (PVC) that is formulated with a high proportion of inert filler material so as to minimize thermal expansion and contraction. Even so, the coefficient of thermal expansion for vinyl is considerably greater than that of other materials used in window manufacture (Figure 11.13). For this reason, only light (non-heat-absorptive) colors are available, and manufacturers specify that installers allow tolerances between the window frame and the building frame to allow for expansion and contraction. Some typical PVC window details are shown in Figures 11.14 and 11.15.

Glass-fiber-reinforced plastics (GFRPs) are new to the window market. GFRP frame sections are produced by a process of pultrusion: Continuous lengths of glass fiber are pulled through a bath of plastic resin, usually polyester, and then through a shaped, heated die in which the resin hardens. The resulting sash pieces are strong, stiff, and relatively low in thermal expansion. Like vinyl, they are fairly good thermal insulators.

In earlier times, because of the difficulty of manufacturing large pieces of glass, sashes were necessarily divided into small *lights* by *muntins* (Figure 11.16). A typical double-hung window had its upper sash and lower sash each divided into six small lights and was referred to as a "six over six." Muntin arrangements changed with changing architectural styles and improvements with glass manufacture. Today's windows, glazed with large, virtually flaw-less lights of float glass, need no muntins at all, but many building owners and designers prefer the look of traditional *true-divided-light* windows. This desire for divided lights is greatly complicated by the necessity of using double glazing. Some manufacturers offer the option of individual small lights of double glazing held in deep muntins. This is relatively expensive and tends to look heavy because the muntins must be wide to cover the edges of the double glazing. The least expensive option utilizes grids of imitation muntin bars, made of wood or plastic, that are clipped into each sash against the interior surface of the glass. These are designed to remove easily for washing the glass. Other alternatives are imitation muntin grids between the sheets of glass, which are not very convincing replicas of the real thing, and grids, either removable or permanently bonded to the glass, on both the outside and the inside faces of the window. Another option is to use a prime window with authentic divided lights of single glazing and to increase its thermal performance with a storm sash. This looks the best of all the options from the inside, but reflections in the storm sash largely obscure the muntins from the outside.

Glazing

The term *glazing* refers to glass incorporated into a building or set within a window sash. A number of glazing options are available for residential windows (Figure 11.17). *Single glazing* is acceptable only in the mildest climates because of its low resistance to heat flow and the likelihood that moisture will condense on its interior surface in cool weather. Sealed *double glazing* or single glazing with storm windows is the minimum acceptable glazing under most building codes. More than 90 percent of all residential windows sold today in North America have two or more layers of glazing. Double glazing with *a low-emissivity (low-e)* coating on one glass surface performs at least as well as triple glazing. The most usual form of

FIGURE 11.16
A pair of 6/1 double-hung windows. The muntins in the divided upper sash can be made in a variety of ways that are only apparent upon close scrutiny. These two windows were "ganged" at the factory into a single unit with a vertical mullion at the center. (*Courtesy of Marvin Windows and Doors*)

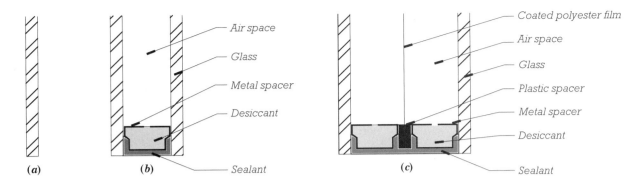

FIGURE 11.17
Comparison of single, double, and triple glazing: (*a*) Single residential glazing is usually double strength, about ⅛ inch (3 mm) thick. (*b*) Double glazing consists of two sheets of glass with an airspace between. A metal spacer at the edge of the glass contains desiccant that absorbs moisture contained in the air between the two sheets of glass. The entire unit is sealed at the edges. If this seal fails, moisture can enter and saturate the desiccant, causing the unit to fog. (*c*) Triple glazing is similar to double glazing but with an extra airspace created by either a transparent polyester film or another layer of glass between the outer layers of glass. Single, double, or triple glazing can be coated to transmit only selected parts of the light spectrum, and the cavity can be filled with argon or other inert gases to increase thermal performance.

Thermal Insulating Values, Center of Glass

	U		R	
	U.S.	**S.I.**	**U.S.**	**S.I.**
Single glass	1.11	6.29	0.90	0.16
Double glass or single glass plus storm window	0.50	2.84	2.00	0.35
Double glass with argon fill	0.46	2.61	2.17	0.38
Double glass with low-e coating on one surface	0.34	1.93	2.94	0.52
Double glass with argon fill and low-e coating on one surface	0.28	1.59	3.57	0.62
Triple glass	0.34	1.93	2.94	0.52
Double glass, polyester film in middle of airspace, argon fill, low-e coating on two interior surfaces	0.17	0.96	5.88	1.03

FIGURE 11.18
Center-of-glass thermal insulating values for various types of glazing.

triple glazing today consists of two outer sheets of glass with a thin, virtually weightless, highly transparent plastic film stretched in the middle of the airspace. Removable glazing panels and storm windows must be removed periodically and cleaned, which is a nuisance that can be avoided by using sealed double glazing.

Figure 11.18 shows the thermal properties of various glazing options. These values apply only to the center area of each light of glass. The edges of the light transmit heat much more rapidly because of the solid strips of material used to space the sheets of glass and hold them together. For this reason, whole-window values for heat transmission that take into account heat transmission through the edge spacers, sash, and frame are the only accurate way of comparing the thermal efficiencies of different windows. Higher whole-window thermal efficiency is achieved by using thermally efficient frames and less conductive spacers around the perimeter of lights of multiple glazing.

In warm climates, the dominant thermal problem in windows is likely to be excessive admission of solar heat to the interior of the building through windows on the east, west, and south sides of the building. Tinted glass can reduce the transmission of solar heat somewhat and make the building easier to cool. More effective, however, are low-emissivity coatings that are designed specifically to reflect the wavelengths of sunlight that contain the most heat, while admitting useful amounts of light to the building and providing good viewing conditions to the outdoors.

Safety Considerations in Windows

To prevent accidental breakage and injuries, building codes require that large window lights that are near enough to the floor or to doors that they might be mistaken for open doorways must be made of breakage-resistant material. *Tempered glass* is most often used for this purpose, but *laminated glass* and plastic glazing sheets are also permitted. Building codes also require that at least one window in each bedroom must open to an aperture large enough to permit occupants of the bedroom to escape through it and firefighters to enter through it. Called an *egress window,* the typical requirement is that the clear opening must be at least 5.7 square feet (0.53 m²), the clear width must be at least 20 inches (510 mm), the clear height must be at least 24 inches (610 mm), and the sill may be no higher than 44 inches (1.12 m) above the floor.

Window Testing and Standards

The designer's task in selecting windows is facilitated by testing programs that allow objective comparisons of the thermal and structural performances of windows of different types and different manufacturers. Performance-based standards for wood and clad windows are published by the Window and Door Manufacturers Association (WDMA), for aluminum and plastic windows by the American Architectural Manufacturers Association (AAMA), and for polyvinyl chloride windows by the American Society for Testing and Materials (ASTM). Each of these standards establishes performance grades of windows. Within each grade, a design wind pressure, water resistance, structural performance, air infiltration, and operating force are specified. The WDMA ratings for wood windows, for example, establish six grades corresponding to satisfactory performance in wind pressures of 15, 20, 25, 30, 35, and 40 pounds per square foot (718 to 1914 Pa). The designer may select a grade of window to correspond to the severity of exposure to weather of the building site and the level of performance that the building owner is

willing to pay for. For an inexpensive residence in a sheltered location, a Grade 15 window, the lowest classification, is likely to suffice. For an expensive custom house at the coast, a Grade 35 or 40 window is likely to be a better match for the more demanding environmental conditions in which it must perform.

With regard to energy efficiency, the National Fenestration Rating Council (NFRC) sponsors a program of testing and labeling that is based on the performance of the whole window, not just the glass. At the present time, the standard label that is affixed to each window shows values for the thermal insulating value (U-value), solar heat gain coefficient, and visible light transmittance (Figure 11.19). Condensation performance, long-term energy performance, and other fac-

tors are likely to be added to this label in the future.

Installing Windows

Some catalog pages for windows are reproduced in Figure 11.11 to give an idea of the information on window configurations that is available to the designer. The most important dimensions given in catalogs are those of the *rough opening*. The rough opening height and width are the dimensions of the hole that must be left in a framed wall for the window's installation. They are slightly larger than the corresponding outside dimensions of the window unit itself, to allow the installer to locate and level the unit accurately. The *frame size* is the size of the unit itself and is used by the designer to calculate the overall size of

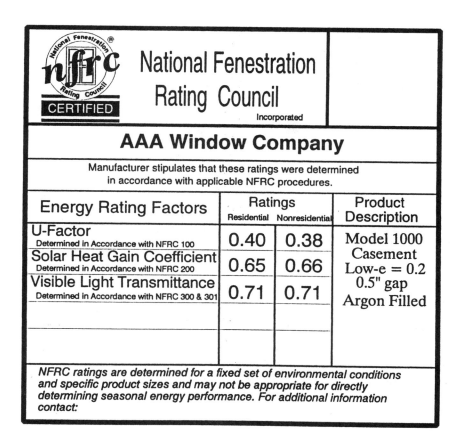

FIGURE 11.19

A typical certification label that is affixed to a window unit so that buyers may compare energy efficiencies. (*Courtesy of National Fenestration Rating Council, Inc.*)

Sheldon Village					Window Schedule			3-Bedroom Townhouse	
Location Key	Type	No. Ea.	Rough Opening	Jamb Depth	Header Height	Header Size	Notes: All Screened
1	XO	1	4-0X4-0	6⅝	6-10-½"	See plan	
2	XO	1	4-0X4-0	6⅝	6-10-½"	See plan	
3	SH-F-SH	3	7-6X4-0	6⅝	6-10-½"	See plan	(3) 2–6 windows
4	SH	1	2-6X4-0	6⅝	6-10-½"	See plan	
5	XO	1	4-0X4-0	6⅝	6-10-½"	See plan	
6	SH	1	2-6X4-0	6⅝	6-10-½"	See plan	Tempered
7	XO	1	4-0X3-0	6⅝	6-10-½"	See plan	
8	SH	1	3-0X5-0	6⅝	6-10-½"	See plan	Egress
9	XO	1	2-0X4-0	6⅝	6-10-½"	See plan	
10	SH	1	3-0X5-0	6⅝	6-10-½"	See plan	Egress
11	AWNING	1	2-0X2-0	6⅝	4-10-½"	See plan	
12	SH	1	3-0X5-0	6⅝	6-10-½"	See plan	Egress
13	SKYLT.	1	1-8X2-10				

FIGURE 11.20
A window schedule from a set of construction documents. Each window has a reference number that is keyed to a plan drawing. Window manufacturers use this schedule to create a bid for the general contractor, and framers use it to coordinate installation of the windows.

(a)

(b)

(c)

(d)

FIGURE 11.21
Installing a wood casement window unit in an existing building. (Installation in new construction is identical, except that the siding is installed after the window.) (*a*) The window unit, with its factory-applied exterior wood casing flange, is placed in the opening and centered. (*b*) A single finish nail is driven through a lower corner of the casing and into the wall framing. (*c*) The unit is plumbed, leveled, squared, and shimmed before inserting nails around the remainder of the unit. (*d*) Sealant is applied between the siding and the exterior casing of the unit to reduce water leakage. The installation is now complete except for the installation of interior casings, which is shown in Chapter 18. Installation procedures for other types of wood windows are similar. (*Courtesy of Marvin Windows, Warroad, Minnesota*)

ganged windows made of individual window units joined together.

Information from the manufacturer's catalog is used by the designer to assemble a *window schedule* for utilization by the contractor (Figure 11.20). The window schedule lists each window shown on the plan and describes its operation, the size of rough opening it requires, its position on the wall (head height), its jamb depth, and any other information pertinent to its complete installation. The contractor uses the window schedule in conjunction with the construction drawings to get bids for windows, to order them, and to frame the window openings.

Most factory-made windows are extremely easy to install, often requiring only a few minutes per window. Figure 11.21 shows a typical procedure for installing a wood window, nailed through the casing into the framing. Windows that are framed or clad in aluminum or vinyl are usually provided with a continuous flange called a *nailing fin* around the perimeter of the window unit (Figure 11.22). When the unit is pushed into the rough opening from the outside, the flange bears against the sheathing along all four edges. After the unit has been located and leveled and squared in the opening, it is attached to the frame by means of nails driven through the flanges. The siding or trim later conceals the flanges (Figure 11.23). It is important to caulk behind casing or flanges or to tape over the flanges in order to control air infiltration at the window opening.

FIGURE 11.22
Most windows are attached to the building by means of a nailing fin. The fin is nailed to the sheathing to hold the window in the plane of the wall, and caulking between the fin and the weather barrier seals against moisture and air infiltration. (*Photo by Rob Thallon*)

FIGURE 11.23
The windows at the top of the photo have a wide casing, whereas those at the bottom of the photo do not, reflecting the intention of the architect to differentiate the two otherwise identical sets of windows. (*Photo by Rob Thallon*)

EXTERIOR DOORS

Exterior doors fall into two general categories, swinging (hinged) and sliding (Figure 11.24). Swinging doors are available in a wide range of styles and are often combined with *sidelights* or *transoms* to make an elegant entry (Figure 11.25). Simple swinging doors are also used in the most mundane of locations such as between the garage and the house. Sliding doors, however, are almost invariably glazed and are typically used as a connection between house and garden (Figure 11.26).

Residential hinged entrance doors almost always swing inward and are mounted at the interior edge of the jamb (Figure 11.27). This arrangement pulls the door back slightly from the weather and exposes the hinges only at the interior, making the door more secure and avoiding the need for

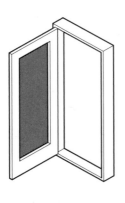

Single Hinged Door

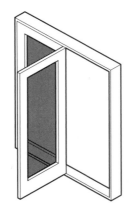

Terrace Door

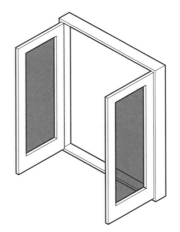

French Door

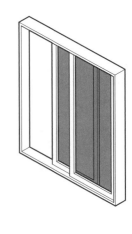

Sliding Door

FIGURE 11.24
Basic exterior door operational types as seen from the inside of the house.

FIGURE 11.25
A six-panel wood entrance door with flanking sidelights and a fanlight above. A number of elaborate traditional entrance designs such as this are available from stock for use in light frame buildings. (*Courtesy of Morgan Products, Ltd., Oshkosh, Wisconsin*)

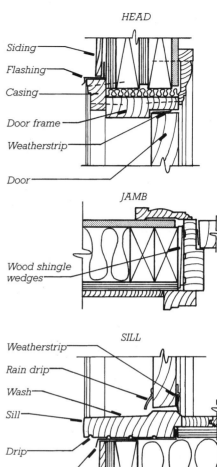

HEAD

Siding

Flashing

Casing

Door frame

Weatherstrip

Door

JAMB

Wood shingle
wedges

SILL

Weatherstrip

Rain drip

Wash

Sill

Drip

Flashing

FIGURE 11.27
Details of an exterior wood door installation. The door opens toward the inside of the building to protect it from the weather, to keep the hinges on the inside for security, and to allow for a screen door on the outside.

FIGURE 11.26
Sliding glass patio doors are common in affordable housing and apartment buildings because a single unit provides both a door and a large area of glass. (*Photo by Rob Thallon*)

nonremovable pin hinges. An in-swinging door also allows a storm door, swinging outward, to be mounted on the outside of the same frame for improved wintertime thermal performance. In summer, a screen door may be substituted for the storm door. A combination door, which has easily interchangeable screen and storm panels, is more convenient than separate screen and storm doors.

The hinged *French door* opens fully and can be used to regulate airflow through the room if both leaves are fitted with doorstops that can

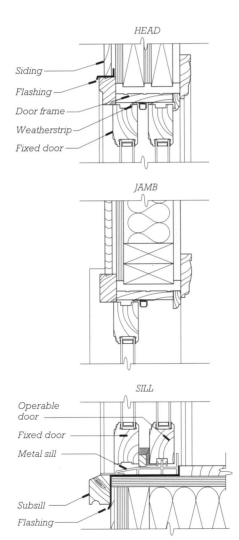

FIGURE 11.28
Details of a wood sliding patio door. A sliding screen door is typically mounted at the exterior of the fixed door panel.

hold them securely in any open position. With its seven separate edges that must be carefully fitted and weatherstripped, the French door is prone to air leakage along these edges. The recently developed *terrace door*, with only one operating door, minimizes this problem but it can open to only half its area.

Sliding *patio doors* are more like overgrown sliding windows than doors in their details and operation (Figure 11.28). These doors are used extensively in affordable housing schemes because they offer both a large glazed area and a door in a single inexpensive unit. The sliding patio door opens to only half of its area and is difficult to secure except from the inside with a site-made stick placed in the track to prevent door movement. An exterior-mounted sliding screen is available.

Weather resistance is an important factor in choosing exterior doors. Exterior doors must be well constructed and tightly weatherstripped if they are not to leak air and water (Figure 11.27). In most climates, doors should be protected with porches or eaves from the brunt of winter storms, and they should be weatherstripped to prevent occasional driving rain from entering the building. Weatherstripping also seals against air infiltration. A tight-fitting threshold at the base of the door completes the weather seal.

Door Construction

At one time, nearly all doors were made of wood (Figure 11.29). In simple buildings, doors made of planks and Z-bracing were once common. In more finished buildings, *stile-and-rail* doors gave a more sophisticated appearance while avoiding the worst problems of moisture expansion and contraction to which plank doors are subjected. In recent decades, while stile-and-rail doors have continued to be used in higher quality buildings, pressed sheet metal doors and molded GFRP doors, usually embossed to resemble wood stile-and-rail doors, have become a popular alternative.

Their cores are filled with insulating plastic foam. Their thermal performance is superior to that of wood doors. They do not suffer from moisture expansion and contraction as wood doors do. The major disadvantage of metal and plastic exterior doors is that they do not have the satisfying appearance, feel, or sound of a wood door. *Flush wood doors*, constructed with a solid core of wood blocks or wood composite material, are also common exterior doors chiefly because they are easier to manufacture and therefore cost less. *Fire doors* have a noncombustible mineral core and are rated according to the period of time for which they are able to resist fire as defined by Underwriters Laboratories Standard 108. In single- and double-family residential work, a 20-minute door between garage and living space is usually the only fire door required.

Installing Exterior Doors

The installation procedure for a door is essentially identical to that for a window (Figure 11.30). The designer selects the exterior doors and prepares a door schedule, showing the dimensions of the rough opening and other information. Prior to installation, the framing around the door is protected with a weather barrier, and caulk is applied around the perimeter of the opening to seal against air infiltration.

Exterior hinged doors are usually furnished *prehung*, meaning that they are already mounted on hinges in a surrounding frame, complete with weatherstripping and casing, ready to install by merely nailing the frame into the wall. When the unit is pushed into the rough opening from the outside, the sill rests on the subfloor and the casing bears against the sheathing along the other three edges. After the unit has been plumbed and squared in the opening, it is shimmed all around and attached to the building frame by means of nails driven through the casing and/or the jamb.

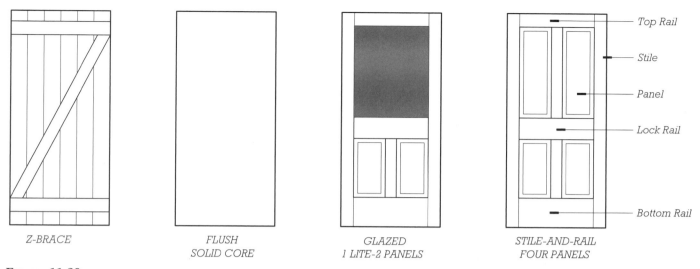

Z-BRACE | FLUSH SOLID CORE | GLAZED 1 LITE-2 PANELS | STILE-AND-RAIL FOUR PANELS

Top Rail
Stile
Panel
Lock Rail
Bottom Rail

FIGURE 11.29
Types of exterior wood doors. Standard sizes range from 2 ft. 6 in. to 3 ft. 0 in. (863 mm to 915 mm) wide by 6 ft. 8 in. (2.0 m) tall. Exterior doors are typically 1¾ inches (44 mm) thick. There are endless variations of the glazed and panel types.

FIGURE 11.30
The base of an exterior hinged door installation. A metal flashing to protect the framing has been installed below the extruded aluminum sill. A terrace set on gravel fill will complete the construction. (*Photo by Rob Thallon*)

Doors are not typically delivered with *lockset* or *deadbolt* installed but may be predrilled for this hardware (Figure 11.31). The locksets, deadbolts and other hardware for all doors should be specified by the designer and included in the door schedule. The carpenter who installs the doors usually installs these items at the site. Screen doors and storm doors may be installed at this time or may be delayed in deference to the painter.

Sliding doors, although heavy, are even simpler to mount than hinged doors because they include fins around the perimeter through which nails are driven into the sheathing and studs to fasten the unit securely in place. These doors are generally furnished complete with hardware and screens.

Overhead garage doors operate differently and are installed differently from either hinged or sliding exterior doors, so they are typically installed by a specialized subcontractor.

FIGURE 11.31
Doors are typically prehung and predrilled for hardware at the distributor. Here, a worker is routing the door and its jamb for hinges. Holes for the door handle have already been drilled by the machine at the bottom edge of the door. (*Photo by Rob Thallon*)

EXTERIOR TRIM

Once the windows and doors have been installed, there is often exterior trim to be installed before the siding is applied. Window and door surrounds, corner boards, handrails, and porch and eave trim are common examples of exterior trim in residential construction (Figure 11.32). This trim can be functional and it can add proportion and interest to the facade. Often, it is shaped to imitate traditional trim. It is generally applied by the same carpenters who install the windows and doors. Flashings must be integrated with the trim at the heads of windows and doors. Trim installation must be coordinated with the painter because the trim is often painted a different color than the surrounding siding.

The materials of choice for exterior trim have traditionally been clear vertical-grain western red cedar or redwood. These woods resist decay, remain stable, work easily, and hold paint well. However, these materials have become costly, so lower grades of these woods and other species as well as composite materials are often used instead. Composite woods include LVL, hardboard, fiber–cement, and finger-jointed trims.

SIDING

The exterior cladding material applied to the walls of a residence is referred to as *siding*. Many different types of materials can be used as siding: wood boards with various profiles, applied either horizontally or vertically; imitation board sidings made of metal, vinyl, or concrete; plywood siding; wood shingles; stucco; imitation stucco; and brick or stone masonry (Figures 11.33 to 11.53).

The primary purpose of siding is to protect the walls from weather while presenting an attractive appearance. Siding materials must be able to endure moisture in the form of water vapor, rain, and melted snow; ther-

FIGURE 11.32
Trim is used effectively on this house to emphasize parts of the composition. The eye is drawn to the central bay, which is made entirely of windows and painted trim. The attic windows and rafters above are painted to complement the bay. Notice that the central windows have less glass area because they contain operable sash for ventilation. (*Photo by Rob Thallon*)

mal expansion and contraction due to extreme heat and cold; freezing temperatures that can cause cracking and splitting in the presence of moisture; and ultraviolet degradation from direct sunlight. Some siding materials are coated with paint or stain to prolong their durability. The effectiveness of siding materials is enhanced by the weather barrier just under them, which provides a second line of defense against the weather.

Virtually all of the sidings used today in North American residential construction are made from one of the six different materials discussed next.

Board Sidings

Horizontally applied board sidings, made either of solid wood or of wood composition board, are nailed in such a way that the nails pass through the sheathing and into the studs, giving a very secure attachment. This procedure also allows horizontal sidings to be applied directly over insulating sheathing materials without requiring a nail-base sheathing. (A nail-base sheathing is a material such as plywood or waferboard that is dense enough to hold nails.) *Siding nails,* whose heads are intermediate in size between those of common nails and those of finish nails, are used to give the best compromise between holding power and appearance when attaching horizontal siding. Siding nails should be hot-dipped galvanized or made of aluminum or stainless steel to prevent corrosion staining. Nailing is done in such a way that the individ-

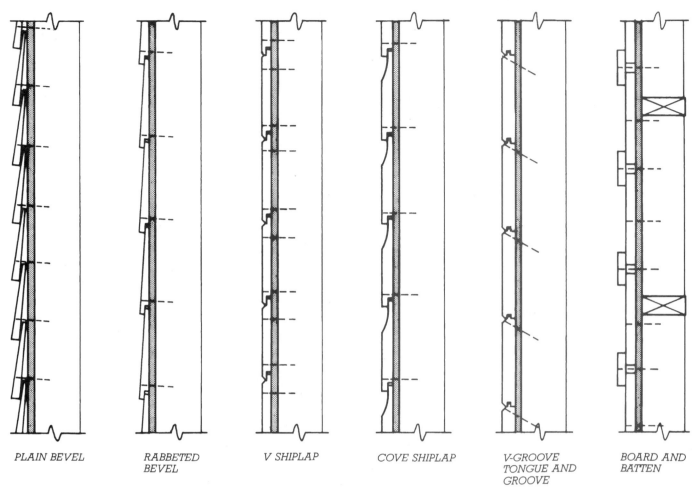

PLAIN BEVEL   RABBETED BEVEL   V SHIPLAP   COVE SHIPLAP   V-GROOVE TONGUE AND GROOVE   BOARD AND BATTEN

FIGURE 11.33
Six types of wood siding, from among many. The four bevel and shiplap sidings are designed to be applied in a horizontal orientation. Tongue-and-groove siding may be used either vertically or horizontally. Board-and-batten siding may only be applied vertically. The nailing pattern (shown with broken lines) for each type of siding is designed to allow for expansion and contraction of the boards. Nail penetration into the sheathing and framing should be a minimum of 1½ inches (38 mm) for a satisfactory attachment.

FIGURE 11.34
Workers installing cement board siding. The worker on the ground cuts the siding to size with a power saw while the two on the scaffold nail it to the wall with pneumatic nailers. The designer has specified alternating board widths to create visual interest. (*Photo by Rob Thallon*)

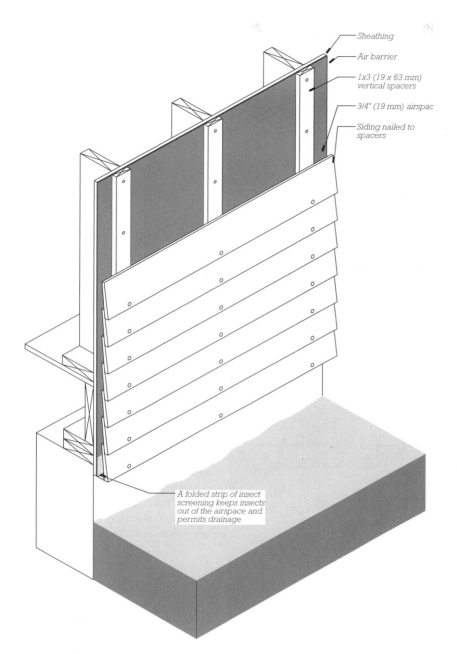

Sheathing

Air barrier

1x3 (19 x 63 mm) vertical spacers

3/4" (19 mm) airspac

Siding nailed to spacers

A folded strip of insect screening keeps insects out of the airspace and permits drainage

FIGURE 11.35
Rainscreen siding application. Any water that penetrates through joints or holes in the siding drains away before it reaches the sheathing.

ual pieces of siding may expand and contract freely without damage (Figure 11.33). The ends of horizontal siding boards are butted tightly to corner boards and window and door trim, usually with a sealant material applied to the joint during assembly.

More economical horizontal siding boards made of various types of wood composition materials are readily available. Caution should be exercised in selecting these materials, however, as some builders have experienced problems in service with excessive water absorption, decomposition, or fungus growth. Composition sidings made from wood fiber, sand, and portland cement appear to be the most

reliable and are increasing their share of the siding market (Figure 11.34). They are nonflammable, impervious to rot, and unattractive to insects. In addition, they hold paint better and are more dimensionally stable than wood or composite wood products. However, because of their weight and hardness, they are more difficult to handle and to cut than wood siding products.

In North America, horizontal siding boards are customarily nailed tightly against the sheathing. In Europe, and increasingly on this side of the Atlantic, siding is nailed to vertical wood spacers that are aligned over the studs (Figure 11.35). This *rainscreen* detailing creates a pressure equalization chamber behind the siding that acts to prevent water penetration in a wind-driven rain. It also provides a free drainage path for any water that leaks through the siding, and it permits rapid drying of the siding if it should become soaked with water, advantages that usually justify the additional expense of this procedure. Special attention must be given to corner boards and window and door casings to account for the additional thickness of the wall.

Vertically applied sidings (Figure 11.36) are nailed at the top and bottom plates of the frame and at one or more intermediate horizontal lines of wood blocking installed between the studs. Rainscreen detailing of vertical sidings is inherently difficult because the horizontal spacers do not allow free drainage of water from the cavity.

Heartwood redwood, cypress, and cedar sidings, which are decay resistant, may be left unfinished if desired, to weather to various shades of gray. The bare wood will erode gradually over a period of decades and will eventually wear thin and have to be replaced. Other woods must be either stained or painted to prevent weathering and decay. If these coatings are renewed faithfully at frequent intervals (usually every 3 to 6 years), the siding beneath will last indefinitely.

FIGURE 11.36
Board-and-batten siding similar to that historically used for barns. The modern version generally uses narrower boards and is applied over sheathing and a weather barrier. (*Photo by Rob Thallon*)

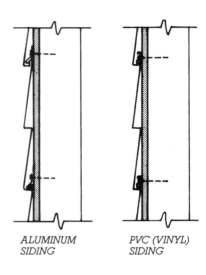

ALUMINUM SIDING   PVC (VINYL) SIDING

FIGURE 11.37
Aluminum and vinyl sidings are both intended to imitate horizontal lap siding in wood. Their chief advantage in either case is low maintenance. Nails are completely concealed in both systems.

Metal and Vinyl Sidings

Sidings formed of prefinished sheet aluminum or molded of vinyl plastic are usually designed to imitate wood sidings. Their primary advantage is that they will not decay and are generally guaranteed against needing repainting for long periods, typically 20 years (Figures 11.37 and 11.38). Such sidings do have their own problems, however, including the poor resistance of aluminum sidings to denting and the tendency of vinyl sidings to shatter on impact, especially in cold weather. Vinyl also expands and contracts much more than other siding materials (Figure 11.13), so installation details that allow for this

movement must be incorporated into its application (Figure 11.39). Although vinyl and aluminum sidings bear a superficial similarity to the wood sidings that they mimic, their details around openings and corners in the wall are sufficiently different that they are usually inappropriate for use in historic restoration projects.

Plywood Sidings

Plywood siding materials have become popular for their economy (Figure 11.40). The cost of the material per unit area of wall is usually somewhat less than for other siding materials, and labor costs tend to be relatively low

because the large sheets of plywood are more quickly installed than equivalent areas of boards. In many cases, the plywood siding can also function as structural sheathing, leading to further cost savings. All plywood sidings must be painted or stained, even those made of decay-resistant heartwoods, because their veneers are too thin to withstand weather erosion for more than a few years. The most popular plywood sidings are those that are grooved to imitate board sidings. The grooves also serve to conceal the vertical joints between sheets.

The largest problem in using plywood sidings for two- or three-story houses is how to detail the horizontal

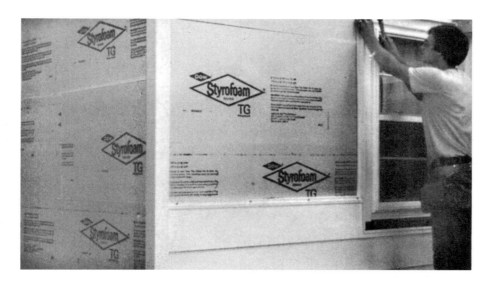

FIGURE 11.38
Installing aluminum siding over insulating foam sheathing. Special aluminum pieces are provided for corner boards and window casings; each has a shallow edge channel to accept the cut ends of the siding. Each horizontal "board" clips to the siding piece below and is nailed at the top edge in a slot that allows for expansion and contraction. (*Courtesy of Dow Chemical Company*)

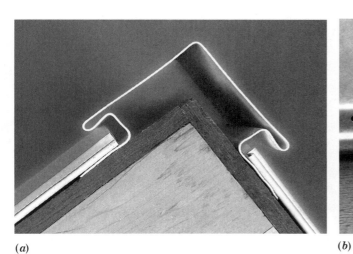

(*a*)

(*b*)

FIGURE 11.39
Vinyl siding details: (*a*) Plan view at a corner, showing slot with room for thermal expansion and contraction of the siding.
(*b*) Elevation view of the siding, showing nailing slots that also allow movement. (*Courtesy of Certainteed Corporation*)

FIGURE 11.40
Plywood siding is often disguised with vertical battens that cover the joints between panels and break up the 4 feet of space between. (*Photo by Rob Thallon*)

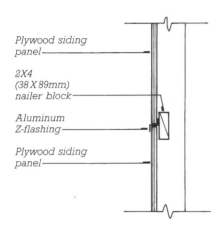

FIGURE 11.41
A detail of a simple Z-flashing, the device most commonly used to prevent water penetration at horizontal joints in plywood siding.

FIGURE 11.42
Applying wood shingle siding over asphalt-saturated felt building paper. The corners are woven as illustrated in Figure 11.44. (*Photo by Edward Allen*)

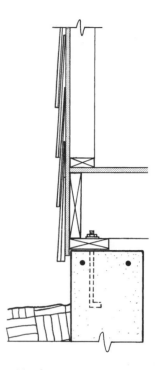

FIGURE 11.43
A detail of wood shingle siding at the sill of a platform frame house. The first course of shingles projects below the sheathing to form a drip, and is doubled so that all the open vertical joints between the shingles of the outside layer are backed up by the under-course of shingles. Succeeding courses are single, but are laid so that each course covers the open joints in the course below.

end joints between sheets. A *Z-flashing* of aluminum (Figure 11.41) is the most common solution and is clearly visible. The designer should include in the construction drawings a sheet layout for the plywood siding that organizes the horizontal joints in an acceptable manner. This will help avoid a random, unattractive pattern of joints on the face of the building.

Shingle Sidings

Wood shingles and shakes (Figures 11.42 to 11.47) require a nail-base sheathing material such as OSB, plywood, or waferboard. Either corrosion-resistant box nails or air-driven staples may be used for attachment. Most shingles are of cedar or redwood heartwood and need not be coated with paint or stain unless such a coating is desired for cosmetic reasons. Figures 11.44 and 11.45 show two different ways of turning corners with wood shingles.

The application of wood shingle siding is labor intensive, especially around corners and openings, where many shingles must be cut and fitted. Several manufacturers produce shingle siding in panel form by stapling and/or gluing shingles to wood backing panels. A typical panel size is approximately 2 feet high and 8 feet long (600 by 2450 mm). Several different shingle application patterns are available in this form, along with prefabricated woven corner panels. The application of panelized shingles is much more rapid than that of individual shingles, which results in sharply lower labor costs and in most cases a lower overall cost.

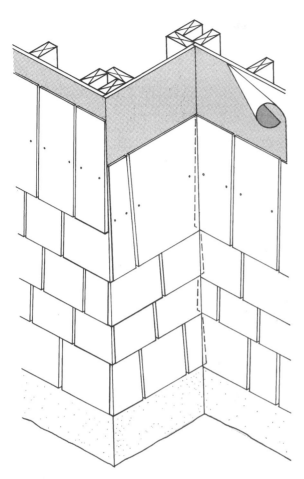

FIGURE 11.44
Wood shingles can be woven at the corners to avoid corner boards. Each corner shingle must be carefully trimmed to the proper line with a block plane, which is time consuming and relatively expensive, but the result is a more continuous, sculptural quality in the siding.

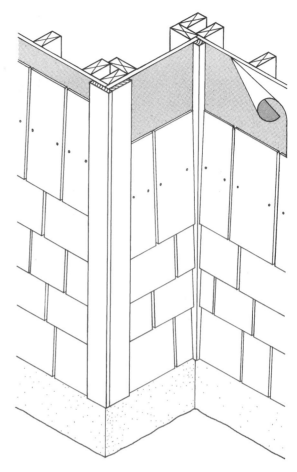

FIGURE 11.45
Corner boards save time when shingling walls, and become a strong visual feature of the building. Notice in this and the preceding diagram how the joints and nailheads in each course of shingles are covered by the course above.

FIGURE 11.46
Both roof and walls of this New England house by architect James Volney Righter are covered with wood shingles. Wall corners are woven. (*Photo © Nick Wheeler/Wheeler Photographics*)

FIGURE 11.47
Fancy cut wood shingles were often a featured aspect of shingle siding in the late 19th century. Notice the fish-scale shingles in the gable end, the serrated shingles at the lower edges of walls, and the sloping double-shingle course along the rakes. Corners are woven. (*Photo by Edward Allen*)

Stucco

Stucco, a portland cement plaster, is a strong, durable, economical, fire-resistant material for siding light frame buildings (Figure 11.48). Stucco is applied over galvanized metal lath, using galvanized steel or plastic accessories to prevent rusting in damp locations. One disadvantage of portland cement stucco is that it shrinks and is therefore prone to cracking. To compensate, stucco walls should be provided with control joints at frequent intervals to channel the shrinkage into predetermined lines rather than allowing it to cause random cracks. The curing reaction in stucco is the same as that of concrete. Stucco must be kept moist for a period of at least a week before it is allowed to dry, in order to attain maximum hardness and strength through full hydration of its portland cement binder.

Stucco can be applied either over sheathing or without sheathing. Over sheathing, asphalt-saturated felt paper is first applied as an air and moisture barrier. Then *self-furring metal lath* is attached with nails or screws. Self-furring metal lath is formed with "dimples" that hold the lath away from the surface of the wall a fraction of an inch to allow the stucco to key to the lath. If no sheathing is used, the wall is laced tightly with strands of line wire a few inches apart, and paperbacked metal lath is attached to the line wire, after which stucco is applied to encase the building in a thin layer of reinforced concrete. Stucco is usually applied in three coats, either with a hawk and trowel or by spraying (Figure 11.48). Pigments or dyes are often added to stucco to give an integral color, and rough textures are frequently used.

Synthetic Stucco

In recent years, the *exterior insulation and finish system (EIFS)*, or synthetic stucco, has rapidly grown in popularity (Figure 11.49). The system combines insulation with a weather barrier and provides a simple way to add texture and relief to the walls of a building. EIFS consists of rigid insulation panels adhered directly to the sheathing and then coated with a two-layer acrylic compound whose finished surface resembles stucco. The acrylic coating is both more water resistant and more flexible than traditional stucco.

FIGURE 11.48
Applying exterior stucco over wire netting. The workers to the right hold a hose that sprays the stucco mixture onto the wall, while the man at the left levels the surface of the stucco with a straightedge. The small rectangular opening at the base of the wall is a crawlspace vent. (*Courtesy of Keystone Steel and Wire Company*)

FIGURE 11.49
EIFS at the corner of a residence. Dimples appear on the surface where mechanical fasteners hold the rigid insulation to the sheathing. A horizontal strip of insulation has been added to create a shadow line on the building. A final coat of synthetic stucco will hide the imperfections now apparent in the surface. (*Photo by Rob Thallon*)

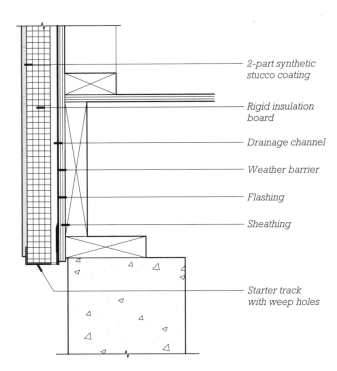

FIGURE 11.50
A detail of water-managed EIFS at the sill of a house. The weather barrier against the sheathing prevents moisture from entering the framing. An open drainage layer at the inner face of the rigid insulation allows gravity to carry the moisture to the base of the wall where it escapes through holes in the starter track.

2-part synthetic stucco coating

Rigid insulation board

Drainage channel

Weather barrier

Flashing

Sheathing

Starter track with weep holes

Early versions of the system relied on the acrylic coating to act as a complete barrier to the penetration of moisture. Called *barrier EIFS*, this system experienced a considerable failure rate from moisture that found its way behind the barrier at imperfections in the seals at windows and other locations. This moisture was trapped beneath the barrier with no way to escape and often caused sheathing, framing, and trim to rot. As a consequence of these failures, a second generation of synthetic stucco, called *water-managed EIFS*, has been developed. Water-managed EIFS assumes that some leaks will occur; accordingly, it incorporates a weather barrier and drainage channels at the inside of the rigid insulation (Figure 11.50). Both barrier EIFS and water-managed EIFS systems are now in use, with each being favored in particular climatic regions.

Masonry Veneer

Light frame buildings can be faced with a single wythe of brick or stone in the manner shown in Figure 11.51. The corrugated metal ties support the masonry against falling away from the building, but allow for differential movement between the masonry and the frame. Masonry materials and detailing are covered in Chapter 5.

Manufactured Stone

A less expensive alternative to masonry veneer is manufactured stone. Because the manufactured imitation is much lighter and thinner than the genuine material, it can be applied directly to the sheathing—much like a tile is applied to an interior wall (Figure 11.52). The thinness of the manufactured stone makes transitions to other siding materials easier than with masonry veneer because the manufactured stone and the siding are often nearly at the same plane, whereas there is usually a considerable offset between masonry veneer and other sidings. Manufactured stone as a material is discussed in Chapter 5.

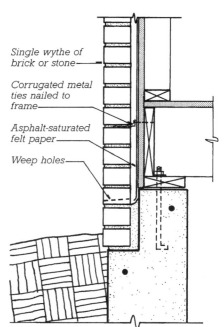

Single wythe of brick or stone

Corrugated metal ties nailed to frame

Asphalt-saturated felt paper

Weep holes

FIGURE 11.51
A detail of masonry veneer facing for a platform frame building. The weep holes drain any moisture that might collect in the cavity between the masonry and the sheathing. The cavity should be at least 1 inch (25 mm) wide. A 2-inch (50-mm) cavity is better because it is easier to keep free of mortar droppings that can clog the weep holes.

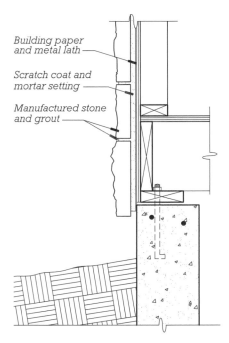

Building paper and metal lath

Scratch coat and mortar setting

Manufactured stone and grout

FIGURE 11.52
A detail of manufactured stone facing for a platform frame building. The thin lightweight ersatz stone panels are adhered to a setting bed that is very similar to a stucco wall. The facing is normally applied first to the top of the wall, and work then proceeds downward so that the work already in place does not have to be protected from falling mortar.

Coating Materials

Most architectural coating materials are formulated to include four basic types of ingredients: vehicles, solvents, pigments, and additives. The *vehicle* provides adhesion to the substrate and forms a film over it. *Solvents* are volatile liquids used to improve the working properties of the paint or coating. The most common solvents used in coating materials are water and hydrocarbons, which are commonly called paint thinners or mineral spirits. *Pigments* are finely divided solids that add color, opacity, hardness, abrasion resistance, weatherability, and gloss control to the coating material. *Additives* modify the coating material by affecting such properties as drying time, ease of application, and resistance to fading.

Coating materials fall into two major groups, water based and solvent based. *Water-based coatings* use water as a solvent, and cleanup is done with soap and water. Water-based coatings are commonly referred to as "latex paints." *Solvent-based coatings* use solvents other than water such as hydrocarbons, turpentine, alcohols, and esters. In everyday use, solvent-based paints are often referred to as "oil paints" or "alkyd paints." Solvent-based and water-based coatings each have many uses for which they are preferred; however, there are also some applications for which they can be used more or less interchangeably.

As a component of their air pollution control programs, many states now place stringent controls on the amount of *volatile organic compounds (VOCs)* that coatings may release into the atmosphere. As a consequence of this, and the fact that many painters sensitive to health issues prefer to work with latex paints, coating manufacturers have undertaken the development and promotion of new types of water-based coatings, including water-based clear coatings, and the production of solvent-based coatings has declined significantly.

FIGURE 11.53
Manufactured stone can make a wood-framed metal flue enclosure resemble a masonry chimney. (*Photo by Michelle Conley*)

EXTERIOR PAINTS AND COATINGS

Once the carpenters and siding subcontractors have finished installing all of the exterior features of the building, it is time to paint. Paints and other architectural coatings (such as stains, varnishes, lacquers, and sealers) protect and beautify these surfaces. The painting subcontractor may be called on to coat siding, windows and doors and their trim, eaves and rakes and various roof trims, gutters and downspouts, and other exterior construction such as porch structure and railings. The painter must work with the contractor to coordinate the painting of all these items so that the work progresses efficiently. Although the painting work may be interspersed with the other aspects of finishing the exterior walls, the painter is generally the last contractor to finish. When the exterior painting is done, the exterior shell of the house is complete (Figure 11.54).

Types of Coating Materials

The various types of architectural coatings can be defined by the relative proportions of vehicle, solvent, pigment, and additives in each.

Paints contain relatively high amounts of pigments. The highest pigment content is in *flat paints,* those that dry to a completely matte surface texture. Flat paints contain a relatively low proportion of film-forming vehicle. Paints that produce more glossy surfaces are referred to as *enamels.* A high-gloss enamel contains a very high proportion of vehicle and a relatively low proportion of pigment. The vehicle cures to form a hard, shiny film in which the pigment is fully submerged. A semigloss enamel has a somewhat lower proportion of vehicle, though still more than a flat paint.

Stains range from transparent to semitransparent to solid. Transparent stains are intended only to change the color of the substrate, usually wood and sometimes concrete. They contain little or no vehicle or pigment, a very high proportion of solvent, and a dye additive. Excess stain is wiped off with a rag a few minutes after application, leaving only the stain that has penetrated the substrate. Usually. a surface stained with a transparent stain is subsequently coated with a clear finish such as varnish to bring out the color and figure of the wood and produce a durable, easily cleaned surface.

Semitransparent stains have more pigment and vehicle than transparent ones and are not wiped after application. They are intended for exterior application in two coats and do not require a clear top coat. Also self-sufficient are the solid stains, which are intended for exterior use and are usually water based. These contain much more pigment and vehicle than the other two types of stains and resemble a dilute paint more than they do a true stain.

The *clear coatings* are high in vehicle and solvent content and contain little or no pigment. Their purpose is to protect the substrate, make it easier to keep clean, and bring out its inherent beauty, whether it be wood, metal, stone, or brick. *Lacquers* are clear coatings that dry extremely rapidly by solvent evaporation. They are employed chiefly in factories and shops for rapid finishing of cabinets and millwork. A slower drying clear coating useful for on-site finishing is known as a *varnish.* Varnishes and lacquers are available in gloss, semigloss, and flat formulations.

FIGURE 11.54
Painters finishing the exterior of a house. The efficiency of exterior painting depends more on the weather than any other factor.
(*Courtesy of Energy Studies in Buildings Laboratory, Center for Housing Innovation, Department of Architecture, University of Oregon*)

Field Application of Architectural Coatings

There is no aspect of painting and finishing work that is more important than surface preparation. Unless the substrate is clean, dry, smooth, and sound, no paint or clear coating will perform satisfactorily. Normal preparation of new wood surfaces involves patching and filling holes and cracks and sanding until the surface is smooth. Preparing metals may involve solvent cleaning to remove oil and grease and, on some metals, chemical etching of the bare metal to improve the paint bond. New masonry, concrete, and plaster surfaces generally require some aging before coating to assure that the chemical curing reactions within the materials themselves are complete.

There are a number of materials designed specifically to prepare a surface to receive paints or clear coatings. Paste fillers are used to fill the small pores in open-grained woods such as pine, fir, and cedar prior to finishing. Various patching and caulking compounds serve to fill nail holes and other larger holes in the substrate. A *primer* is a pigmented coating especially formulated to make a surface more paintable. A wood primer, for example, improves the adhesion of paint to wood. It also hardens the surface fibers of the wood so it can be sanded smooth after priming. Other primers are designed to be used as first coats for various metals or masonry materials. A *sealer* is a thin, unpigmented liquid that can be thought of as a primer for a clear coating. It seals the pores in the substrate so that the clear coating will not be absorbed.

Most finish coatings adhere better if the substrate is primed first. Specialized primers are produced for both interior and exterior surfaces of wood, various metals, plaster, gypsum board, and a range of masonry and concrete surfaces. Wood trim, casings, and siding in high-quality work are back primed by applying primer to their back surfaces before they are

FIGURE 11.55
To avoid the complication of masking, all of the roof overhangs of this house have been spray painted before the roofing or siding is applied. (*Photo by Rob Thallon*)

installed; this helps equalize the rate of moisture change on both sides of the wood during periods of changing humidity, which reduces cupping and other distortions.

Coatings should be applied only to dry surfaces, or they may not adhere. To prevent premature drying, freshly painted surfaces should not be exposed to direct sunlight, air temperatures during application should be between 50 and 90 degrees Fahrenheit (10 to 32°C), and wind speeds should not exceed 15 miles per hour (7 m/s). The paint materials themselves should be at normal room temperature.

A single coat of paint or varnish is usually insufficient to cover the substrate and build the required thickness of film over it. A typical requirement for a satisfactory paint coating is one coat of primer plus two coats of finish material. Two coats of varnish are generally required over raw wood, sanding the surface lightly to produce a smooth surface after the first coat has dried.

Paint and other coatings may be applied by brush, roller, pad, or spray. Brushing is the slowest and most expensive, but best for detailed work and for applying many types of stains and varnishes. Spraying is the fastest and least expensive, but also the most difficult to control. Roller application is both economical and effective for large expanses of flat surface. Many painters prefer to apply transparent stains to smooth surfaces by rubbing with a rag that has been saturated with stain.

If the painting is carefully coordinated with the installation of windows, doors, siding, and trim, the painter's work will be much easier (Figure 11.55). Soffit and eave trim, for example, can be prepainted on the ground before it is applied, saving the time and effort of erecting scaffolding and the difficulty of painting overhead. Window and door trim can be painted in place before the siding is installed, avoiding the necessity of masking the siding.

Deterioration of Paints and Finishes

Coatings are the parts of a building that are exposed to the most wear and weathering. Over time, they deteriorate and require recoating. The ultraviolet component of sunlight is particularly damaging, causing fading of paint colors and chemical decomposition of the paint film. Clear coatings are especially susceptible to ultraviolet (UV) damage, often lasting no more than a year before discoloring and peeling, which is why they are generally avoided in exterior locations. Some clear coatings are manufactured with special UV-blocking ingredients so that they may be used on exterior surfaces.

The other major force of destruction for paints and other coatings is water. Most peeling of paint is caused by water getting behind the paint film and lifting it off. The most common sources of this water in wood sidings are lumber that is damp at the time it is coated, rainwater leakage at joints, and water vapor migrating from damp interior spaces to the outdoors during the winter. Good construction practices and airtight vapor retarders can eliminate these problems.

Other major forces that cause deterioration of architectural coatings are oxygen, air pollutants, fungi, dirt, degradation of the substrate through rust or decay, and mechanical wear. Most exterior paints are designed to "chalk" slowly in response to these forces, allowing the rain to wash the surface clean at frequent intervals.

C.S.I./C.S.C. MasterFormat Section Numbers for Finishing the Walls

07400	**ROOFING AND SIDING PANELS**
07460	**Siding**
08200	**WOOD AND PLASTIC DOORS**
08210	**Wood Door**
08220	**Plastic Door**
08250	**Preassembled Wood and Plastic Door and Frame Unit**
08260	**Sliding Wood and Plastic Door**
08280	**Wood and Plastic Storm and Screen Door**
08500	**WINDOWS**
08550	**Wood Window**
08560	**Plastic Window**
08570	**Composite Window**
08600	**SKYLIGHTS**
08610	**Roof Window**
08620	**Unit Skylight**
08700	**HARDWARE**
08710	**Door Hardware**
08720	**Weatherstripping and Seal**
08750	**Window Hardware**
08800	**GLAZING**
08810	**Glass**
09900	**PAINTS AND COATINGS**
09910	**Paint**
09930	**Stains and Transparent Finishes**
09980	**Coatings for Concrete and Masonry**

SELECTED REFERENCES

1. In addition to consulting the reference works previously listed at the end of Chapter 9, the reader should acquire current catalogs from a number of manufacturers of residential windows. Most lumber retailers also distribute unified catalogs of windows, doors, and millwork that can be an invaluable part of the designer's reference shelf.

KEY TERMS AND CONCEPTS

weather barrier
asphalt-saturated felt
housewrap
glazed area
sash
jamb
casing
sill
mullion
muntin
prime window
storm window
combination window
replacement window
single-hung window
double-hung window
sliding window
projected window
casement window
awning window
tilt/turn window
skylight
roof window
weatherstripping
insect screen
wood window
clad window
vinyl window
glass-fiber-reinforced plastic window

light
true divided-light window
glazing
single glazing
double glazing
triple glazing
low-emissivity (low-e) coating
tinted glass
tempered glass
laminated glass
egress window
rough opening
frame size
window schedule
nailing fin
sidelight
transom
French door
terrace door
patio door
stile-and-rail door
flush door
fire door
prehung door
lockset
deadbolt
siding
board siding
siding nail

rainscreen principle
aluminum siding
vinyl siding
plywood siding
Z-flashing
shingle siding
stucco
self-furring lath
exterior insulation and finish system (EIFS)
barrier EIFS
water-managed EIFS
masonry veneer
manufactured stone
exterior paints and coatings
vehicle
solvent
pigment
additive
water-based coating
solvent-based coating
volatile organic compound (VOC)
flat paint
enamel
stain
clear coating
lacquer
varnish
primer
sealer

REVIEW QUESTIONS

1. In what order are exterior finishing operations carried out on a platform frame building, and why?

2. Which types of siding require a nail-base sheathing? What are some nail-base sheathing materials? Name some sheathing materials that cannot function as nail bases.

3. What are the reasons for the relative economy of plywood sidings? What special precautions are advisable when designing a building with plywood siding?

4. How are corners made when siding a building with wood shingles?

5. Specify two alternative exterior coating systems for a building clad in wood bevel siding.

6. What are the usual reasons for premature paint failure on a wood-sided house?

EXERCISES

1. For a completed wood frame building, make a complete list of the materials used for exterior finishes and sketch a set of details of the eaves, rakes, corners, and windows. Are there ways in which each could be improved?

2. Visit a lumberyard and look at all the alternative choices of sidings, windows, doors, trim lumber, and roofing. Study one or more systems of gutters and downspouts. Look at eave vents, gable vents, and ridge vents.

3. For a wood light frame building of your design, list precisely and completely the materials you would like to use for the exterior finishes. Sketch a set of typical details to show how these finishes should be applied to achieve the appearance you desire, with special attention to the roof edge details.

PLUMBING

Running water, both hot and cold, which we take for granted in a modern house, is a relatively recent development. It wasn't until the latter part of the 18th century that the flush toilet was invented, and the mass-produced pipes to carry water and waste did not precede the toilet by many years. Only in the past 50 years or so—since copper and plastic have replaced galvanized steel piping—have we been able to install plumbing that can be expected to last for the life of the building.

PLUMBING BASICS

Plumbing consists of two basic systems—a *supply system* of pressurized pipes that carry water into the building and a *waste system* of unpressurized pipes that drain sewage from the building by gravity. Water for the supply system is usually drawn from either a municipal water system or a well. The waste system drains to either a municipal waste facility or an on-site septic system. Residences within urban boundaries are usually connected to public water mains and sewers, whereas a rural property often has a private well and septic system.

The pipes within a house, whether for supply, waste, or vent, are connected to each other with *fittings,* primarily right-angle *elbows* and T-shaped *tees.* These connectors slip over the ends of adjacent pipes. In copper piping, the fittings are made of copper and are soldered to the pipes. In plastic piping, the fittings are glued or clamped to the pipes, depending on the piping system used. Cast iron waste and vent piping is connected with stainless steel clamps that are lined with neoprene gaskets for a tight seal.

The pressurized supply system is equipped with *shutoff valves* so that all or part of it can be deactivated for repair or replacement work. These include a main shutoff valve where the water service enters the house and small shutoff valves at each fixture.

Waste piping and *vent piping* must slope at code-specified angles that range from $\frac{1}{16}$ inch per foot to $\frac{1}{4}$ inch per foot, the exact angle depending on the size of the pipe. These slopes assure that the waste piping will drain efficiently and that any condensate that accumulates in vent piping will drain into the waste system. Removable cleanout plugs must be provided at strategic locations so that a plumber can gain access to the line by unscrewing one or more of these plugs if a line should become clogged.

Supply Piping

Water enters the house through a single *supply pipe* (usually 1 inch or ¾ inch inside diameter) that is buried deeply enough in the ground to prevent it from freezing. When connected to a public water supply, the supply pipe passes through a *water meter* that measures the residence's water usage. The meter is usually at the street in temperate or warm regions and inside the house with an electronic readout on the outside in cold regions. Once within the house, the supply pipe is not reduced in size until it reaches the water heater, after which it may be divided into branch lines to supply plumbing fixtures. The water supply to individual fixtures consists of a pair of supply lines—one hot, one cold—usually of ½ inch diameter, depending on the number of fixtures that are fed by the same pair of pipes (Figure 12.1).

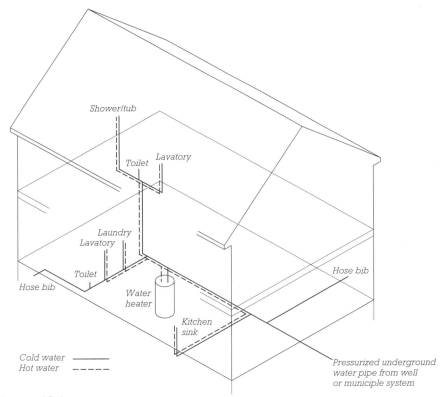

FIGURE 12.1

A typical residential water supply distribution system. Water enters the house through a buried line and branches into two parallel lines. One is for cold water, and the other passes through the water heater and supplies hot water. Most plumbing fixtures require both cold and hot water supplies, but some fixtures such as toilets (water closets) require only cold water.

Supply piping can consist of *copper pipes* with soldered fittings, *polyvinyl chloride plastic (PVC) pipes* with glued fittings, or flexible *cross-linked polyethylene plastic (PEX) tubing* with crimped fittings (Figure 12.2). Copper has been the standard of quality since the 1950s, but PVC, introduced in the 1970s, is seen by some to be an improvement because it is faster and safer to install (no flame to melt solder is required) and is not corroded by minerals in the water. PEX, the most recent introduction, is even simpler to install than PVC because it can bend like a garden hose, is quieter with water running through it, and it can withstand expansion when freezing.

The *water heater* is usually a large, insulated storage tank (typically 40 to 55 gallons) heated by either electricity or gas (Figure 12.3). Electric water heaters are simpler and therefore less expensive to install, whereas gas heaters heat water faster and can therefore be smaller, but their installation is complicated by the requirement to vent combustion products to the outdoors. Hot water may also be supplied by a *demand system*, which passes water through a heating element to heat it instantaneously as it is needed and does not have a storage tank. The advantages of a demand system are that the system is compact and that heat is not lost from stored hot water. Demand systems use either electricity or gas as an energy source and are more expensive than conventional water heaters. The most common use of a demand system is for the "instant hot" faucets found at many kitchen sinks.

FIGURE 12.2
The plumbing department at this home improvement center displays three generations of piping. The machine on the floor cuts and threads galvanized iron pipe. The racks hold a variety of sizes and types of rigid copper and plastic pipe. Out of view is the most recent generation of flexible plastic pipe. (*Photo by Greg Thomson*)

FIGURE 12.3
A standard water heater stands next to a forced-air furnace in a garage. This example uses natural gas as a fuel. One of the flexible copper pipes at the top of the heater supplies cold water to the heater, and the other conveys hot water to the house. The large vertical pipe at the top is a vent to exhaust combustion gases to the outdoors. The vertical pipe at the right side is connected to a pressure relief valve at the top of the heater tank and drains to the outside. The gas supply piping can be seen at the lower left of the heater. The heater is strapped to the wall to avert catastrophic damage in the event of an earthquake. The efficiency of the heater can be increased with the addition of an insulating jacket. (*Photo by Rob Thallon*)

Supply piping should be wrapped in thermal insulation. This usually takes the form of either a plastic foam jacketing or slit tubes of closed-cell rubber, which are slipped over the pipes, or a wrapping of glass fiber insulation encased in a vapor-resistant foil or plastic film (Figure 12.4). This insulation has several functions: It helps to keep cold water cold in the pipes, and hot water hot. It helps conserve the energy used to heat the water. In humid conditions, it prevents condensation of moisture on cold water pipes.

In locations that are subject to temperatures that will never drop more than several degrees below the freezing mark, pipe insulation will prevent water from freezing in the pipes. However, no amount of pipe insulation can prevent a pipe from freezing if it is exposed to very low temperatures for a period of hours. This is because a pipe contains only a very small volume of water that does not contain enough heat to overcome heat losses from its large surface area. In cold climates, it is best to design every building so that no supply pipes are located in exterior walls or unheated spaces. If this is not possible and piping must run in crawlspaces, exterior walls, attics, or other unheated locations, it is imperative that the pipes be located on the warm side of the house insulation, next to the interior finish material. The designer should highlight this requirement in the plumbing specification and follow up with careful on-site inspection to be sure that it is implemented. As long as the house itself is heated, a pipe in this location will never freeze. If a pipe falls within the house's insulation or outside it, it is likely that it will freeze, even if it is jacketed with its own insulation. When a pipe freezes, the expansion of the water when it turns to ice will usually split the pipe open. As soon as warm weather returns and the pipe thaws, large quantities of water will flood the house.

Waste Water Piping

Waste water piping works much like a natural watershed, in which numerous small streams feed into a river that ultimately empties into the sea. In a residential wastewater system, the flow starts at the individual fixtures (toilets, sinks, tubs, etc.). From those fixtures, the waste flows through branch lines to a main drain that passes through the foundation to the building sewer. The building sewer then empties into either a municipal sewer or a septic system.

The pipes in this system have specific names, depending on their use (Figure 12.5). Lines connected to toilets are at least 3 inches in diameter and are called *drain piping* or *soil lines*. Lines free of solids, called *waste piping*, can be smaller—usually 1½ or 2 inches in diameter. The line at the base of the system that is the receptor for all the drain and waste piping is the *building drain*. At a distance of 2 feet beyond

FIGURE 12.4
A slotted foam insulation jacket is installed on this copper supply piping to help conserve energy. When insulation is used on pipes to prevent freezing or condensation, the entire length of pipe must be covered. (*Photo by Greg Thomson*)

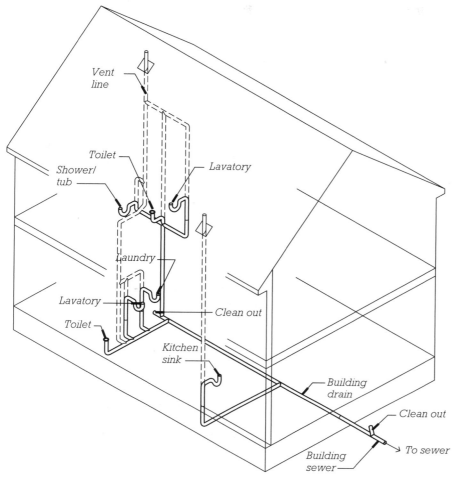

FIGURE 12.5
A typical residential wastewater system. All fixtures drain to the building drain through sloping or vertical branch lines. At the exterior of the foundation, the building drain becomes the building sewer. All fixtures are vented to the exterior through a network of vent pipes (shown with broken lines).

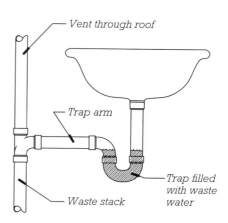

FIGURE 12.6
A typical plumbing trap prevents noxious waste line gases from escaping into the living space.

Fixture	Drain/Waste Pipe Diameter (in.)
Kitchen sink	2
Bar sink	1½
Washer	2
Laundry tub	2
Lavatory	1½
Shower	2
Tub	2
Toilet	3

FIGURE 12.7
Typical drain/waste pipe diameters for the plumbing fixtures found in the average house. Actual sizes are determined by the plumber based on building code formulas that assign drainage fixture units (DFU) to various fixtures. A 2½ bathroom house with a dishwasher and laundry facilities usually has fewer than 20 DFU. Thus, all the fixtures in a residence typically discharge into a 3-inch building drain, which can hold 42 DFU when sloping at ¼ inch per foot.

the foundation wall, the building drain becomes the *building sewer*.

The waste water system contains noxious, flammable sewer gases. These gases are kept out of the house by means of a *P-trap* at each fixture that seals the end of the pipe with water (Figure 12.6). When wastewater is released from the fixture, it displaces water held in the trap. A vent on the sewer side of each trap supplies air so that draining water cannot create a suction to pull water out of the trap. The *vent piping* extends through the roof of the house. *Cleanouts* are required at strategic points in the system to allow reasonable access for the cleaning of clogged lines.

The entire system, consisting of drain pipes, waste pipes, and vent pipes, is called a *drain–waste–vent (DWV) system*. Materials for residential DWV systems generally consist of black *ABS pipe* (composed of an acrylonitrile–butadiene–styrene copolymer) that is glued at the fittings. This pipe is the least expensive of all approved alternatives, and its only significant drawbacks are that it transmits the sound of draining water and expands and contracts considerably with thermal changes. Other waste pipe materials include *PVC pipe*, used because of its lighter weight, and *cast iron pipe*, used because its mass makes it quieter than plastic. Pipe diameters for resi-

dential DWV systems usually include 1½ inches, 2 inches, 3 inches, and sometimes 4 inches (Figure 12.7). All DWV components, including vents, must be sloped to drain. All plastic pipes require special attention to thermal expansion and contraction.

PLANNING FOR PIPES

Planning for plumbing begins early in the design phase of a residential project. The location of the water main or the well determines the direction from which the supply pipe will enter the building. The location of the public sewer or the septic system establishes the approximate route for the building drain. Once the basic routes for plumbing to enter and exit the house have been established, sleeves may be designed into the foundation to allow for their passage.

The framing of the building must be planned by the designer to accommodate plumbing. Supply pipes and the smaller diameters of waste and vent pipes will fit within the stud or joist cavities, but special provisions must be made for the large drain pipes. A toilet drain pipe, for example, must be in a particular location, must have a prescribed slope, and can be difficult to route through framing—particularly from upper floors (Figure 12.8). As framing proceeds, the framer can also make the plumber's job more efficient by paying attention to the layout of studs and joists around the plumbing fixtures (Figure 12.9).

Outside the boundaries of the house, all pipes are buried. Usually, the same trench may be used for more than one utility (i.e., water supply, electrical and phone lines), so coordination of plumbing with other utilities is beneficial.

Although the clustering of plumbing fixtures around a plumbing wall (Figure 12.10) does minimize the amount of pipe that must be used and can make the plumber's job easier, it is not usually a high priority except in multiunit projects where savings are multiplied. In custom houses, plumbers generally base their bid on the number of fixtures in any case, ignoring the relative efficiency of the layout. It is always wise to locate the water heater centrally so as to deliver hot water to each fixture as quickly as possible and to specify that vents be tied together in the attic to avoid multiple roof penetrations.

FIGURE 12.8
A typical water closet flange measures 6 inches (150 mm) from the bottom of the subfloor to the bottom of the bend. The drain line must slope at ¼ inch per foot (1:50). The joists in this example must therefore be a minimum of 7 inches deep for the waste line to travel the 4 feet across the bathroom.

FIGURE 12.9
The framer, knowing that the plumber will need to locate tub and shower valves centered on the tub, has located studs accordingly. Joists below the tub have also been located to allow for the tub trap. The plumber has installed the tub and rough-in supply and waste piping. The framer will now add blocking around the tub edge to provide nailing for finish wall surfaces. (*Photo by Rob Thallon*)

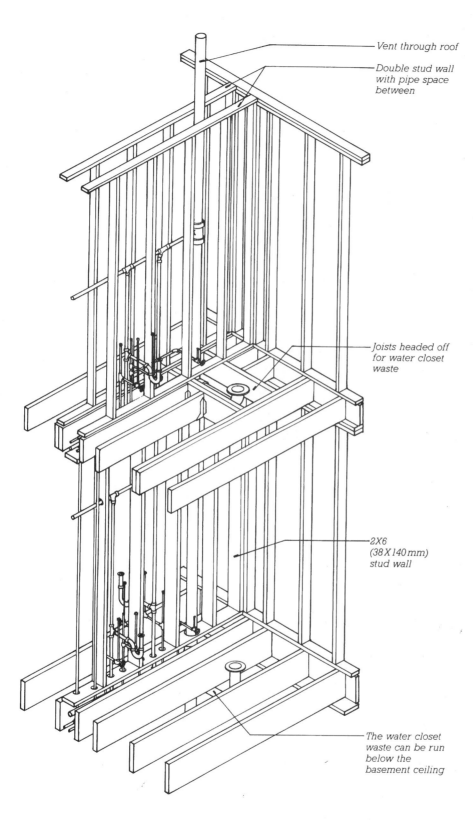

Vent through roof

Double stud wall with pipe space between

Joists headed off for water closet waste

2X6 (38 X 140 mm) stud wall

The water closet waste can be run below the basement ceiling

FIGURE 12.10
The plumber's work is easier and less expensive if the building is designed to accommodate the piping. The "stacked" arrangement shown here, in which a second-floor bathroom and a back-to-back kitchen and bath on the first floor share the same vertical runs of pipe, is economical and easy to rough in, as compared with plumbing that does not align vertically from one floor to the next. The double wall framing on the second floor allows plenty of space for the waste, vent, and supply pipes. The second-floor joists are located to provide a slot through which the pipes can pass at the base of the double wall, and the joists beneath the water closet (toilet) are headed off to house its waste pipe. The first floor shows an alternative type of wall framing using a single layer of deeper studs, which must be drilled to permit horizontal runs of pipe to pass through.

ROUGH-IN PLUMBING

The scheduling of the plumber varies depending on the type of construction. With slab-on-grade construction, the plumber arrives on the site with the foundation crew because all subfloor piping must be in place before the slab is poured (see Figure 12.11). In houses with a crawlspace, the plumber should complete work in the crawlspace before the subfloor sheathing has been installed (Figure 12.12). If the house has a basement, the plumber may wait to start work until after the roof has been framed; then the workplace will be protected from the weather and all the rough plumbing can be installed at the same time (Figure 12.13).

FIGURE 12.11
The plumber has roughed in the drains for this residence to be built on a slab-on-grade foundation. Set at the specified level and sloped to drain, the pipes are supported by the gravel under the future slab. More gravel will be added before the concrete is poured. The batt insulation taped to the vertical sections of pipe will prevent the concrete from making direct contact with the pipes and permit independent movement of plumbing and slab. (*Photo by Rob Thallon*)

FIGURE 12.12
Both supply and waste piping are typically installed in a crawlspace floor structure before the sheathing is applied. The supply piping is located at the center of the joists, while the waste piping hangs below. Notice the plastic straps holding the waste piping in place. (*Photo by Rob Thallon*)

This first phase of the plumbing work is called *rough-in plumbing*. It consists of installing virtually all the supply and waste piping within the building. Bathtubs and manufactured showers—because they are very large and will not fit through doors and because they must be connected below the floor—are typically set during the rough-in plumbing phase as well (Figure 12.14). Generally, the waste pipes are installed first because they are the largest and must slope properly to drain. The plumber will drill holes in framing members for these pipes according to guidelines established by the building codes. The pipes are then installed and glued together with *fittings*. The plumber must also install supports for the pipes, which are placed at code-required intervals (Figure 12.15). The supports consist of plastic pipe hangers or a combination of blocking and *plumber's tape*—a flexible galvanized metal strap with holes for fasteners. Sometimes, framing must be modified slightly to accommodate the plumbing. This may be done by either the plumber or the framer.

The supply pipes are installed during the rough-in phase as well. Because they are much smaller than waste pipes and do not rely on gravity, supply pipes can be routed through the framing without much planning.

FIGURE 12.13
At the basement ceiling, the plumber installs the plastic waste pipes first to be sure that they are properly sloped to drain. The copper supply pipes for hot and cold water are installed next. (*Photo by Edward Allen*)

FIGURE 12.14
Tubs and preformed shower enclosures are installed during the rough-in plumbing phase because their trap is connected below the floor and because they are often too large to maneuver through a fully framed house. (*Photo by Rob Thallon*)

Small holes are drilled for pipes to pass through members, and clamps are used to secure pipes running parallel to framing. A variety of manufactured clips and brackets are used to hold the fixture end of the pipes in place, and the pipes are temporarily capped to keep out dirt and to allow them to be pressure tested. Where pipes must pass close to the surface of the framing, a steel *nailing plate* must be installed on the face of the framing to prevent nails driven during the finish phase of construction from penetrating the pipes.

Once the rough plumbing is installed, it must be tested to detect any leaks, which would be costly to repair once the plumbing is covered with finish materials. The plumber performs the test, and the plumbing inspector verifies it. The supply pipes are tested under pressure with either water or air. The waste system is tested by plugging the lowest part of the drain with an inflatable test plug, capping all branch lines, and filling the system with water until it emerges from a vent on the roof. At the same time, the plumbing inspector confirms that the pipes do not leak; all pipes are checked for proper sizing, for adequate support, and for protection by nailing plates where they are near the surface of the framing.

FINISH PLUMBING

After the floors are finished, the cabinets installed, and the appliances in place, the plumber returns to the job site to install the plumbing *fixtures* (Figure 12.16). The water heater is installed at this time. Toilet(s) are bolted to the floor and their cold water supply connected. Sinks are fastened and sealed to countertops, connected with traps to the waste pipes in the walls, and their faucets installed and connected to hot and cold water supply. The tub fillers, shower valves, and shower heads are installed and adjusted in the bathroom(s). The dishwasher is supplied with hot water and its waste line is connected to the kitchen sink drain. When all the fixtures have been installed and tested, the plumber calls the plumbing inspector for a final inspection. The inspector looks to see that code-approved fixtures have been installed, that traps have been installed properly, and that there is adequate vertical distance (an air gap) between spouts and the flood level of the receptacle into which they flow.

PLUMBING CODES

The safety and quality of plumbing is governed by the *Uniform Plumbing Code*. This code has been adapted and incorporated into the International Residential Code. Residential plumbing systems must pass at least two on-site inspections: the rough-in inspection and the final inspection, as

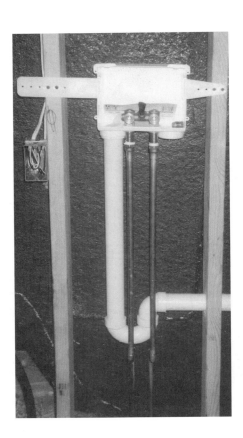

FIGURE 12.15
Mounting brackets for piping are now standard materials for plumbers during the rough-in phase of their work. This washer assembly has hot and cold supply valves and a drain for the washer discharge pipe integrated with the mounting bracket. (*Photo by Edward Allen*)

FIGURE 12.16
During the finish plumbing phase, the plumber installs fixtures and connects them to supply and waste lines. In the case of a sink, (1) the strainer(s) and trap are installed on the sink, (2) the sink is fastened to the countertop and sealed, (3) the trap is connected to the waste line, (4) the valves are installed on the sink, (5) the hot and cold water are connected to the valves, and (6) the system is tested for leaks. (*Photo by Rob Thallon*)

described previously in the Rough-in Plumbing and Finish Plumbing sections of this chapter.

Plumbers are intimately aware of the requirements of the code, but architects and builders must also be familiar with specific provisions in the code that will impact the design and construction of the residence. These include the following:

- The minimum number of fixtures in a dwelling

- The materials that can be used for pipes and fittings

- The minimum size of pipes based on how many fixtures are connected to them

- The location and maximum spacing of cleanouts

- The slope of waste drainage piping

- The support of piping

- The minimum distance of vents from operable windows to prevent sewer gas from entering the house

OTHER PIPING SYSTEMS

In addition to the supply and waste systems supplied and installed by the plumbing subcontractor, there are a number of other piping systems installed in a residence by the plumber or other subcontractors. The installation of these systems is coordinated with the installation of rough plumbing, wiring, and heating.

The Building Sewer

The plumbing subcontractor is generally responsible for extending the supply and waste pipes only about 2 feet from the house. In the case of the waste system, this is the end of the building drain and the beginning of the building sewer. The installation of the building sewer that connects the building drain to a municipal sewer or septic system is generally performed by either the excavator (sometimes in collaboration with the plumber) or the septic system subcontractor. Both have heavy equipment to excavate ditches and both have licenses to do this work (Figure 12.17).

Water Supply Line

The water line between the house and the municipal water system or the well may or may not be installed by the plumber. With municipal systems, this line may be installed by the water utility itself. When the supply comes from a well, the well driller will often install the supply line to the house. In either case, the water supply line is often laid in a common trench with other utilities such as electricity, gas, or phone.

Gas Piping

Piping for natural gas or propane is common in residences. This must be

FIGURE 12.17
A building sewer that serves 10 apartments is being installed in this trench. The vertical pipe at the far end of the sewer is a cleanout and will be cut flush with the ground and capped when the finish grade level is established. The sewer is sloped to drain and is supported by a gravel bed. Other utilities will also be installed in the same trench before it is backfilled. (*Photo by Rob Thallon*)

resistant to heat in order to safely contain its combustible contents in the event of a house fire. Until recently, all gas piping was made of *black pipe*—a combination of steel and wrought iron that is resistant to heat and to the corrosive actions of both gas and moisture. However, black pipe has to be threaded at the site wherever it is joined—a very costly operation. A plastic-coated corrugated stainless steel pipe that bends without fittings has now replaced black pipe because its installation costs are much lower. The few connections that are required with this new pipe are made with pressure fittings. Plastic piping may be used for underground gas lines. All gas piping must be installed by a plumber licensed to do so.

Fire Sprinkler System

Residential sprinkler systems that automatically extinguish incipient fires are becoming more common in both multifamily and single-family hous-

ing. Like their big brothers in commercial and industrial buildings, these systems minimize dollar losses and save many lives. Fed by a supply line connected to the water supply near its source, the system provides at least one sprinkler in the ceiling of each room. The piping is generally made of either copper or plastic and can be installed by a plumber or a specialized subcontractor. The system is typically coordinated with and installed at the same time as the plumbing (Figure 12.18).

FIGURE 12.18
A roughed-in residential fire sprinkler system. A ceiling sprinkler head will be connected to the pipe in the left foreground. At the right, a pipe projects through the wall where an exterior sprinkler head will be installed over a window. (*Photo by Rob Thallon*)

Solar Water Heating

Piping for active solar heating applications is usually installed concurrently with the rough plumbing. Piping is typically similar in size and material to the domestic supply system. The system may be installed by a plumber or a licensed subcontractor specializing in solar heating. It is not uncommon to install pipes for future use in a solar system, such as pipes that extend from attic to basement, which could later connect an active solar panel on the roof with a water heater in the basement.

Hydronic Heating System

Hydronic heating systems heat water in a boiler and convey this heat to the rooms of a house through pipes. These systems may operate using high pressure equivalent to supply piping or use lower pressure. In the former case, the system is typically installed by a licensed plumber. In the latter case, the system may be installed by a separate specialized subcontractor. In both cases, the piping is similar to residential supply piping and is installed concurrently. These systems are discussed further in Chapter 13.

Central Vacuum System

A central vacuum system is powered by a central vacuum/collector unit that is connected by means of 2-inch rigid PVC piping to outlets distributed throughout the house (Figure 12.19). Low voltage wiring is run with the pipe and is connected to a switch at each outlet that turns on the vacuum when a vacuum hose is inserted.

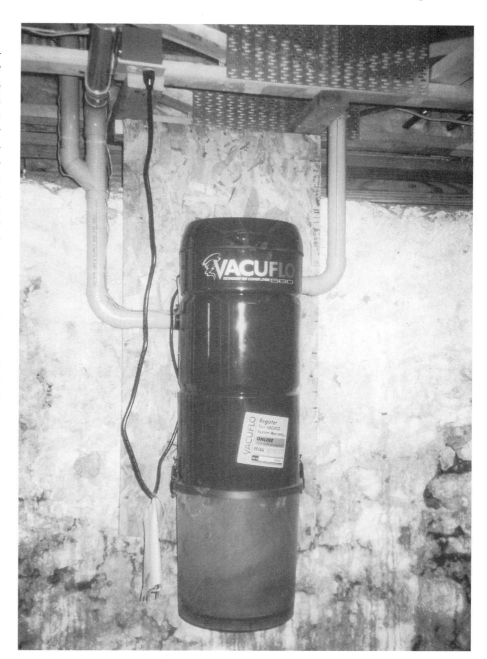

FIGURE 12.19

A central vacuum system mounted in the basement of a house. Dirt and dust are sucked into the machine through the 2-inch PVC pipes at the left, and air is exhausted to the outside through the pipe at the right. (*Photo by Greg Thomson*)

```
C.S.I./C.S.C.
MasterFormat Section Numbers for Plumbing
15100            BUILDING SERVICE PIPING
15140            Domestic Water Piping
15150            Sanitary Waste and Vent Piping
15180            Heating and Cooling Piping
15300            FIRE PROTECTION PIPING
15400            PLUMBING FIXTURES AND EQUIPMENT
15410            Plumbing Fixture
15480            Domestic Water Heater
```

SELECTED REFERENCES

1. Hemp, Peter. *Plumbing a House.* Newtown, CT: Taunton Press, 1994.

A clearly written and illustrated guide to the design and installation of supply and DWV piping. Sections on tools and materials complete the coverage of rough plumbing.

2. Hemp, Peter. *Installing and Repairing Plumbing Fixtures.* Newtown, CT: Taunton Press, 1994.

The companion to the previous reference. This one describes the materials and process of finish plumbing.

KEY TERMS AND CONCEPTS

supply system
waste system
plumbing fitting
elbow
tee
shutoff valve
supply pipe
water meter
branch line
copper pipe
PVC pipe
cross-linked polyethylene (PEX) tubing

water heater
demand water heater
drain–waste–vent (DWV) system
drain piping
soil line
waste piping
building drain
building sewer
P-trap
vent piping
cleanout
drain–waste–vent (DWV) piping
ABS pipe

cast iron pipe
rough-in plumbing
plumber's tape
nailing plate
plumbing fixture
Uniform Plumbing Code
gas piping
black pipe
fire sprinkler system
solar water heater
hydronic heating system
central vacuum system

REVIEW QUESTIONS

1. What are the three most commonly used types of supply piping? What are the advantages and disadvantages of each?

2. What are the reasons for insulating supply piping?

3. What is the difference between a drain line and a waste line?

4. Explain in detail the reason for vent piping.

5. How does a trap prevent sewer gases from entering the house?

6. Why does the plumber usually have to schedule at least three phases for the completion of the plumbing of a new house?

EXERCISES

1. Design a new bathroom for an existing house. Make a three-dimensional drawing showing all of the supply and waste piping required to serve the fixtures in the new room. If possible, indicate how to connect the new proposed piping to the existing plumbing in the house.

2. Go to a home improvement store and make a list of all the materials required to add a lavatory in your bedroom.

3. Call a local plumbing contractor and ask permission to visit a residential construction project with the plumber. Describe the phase of construction under way and take notes about exactly what the plumber will be doing that day.

HEATING AND COOLING

Modern heating and cooling systems, designed to maintain temperature and humidity within a zone of human comfort, are often the most sophisticated system of a house. They feature efficient equipment, complex distribution systems, and sensitive controls. Their installation can involve the trades of plumbing, sheet metal work, electrical work, insulation, and carpentry. The same system may be called on to heat during the winter months, cool during the summer, control humidity all year long, and remove dust and pollen from the air. As a category, systems that condition the interior environment for human comfort are referred to as *heating, ventilating, and air conditioning (HVAC) systems.*

The design of an efficient HVAC system should take place simultaneously with the design of the building itself. Simple systems such as baseboard heaters may be designed by rules of thumb, but more sophisticated systems that both heat and cool are often designed by an engineer in collaboration with the building designer. In a typical scenario, the design process starts with a proposed preliminary design by the building designer based on the needs of the client. The engineer or heating system specialist then develops the proposal sufficiently to make suggestions about the size and location of equipment and a heat distribution system. Next, the engineer and designer working together refine the design sufficiently to specify its details in drawings and specifications. In complicated buildings, there may be several cycles of discussion between engineer and designer, but in simpler buildings, one round of discussions may be adequate. In all cases, the building designer must integrate the system into the planning, the structure, and the finish details of the house, and the engineer will refine the details of the system as it is being installed.

A well-designed system takes into account the many energy conservation principles now well known in the house building industry. With a coordinated approach between building designer and heating engineer, the system can be both efficient and effective. Extra insulation in the envelope of the building can result in a smaller and less expensive system. For heating systems, the potential for passive solar assistance should be considered. For cooling systems, the simple principles of shading and natural cross ventilation can have a significant beneficial effect.

Choosing a System

The ideal HVAC system effectively provides for human comfort, is efficient in terms of energy consumption, and can be upgraded or extended over time. There are so many factors to consider that the choice is difficult, but consumers have first-hand experience with most systems, and design professionals are becoming more skillful in explaining the strengths and weaknesses of each (Figure 13.1). Recently, designers and contractors have begun to use life-cycle costing, which examines all energy-related costs over the life of the building, as an effective tool for selling responsible investment in long-term energy savings.

There are four basic categories of HVAC systems from which to choose:

- *Forced-air systems,* which heat (and cool) air at a central location and force it through ducts to all the rooms of a house. The process of moving heat by means of air is called *convection.*
- *Radiant panel systems,* which employ water or electricity to heat large areas such as floors or ceilings. The heat travels directly from the heated surfaces by means of *radiation* to reach the objects and occupants of a house.
- *Hydronic baseboard systems,* which operate with a centralized boiler that supplies heated water to small, locally controlled radiators. These systems employ both convection and radiation.
- *Local source systems,* which heat (or cool) by means of small local units such as baseboard heaters or wall heater/air conditioners built into each room of a house. These systems also employ both convection and radiation.

Each of these categories has several variations. The selection of the system most suited to the particular circumstances of each building should take all of the following into consideration:

	Advantages	Disadvantages
Forced air	• Rapid response time • Ability to filter air and control humidity • Ability to both heat and cool	• Relatively noisy • Ducts are bulky • Difficult to zone
Radiant panel	• Quiet • Invisible • Zoning is simple	• Slow response time • No air filtration
Hydronic baseboard	• Zoning is simple	• Baseboard placement can limit furniture arrangements • No cooling or air filtration
Local source	• Zoning is automatic • Low first cost	• Most are relatively noisy • No air filtration

Figure 13.1
A comparison of the principal types of heating systems. Local source systems include electrical baseboard heaters, wall heaters, and radiant stoves.

Climate. The difference between a cold climate where heating is the dominant need and a warm climate where cooling is required will have a significant impact on the selection of a system. More subtle climatic differences such as high or low humidity, wind, and other factors will also impact system selection.

Construction system. If a house is built on a slab, the options for heating systems are quite different than those for a house with a basement or crawlspace. Radiant systems are easily incorporated into a slab, for example, whereas forced-air systems are not.

Construction cost. There is a wide range of costs associated with various heating systems. As a rule, the first cost of electric baseboard heaters is the lowest, whereas a sophisticated heat pump system can cost many times more.

Energy source. The availability of inexpensive energy often affects the selection of a heating system. Inexpensive electric baseboard heaters and wall heaters are common in areas with low electrical rates, but rare in areas where electricity is expensive.

Zoning. It is often desirable to create distinct *heating zones* in a house. There may be a desire to have sleeping areas cooler, bathing and dressing areas warmer, or rooms that are seldom occupied at different temperatures. The desire to create distinct heating zones in a house will lead to the selection of systems that are reasonably simple to zone.

Air quality. Humidity control and air filtration are possible with some systems, but not with others.

Aesthetics. Some systems such as the radiant floor are free of grills, registers, or appliances and allow total flexibility of furniture arrangement. Other systems such as freestanding heaters can become the focus of a room.

Distribution system. Centralized systems can have heating and cooling distributed throughout the house by air (forced-air systems) or water *(hydronic systems).* Forced-air systems require relatively large ductwork, while hydronic systems require only small pipes.

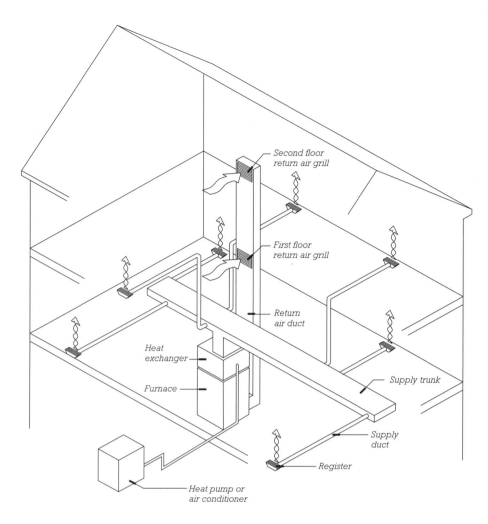

Second floor
return air grill

First floor
return air grill

Return
air duct

Heat
exchanger

Furnace

Supply trunk

Supply
duct

Register

Heat pump or
air conditioner

FIGURE 13.2
A forced-air system in a two-story house with a basement. The furnace is in the basement and blows warm air through supply ducts to registers in the floor near the exterior wall of each room. The air returns to the furnace through a centrally located return air duct with a return air grill near the ceiling of each floor. With the addition of a heat pump or an air conditioner, the system is capable of delivering cool air as well.

FORCED-AIR SYSTEMS

Forced-air heating is by far the most prevalent heating system in North America, accounting for over 60 percent of all residential heating. The *forced-air system* is composed of a *central furnace* that heats air and forces it through small *supply ducts* to each room. From the room, the air travels back to the furnace via a *return-air duct*.

A fan located in the furnace moves the loop of heated air. The ducts that conduct the air between the furnace and the rooms are usually insulated and are sized for the volume of air passing through them (Figure 13.2).

Advantages of forced-air systems include the ability to heat a house rapidly and the potential to integrate other climate control devices with it. The most common addition

to the system in moderate climates is a *heat pump* (Figure 13.3), which can increase heating efficiency by a factor of about 3. A typical heat pump works like a refrigerator in reverse and will extract heat from outdoor air even though the air is cool. Air-to-air heat pumps commonly work efficiently in air as cold as 30 degrees Fahrenheit. Below this temperature, an auxiliary source of heat must be used to warm the air in the system. The same heat pump that provides heating will also provide cooling; however, cooling ducts need to be slightly larger since more air volume is required. Cooling can also be added to the system with a simpler and less expensive electric *air conditioner* (Figure 13.4). Both humidity and air quality can be controlled at the furnace with specialized equipment added to the main supply duct. A *humidifier* can add moisture to the air, and a *dehumidifier* can lower the moisture content. Both are controlled by a humidistat located strategically in the house. Every furnace has a simple filter located in the central return air duct in order to remove dust from the air for the benefit of the occupants of the house and the longevity of the furnace. An *electronic air filter* that will eliminate much smaller particles including pollen can be added to the system at this same location.

The efficiency of any heating device is measured by its *coefficient of Performance (COP)*. The COP is the ratio of the energy output of a heating device to its energy consumption. The COP of an electric resistance furnace, for example, is 1.0, meaning that all of the electricity consumed to produce heat is returned in the form of heat. Air-to-air heat pumps typically operate in the 3.0 COP range when outdoor temperatures are above 32 degrees Fahrenheit (0°C), meaning that they produce three times as much heat as an electric resistance furnace consuming the same amount of energy. There are

FIGURE 13.3
The residential air-to-air heat pump extracts heat from outdoor air as cold as 32 degrees Fahrenheit (0°C), and this heat is transported through refrigerant lines to a heat exchanger in the furnace. The compressor unit shown here is the exterior component of a heat pump. The fan moves outdoor air across coils containing refrigerant to change the temperature of the refrigerant. The heat pump can work in reverse to cool during warm summer months. These units are generally noisy and should be located as far as possible from outdoor social spaces or bedroom windows. (*Courtesy of Carrier Corporation*)

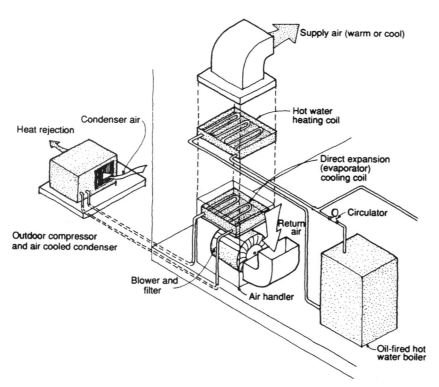

FIGURE 13.4
An air-conditioning unit sends cooled refrigerant to a coil in the supply duct that cools the circulated air. When heat is called for, it is introduced into the same supply duct with a heating coil. In this case, heat is from hot water, but electricity or a gas-fired burner can also be used. (*From Benjamin Stein and John Reynolds.* Mechanical and Electrical Equipment for Buildings, *9th ed., John Wiley & Sons, New York, 2000*)

also water-to-air heat pumps and ground-source heat pumps that have COPs up to 4.0. These heat pumps extract heat from groundwater or from the ground itself. The ground is usually at approximately 55 degrees Fahrenheit (13°C), making it a practical source for both heating and cooling.

A principal disadvantage of the forced-air system is that it is difficult to zone because the furnace supplies air to the supply ducts at a single temperature. Further, the moving air makes noise as it passes through the ducts, the supply registers, and the return grills. In addition, the heat of the furnace tends to dry out the air as it passes through, and the air passing through rooms can create drafts. Furniture placement is somewhat limited by the location of supply registers, and the potential for heat loss through leaky ductwork is significant. When a forced-air system incorporates a heat pump, care must be taken to keep the noisy outdoor units away from bedroom windows and other areas where tranquility is desired.

Components of the System

The central furnace contains two principal components—the *heating element* and the *fan* (Figure 13.5). The heating element can operate with a variety of fuels, including electricity, heating oil, kerosene, natural gas, or propane. A cooling element may be added in line with the furnace for warm summer months. The fan is invariably powered by electricity; thus the system will be inoperable during a power failure regardless of the source of heating fuel.

Many central furnaces are designed to accept the cooling coils of a heat pump. In fact, if provisions are made for refrigerant lines, it is relatively simple to add the outdoor unit of the heat pump at a later date. However, cooling typically requires a higher volume of air than does heating, so the ducts need to be sized accordingly.

Planning a Forced-Air System

The location of the furnace has a notable effect on the cost and efficiency of a forced-air system. The most important considerations are that the furnace be centrally located and accessible for service. The center of a basement is an ideal place for a furnace because supply ducts can all be relatively small and the return-air duct can be short because of its central position. Many furnaces are located in the garage, which is good for access but requires longer and larger ducts than a central furnace location, so duct heat loss becomes an issue. Provided that there is adequate height and that the furnace may be positioned on its side, the crawlspace may also be used to harbor a furnace so long as maintenance and replacement of the furnace are considered. Furnaces may also be located in a closet or in the attic. A furnace with a combustion-fired heat source must be supplied with outdoor air and vented to the outdoors to remove the byproducts of combustion. The type and size of the required vent can play a role in the location of the furnace.

Modern furnaces capable of heating a 2500-square-foot house are generally 18 to 24 inches square by 3 to 4 feet long. They are classified according to their orientation into *upflow, downflow,* or horizontal units (Figure 13.6). Upflow units blow supply air from the top of the unit and downflow from the bottom. The type of unit needs to be matched with its position in a house. Upflow units, for example, are commonly used in a basement, and horizontal units are typical in crawlspaces or attics where height is limited.

Coincident with planning the furnace location and of equal importance is planning the location of the return-air duct. This large duct, which draws warm air from the ceiling or floor level, must be planned for carefully, especially if it extends between floors, because it cannot fit within the standard thickness of a wall. (A return-air duct for a 2500-square-foot house is approximately 10 inches by 20 inches.) Too often this duct is not accommodated in the design and must be shoehorned in during the framing phase of construction. In economical forced-air systems, the return duct draws from one or two central locations in a hallway or at the top of a stair (Figure 13.2). In this case, the supply air reaches the return-air duct by passing under doors that have been shortened (undercut) at their base to allow for the passage of air. For improved air distribution, the return duct branches out to draw air from each room supplied with warm air.

The *main supply duct* is approximately the same size as the return duct as it leaves the furnace, but it may branch immediately into two or more ducts that are proportionately smaller. Planning a route for supply ducts to individual rooms is a reasonably simple task because these ducts are small. (A supply duct for a bedroom, for example, can generally fit within the stud space of a 2 ×4 wall.) As a rule, it is most efficient to locate supply registers in the floor below a window. The rising warm air from the supply

duct will counteract falling cool air from the window to reduce air movement across the glass, thus reducing heat loss (Figure 13.7). In two-story residences, supply ducts must rise vertically (on an interior wall if possible) to the second floor and often must also travel horizontally through a joist space to reach their destination. Their route must be carefully planned in order to minimize bends and to avoid structural members, plumbing, and other possible obstructions.

FIGURE 13.5
A forced-air furnace. Return air enters at the base of the unit where it is filtered. From here, a fan forces the air through the gas-burning elements and a cooling coil at the top of the unit. Units designed for horizontal mounting are also available. (*Courtesy of American Furnace Division, Singer Company*)

Supply (or return) ducts may be eliminated altogether by using the crawlspace as a *plenum*. This is accomplished by eliminating ventilation to the crawlspace and insulating its walls and ground surface while also sealing it carefully against air infiltration and moisture from the ground. The warm supply air is then blown directly from the furnace into the crawlspace and registers are cut into the floor (Figure 13.8). The crawlspace has effectively been made into a large supply duct.

Rooms on upper floors can be supplied via ducts running from the crawlspace to floor registers. Return air is ducted to the furnace as it would be in a normal fully ducted system. Advantages of this system include warm floors on the ground level and the elimination of ground-floor supply ducts from the heating budget. (The savings accrued by not having to insulate the joist floor above the crawlspace are offset by the need for crawlspace wall and floor insulation.)

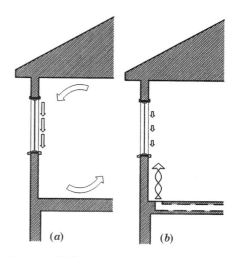

FIGURE 13.7
(*a*) Air cooled by cold glass will fall and induce a convective air current in a room, which brings more air into contact with the cold glass. (*b*) Heat released by a heat source located under a window will rise, counteracting the downward flow of cold air at the window. This reduces the flow of air across the cold glass, thereby reducing heat loss.

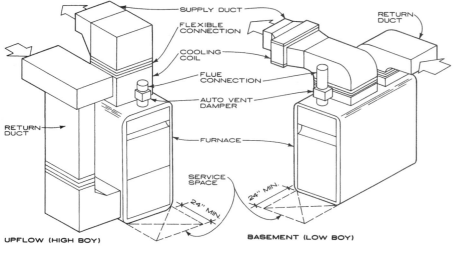

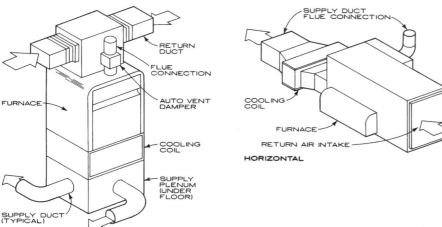

FIGURE 13.6
Warm-air furnaces are classified by orientation. Upflow furnaces are most common for basements, while downflow furnaces are most common when located on the first floor. Horizontal units are used in crawlspaces. (*Reprinted by permission from Ramsey/Sleeper, Architectural Graphic Standards, 8th ed., John Ray Hoke, Jr., A.I.A., Editor, John Wiley & Sons, New York, 1988.*

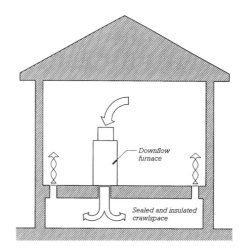

FIGURE 13.8
The downflow furnace delivers warm air to the crawlspace, which is sealed and insulated. The crawlspace acts as a huge supply air duct (plenum) with supply registers in the floor of the various rooms in the house. Return air is collected at the ceiling level and ducted to the furnace.

Forced-Air Rough-in Construction

Installation of ducts by the forced-air heating contractor is generally scheduled just after the plumber. Both plumbing waste lines and heating ducts are large, but plumbing usually takes precedence because its location is less flexible and waste lines must have a specified slope (Figure 13.9). As with plumbers, forced-air heating contractors will want to complete their work in a crawlspace before the subfloor is applied. The remaining ductwork is installed later when the framing is complete. If the house has a basement (and thus good access to the area under the first floor), all the ductwork will likely be installed at one time after the roof has been dried in.

Large ducts are customarily fabricated in the heating subcontractor's shop. Made of sheet metal lined with sound-absorbing insulation or with insulated fiberglass *duct board,* these fabricated items typically include *trunk ducts* such as return-air ducts and main supply ducts and *adapters* to fit the central supply and return ducting to the furnace. Premanufactured fittings are available to the heating subcontractor, including branch duct adapters such as angles, elbows, and transitions from rectangular to round and *boot fittings* for the termination of ducts at the room (Figure 13.11).

Between the main supply duct and the boots are the smaller *branch ducts.* The most common material for these ducts is *flex-duct,* a flexible tube made of insulation wrapped around coiled wire and sandwiched between layers of plastic. Although most heating and cooling contractors like to use flexible duct, rigid ducts are preferred because they have smooth walls that allow the air to flow efficiently, without turbulence. Also, flexible duct is too tempting to crimp to solve a clearance problem. Rigid branch ducts are connected to the main supply duct and to the boot with sheet metal screws and sealed with *duct tape.* Flexible ducts are both con-

FIGURE 13.9
Both plumbing pipes and heating ducts are installed after the framing of a house but before finish materials are in place. The plumber is typically scheduled before the heating/cooling contractor because of the limited number of ways drain and waste piping can be installed. Good communication among the framer, the plumber, and the heating contractor is essential during this phase of construction to make their jobs run efficiently. (*Photo by Rob Thallon*)

FIGURE 13.10
A heating contractor specializing in forced-air systems typically has a sheet metal shop where customized ductwork is manufactured. Many standard fittings and ducts are also kept in stock. (*Photo by Rob Thallon*)

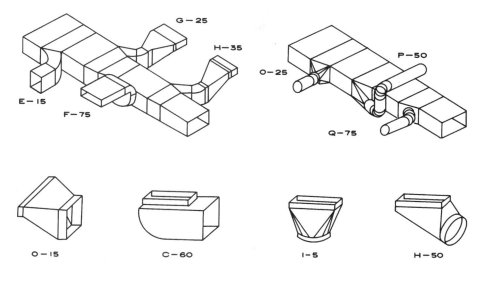

FIGURE 13.11
Trunk ducts are reduced in size in relationship to the volume of air that they carry. Both supply and return ducts are largest at the furnace. Premanufactured boot fittings are at the ends of ducts where they enter or leave a room.

nected and sealed with duct tape. When return-air ducting extends to individual rooms, boots and flex-duct are used here also.

The heating contractor installs the largest components of the system first. The largest ducts are supported by sheet metal straps, which are nailed or screwed to the framing. The furnace is best supported on a concrete pad on the ground because this minimizes the transfer of fan vibrations to the framing. When the framing supports the furnace, rubber vibration dampers should be located between furnace and framing. In all cases, isolation of the furnace from the ducts by means of a flexible connection is recommended to minimize noise and vibration. Gas furnaces should be tied down to minimize catastrophic earthquake damage.

FIGURE 13.12
The installation of this warm-air furnace and air-conditioning unit is complete. The white PVC pipe running vertically from the top of the furnace carries the exhaust gases to the outside. The ductwork to the left carries conditioned air to three different zones. (*Photo by Greg Thomson*)

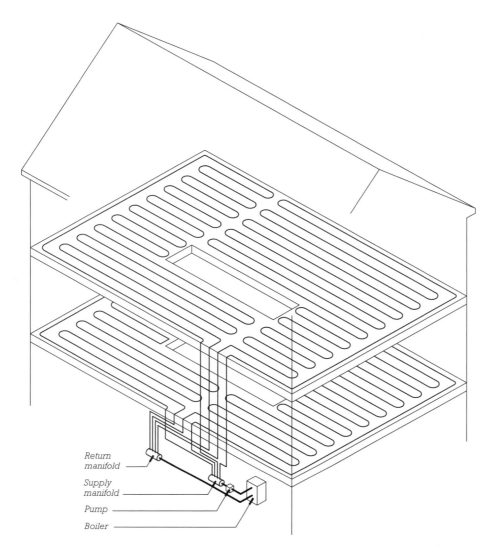

FIGURE 13.13
The radiant heating system in this two-story house uses hydronic tubes in the floor. There are four zones—two upstairs and two downstairs, each heating about a 250-square-foot (23 m²) area. The water is heated in a boiler and pumped through a supply manifold to each of the four zones. The warmest water is routed first to the perimeter of each zone. After passing through the floor and losing heat, the water passes through a return manifold to the boiler. Zones can be controlled independently, each with a thermostat connected to a valve at the supply manifold.

Labels in figure:
Return manifold
Supply manifold
Pump
Boiler

RADIANT PANEL SYSTEMS

Radiant panel heating, introduced into house construction in the 1920s, heats spaces by producing large heated surfaces (floors, walls, or ceilings) that radiate heat to other surfaces, objects, and people in the space. The most common form of radiant panel heating is the *radiant floor*—a form of which was developed over 2000 years ago by the early Romans. Radiant floors were popular in the 1950s when they were used extensively in single-story tract houses. Interest in radiant floors waned as the joints in the copper or wrought iron hot water pipes embedded in the slabs began to fail due to corrosion caused by impurities in the water. However, there has been a dramatic resurgence of interest in radiant panel systems recently with the introduction of noncorrosive flexible plastic tubing, which allows piping to be placed in long continuous runs without joints.

Radiant heating panels operate by emitting heat in the form of infrared waves that pass through the air without heating it and are absorbed by solid objects that are within "view" of the radiant source (Figure 13.13). A fire, the simplest example of a radiant source, heats our bodies, our furniture, and the surfaces of our rooms without directly heating the air in the space. The air is ultimately heated indirectly, however, as it passes over the warm surfaces. A radiant floor operates like a fire except that the heat is spread over a much larger area at a much lower temperature.

The advantages of radiant panel systems are numerous as long as it is the floor, not the ceiling, that is heated. The comfort of radiant systems is mentioned most frequently by those who have experienced them. Radiant floor systems keep the feet warm while the air within a room remains cool and fresh. Stratification of heated air caused by the natural tendency of warm air to rise is negligible with radiant panels because the air is not heated directly. In addition, because radiant heat is absorbed directly by the body, radiant systems can operate at air temperatures 6 to 8 degrees Fahrenheit (3 to 4°C) lower than convective systems. Radiant heating systems are more easily zoned than forced-air systems, adding both comfort and efficiency. When the energy-saving features of radiant panels mentioned previously are combined with the facts that glass (especially low-e glass) reflects radiant heat back into the building and air infiltration is reduced due to lower operating temperatures, an overall reduction in heat loads of 15 to 20 percent can sometimes be achieved with radiant panel systems.

There are architectural benefits of radiant panel systems as well. The panels are essentially invisible—buried within the floors, ceilings, or walls of a space—so that no registers, grills, radiators, or baseboards are visible (Figure 13.14). There is also more flexibility in the placement of furniture because there are no localized heat sources to obstruct. Finally, the systems are silent, eliminating the sounds of moving air or expanding metal that occur with other common systems.

The primary disadvantage of radiant panel systems is that most do not include a cooling option. Coupled with the fact that the initial cost of the system is often higher than that of a comparable forced-air system, the need for cooling often leads away from a radiant panel system. However, hybrid systems with radiant panel heat and a simple centralized cooling system are not uncommon. Another disadvantage is the fact that radiant systems have a slow response to thermostatic changes as compared to other systems. Radiant systems are therefore recommended for situations that demand a reasonably constant temperature and are not affected by sudden exterior temperature swings.

Types of Radiant Panel Systems

There are two basic methods of supplying heat to radiant panel systems —*electric heating cables* and (hot water) *hydronic tubing*. The most common installations for new construction are hydronic floors. Electric radiant ceiling systems were recently popular for low-cost work, but today, electric radiant panels are primarily employed in floors for remodeling work. Both electric and hydronic systems can be used in walls, but this is uncommon, primarily because occupants frequently drive nails and screws into walls to hang pictures and other items. Radiant ceilings are also highly susceptible to damage by remodelers or owners unaware of cables just beneath the surface.

Hydronic floor systems may be located in either a concrete slab on grade or a wood-framed floor (Figure 13.15). For a slab-on-grade installation, tubing carrying the hot water is positioned in the slab itself—about 2 inches below the surface. For framed floors, the tubing may be located either above the subfloor or below it. When

FIGURE 13.14
One advantage of radiant floor heating is that there are no registers in the floors or radiators on the walls to limit the placement of furniture. A staple-up system was installed in this house without any adjustments to the framing, and hardwood maple floors were employed as a finish floor. (*Courtesy of The Energy Service Company, Eugene, Oregon*)

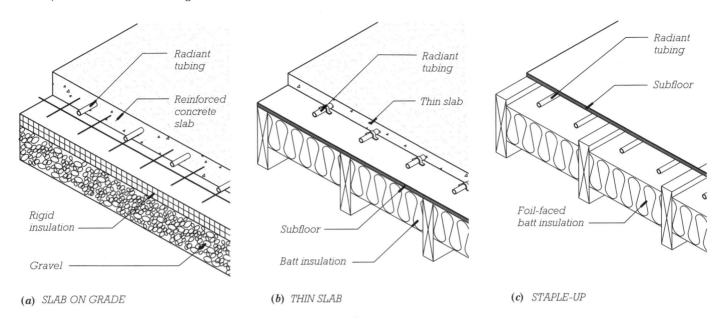

(a) SLAB ON GRADE (b) THIN SLAB (c) STAPLE-UP

FIGURE 13.15
Three hydronic radiant floor systems. (a) Radiant tubes are embedded in a concrete slab on grade. (b) Radiant tubes are fastened to the top of a subfloor and covered with lightweight concrete or gypsum. (c) Radiant tubes are stapled up to the underside of a subfloor.

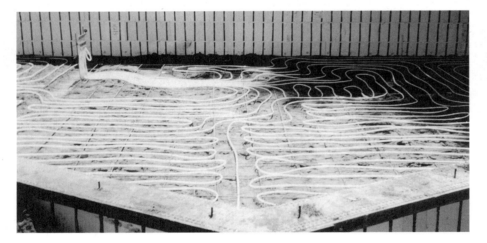

FIGURE 13.16
Hydronic tubes for heating a slab on grade are usually attached to reinforcing bars. The bars hold the tubes in place while the concrete slab is being poured. (*Courtesy of The Energy Service Company, Eugene, Oregon*)

above the subfloor, the hydronic tubes are typically installed in what is called a *thin slab*. Here, the tubes are stapled to the subflooring and covered with a thin layer of lightweight concrete or gypsum. When below the subfloor, the tubes are fastened to the underside of the subfloor with clips or within aluminum plates. This is called a *staple-up* application. In all framed floor applications, a layer of foil-faced insulation is installed below the subfloor to prevent heat from escaping downward. In all cases, the radiant floor may be covered with a finish floor

of tile, wood, resilient flooring or even some carpets with low insulative values. The thinner and denser the finish floor, the better.

Ceiling systems are economical but suffer from the fact that warmed air rises and tends to stratify at the ceiling where the heat source lies, so there is no natural convection loop as there is in a radiant floor system. *Ceiling radiant panel* systems constructed with electric resistance cables stapled to the underside of drywall and then covered with plaster used to be common, but are now rare. Radiant elec-

tric cables and mats for use on floors are now available and are often used in remodeling work because their thin profile allows the buildup over existing floors to be minimal.

Radiant Rough-in Construction

The radiant slab is the most economical of all radiant panel systems and the most easily integrated with standard residential construction. The *cross-linked polyethylene (PEX) tubing* that distributes the hot water in the slab is simply wired to reinforcing steel located just below the surface of the slab (Figure 13.16). Perimeter insulation will likely need to be thicker than that required for a nonradiant slab and will need to extend at least 4 feet (1.2 m) into the building. Insulation under the entire slab is recommended to reduce the response (lag) time of the floor and where groundwater or conductive soils

would continuously siphon off the heat of the floor. This increase in the amount of insulation material does not require a scheduling adjustment or the addition of a subcontractor.

For framed floors, the staple-up method is least disruptive of standard construction practice. Aside from taking precautions not to nail down through

the PEX tubing, the general contractor needs to make very few adjustments. Although not the easiest installation for the radiant heating contractor, the system is reasonably straightforward if there is good access to the underside of the floor and floor construction is not cluttered with closely spaced joists or excessive blocking.

The thin-slab system is simplest for the radiant heating subcontractor, but for the general contractor it represents a number of complications to the construction process. First, it requires the installation of double sole plates to provide a continuous edge for the slab. The construction schedule will have to be adjusted to allow the slab to be poured and to cure, and it may take weeks before the moisture from the wet concrete has stabilized such that flooring or cabinets may be installed on it (Figure 13.17).

The boiler and control valves for hydronic systems are often mounted on a wall and cannot be installed until the wall has been covered with gypsum wallboard, which is a late stage of construction. The boiler supplies hot water to manifolds that, in turn, connect the loops of tubing for each zone. (For small areas, a standard water heater can serve as a heat source, but heating tubes operate under 60 psi in this case instead of 20 psi.) Zone valves controlled by thermostats in each zone control the flow of hot water to the manifolds and, ultimately, to each zone (Figure 13.18). All hydronic tubes must be pressure tested and inspected before they are covered by concrete or other construction.

For electric heating cables, the timing of the installation is similar to that of hydronic systems. Cables are stapled and/or taped on 15-lb felt laid on the subfloor for floor applications. These floor cables are then covered with 1½ inches (40 mm) of lightweight concrete similar to that used in the thin-slab system for hydronic tubes. For electric radiant ceilings, the cables are stapled up to the gypsum wallboard ceiling and are then covered with plaster or sprayed texture.

FIGURE 13.17
A thin-slab hydronic tube installation before the installation of the poured gypsum thin slab. A polyethylene moisture barrier has been installed on the subfloor to contain the moisture from the poured gypsum. (*Courtesy of The Energy Service Company, Eugene, Oregon*)

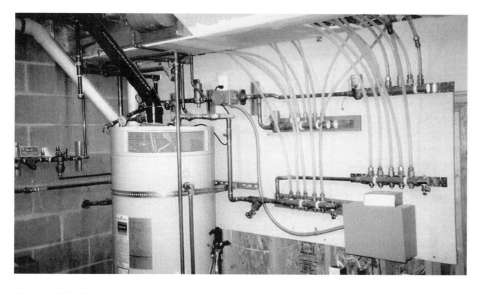

FIGURE 13.18
A typical radiant hydronic system heater and controls. In this case, the water is heated by a factory-installed heat exchanger within the water heater. The heated water is pumped through the two upper manifolds with zone valves to ten separate zones in the house. The water returns to the heater through the two lower manifolds. (*Photo by Rob Thallon*)

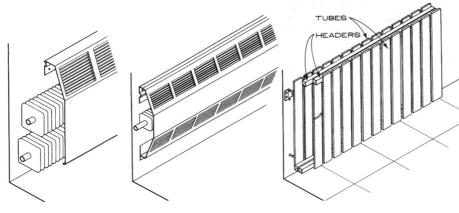

FIGURE 13.19
Hydronic baseboard heaters come in many shapes and sizes. (*Reprinted by permission of John Wiley & Sons, Inc. from Ramsey/Sleeper,* **Architectural Graphic Standards,** *10th ed., John Ray Hoke Jr., A.I.A., Editor © 2000 by John Wiley & Sons, Inc.*)

HYDRONIC BASEBOARD HEATERS

In *hydronic baseboard heating systems,* a central boiler and circulating pumps supply hot water to *baseboard fin-tube convectors,* which are horizontal pipes fitted with closely spaced vertical aluminum fins and mounted in a simple metal enclosure with inlet louvers below and outlet louvers above. These convectors are designed to replace wood baseboards around the perimeters of rooms (Figure 13.19). The heated fins set up convection currents that draw cool room air into the enclosure from below, heat it, and discharge the warm air out the top (Figure 13.20). Instead of baseboard units, *fan–coil units* may be used. A fan-coil unit is a compact sheet metal housing with a coil of hot water piping inside and a small electric fan to blow room air past the coil. Such units furnish relatively large amounts of heat from a small fixture, which makes them especially useful in kitchens, baths, and other rooms where there may not be enough baseboard length available to supply the required amount of heat. They are available in several basic forms, including wall-mounted cabinets and flat units that may be recessed in the toe space of a base cabinet in kitchens or baths.

Hydronic baseboard systems are quiet, efficient, and provide excellent wintertime comfort. They are easy to zone, using separate thermostats in different parts of the house to control either zone pumps or zone valves at the boiler. Their chief disadvantage is their inability to provide summertime cooling or dehumidification. Also, baseboard convectors can interfere with furniture placement.

LOCAL SOURCE HEATERS

There are a number of heating devices that produce heat or cool locally, independent of a central system or boiler. These devices are called *local source* heaters and include baseboard heaters, wall heaters, in-floor heaters, through-the-wall air conditioners and heat pumps, radiant stoves, and fireplace heaters. They heat and/or cool by means of radiation (radiant stoves), convection (wall heaters and through-wall units), or a combination of the two (electric baseboard heaters). A wide variety of fuels, including electricity, natural gas, heating oil, and wood, operate these appliances. Because they are not connected to a central source of heat, they act independently to heat one room at a time.

The advantage of having a heating device with its own thermostat in each room is that the temperatures of the various rooms may be controlled independently. Parts of the house such as bedrooms that are used only occasionally can be heated to a lower temperature, and considerable energy savings can be realized overall. Other rooms used for specialized purposes can have individualized settings. This kind of zoning can be provided by other types of heating systems (radiant panels), but not usually with the ease or the economy afforded by direct source heaters. Especially in superinsulated and passive solar houses, direct source heaters can be the most efficient long-term way to provide heat.

The most common local source heater is the ubiquitous *electric baseboard heater.* These heaters are extremely popular even where electricity rates are high because they are the least expensive to install of all types of heating devices. They heat with a combination of direct radiation and convection, which is induced by openings at the top and base of the device. Warm air rising from the baseboard

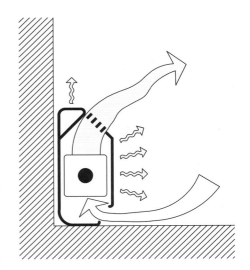

FIGURE 13.20
A baseboard heater heats a room by both radiation and convection. The heater radiates directly into the room, heating objects and surfaces within its "view." The heater also heats local air, which rises, creating convection currents that carry the heat into the room.

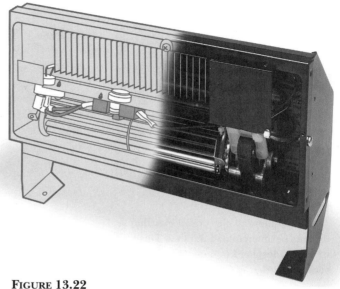

FIGURE 13.22
An electric wall heater warms by convection as a small fan forces air over a heated electric element and out into the room. Most electric wall heaters are designed to fit within a framed wall cavity, so it is best to avoid locating them in exterior insulated walls because their presence would interrupt the insulation. (*Photo courtesy of Cadet Manufacturing*)

FIGURE 13.21
The best place to locate a baseboard heater is under a window where heat rising from the unit will counteract the flow of cold air adjacent to the glass. (*Photo courtesy of Cadet Manufacturing*)

heater is replaced by cooler air that is allowed to flow in at the bottom. Fins that enlarge the heated area over which the convection current passes enhance this induced convection current (Figure 13.20). The convection loop from electric baseboard heaters can counteract the cool air falling from windows. This is why these heaters are located below windows when practical (Figure 13.21).

Local source heaters located in walls are also common. These devices, called *wall heaters,* generally have a fan-forced convection air current to distribute the heat (Figure 13.22). The most prevalent fuel sources for these heaters are electricity, natural gas, or propane —all of which are easily distributed to the heating devices. Most combustion-type primary heaters (all except electrically powered) must be vented to the exterior of the building, which can add considerably to the installation cost. Wall-mounted air-conditioning units are also common (Figure 13.23), and in-floor heaters similar to the wall heaters are also available.

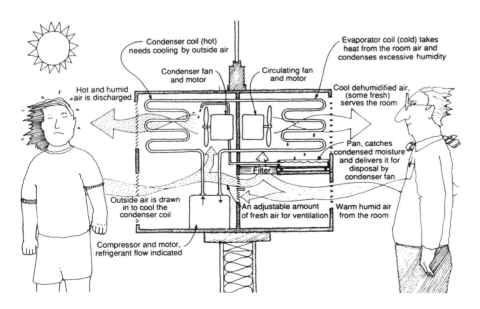

FIGURE 13.23
A self-contained through-the-wall air conditioner works on the same principles as a larger air conditioner, but the cold air is discharged into the room on the wall directly opposite the condenser. (*From Benjamin Stein and John Reynolds.* **Mechanical and Electrical Equipment for Buildings,** *9th ed., John Wiley & Sons, New York, 2000*)

FIGURE 13.24
A through-the-wall air conditioner that mounts permanently to the wall. This type of unit is designed to cool an individual room. (*Courtesy of Carrier Corporation*)

Another type of local source heat is the freestanding *radiant stove*. Descendants of wood- or coal-fired potbelly stoves, these devices now operate on almost every variety of heating fuel. Common fuels include natural gas, propane, heating oil, and wood. Advantages of the freestanding stove are that the radiant source can be closer to the center of a room where it is most effective and that the stoves themselves are often quite handsome —revealing the soothing warmth of a flame inside (Figure 13.25). For wood-burning stoves, advantages include the low cost of fuel, but there are disadvantages of the mess (bark, sawdust, insects, and ashes), the effort of tending the fire, and the increased air pollution. All radiant stoves take up floor space, and because they involve combustion, most must be vented to the exterior. Radiant stoves and fireplaces are discussed further in Chapter 16.

Disadvantages of local source heaters are that they take up space and can make furniture placement difficult. Electric baseboard heaters cannot have furniture placed against them because of fire danger, and radiant stoves also require clearances from combustible materials. Injuries can also occur from contact with hot stoves.

FIGURE 13.25
A freestanding radiant stove that uses propane as a fuel. The flue pipe is a double chamber that vents combustion gases to the outside and provides combustion air for the fire. Clearances from this appliance to combustible surfaces are stipulated by the manufacturer based on code-approved testing. (*Courtesy of Midgley's, Eugene, Oregon*)

Installation of Local Source Heating Devices

With the exception of electric heaters, which are usually installed by an electrician, local source heaters are installed by a specialized heating contractor affiliated with the vendor or by the general contractor. Most heaters may be installed once the house is dried in (tight to the weather), but often the installation will be delayed in order to coordinate with finishing operations. When the heating appliance serves as the sole source of heat, it is installed as early as possible in cold climates in order to furnish comfortable working conditions, to provide a stable environment for finish materials, and to promote drying of plaster and paint.

FINISHING A HEATING/ COOLING SYSTEM

The equipment for most heating/ cooling systems is installed during the rough-in phase of construction. This includes the furnace, the heat pump, and all the ducts in forced-air systems. For radiant systems, equipment to be installed would include the boiler, the valves, the tubing, and all the insulation. Heat-resistant wires would be installed for electric radiant panel systems. Local source heaters, depending on their type, may be installed at the rough-in phase or the finish phase.

To complete the installation of a heating system, the *registers,* grills, convector covers, and *thermostat(s)* must be installed, and the system must be *balanced.* This is done by the heating contractor during the finish phase of construction. It is common practice to provide a temporary thermostat during construction because the permanent thermostat is a sensitive instrument that tends to get clogged with dust and sprayed with paint if installed at the same time as the equipment it controls. The temporary thermostat provides crude control over the system for the purpose of controlling temperature during construction.

Centralized systems usually need to be balanced once construction is complete. This is best done after the users of the space have had a chance to experience the system. It is daily use that will indicate if a particular room or part of a room is too warm or too cool. In forced-air systems, balancing consists of adjusting airflow through each supply register. This can be accomplished by adjusting fan speed at the furnace, by adjusting *air volume dampers* in the ducts, or by adjusting the openings at the supply registers. In hydronic radiant systems, balancing is accomplished by adjusting the flow of water through each zone by means of *balancing valves.*

HEATING AND COOLING SYSTEMS AND THE BUILDING CODES

The 2000 International Residential Code devotes 13 chapters and one-sixth of its pages to heating and cooling systems. The code relies heavily on approved testing agencies such as the American National Standards Institute (ANSI) and Underwriters Laboratories to provide specifications for the installation of equipment. The chapters in the IRC are based on the International Mechanical Code and the International Fuel Gas Code, both of which are cited as references for equipment and systems not addressed in the IRC.

The IRC requires minimum clearances to equipment for inspection, service, repair, and replacement. It also specifies minimum distances from equipment to combustible materials and requires that the equipment be labeled with the manufacturer's name, the model number, the serial number, and the approved testing agency. Each type of equipment must be labeled with information about its capacity, its output, and required clearances. Fuel-burning units, for example, must list hourly rating in British thermal units (Btu)/hour, type of approved fuel, and required clearances.

C.S.I./C.S.C MasterFormat Section Numbers for Heating and Cooling	
15500	HEAT GENERATION EQUIPMENT
15530	Furnace
15540	Fuel-Fired Heater
15700	HEATING, VENTILATING, AND AIR-CONDITIONING EQUIPMENT
15710	Heat Exchanger
15720	Air-Handling Unit
15730	Unitary Air-Conditioning Equipment
15740	Heat Pump
15750	Humidity Control Equipment
15760	Terminal Heating and Cooling Unit
15770	Floor-Heating and Snow-Melting Equipment
15800	AIR DISTRIBUTION
15810	Duct
15820	Duct Accessories
15830	Fan
15850	Air Outlets and Inlet
15860	Air-Cleaning Device
15900	HVAC INSTRUMENTATION AND CONTROL
15910	Direct Digital Control
15915	Electric and Electronic Control
15950	TESTING, ADJUSTING, AND BALANCING

SELECTED REFERENCES

1. Bobenhausen, William. *Simplified Design of HVAC Systems.* New York: John Wiley & Sons, 1994.

Starting with a practical overview of what to consider, this book describes the theory and concepts of HVAC design. There are chapters on warm-air systems, hot-water systems, and cooling systems—all with an emphasis on sizing of the system.

2. Stein, Benjamin, and Reynolds, John S. *Mechanical and Electrical Equipment for Buildings,* 9th ed. New York: John Wiley & Sons, 2000.

Chapter 7 of this monumental work contains an excellent explanation of both the principles behind heating and cooling systems and the details of their design and operation.

3. In Chapter 6 of *Builder's Guide,* listed in reference 1 at Chapter 15, there is a good discussion of the principles of heating and cooling with an emphasis on environmental responsibility.

KEY TERMS AND CONCEPTS

HVAC systems
life-cycle cost
convection
radiation
heating zones
hydronic heating system
forced-air heating system
central furnace
supply duct
return-air duct
heat pump
air conditioner
humidifier
dehumidifier
electronic air filter
coefficient of performance (COP)

heating element
fan
upflow furnace
downflow furnace
main supply duct
crawlspace plenum
duct board
trunk duct
branch duct
duct adaptor
boot fitting
flex duct
duct tape
radiant panel
radiant floor
electric heating cable

hydronic tubing
thin-slab radiant floor
staple-up radiant floor
radiant ceiling panel
hydronic baseboard system
baseboard fin-tube convector
fan–coil unit
local source heaters
electric baseboard heater
wall heater
radiant stove
register
thermostat
system balancing
air volume damper
balancing valve

REVIEW QUESTIONS

1. What are the differences between a central forced-air fuel-burning furnace and a forced-air heat pump system?

2. Explain why a hydronic radiant floor system is more practical than a forced-air system for a single-story house built on a slab on grade. What other heating systems are practical for this type of foundation?

3. What are the advantages and disadvantages of using a crawlspace as a supply plenum for a forced air system?

4. Discuss the advantages of zoning for a heating system.

5. Explain in detail the differences between a hydronic radiant panel system and a hydronic baseboard system.

6. Discuss the reasons for selecting a thin-slab radiant floor as opposed to a staple-up radiant floor.

EXERCISES

1. Call a local heating contractor and try to get a sense of the relative percentages of the various types of heating and cooling systems installed in new houses in your area. What do you suppose are the principal reasons for the percentage breakdown?

2. Call a local architect with a reputation for incorporating passive solar heating and/or cooling into residential designs. Try to elicit a list of principles that can be applied in your geographical area.

3. Visit the site of a house under construction and make a diagram of the heating/cooling system. Compare this system with other systems studied by your classmates. What components of the system were not yet installed when you visited the site?

14

ELECTRICAL WIRING

Although we take it for granted, electricity has been an important part of people's lives only for about a century. In the 1890s, when electricity was first being introduced into the home, light fixtures were designed to operate on either gas or electricity because of the unreliability of electric power. Now virtually every appliance and apparatus in the house operates on electricity. Houses are so reliant on electricity that backup generators to supply energy during the occasional power outage are commonplace in some regions.

WIRING BASICS

How Electricity Works

Electricity is the flow of electrons through a conductor. Because it cannot be seen, it is best described with the analogy of water under pressure in a hose. When the pressure or the hose diameter is increased, more water will pass through the hose, thus creating more force (power) as the water exits the end of the hose. An electrical system works similarly, with wire substituting for the hose and electrons for the water. In an electrical system, the pressure is called *voltage (V)*, the flow of electrons is called *current (I)*, and the restriction to flow is *resistance (R)*. The relationship among these three elements can be expressed in a basic equation called *Ohm's law, V=IR*. With a constant volt-age, less resistance (a larger diameter wire) will produce a greater current. Current is measured in *amperes,* commonly referred to as *amps.*

A flashlight is an example of the simplest electrical system and can be represented by a battery connected with wires to a light (Figure 14.1). Electrical current flows in one direction through the wires from the battery through the light and back to the battery. When either piece of wire is disconnected, the circuit is broken, the electricity ceases to flow, and the light stops glowing. The electricity now exists only as potential (voltage) within the battery. The *power (P)* produced at the electrical device (the light bulb in this example) when the current is flowing, is the product of the current and the voltage, *P=VI*. This type of system, with a current that flows constantly and in one direction, is called *direct current (DC)*.

The *alternating-current (AC)* system used to power buildings is not so simple as direct current. Alternating current in North America is driven by voltage that fluctuates between the positive and the negative at a rate of 60 cycles per second (the rate varies among different regions of the world) (Figure 14.2). The reason that we use alternating current is that its voltage is easily increased or decreased with transformers. Electricity is transmitted most efficiently over long distances by wiring at as high a voltage and as low an amperage as possible. Thus, high voltage originating at the power plant can pass through long transmission lines with minimal voltage drop (power loss) and then be stepped down with a transformer to lower, safer voltage when it reaches its destination (Figures 14.3 and 14.4). As it enters the house, electrical power is rated as *110/220V AC*. This means that it is alternating current that can be utilized at either 110 or 220 volts.

Residential voltage actually varies from region to region, depending on the electrical utility company. East Coast voltage is generally 110/220V or 115/230V, while West Coast utilities deliver power closer to 120/240V. In addition, voltage fluctuates locally throughout the day as well, so that voltage to a residence might range from 118 to 122V, for example.

There are three wires that carry the electricity to a house—two *hot wires* and one *neutral wire.* Assuming a 110/220V system, the hot wires each carry 110V and the neutral wire provides the return path to complete the circuit. The potential between either of the hot wires and the neutral wire is 110 volts. The potential between the two hot wires is 220 volts (Figure 14.5). This high potential between the two hot wires is possible because the two are 180 degrees out of phase so that while one is positive, the other is negative (Figure 14.6).

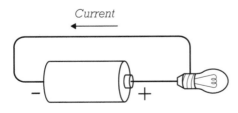

Current

FIGURE 14.1
In the simplest circuit, direct current (DC) flows in a loop, starting from a source (battery), through wires to a device (light bulb), and back to the source.

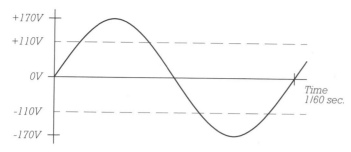

FIGURE 14.2
Alternating current (AC) cycles in a sine curve from positive to negative and back again at a rate of 60 cycles per second. The graph shows one cycle in which the voltage reaches both positive and negative peaks of 170 volts (positive and negative) but averages 110 volts.

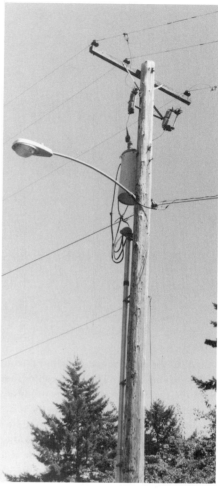

FIGURE 14.4
Typical local transmission lines carry direct-current voltage of several thousand volts that is converted with a transformer to 110/220-volt alternating current. Communication lines are often strung between the same poles. (*Photo by Rob Thallon*)

FIGURE 14.3
Electric substations such as this use step-down transformers to convert very high voltage from long-distance transmission lines to a few thousand volts for distribution to residential neighborhoods. (*Photo by Rob Thallon*)

FIGURE 14.5
The voltage between each hot wire and the neutral wire is 120 volts. Because the sine curves of the hot wires are out of phase (Figure 14.6), the potential between these two wires is 240 volts.

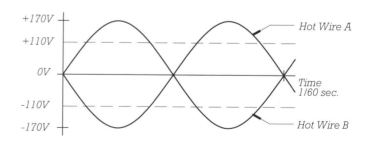

FIGURE 14.6
In a 110/220 AC electrical system, the two hot wires are out of phase so that the average potential between them is 240 volts. The potential between each individual hot wire and the neutral wire averages 110 volts.

FIGURE 14.7

A typical service entrance as viewed from the inside of the house. The large cables that connect the residence to the power utility enter the house through the conduit at the left, which is connected to the meter base. From the meter base, the cables will pass through another conduit, this one very short, through the studs to the 200-amp service panel at the right. The small cables at the lower left are telephone and television cables, each housed in a separate conduit. (*Photo by Rob Thallon*)

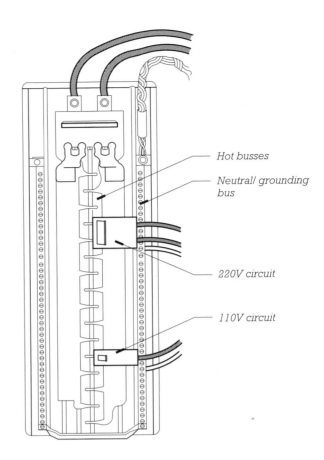

FIGURE 14.8

A typical 200-amp main panel box. The box is made of steel and receives the three main wires from the meter base. The breakers clip onto the hot busses to which the hot cables from the meter base are screwed. The neutral/grounding bus is connected to the neutral wire from the meter base and also to a ground rod driven into the soil. A 110-volt circuit is created by connecting the black (hot) wire of a cable to a breaker and the white (neutral) wire to the neutral/grounding bus. The bare grounding wire is also connected to the neutral/grounding bus. A 220-volt circuit is created by connecting two hot wires to a double breaker, which is connected to each of the two hot busses.

FIGURE 14.9

This 200-amp panel is typical of the average 2300-square-foot residence. The panel has a 200-amp main breaker and smaller breakers for each circuit. There are three 220-volt breakers just below the main breaker, and the remainder are 110-volt breakers. In a properly sized panel, there is capacity for the addition of future circuits. (*Photo by Rob Thallon*)

The three wires (two hot, one neutral) enter the house through a *service entrance* and typically pass first through a *meter base* that holds the meter to measure power usage. From the meter base, the wires are routed to the *main panel,* called a panelboard in the code (Figure 14.7). The main panel is the control center for electrical circuits in the house. It contains a *main disconnect* switch and *circuit breakers* that act as fuses for each electrical circuit (Figures 14.8 and 14.9).

Circuits are branches of electrical service within the house. All circuits originate at the main panel (Figure 14.10). A circuit may serve a single piece of equipment such as a water heater (equipment circuit), a set of receptacles designed for appliances (appliance circuit), or a collection of lights or receptacles in a number of rooms (general-purpose circuit). Each circuit is supplied by a single cable connected to a circuit breaker in the main panel. The cable for a 110V cir-

cuit contains three wires—a hot, a neutral, and a *grounding wire.*

The grounding wire is a safety device that provides a route for electricity other than the human body should a *short circuit* occur anywhere in the system. A short circuit results from a hot wire touching (and thus energizing) any conductor other than the part of the electrical equipment to which it is designed to supply power. Short circuits can occur when electrical motors fail, when wire insulation fails, or for

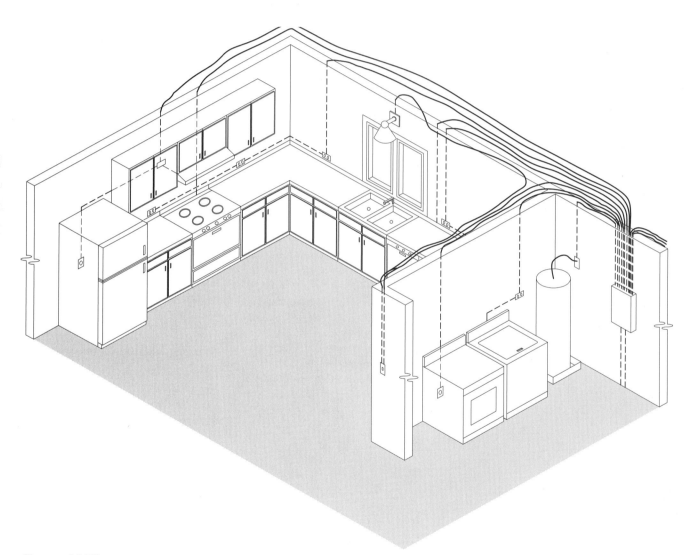

FIGURE 14.10
Circuits in a small house. Each circuit originates with a single cable at the main panel. Circuits are most concentrated in the kitchen where there are numerous appliances. Some circuits serve only one appliance, while others may connect together many receptacles, switches, and lights. In a larger residence, there would be more circuits, but each circuit would be limited to the same maximum electrical load as in the smaller house.

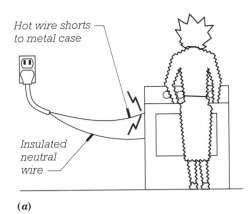

Hot wire shorts to metal case

Insulated neutral wire

(*a*)

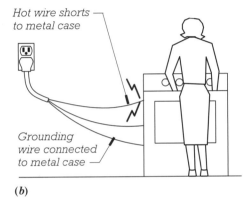

Hot wire shorts to metal case

Grounding wire connected to metal case

(*b*)

FIGURE 14.11
An ungrounded appliance (*a*) is deadly dangerous. Current from a shorted hot wire will find a path to the ground through a person, if possible. A grounded appliance (*b*), on the other hand, will pass the current from a shorted hot wire through the grounding wire, causing the circuit breaker to trip.

FIGURE 14.12
A ground-fault interrupter (GFI) receptacle provides an extra degree of safety for electrical power outlets that must be located near sinks, outdoors, or in other dangerous locations. The GFI receptacle will shut down instantly upon detecting a small change in the current flowing through it. (*Photo by Greg Thomson*)

any number of other reasons. In a 110V circuit, the electrical potential between the shorted equipment and the ground becomes 110 volts, and any electrical conductor touching both the equipment and the ground will become the conduit for 110 volts. This much voltage can easily kill a person, so a grounding wire is connected to the metallic cases of electrical equipment. The grounding wire leads back to the main panel where it is connected directly into the ground. If a short circuit occurs, the grounding wire rather than a person will carry the 110 volts to the ground (Figure 14.11).

The danger to humans of short circuits is especially high near kitchen and bathroom sinks, where plumbing provides a strong potential electrical conductor to the ground. If a person were touching both a shorted appliance and a faucet, it could take several seconds before the breaker in the main panel would trip to shut off the power to that circuit. In these especially dangerous locations, therefore, codes require special *ground-fault interrupter (GFI)* receptacles (Figure 14.12) that detect small current changes such as occur when a person becomes an electrical conductor and respond by instantly shutting down the power to the circuit.

Wiring Materials

The wires that carry the electricity are made of copper or aluminum—the most highly conductive metals that are affordable. (Silver is a better conductor than copper or aluminum but is too costly to be used in wiring.) Aluminum is usually used for the larger diameter conductors leading to the house, whereas copper is used within the house for normal wiring. Aluminum has a cost advantage over copper, especially in large-diameter wire, but it corrodes when exposed to air, so it must be carefully coated with an anticorrosion agent wherever a connection is made. This makes it impractical for local circuits within a house, where multiple connections are required.

FIGURE 14.13

Nonmetallic (NM) cable is virtually the only cable used in residential wiring today. It consists of two or more insulated wires plus a bare grounding wire, all wrapped in a paper insulation and sheathed in a waterproof thermoplastic jacket. Early versions of this cable without the grounding wire have been around since before 1920.

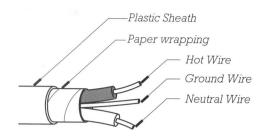

- Plastic Sheath
- Paper wrapping
- Hot Wire
- Ground Wire
- Neutral Wire

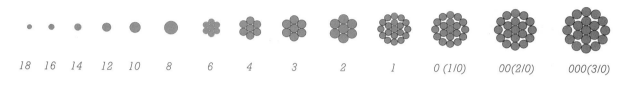

| 18 | 16 | 14 | 12 | 10 | 8 | 6 | 4 | 3 | 2 | 1 | 0 (1/0) | 00(2/0) | 000(3/0) |

Size, AWG

FIGURE 14.14

The chart shows the American Wire Gauge (AWG) size of copper and aluminum wires commonly used in residential wiring. Typically, 18-gauge and 16-gauge are used for low-voltage wiring; 14-, 12-, and 10-gauge for common lighting, receptacle, and appliance circuits; and the larger gauges for the service entrance and major equipment and appliances.

Wire Gauge	Allowable Ampacities	
	Copper	Aluminum
14	20	—
12	25	—
10	35	—
8	50	40
6	65	50
4	—	65
3	—	75
2	—	90
1	—	100
1/0	—	120
2/0	—	135
3/0	—	155
4/0	—	180

FIGURE 14.15

The relationship of (AWG) wire gauge to allowable ampacities of insulated conductors at 167 degrees Fahrenheit (75°C). For safety, circuits using 12-gauge cable typically are connected to a 20-amp breaker, and circuits using 14-gauge cable to a 15-amp breaker. (*Source: 1999 National Electric Code, Table 310-16*)

The wires are sheathed in an insulative coating that protects them from their surroundings and vice versa. Until the 1930s, each wire (hot and neutral) was run through the building independently, isolated from the structure by porcelain knob and tube fasteners. Now the most common wiring is a single *nonmetallic (NM) cable*—a thermoplastic jacket containing insulated hot, insulated neutral, and an unsheathed grounding wire (Figure 14.13). There are two types of NM cable—NMB for common use and UF with a plastic-sheathed grounding wire for underground or wet locations. In some locations such as a water heater or electrical range, wires require more protection from physical abrasion than NM cable can provide. In these situations, a flexible *metal-clad cable* or metal or plastic conduit is employed.

Wire size is described by *gauge*—the larger the gauge designation, the smaller the wire (Figure 14.14). The most common wire sizes in residential construction are 12-gauge, used for most receptacle circuits, and 14-gauge, used for lighting circuits. 10-gauge and 8-gauge wire are required for many large appliances. The main wires connecting the residence to the power grid are the largest, the most common size being 3/0 for a 200-amp service. The relationship of wire size to amp ratings is shown in Figure 14.15.

ELECTRICAL SYSTEM DESIGN

The Electrical Plan

The design of the electrical system is primarily the responsibility of the building designer. A plan drawing of the electrical system prepared by the designer is required in order to obtain a building permit and provides the basic information for the electrical contractor to install the electrical system (Figure 14.16). Using standardized symbols, the plan shows the location of *receptacles, switches, lighting fixtures,* the electric meter, and the main panel. In addition, the locations in the house of electrical equipment such as furnace, water heater, space heaters, washer, dryer, and kitchen appliances are noted in the electrical plan because they require wiring to supply electrical power. A good designer will cross-reference electrical plans with framing plans to ensure that joist locations do not conflict with the placement of lighting fixtures or other electrical devices. When combined with the electrical specifications, the electrical drawings allow the electrical contractor to estimate the cost of construction.

The code specifies that receptacles must be spaced along each wall such that an electrical device with a 6-foot-long (2 m) cord can always be plugged in no matter where the device is positioned along the wall. Effectively, this means that receptacles can be located no farther than 6 feet from the end of a wall and no more than 12 feet apart along the wall (Figure 14.17). With increasingly widespread use of computers and electronic entertainment devices in the home, many more receptacles than this are advisable. More receptacles are required in a kitchen where electrical usage can reasonably be expected to be higher than in a typical room.

One of the most challenging tasks for the designer of a house is the lighting design. Lights can be *surface mounted* or *recessed,* incandescent or fluorescent, line voltage or low voltage (Figure 14.18). Beyond these basic decisions, there is a vast array of fixture designs and light bulb types. With so many variables, the selection of appropriate fixtures as well as their sensitive location within a room and in relation to other fixtures has become an art. In commercial work, lighting design is a specialized field.

Switching for lighting must be indicated on the electrical plan as well.

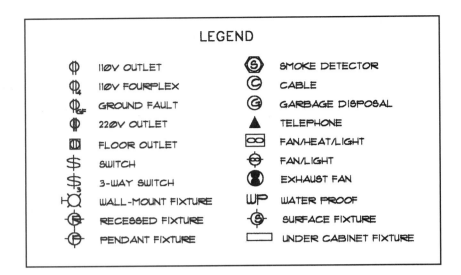

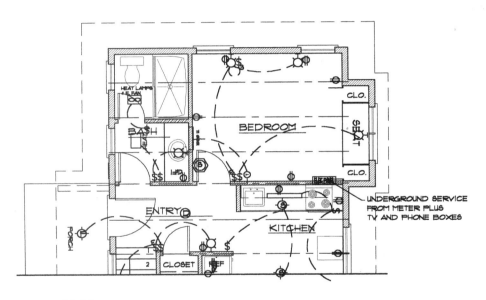

FIGURE 14.16
The electrical plan is drawn by the designer and serves as a guide for the electrical wiring. Standardized symbols show the types of electrical devices and how they are wired. (*Courtesy of David Edrington, Architect*)

The basic toggle switch next to the door to turn on the light in a room is simple to locate (Figure 14.17). *Three-way switches* that allow a light to be turned on or off from two locations are useful at the top and bottom of a stair, the ends of a corridor, or two entrances to a room. Finally, *four-way switches* that allow switching from multiple locations offer even more options.

The location and size of the main panel are also determined before plans are submitted for approval. Because the wires between the meter base and the main panel are large and expensive, these two elements are located back to back on an exterior wall when possible. Most residential panels are rated at 200 amps, although smaller (125 amps) and larger (400 amps) pan-

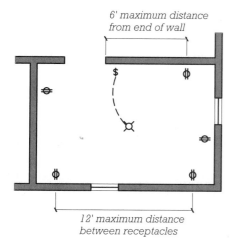

6' maximum distance from end of wall

12' maximum distance between receptacles

FIGURE 14.17
Plan drawing showing typical electrical requirements for a simple room. The light fixture is not required by code but is standard.

els are not uncommon. The designer may consult with an electrical subcontractor to select a practical size based on the present needs of the proposed building plus room for future expansion.

Sometimes, it is advisable to install a *subpanel*. A subpanel is a separate panel with its own circuit breakers and is connected to the main panel with a single large wire. It is useful when there are a number of circuits required at a location remote from the main panel. A subpanel might be used, for example, for the circuits in the top floor of a two-story house with the main panel in the basement. A subpanel might also be used for all the circuits in a house when the main panel is located in a detached garage.

Deciding on overhead or underground service and locating the meter base and the service entrance, where the electric service enters the building (Figure 14.7), are responsibilities of the designer. These decisions are best made in consultation with an engineer from the local utility company. The utility company typically provides

FIGURE 14.18
A wholesale electrical distributor displays a few of the hundreds of surface-mounted light fixtures available. These fixtures mount to standard electrical boxes located in the ceiling or the wall. Recessed fixtures are also available for ceiling locations. (*Photo by Rob Thallon*)

the labor and materials to run wires from the local power line to the building, but trees may have to be cut down or trimmed to provide clearance for overhead service, and trenching for underground service may be complicated by rugged terrain, groundwater, or rocky soil. Locating the meter base

for easy monthly inspection is also something that should be discussed with a utility company representative.

Planning the Circuits

Before the electrical system can be installed, the circuits must be planned.

The intent of a circuit is to divide the electrical system into subsystems, each of which can be safely served with a single reasonably sized cable (Figure 14.10). Each circuit is protected against electrical overload with a circuit breaker at the main panel. Circuit breakers are designed to disconnect the circuit when amperage exceeds normal designed usage. Excessive current (amperage) can occur when too many devices are operating simultaneously on a single circuit, when there is a *direct short* (the hot and neutral wires are touching), or when there is a *ground fault* (the hot wire touches a ground or grounded conductor). Any of these conditions could cause the cable to overheat and ignite a fire if allowed to continue.

A typical set of circuits for an all-electric 2000-square-foot (200 m²) house might include the following:

Dryer	30A	240V
Range	50A	120/240V
Dishwasher	20A	120V
Water heater	30A	240V
Heat pump	100A	240V
Kitchen plugs, 2 or 3 circuits at	20A	120V
Bath heater	20A	120V
Lights and receptacles, 6 or 7 circuits at	15A	120V
Garage	20A	120V

All these circuits can be wired to a 200-amp main panel even though their total amperage adds to over 400 amps because the code recognizes that only a fraction of the capacity will ever be used simultaneously.

With knowledge of the electrical code and practical experience in the field, it is the electrician who generally determines the number of circuits and their organization. This is done during the bidding process and refined when the system is being installed. The designer usually only specifies circuits when they are for a particular purpose such as a designated computer circuit.

ROUGH-IN ELECTRICAL INSTALLATION

The rough-in electrical work begins when the house is framed and dried in. The electrician is generally scheduled after the plumber (and the heating subcontractor, if there is one) because electrical cable is small and easily routed around obstructions. The small cables are also numerous and would likely interfere with plumbing and other utilities if they were to be installed first. All electrical wiring must be performed by a licensed electrician unless it is done personally by the owner.

Setting the Boxes

The first stage of the rough-in electrical work involves attaching the *boxes* (including the main panel box) to the framing. The main panel is made of steel and is designed to fit between studs, to which it is typically screwed (Figure 14.7). Boxes for receptacles, switches, and lights are made of steel or plastic and are nailed to the framing (Figure 14.19).

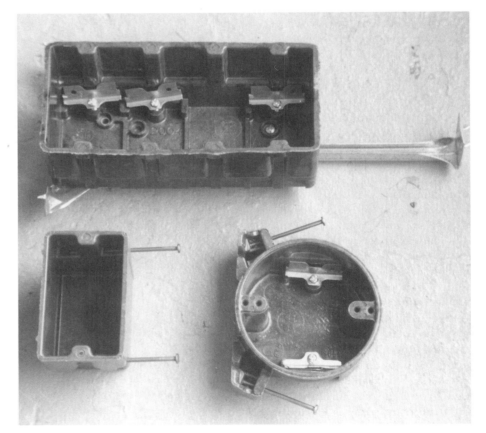

FIGURE 14.19
The most common electrical boxes: At bottom left, a receptacle box is nailed to studs and accommodates a single duplex receptacle or a switch. At bottom right, a light fixture box designed to nail to a joist. Light fixture boxes with adjustable brackets to position the light fixture between ceiling joists are also available. At the top, a four-gang box for four switches. Multiple-gang boxes for two or three switches (or duplex receptacles) are also common. (*Photo by Rob Thallon*)

There are accepted standards for the heights of receptacles and switches, but these are interpreted loosely by electricians whose principal goal in setting boxes is to do it rapidly. The size of each box is prescribed by code and depends on the number and size of wires that enter the box and the type of electrical device that it houses. Decisions about box size are made by the electrician at the site.

There generally needs to be a dialogue among the electrician, the designer, and the owner at the time the boxes are set. The electrician will locate boxes according to the electrical plan, but will have to make some adjustments because the boxes are attached to framing, which cannot be entirely laid out to accommodate electrical boxes. Occasionally, the electrician will have to add more receptacles than shown on the electrical plan in order to meet code. The designer must inform the electrician of the dimensions of trim, backsplashes, and other finish details that might interfere with electrical boxes. The faces of the boxes are typically set to project ½ inch (12.7 mm) from the framing in order to be flush with the finish wall, so the electrician needs to be informed if any finish wall thickness will vary from this standard.

Boxes on exterior walls are potential sources of air infiltration (discussed in Chapter 15) because they penetrate both the gypsum board finish and the polyethylene vapor barrier —both of which are used commonly as an air infiltration barrier. There are various schemes for sealing electrical boxes against air infiltration, and the responsible designer in cold climates will specify one of these schemes or locate the air infiltration barrier at the exterior of the wall.

Running the Cable

After the panel and boxes have been attached to the framing, the cable is strung between them. The cable for each circuit starts at the main panel and extends from box to box, linking all the boxes in a circuit together. Small holes are drilled in framing members to allow cable to pass through them. When running parallel to a framing member, the cable is stapled to it at frequent intervals as specified by code. Cable must also be stapled to a stud or other support just before it enters a box. At the box, the cable is cut—leaving sufficient length projecting from the box to connect the electrical device (receptacle, switch, light, or other device) at a future date (Figure 14.20).

Making-Up the Boxes

When all the cables have been run, the wires are connected to one another and tucked into the boxes in preparation for the later installation of the receptacles, switches, and lights. This phase of the electrical work is referred to as *making-up* the boxes. The first step in making-up a box is to strip the plastic sheathing off of the cables that project from the box, exposing the individually insulated wires. The wires are then cut to length according to their function, the insulation is stripped from them where required, and wires are connected to one another with *wire nuts*. The grounding wires of each cable entering a box are always connected during the make-up phase.

After the connections have been made, wires are folded back into each box to complete the process of

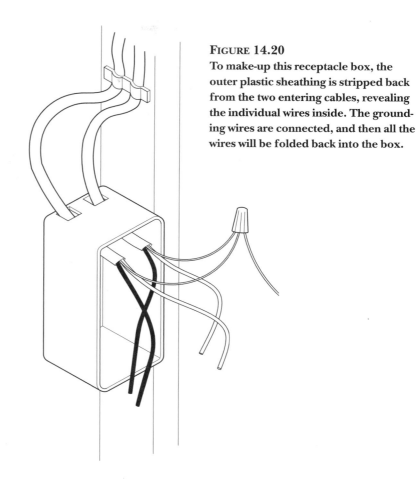

FIGURE 14.20
To make-up this receptacle box, the outer plastic sheathing is stripped back from the two entering cables, revealing the individual wires inside. The grounding wires are connected, and then all the wires will be folded back into the box.

FIGURE 14.21
The made-up box has all of the wiring complete except for the connection of the
electrical device. The wires are folded back into the box until the device is connected.
Notice that the face of the box is set forward from the face of the stud so it will be
flush with the surface of the (future) finished wall. (*Photo by Edward Allen*)

making-up the circuits (Figure 14.21). At this time, the temporary power is usually disconnected, and the main panel is connected to the permanent service entrance. Circuit breakers are installed in the panel, and some circuits are usually activated in order to provide electricity for construction tools and lights.

Rough Inspection

A rough-in electrical inspection is scheduled when the rough-in wiring is complete. At this time, the electrical inspector verifies clearances from the ground to overhead service wires and from the service entrance to the building. If there is an underground service, the depth of the cable is scrutinized. Connections to the power supply at the meter base and the main panel are also checked. The main panel is checked to see that breakers are properly sized, located, and connected and that there is adequate clearance around the panel. For local wiring, the inspector looks to see that there are an adequate number of circuits, an adequate number of receptacles, and that the cables are properly supported, fastened, and protected.

FINISH ELECTRICAL

The electrician returns near the end of construction to finish the system. At this point, the cabinets have been installed, the trim is done, and all the painting is complete. The finish electrical work involves installing all of the receptacles and switches, along with their cover plates; installing the lighting fixtures; and connecting appliances and equipment. The electrician tests all circuits and equipment before their final inspection.

General-Purpose and Appliance Circuits

Receptacles and switches are the most straightforward devices to install. Each is connected to the hot, neutral, and grounding wires; screwed into the box; and, finally, covered with the protective *cover plate* (Figure 14.22). Receptacles and switches are available in several colors and in a wide variety of styles and range of quality. The designer stipulates in the specifications the type to use.

Lighting fixtures present somewhat more of a challenge to the electrician, because there is such a wide variety of types, each with its own idiosyncrasies. Surface-mounted fixtures are easier to install than recessed ones. Heavy fixtures need to be attached to framing adequate for their support. Recessed fixtures fit into housings

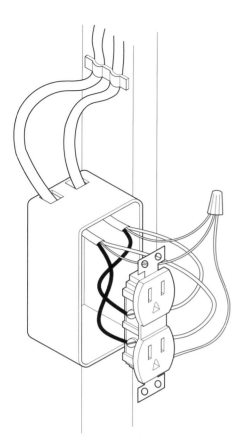

FIGURE 14.22
A receptacle that has been wired but not yet screwed to its box. Connections may be made at push-in terminals on the back of the receptacle or screw terminals on the sides. The screw terminals are safer and are used in all high-quality work because the contact area between wire and receptacle is much greater. Once connected electrically, the receptacle is screwed to the face of the box and the cover plate is attached to its face to trim the device.

that have to be installed during the rough-in phase. Recessed lighting fixtures located in insulated ceilings must be of a design that is approved for such locations to prevent overheating. Low-voltage lights must have a *transformer*, ideally concealed but accessible, to step down the voltage.

Equipment Circuits

Heat pumps, furnaces, water heaters, dryers, ranges and ovens, range hoods, dishwashers, garbage disposals, submersible pumps, and garage door openers are among the most frequently installed electrical equipment. Each may be supplied by a different manufacturer, and each may have different requirements for electrical connection. For this reason, wiring is most efficient if this equipment is specified before the rough-in wiring is installed. When this is not possible (which is frequently the case), the electrician

can allow for a range of possibilities by leaving extra length of wire at the equipment location. Connections to appliances such as the water heater, dishwasher, furnace, heat pump, and garbage disposal often require flexible metal tubing to protect the wires. Coordination with the plumber and the cabinetmaker may also be necessary for some equipment.

Final Inspection

When the wiring is entirely complete, a final inspection is arranged. The municipal electrical inspector will spot-check some of the receptacles to be sure that they are operational and will flip some switches to assure that they work. In addition, all GFIs are checked, some equipment is spot-checked, and the main panel is examined to verify its completion, including the labeling of circuits. The successful completion of a

final electrical inspection is required as a part of the final building inspection and is necessary for the granting of the occupancy permit.

ELECTRICAL WIRING AND THE BUILDING CODES

Regulations governing the installation of electrical systems in residences are currently contained in the International Residential Code, Chapters 33–42. The regulations are derived from the National Electric Code, a more comprehensive code covering electrical work in buildings of all types and sizes.

The electrical codes are primarily concerned with safety. For example, clearances around the main panel are dictated to allow access in emergencies, and a maximum spacing between receptacles is specified to eliminate extension cords and dangerous long runs of small wire. The code-writing organizations work with approved testing agencies to set minimum standards for electrical equipment and devices. These organizations also work with licensing agencies to train qualified electricians and to require their supervision of electrical work at the site.

LOW-VOLTAGE WIRING

In addition to electrical power, several other residential systems require wiring that runs through the framing. The most common of these are telephone, cable television, and satellite television. Other such systems include Internet cabling, low-voltage lighting, security, sound, intercom, heating and cooling thermostats, and the old-fashioned doorbell or chime. Many require installation by specialized subcontractors.

Each of these systems operates on electricity of such low voltage that it cannot give a fatal shock or start a fire. Therefore, they need not be installed by a licensed electrician. None is regulated by code, and thus no inspections are required. Telephone, cable television, and Internet systems are powered by their service providers, but low-voltage electricity must be produced for other systems by a small step-down transformer within the house. Often, this transformer, generally about the size of a single-serving package of breakfast cereal, is located inside a piece of equipment such as a central control panel for a security system.

Low-voltage wiring systems are roughed-in and finished at approximately the same time as the conventional wiring. In addition to the conventional electrical contractor, a high end house may require the services of a home theater installer; a specialized lighting contractor; and installers of a security system, cable television, Internet wiring, and intercom.

C.S.I./C.S.C MasterFormat Section Numbers for Wiring

16100	**WIRING METHODS**
16120	Conductors and Cable
16130	Raceway and Boxes
16140	Wiring Device
16150	Wiring Connection
16400	**LOW-VOLTAGE DISTRIBUTION**
16500	**LIGHTING**
16510	Interior Luminaires
16120	Exterior Luminaires
16700	**COMMUNICATIONS**
16800	**SOUND AND VIDEO**

SELECTED REFERENCES

1. Cauldwell, Rex. *Wiring a House*. Newtown, CT: Taunton Press, 1996.

A collection of just about all the knowledge required to wire a new house. The book is clearly illustrated with drawings and photos and covers topics from the nature of electrical power to tools, wiring methods, and materials.

2. Mullin, Ray C. *Electrical Wiring Residential 1999*. Chicago: Delmar, 1999.

A widely used textbook explaining how residential electrical systems work, how they are designed, and how they are installed.

KEY TERMS AND CONCEPTS

voltage
current
resistance
Ohm's law
amperes
amps
power
direct current (DC)
alternating current (AC)
hot wire
neutral wire
service entrance
meter base
main panel
main disconnect
circuit breaker
circuit
grounding wire
short circuit
ground-fault interrupter (GFI)
nonmetallic cable
metal-clad cable
conduit
gauge
electrical plan
receptacle
switch
lighting fixture
surface-mounted fixture
recessed fixture
three-way switch
four-way switch
subpanel
direct short
ground fault
electrical box
making-up a box
wire nut
cover plate
transformer
low-voltage wiring

REVIEW QUESTIONS

1. Explain the relationship among voltage, current, and resistance.

2. What are the differences between a neutral wire and a grounding wire?

3. Trace the path of electricity from the municipal power supply at the street to a light bulb in a house. Use specific terms for all the cables and devices through which the electricity flows.

4. What are the steps involved in rough wiring a house? Discuss why each step is in its particular sequence.

5. What are the typical wire sizes used in residential construction? For what purpose is each employed?

6. Explain the differences between line voltage and low-voltage systems. Make a list of typical uses of low-voltage systems.

EXERCISES

1. Obtain a residential wiring plan from a local architect or builder and discuss with your class the symbols used to describe the work of the electrician. Can you think of improvements to the drawing or to the design?

2. Make a wiring plan for a simple house plan taken from a newspaper or magazine.

3. Visit a house under construction with the rough wiring exposed. Sketch or photograph installed components such as a switch box, subpanel, main panel, and meter base and annotate the illustrations with procedures performed by the electrician in order to install the component. Include such procedures as the drilling of holes, stapling of cables, stripping of wires, and so on.

15

THERMAL INSULATION

Thermal insulation helps keep a building cooler in summer and warmer in winter by retarding the passage of heat through the exterior surfaces of the building. It helps keep the occupants of the building more comfortable by moderating the temperatures of the interior surfaces and reducing convective drafts. It also reduces the energy consumption of the building for heating and cooling to a fraction of what it would be without insulation.

However, insulation is only one part of a balanced approach to energy efficiency that must also include the control of water vapor and air infiltration. If water vapor enters the insulated exterior structure and condenses there, it can render the insulation ineffective and cause decay of the structure of the house. Air filtering into the house through cracks and breaks in the construction can introduce large volumes of heat or cold that counteract the effects of insulation. All these issues must be addressed simultaneously in order to achieve thermal efficiency and comfort.

Thermal insulation is any material that is added to a building assembly for the purpose of slowing the conduction of heat through that assembly. Insulation is almost always installed in new roof and wall assemblies in North America and often in floors and around foundations and concrete slabs on grade, anywhere that heated or cooled interior space comes in contact with unconditioned space or the outdoors.

The effectiveness of a building assembly in resisting the conduction of heat is expressed in terms of its *thermal resistance*, abbreviated as R. R is expressed either in English units as square foot–hour–degree Fahrenheit per Btu or in metric units of square meter–degree Celsius per watt. The higher the *R-value*, the higher the insulating value.

Every component of a building assembly contributes in some measure to its overall thermal resistance. The magnitude of the contribution depends on the amount and type of material used. Metals and glass have very low R-values, and concrete and masonry materials are only slightly better. Wood has a substantially higher thermal resistance, but not nearly as high as that of an equal thickness of any common insulating material. Most of the thermal resistance in an insulated building assembly is attributable to the insulating material.

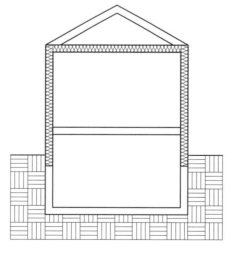

FIGURE 15.1
The thermal envelope surrounds the conditioned spaces of the house, protecting them from the temperature extremes of the outdoor air.

THE THERMAL ENVELOPE AND ITS COMPONENTS

Thermal insulation forms a container around the conditioned (heated and/or cooled) spaces in the house. This container is commonly referred to as the *thermal envelope* (Figure 15.1). The most critical parts of the thermal envelope are those parts in contact with the atmosphere —the walls, the ceiling, and sometimes the foundation of the house— because the temperature of the atmosphere varies much more than the ground upon which the house rests. A ceiling requires more insulation than a wall because heat flow upward through ceiling insulation is substantially faster than heat flow horizontally through wall insulation. Also, under summer conditions, the sun beating down on a roof can raise the surface temperature of the roof 30 to 40 degrees Fahrenheit (20 to 25°C) higher than that of the exterior surface of the walls. A floor over an unheated space requires about the same amount of insulation as a wall. Energy codes reflect these facts in their requirements for various components of the house (Figure 15.2). A cross section of the most common type of thermal envelope is shown in Figure 15.3.

Table N1102.1
Simplified Prescriptive Building Envelope Thermal Component Criteria Minimum Required Thermal Performance (*U*-Factor and *R*-Value)

HDD	Maximum Glazing *U*-Factor [Btu/(hr-ft² ·°F)]	Minimum Insulation *R*-Value [(hr · ft² · F)/Btu]					
		Ceilings	Walls	Floors	Basement Walls	Slab Perimeter *R*-value and Depth	Crawl Space Walls
0–499	Any	R-13	R-11	R-11	R-0	R-0	R-0
500–999	0.90	R-19	R-11	R-11	R-0	R-0	R-4
1,000–1,499	0.75	R-19	R-11	R-11	R-0	R-0	R-5
1,500–1,999	0.75	R-26	R-13	R-11	R-5	R-0	R-5
2,000–2,499	0.65	R-30	R-13	R-11	R-5	R-0	R-6
2,500–2,999	0.60	R-30	R-13	R-19	R-6	R-4, 2 ft.	R-7
3,000–3,499	0.55	R-30	R-13	R-19	R-7	R-4, 2 ft.	R-8
3,500–3,999	0.50	R-30	R-13	R-19	R-8	R-5, 2 ft.	R-10
4,000–4,499	0.45	R-38	R-13	R-19	R-8	R-5, 2 ft.	R-11
4,500–4,999	0.45	R-38	R-16	R-19	R-9	R-6, 2 ft.	R-17
5,000–5,499	0.45	R-38	R-18	R-19	R-9	R-6, 2 ft.	R-17
5,500–5,999	0.40	R-38	R-18	R-21	R-10	R-9, 4 ft.	R-19
6,000–6,499	0.35	R-38	R-18	R-21	R-10	R-9, 4 ft.	R-20
6,500–6,999	0.35	R-49	R-21	R-21	R-11	R-11, 4 ft.	R-20
7,000–8,499	0.35	R-49	R-21	R-21	R-11	R-13, 4 ft.	R-20
8,500–8,999	0.35	R-49	R-21	R-21	R-18	R-14, 4 ft.	R-20
9,000–12,999	0.35	R-49	R-21	R-21	R-19	R-18, 4 ft.	R-20

For SI: 1 Btu/(hr • ft² • °F)=5.68W/m² • K, 1 (hr-ft² °F)/Btu = 0.176m² • K/W.

FIGURE 15.2

This table shows the prescriptive building envelope requirements from the International Residential Code. The zones in the first column refer to climatic regions from warmest at the top to coldest at the bottom. HDD represents heating degree days for each zone. Type A-1 refers to detached one- and two-family dwellings, and Type A-2 refers to multiple dwellings up to three stories in height. The U-factor is the inverse of the R-value so that a U-factor of 0.40 translates to an R-value of 1/0.40 or 2.5.

The Roof

The roof assembly may be insulated at the ceiling to leave an uninsulated attic or at the roof itself if there is no attic. (Figure 15.5). The most effective strategy is to insulate the minimum perimeter around the inhabited space of the house. Ceiling insulation usually must be compressed somewhat to fit into the diminishing space under the roof sheathing where it meets the exterior wall. This compression, which can be seen in Figure 15.3, reduces the thermal resistance of the ceiling insulation at this point. A raised-heel roof truss (Figure 15.6) overcomes this problem and also facilitates roof ventilation.

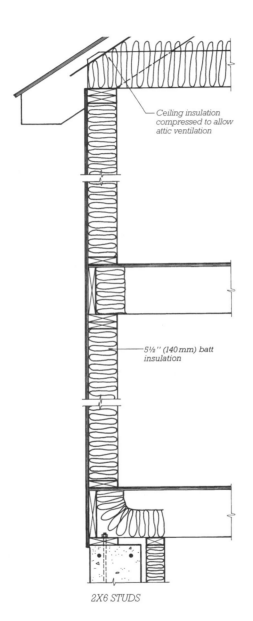

Ceiling insulation compressed to allow attic ventilation

5½″ (140 mm) batt insulation

2X6 STUDS

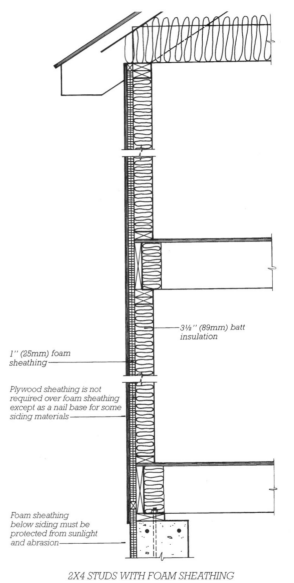

3½″ (89mm) batt insulation

1″ (25mm) foam sheathing

Plywood sheathing is not required over foam sheathing except as a nail base for some siding materials

Foam sheathing below siding must be protected from sunlight and abrasion

2X4 STUDS WITH FOAM SHEATHING

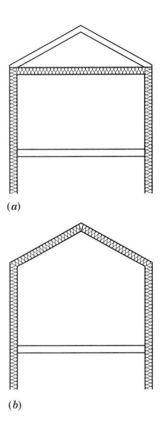

(*a*)

(*b*)

FIGURE 15.3
Insulation levels in this 2×6 stud wall are R-19 (R-132 in SI units). This is an increase of approximately 50 percent over the standard insulation in a 2×4 wall. The thickness of insulation in the ceiling and the basement or crawlspace is not limited by wall thickness.

FIGURE 15.4
2×4 framing with plastic foam sheathing in combination with batt insulation can achieve levels greater than 2×6 framing alone. The foam sheathing insulates the wood framing members as well as the cavities between, but can complicate the process of installing some types of siding.

FIGURE 15.5
Insulation at the ceiling: (*a*) can be located at the ceiling, creating an uninsulated attic, or (*b*) can follow the slope of the roof with no attic. The first condition allows for greater insulation depth at lower cost and is common with roof trusses since the attic is not usable as living space. The second creates a more interesting living space below the ceiling.

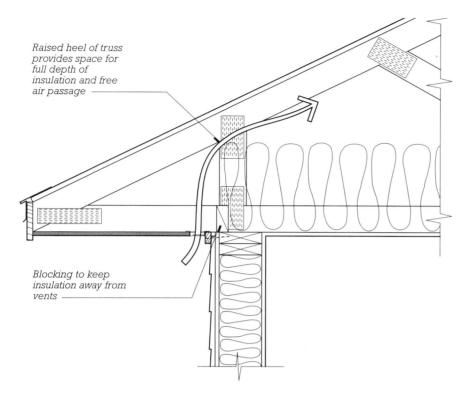

Raised heel of truss provides space for full depth of insulation and free air passage

Blocking to keep insulation away from vents

FIGURE 15.6
A raised-heel roof truss. A photograph of a raised heel truss is shown in Figure 9.43.

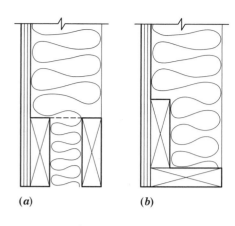

(a)   (b)

FIGURE 15.7
Window and door headers in 2 × 6 framing require special detailing. Two alternative header details are shown here in section view: (a) The header members are installed flush with the interior and exterior surfaces of the studs, with an insulated space between. This detail is thermally efficient, but may not provide sufficient nailing for interior finish materials around the window. (b) An efficient and simply-built header that provides nailing around the opening except, possibly, for wide casings.

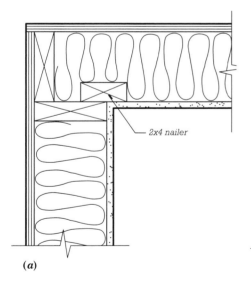

2x4 nailer

(a)

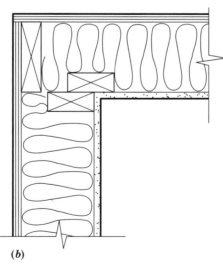

(b)

FIGURE 15.8
Two alternative corner post details for 2 × 6 stud walls, shown in plan view: (a) Each wall frame ends with a full 2 × 6 stud, and a 2 × 4 nailer is added to accept fasteners from the interior wall finish. Metal or plastic drywall clips can replace the nailer, improving insulation further. (b) For maximum thermal efficiency, one wall frame ends with a 2 × 4 stud flush with the interior surface, which eliminates any thermal bridging through studs.

The Walls

The framed outside walls of a house are typically insulated by entirely filling the cavities between studs with insulation. Special consideration is given to headers and wall intersections where framing members compete for space with the insulation (Figures 15.7 and 15.8). Electrical cables are usually present in the wall cavity, and sometimes plumbing pipes as well. These tend to compress the customary batt insulation as it is installed, making it less effective. This problem is best solved by running the cables at the base of the wall through notches made by the framers (Figure 15.9). An alternative is to split the insulation as it is installed so that it fully surrounds the cable (or pipe).

Windows and doors occupy a considerable portion of the wall area in a residence. These elements are very conductive as compared to the framed wall, so their maximum total area and their minimum levels of thermal resistance are mandated by the energy sections of the building code. Generally, colder climates have more stringent requirements. The National Fenestration Rating Council develops standards and testing procedures for insulative values and resistance to air infiltration of windows that are incorporated into the codes.

The Crawlspace

A crawlspace is usually an unconditioned space, much like an uninhabited attic. The temperature of the space can vary considerably with changes in atmospheric conditions; therefore, the floor above it must be thermally insulated to protect the inhabited space (Figure 15.10).

The control of moisture in a crawlspace will be discussed later in this chapter.

The Basement

A basement can be either a conditioned (living) space that is used like all the other spaces in a house, or it

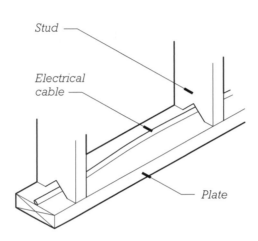

FIGURE 15.9
The notch at the base of the stud is made by the framer who lines up a series of studs and cuts them with two passes of a portable circular saw. The notch allows electrical cables to be run at the base of the stud cavity, eliminating compression of batt insulation by the cable.

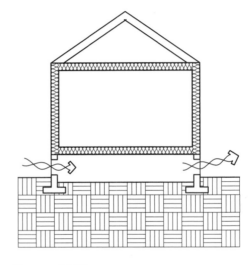

FIGURE 15.10
The ground floor of a house above a ventilated crawlspace is insulated in the floor.

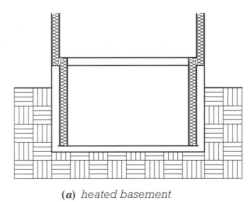

(a) heated basement

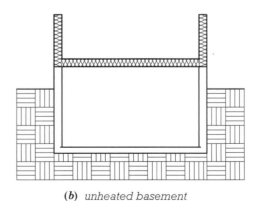

(b) unheated basement

FIGURE 15.11
(a) **A heated basement is insulated at the perimeter foundation walls.** *(b)* **An unheated basement is not insulated, but the conditioned space above is insulated at the floor.**

can be unconditioned and used primarily for storage and mechanical equipment. In a fully conditioned basement, the basement walls are insulated (Figure 15.11a), either on the outside of the foundation wall (Figure 8.36 *A*, *B*) or on the inside (Figure 8.36*C*). A basement used only for storage need not be insulated. The thermal insulation of the living space above is accomplished by insulating the floor, which is also the ceiling of the basement (Figure 15.11*B*).

The Slab Foundation

With a slab foundation, the inhabited space is in direct contact with the earth. The temperature of the earth near the surface remains at a constant temperature of approximately 55 degrees Fahrenheit (13°C). Only within the top few feet is the earth affected by climate—it will freeze in the winter and warm up with the summer sun. (Frost lines are discussed in Chapter 8.) When covered by a house, however, the earth will remain at 55 degrees, which is only about 15 degrees below the ideal indoor temperature. Thus, there is very little need to insulate under the slab itself, and it is rarely done except when the slab is heated or to prevent moisture from condensing on the slab in climates with warm, humid summers.

At the perimeter of the slab, where it is in contact with the atmosphere, there is a great need to insulate because the differential between atmospheric and indoor temperatures can be extreme. The insulation in this zone is called *perimeter insulation*, and it can be installed in either of two basic positions, depending on the footing details (Figure 15.12).

THERMAL INSULATION MATERIALS

Figure 15.13 shows the most important types of thermal insulating materials for residential construction and gives some of their characteristics.

Glass fiber batts are by far the most popular type of insulation for wall cavities in new construction. They are also widely used as attic, roof, and floor insulation. Some details of their installation are shown in Figure 15.14.

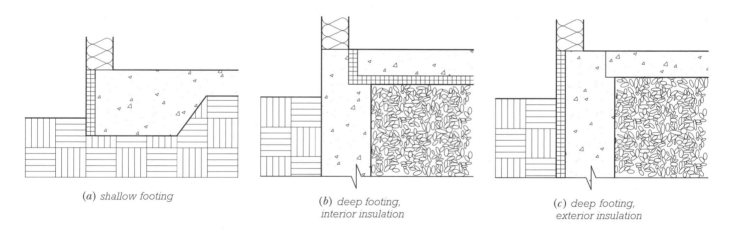

(*a*) shallow footing

(*b*) deep footing,
interior insulation

(*c*) deep footing,
exterior insulation

FIGURE 15.12
A slab edge must be insulated in all but the warmest climates. (*a*) A slab with a shallow turned-down footing is insulated at the outside edge with rigid insulation. (*b, c*) A slab with a deep footing may be insulated inside or outside the foundation with rigid insulation.

Type	Materials	Installation	R-Value[a]	Combustibility	Disadvantages
Batt or blanket	Glass wool; rock wool	The batt or blanket is installed between framing members and is held in place either by friction or by a facing stapled to the framing	3.2–3.7 *22–26*	The glass wool or rock wool is incombustible, but paper facings are combustible	Low in cost, fairly high R-value, easy to install
High-density batt	Glass wool	Same as above	4.3 *30*	Same as above	Same as above
Loose fill	Glass wool; rock wool	The fill is blown onto attic floors, and into wall cavities through holes drilled in the siding	2.5–3.5 *17–24*	Noncombustible	Good for retrofit insulation in older buildings; may settle somewhat in walls
Loose fibers with binder	Treated cellulose; glass wool	As the fill material is blown from a nozzle, a light spray of water activates a binder that adheres the insulation in place and prevents settlement	3.1–4.0 *22–27*	Noncombustible	Low in cost, fairly high R-value
Foamed in place	Polyurethane	The foam is mixed from two components and sprayed or injected into place, where it adheres to the surrounding surfaces	5–7 *35–49*	Combustible; gives off toxic gases when burned	High R-value, high cost, good for structures that are hard to insulate by conventional means
Foamed in place	Polyicynene	Two components are sprayed or injected into place, where they react chemically and adhere to the surrounding surfaces	3.6–4.0 *25–27*	Resistant to ignition combustible self-extinguishing	Fairly high R-value; seals against air leakage
Rigid board	Polystyrene foam;	The boards are applied over the wall framing members, either as sheathing onthe exterior or as a layer beneath the interior finish material	4–5 *27–35*	Combustible, but self-extinguishing in most formulations	High R-value, can be used in contact with earth, moderate cost
Rigid board	Polyurethane foam; polyisocyanurate foam	Same	5.6 *39*	Combustible; gives off toxic gases when burned	High R-value, high cost
Rigid board	Glass fiber	Same	3.5 *24*	Noncombustible	Moderate cost, vapor permeable
Rigid board	Cane fiber	Same	2.5 *17*	Combustible	Low cost

[a]R-values are expressed first in units of hr-ft²-°F/Btu-in., then, in italics, in m-°K/W.

FIGURE 15.13
Thermal insulation materials commonly used in residential construction. The R-values offer a direct means of comparing the relative effectiveness of the different types.

(a)

(b)

(c)

(d)

(e)

FIGURE 15.14

(a) Pressure fitting unfaced glass fiber batt insulation between studs in a wall, the most common use of batts. The batts are unfaced and, because they are slightly wider than the cavity they fill, they stay in place between the studs by friction. (b) Stapling faced batts between roof trusses. The R-value of the insulation is printed on the facing. (c) Placing unfaced batts between ceiling joists in an existing attic. Vent spacers should be used at the eaves (see Figure 15.25). (d) Insulating crawlspace walls with batts of insulation suspended from the sill. The header space between the joist ends has already been insulated. (e) Installing a polyethylene vapor retarder over unfaced glass fiber batt insulation using a staple hammer, which drives a staple each time it strikes a solid surface. (*Photo by Edward Allen. Other photos courtesy of Owens–Corning Fiberglass Corporation*)

Loose-fill glass fiber, rock wool fiber, and fire-retardant-treated cellulose are commonly used in attic floors because they are inexpensive and leave few voids and no gaps between pieces. The material is blown into place through a large hose. Levels of insulating value can be adjusted by regulating the depth of the insulation (Figure 15.15). Because of its method of delivery, this material is also used commonly to retrofit the walls of uninsulated houses by blowing it through holes drilled in the siding.

Rigid board insulation is commonly used where a high R-value per unit thickness is required or where insulation is used to support another material. Its use is common in some roof assemblies, in walls or ceilings where insulation thickness is limited, and for superinsulated walls. Some types whose composition is water resistant are also used below grade (Figure 15.16). However, problems with termites and carpenter ants in subsurface rigid foam have prompted code officials in regions with high concentrations of these pests to require that below-grade rigid insulation have both an insect-proof break such as flashing and a means of inspection for termite tubes. Most rigid foam board insulation is available in 2- or 4-foot-wide sheets, 8 feet long (0.6 or 1.2 m wide by 2.4 m long). Thickness varies in ½-inch increments from ½ inch to 2 inches (13 to 50 mm).

Other insulation such as *foamed-in-place* insulation (Figure 15.17) and dense-packed fire-retardant-treated cellulose provide high R-values but can be relatively expensive to install.

In warmer regions, *radiant barriers* are increasingly used in roofs to reduce the flow of solar heat into the building. These are thin sheets or panels faced with a bright metal foil that reflects infrared radiation. Most types of radiant barriers are made to be installed over the rafters and beneath the sheathing with an airspace between the barrier and the sheathing. They are therefore put in place during the framing of the building. Radiant barriers are used in combination with conventional insulating materials to achieve the desired overall thermal performance.

FIGURE 15.15
Blowing loose-fill glass fiber insulation into an attic space. A vapor retarder has been installed above the drywall ceiling, which supports the insulation. Vent spacers have been installed at the eaves to prevent the insulation from blocking eave vents (Figure 15.26). (*Courtesy of Owens-Corning Fiberglass Corporation*)

FIGURE 15.16
Following completion of the exterior siding, a worker staples a glass fiber reinforcing mesh to the foam insulation on the exposed portions of the basement wall. Next he will trowel onto the mesh two thin coats of a cementitious, stuccolike material that will form a durable, attractive finish coating over the foam. (*Courtesy of Dow Chemical Company*)

(*a*)

(*b*)

FIGURE 15.17
(*a*) Spraying polyicynene foam insulation between studs. At the time of spraying, the components are dense liquids, but they react immediately with one another to produce a low-density foam, as has already occurred at the bottom of the cavity adjacent to the corner. The foam in the cavity to the right has already been trimmed off flush with the studs. (*b*) Trimming off excess foam with a special power saw. (*Courtesy of Icynene, Incorporated, Toronto, Canada*)

HOW MUCH INSULATION?

Most houses built in the 1950s and earlier were not insulated or were insulated to very low standards. From the 1950s through the 1970s, there was an increasing awareness of the value of insulation, and many codes incorporated insulation standards. Before the rapid rise in energy costs in the mid-1970s, code-specified insulation values generally did not exceed the amount of insulation that could fit within the thickness of a standard 2×4 stud wall. With rising energy costs, however, economic analyses indicated the strong desirability of insulating buildings to R-values well beyond those that could be achieved by merely filling the available voids in the framing. In an effort to reduce consumption of nonrenewable fossil fuel resources, building codes began to specify higher minimum insulating values for walls, ceilings, windows, and basement walls. Stringent energy codes are now incorporated into most building codes within the United States and Canada. Generally, higher levels of insulation are required in more severe climates (Figure 15.18), although political boundaries can also affect code specifications. California, for example, has a milder climate but more stringent insulation requirements than any of the states bordering it.

The widespread concern for energy efficiency has led to experimentation at the building site and the manufacturing plant alike. Insulating sheathing materials and 2×6 (38×140 mm) exterior wall studs have become widely accepted in the building industry in North America (Figures 15.3 and 15.4), as have exterior and interior insulation systems for basements. Energy-efficient framing techniques that maximize the performance of the thermal envelope (Figures 15.7 to 15.9) are becoming common. Preservative-treated wood foundations (Figures 8.31 and 8.32), which are easily insulated to high R-values, are gaining increasing acceptance. Other methods of adding extra insulation to platform frame buildings such as the double-framed walls illustrated in Figure 15.19 are not yet standard practice but are representative of current experimentation by designers and building contractors in the colder regions of the United States and Canada.

Manufacturers of glass fiber insulating batts have responded to the call for better insulated buildings by producing high-density batts whose R-value per inch of thickness is substantially higher than that of conventional batts (Figure 15.13). A 3½-inch (89-mm) batt of this material has a total R-value of 15 (104). When this resistance is added to that of the inte-

rior wall finish, sheathing, and siding, it is often possible to meet a legal requirement for an R19 wall while using ordinary 2×4 studs. Other high-efficiency methods for insulating framing cavities include foamed-in-place plastics (Figure 15.17) and high-density blown cellulose, which insulate at the same time as they seal any holes in the framing created by wires, pipes, ducts, and material irregularities.

INSTALLATION OF INSULATION

Insulation is usually installed in the house by a subcontractor after the plumbing, heating, wiring, and all other utilities are in place.

Batt Insulation

Common batt insulation for an entire house can be installed in a day or two by a small crew. Unfaced batts are pushed between framing members and remain in place by friction. If the insulation is faced, the facing is stapled to the framing (Figure 15.14b–d). Batts are cut with a utility knife to fit into irregular spaces and around utilities such as electrical boxes. When the insulation is not faced with a vapor retarder, the insulation subcontractor may also be responsible for

FIGURE 15.18
United States, January mean minimum temperature (°F). Maps like this that indicate the coldest temperature during an average winter are used by building code officials to determine the required levels of thermal insulation according to geographic location. (*Source: National Atlas compiled by U.S. Geological Survey in 1965*)

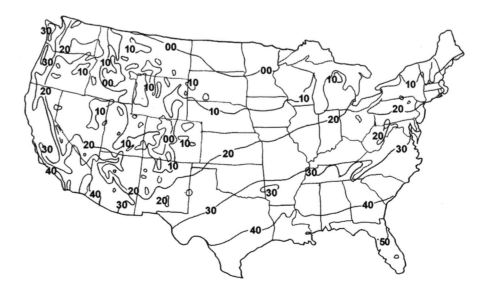

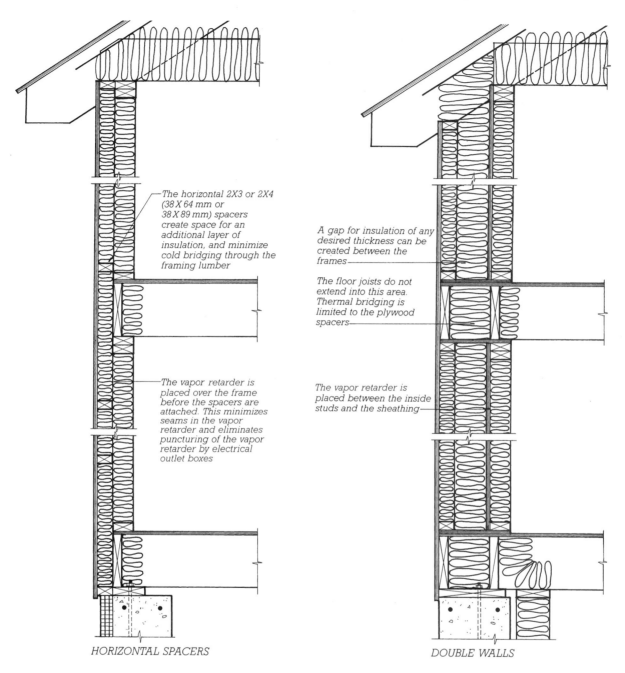

The horizontal 2X3 or 2X4 (38 X 64 mm or 38 X 89 mm) spacers create space for an additional layer of insulation, and minimize cold bridging through the framing lumber

The vapor retarder is placed over the frame before the spacers are attached. This minimizes seams in the vapor retarder and eliminates puncturing of the vapor retarder by electrical outlet boxes

A gap for insulation of any desired thickness can be created between the frames

The floor joists do not extend into this area. Thermal bridging is limited to the plywood spacers

The vapor retarder is placed between the inside studs and the sheathing

HORIZONTAL SPACERS

DOUBLE WALLS

FIGURE 15.19
With the two framing methods shown here, the walls can be insulated to any desired level of thermal resistance.

adding a polyethylene vapor retarder over the batts (Figure 15.14e).

Attic Insulation

It is very common in both new and older buildings to insulate ceilings from the attic above using blown-in loose-fill insulation. The popcorn-sized bits of glass fiber or rock wool insulation fill in around framing members, wiring, and vent pipes to minimize the voids that would be inevitable if batt insulation were used in this location. Fire-retardant-treated cellulose is an effective alternative to mineral wool insulation. The depth of insulation and there-fore the R-value of the installation can be easily controlled. The loose fill material is blown through a large hose connected to the delivery truck (Figure 15.15). Because this insulation is supported by the finish ceiling, it is installed later in the construction process than the batt insulation in walls.

FIGURE 15.20
To avoid the discomfort and inefficiency of working overhead in a cramped crawlspace, first-floor insulation over a crawlspace is usually installed from the top, before subfloor sheathing is applied. Insulation is supported on netting or strips attached to the bottoms of the floor joists.

Floor Insulation Over a Crawlspace

Floor insulation over crawlspaces is often installed before the subflooring is installed, when the joist spaces are easily accessible from above (Figure 15.20). After the plumbing and heating ducts are installed in the floor, one of a variety of support mechanisms (wires, lath, netting) is attached to the underside of the floor joists, and the batts are dropped in from above. This operation requires an extra trip to the job site for the insulation subcontractor, but so much labor and discomfort are saved by installing the insulation from above rather than below that it is generally worth the tradeoff.

Slab and Basement Rigid Insulation

Rigid insulation at the perimeter of slabs or at the exterior of basement walls is installed at the time the foundation work is done—often by the foundation subcontractor or the general contractor (Figures 15.12 and 8.37). This is because the insulation must be in place before the foundation is backfilled, and its installation is a relatively simple task. The framing crew will occasionally install rigid insulation in special situations as well.

CONTROL OF WATER VAPOR

Water Vapor and Condensation

Water exists in three different physical states, depending on its temperature and pressure: solid (ice), liquid, and vapor. Water vapor is an invisible gas that is always present in air. The higher the temperature of the air, the more water vapor it is capable of containing. At a given temperature, the amount of water vapor the air actually contains, divided by the maximum amount of water vapor it could contain, is the *relative humidity* of the air. Air at 50 percent relative humidity contains half as much water vapor as it could contain at the given temperature.

If a mass of air at 50 percent relative humidity is cooled, its relative humidity rises. The amount of water vapor in the air mass has not changed, but the ability of the air mass to contain water vapor has diminished because the air has become cooler. If the cooling of the air mass continues, a temperature will be reached at which the humidity is 100 percent. This temperature is known as the *dew point*. A roomful of very humid air has a high dew point, which is another way of saying that the air in the room would not have to be cooled very much before it would reach 100 percent humidity. A roomful of dry air has a low dew point. It can undergo considerable cooling before it reaches saturation.

When a mass of air is cooled below its dew point, it can no longer retain all its water vapor. Some of the vapor is converted to liquid water, usually in the form of fog. The farther the air mass is cooled below its dew point, the more fog will be formed. The process of converting water vapor to liquid by cooling is called *condensation*. Condensation takes place in buildings in many different ways. In winter, room air circulating against a cold pane of glass is cooled to below its dew point, and a fog of water droplets forms on the glass. If the air is very humid, the droplets will grow in size, then run down the glass to accumulate in puddles on the window sill. On a hot, humid summer day, the moisture in the air in the vicinity of cold water pipe or a cool basement wall will condense in a similar fashion.

In an insulated wall or roof assembly, condensation can become a serious problem under wintertime heating conditions. The air inside the building is at a higher temperature than the air outside and usually contains much more water vapor, especially in rooms where cooking, bathing, or washing takes place. Indoor air leaking through the assembly toward the outside becomes progressively cooler and reaches its dew point somewhere inside the assembly, almost always within the thermal insulation (Figure 15.21). Even in places where the air itself does not leak through the assembly, water vapor can still

migrate from indoors to outdoors, driven by the difference in vapor pressure between the moist indoor air and the drier air outdoors. When that occurs, the water vapor will reach its dew point somewhere within the insulation. The result is that the insulation becomes wet, and portions of it may become frozen with ice. Under these conditions, the insulating value is lost, and the structural materials of the roof or wall may become wet and subject to decay. Water may accumulate to such an extent that it runs or drips out of the assembly and spoils interior finishes or the contents of the building. Furthermore, when the outside of the assembly is later heated by warmer outdoor air or bright sunlight, the water begins to vaporize again. As it does so, its vapor pressure can raise blisters in paint films or roof membranes. (Most cases of peeling paint on wooden buildings are caused not by a poor painting job, but by moisture in the wood or lack of a vapor retarder on the warm side of the wall.)

Under summertime cooling conditions in hot, humid weather, the flow of water vapor through building assemblies can be reversed, as moisture moves from the warm, damp outside air toward the cooler, drier air within. This condition is not usually as severe as the winter condition because the differences in temperature and humidity between indoors and outdoors are not as great. Furthermore, in most areas of North America, the cooling season is short compared to the heating season, allowing the designer to neglect the summer vapor problem and concentrate on the winter problem. In areas of the southern United States, however, the summer problem is more severe than the winter problem and must be solved first.

The Vapor Retarder

To prevent condensation inside building assemblies, a *vapor retarder* (often

called, less accurately, a vapor barrier) is installed on the warmer side of the insulation layer. This is a continuous sheet, as nearly seamless as possible, of plastic sheeting, aluminum foil, kraft paper laminated with asphalt, or some other material that is highly resistant to the passage of water vapor. The effect of the vapor retarder is to diminish the flow of air and vapor through the building assembly, preventing the moisture from reaching the point in the assembly where it would condense. For buildings in most parts of North America, the vapor retarder should be placed on the inside of the insulation. In humid areas where warm weather cooling is the predominant problem, these positions should be reversed. In some mild climates, a vapor retarder may not be required at all.

Like insulation, vapor retarders began to receive increased attention from designers, builders, and building scientists as the need for energy conservation grew. There is agreement that their effectiveness and energy-saving value increase as thermal insulation levels increase, but there is little consensus about their use in warm climates.

Many batt insulation materials are furnished with a vapor retarder layer of treated paper or aluminum foil already attached. However, because a vapor retarder attached to batts has a seam at each stud that can leak significant quantities of air and vapor, most designers and builders in cold climates prefer to use unfaced batts and to apply a separate vapor retarder of polyethylene sheet with very few seams that can be easily taped (Figure 15.14e). Alternatively, a vapor retarder that is incorporated into paint and applied to the finished drywall surface is becoming popular because it has the least first cost. However, this method takes responsibility for vapor control out of the hands of the insulation subcontractor and places it with the painting subcontractor.

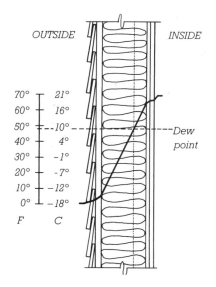

FIGURE 15.21
All of the components of the wall assembly contribute to its overall insulative value. The insulation contributes the most and is invariably the component within which the dew point falls.

	U.S. Perms	CSI Perms
Aluminum foil 1 mil (0.025 mm)	0.0	0.0
Built-up roofing	0.0	0.0
Polyethylene 6 mil (0.15 mm) 4 mil (0.10 mm)	0.06 0.08	0.0010 0.0014
Hot melt asphalt 3.5 oz/ft² (1.1 kg/m²) 2 oz/ft² (0.6 kg/m²)	0.1 0.5	0.002 0.009
Exterior oil paint, three coats on wood	0.3–1.0	0.005–0.017
Oriented strand board (OSB) 7/16" (12 mm) thick	0.7	0.012
Brick masonry 4" (100 mm) thick	0.8	0.014
Extruded polystyrene (XPS) 1" (25 mm) thick	0.8–0.11	0.014–0.019
Expanded polystyrene (EPS) 1" (25 mm) thick	1.1–5.0	0.019–0.9
Plywood, exterior glue 7/16" (12 mm) thick	1.07	0.018
Interior primer plus one coat flat oil paint on plaster	1.6–3.0	0.028–0.052
Plaster on metal lath	15	0.26
Gypsum wallboard 1/2" (12 mm) thick	38	0.65

FIGURE 15.22
Perm ratings for materials commonly used in residential roof and wall assemblies. The higher the number, the more permeable the material.

The performance of vapor retarder materials is measured in perms. In the United States, a *perm* is defined by ASTM E96 as the passage of 1 grain of water vapor per hour through 1 square foot of material at a pressure differential of 1 inch of mercury between the two sides of the material. In Canada, the CSI perm is measured in terms of 1 nanogram per second per square meter per pascal of pressure difference. One CSI perm is equal to 57.38 U.S. perms (Figure 15.22).

When possible, the opposite side of the building assembly from the vapor retarder should be allowed to "breathe," to ventilate freely by means of attic ventilation or topside roof vents. This helps prevent stray mois-ture from becoming trapped between the vapor retarder and another near-impermeable surface such as a ply-wood roof deck. In a wall, where ventilation is not practical, the mate-rials opposite the vapor retarder are selected to be more permeable than the vapor retarder.

VENTILATION OF CAVITIES

Historically, the cavities below, above, and within the structure of a house were passively ventilated by gaps in the construction such as the seams between sheathing boards. Modern materials such as plywood and OSB eliminate most of the gaps, and mem-branes can be used to seal cavities to keep out moisture and other unwanted elements. This tightness, however, can also trap moisture that finds its way into the cavities. In wall cavities filled with insulation, a vapor retarder combined with a highly permeable material at the opposite side of the assembly is generally sufficient. Larger cavities such as crawlspaces and attics, however, contain large amounts of air with the potential to hold a great deal of moisture. It is therefore generally necessary to ventilate these spaces and cavities deliberately.

Attic Ventilation

Attics are ventilated not only to allow water vapor to escape, but also to keep the house cooler in summer by pre-venting the buildup of solar heat con-ducted through the roofing and roof sheathing. For attic ventilation, it is advantageous to have both low (intake) vents and high (exhaust) vents. The difference in height of these vents causes natural convection currents that draw air through the spaces, ven-tilating them. Building codes recog-nize this effect by allowing the *net ventilation area* of roofs to be reduced from 1/150 of the area ventilated to 1/300 of the area, if vents are sepa-rated vertically by at least 3 feet (90 cm). Under extreme conditions, attic ventilation can be increased with a thermostatically controlled fan.

At the eave, the required intake ventilation is created by *soffit vents* or *frieze vents*. Soffit vents usually take the form of a continuous slot covered either with insect screening or with a perforated aluminum strip made espe-cially for the purpose (Figure 15.23). Frieze vents are most commonly drilled or cut into the frieze blocks between rafters and screened with galvanized insect screening (Figure 15.24). Care should be taken to assure that a con-tinuous air path is maintained between the vents at the eave and the exhaust vents higher in the roof. Where insu-lation may shift or expand to block the air passage, a *vent spacer* may be required (Figure 15.25).

FIGURE 15.23
A continuous soffit vent made of perforated sheet aluminum or galvanized steel permits generous airflow to all the rafter spaces but keeps out insects.

FIGURE 15.24
Frieze block vents are drilled or notched into each frieze block and screened on the back side to keep out insects. Vents are located near the top of frieze blocks to connect with the airspace just below the roof sheathing. The collective area of these vents is frequently too small and needs to be calculated based on formulas stipulated in the building codes. (*Photo by Rob Thallon*)

FIGURE 15.25
To maintain clear ventilation passages where thermal insulation materials come between the rafters, wood blocking may be inserted, or vent spacers of plastic foam or wood fiber may be installed as shown here. The positioning of the blocking or vent spacers is shown in Figures 10.38 and 15.28. Vent spacers are generally more economical than blocking and can be installed along the full slope of a roof to maintain free ventilation between the sheathing and the insulation where the interior ceiling finish material is applied to the bottom of the rafters. (*Courtesy of Poly-Foam, Inc.*)

FIGURE 15.26
This building has both a louvered gable vent and a continuous ridge vent to discharge air that enters through the soffit vents or frieze block vents. The gable vent is of wood and has an insect screen on the inside. The aluminum ridge vent is designed to prevent snow or rain from entering even if blown by the wind. (*Photo by Edward Allen***)**

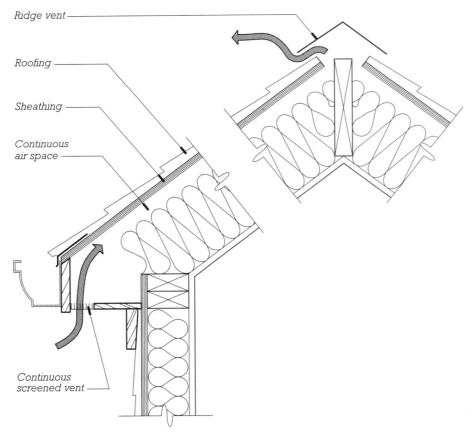

Ridge vent

Roofing

Sheathing

Continuous air space

Continuous screened vent

FIGURE 15.27
Section view of the vent space required by code for roof assemblies insulated with fiber-glass batts. Ideally, the space should connect with outside air at both top and bottom.

At the high part of the roof, there are several common forms of venting (Figure 15.26). The continuous *ridge vent* is very effective because it is located at the highest point of the roof and is distributed evenly across the entire length of the roof. *Gable end vents,* located high in the wall at opposite ends of an attic, can help to create a cross draft in a wind. Gable end vents are made in a wide variety of shapes and sizes to add a decorative feature to the end wall. *Through-roof vents,* which resemble inverted cake pans and are flashed into the roofing material, offer a third alternative. These vents are commonly used in situations such as hip roofs when the other venting types are impractical.

Roof Ventilation

Roof assemblies with insulation between the rafters are similar to walls with insulation between studs, but roofs do not work under the same set of circumstances. Insulated roof assemblies can have both a higher temperature differential between the inside and the outside and a highly impervious roofing material such as preformed metal at the outside of the assembly. A roof cavity is also subject to the passage of moisture-laden interior air through openings such as lighting fixtures. This combination makes the likelihood of condensation within the assembly so great that a ventilation space is required to prevent it (Figure 15.27). Insulated roof assemblies require a 1-inch (25-mm) airspace between insulation and sheathing. This airspace needs to be connected to the outside. With low and high vents to promote air movement, the air moving through this space evaporates any moisture that may condense in the assembly and helps to prevent heat buildup in the summer as well.

Ice Dams

Snow on a sloping roof tends to be melted by heat escaping through the roof from the heated space below. At the eave, however, the shingles and gutter, which are not directly above heated space, can be very cold, and the snow melt can freeze and begin to build up layers of ice. Soon an *ice dam* forms, and melt water accumulates above it, seeping between the shingles, through the roof sheathing, and into the building, causing damage to walls and ceilings (Figure 15.28). Ice dams can be prevented by keeping the entire roof cold enough so that snow will not melt. This is done by ventilating the roof continuously as described previously with low and high vents, or it may be accomplished by filling the rafter cavities with rigid insulation sufficient to prevent escaping heat from melting snow during the most extreme climatic circumstances. An R-value of 50 is generally adequate for this purpose.

Crawlspace Ventilation

The crawlspace is a unique cavity, trapping air between the insulated building above and the moisture-laden ground below. Until recently, the conventional wisdom for minimizing condensation in this cavity has been to provide lots of vents around the perimeter of the space. This strategy worked in most cases except that it did not inhibit moisture at its main source, the soil, and it allowed the introduction of additional moisture from humid air entering through the vents. In the 1970s, vapor retarders on the floor of the crawlspace were introduced, with a corresponding code

FIGURE 15.28
Ice dams form because of inadequate insulation combined with a lack of ventilation, as shown in the top diagram. The lower two diagrams show how insulation, attic ventilation, and vent spacers are used to prevent snow from melting on the roof. Many building codes also require the installation of a strip of water-resistant sheet material under the shingles in the area of the roof where ice dams are likely to form.

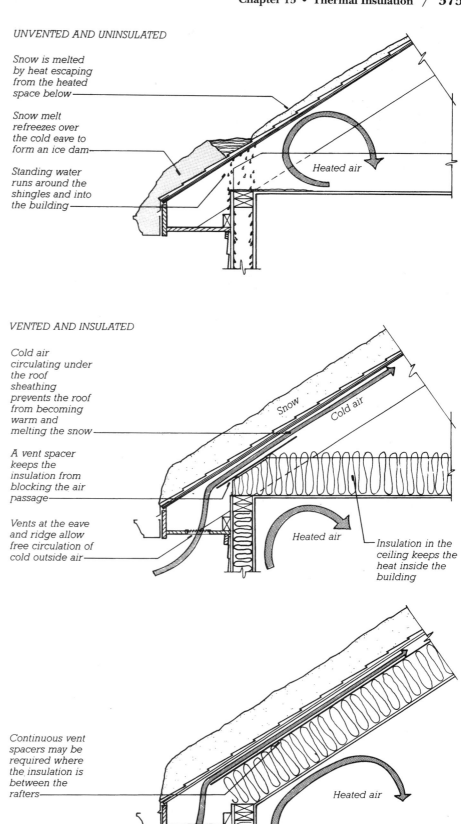

UNVENTED AND UNINSULATED

Snow is melted by heat escaping from the heated space below

Snow melt refreezes over the cold eave to form an ice dam

Standing water runs around the shingles and into the building

Heated air

VENTED AND INSULATED

Cold air circulating under the roof sheathing prevents the roof from becoming warm and melting the snow

A vent spacer keeps the insulation from blocking the air passage

Vents at the eave and ridge allow free circulation of cold outside air

Snow

Cold air

Heated air

Insulation in the ceiling keeps the heat inside the building

Continuous vent spacers may be required where the insulation is between the rafters

Cold air

Heated air

FIGURE 15.29
In some climates, it is best to seal a crawlspace shut after it has been thoroughly dried out. All seams in and between the floor vapor barrier and wall insulation are taped. A sealed crawlspace should be inspected once or twice a year to verify that it remains dry. (*Courtesy of The Healthy Building Company, Bon Lee, North Carolina*)

sealing around electrical outlets and other penetrations of the exterior wall. With reasonable care given to the control of infiltration, a dedicated contractor can cut the rate of infiltration to approximately one-half air change per hour (now the average for new residential construction). Even this reduced amount, however, can account for 30 percent of all heating (or cooling) energy costs in severe climates (Figure 15.30). By taking special measures to control infiltration, the rate can be cut further to 0.2 air change per hour.

Air Barriers at the Exterior of Wall Framing

One standard approach to controlling infiltration is to locate a continuous air barrier membrane under the siding. This is a logical place to control infiltration because it has few penetrations and because the weather barrier and *air barrier* can be one and the same. The most common air barrier in use today is a white polyolefin sheet material, marketed under a variety of trade names as "housewrap." This material, which is permeable to water vapor but not to liquid water or air, comes in very wide rolls—allowing coverage of a single-story house in one pass. When carefully taped at joints and sealed at edges, it makes a very effective air barrier (Figures 11.3 and 15.31).

reduction in the number of required vents. Crawlspace moisture continues to be a problem in many regions, however, primarily due to the condensation of moisture from warm, outside air that enters the crawlspace. Many builders and building scientists now advocate crawlspaces that are totally sealed and fully insulated. The vapor retarder is sealed to insulated foundation walls, and a small warm-air supply duct turns the crawlspace into a conditioned mini-basement (Figure 15.29).

can be very expensive for the homeowner because heating or cooling this air will amount to about 40 percent of the energy bill.

Accordingly, much attention is being given by researchers and conscientious builders and designers to reducing air infiltration between the outdoors and the indoors of residential buildings. This has led to higher standards of window and door construction, increased use of vapor-permeable air barrier papers under the exterior siding, and greater care in

CONTROL OF AIR INFILTRATION

The air leaking into a building through cracks around windows and doors and through gaps in the construction is called *infiltration*. In a house designed and built to code, but with little concern for the tightness of the envelope, infiltration can account for as much as one *air change per hour*, equivalent to changing the entire air volume of the house in 1 hour. This

Component	Heat Loss	Percentage of Total
Floor/foundation	86	15.4
Infiltration	171	30.5
Roof	35	6.2
Wall	98	17.5
Window conduction	170	30.4

FIGURE 15.30
Winter heat loss in an average-sized new two-story house. The figures are based on a design temperature differential of 70 degrees Fahrenheit, so the Btu/hour loss will change depending on outside temperature, but the percentage of Total figure will remain constant. (*Source:* **Residential Heating and Cooling Loads Component Analysis Final Report,** *Lawrence Berkeley National Laboratory for U.S. Department of Energy, November 1999*)

FIGURE 15.31
Building "wrap" is designed to act as both weather barrier and air barrier. All seams, edges, and window and door perimeters must be taped to make an effective air barrier. (*Photo by Donald Corner*)

Air Barriers at the Interior of the Framing

The air barrier may also be located at the inside of the framing. Here, the logical candidates to use as an air barrier are the polyethylene vapor retarder and the drywall. An advantage to this approach is that these materials are applied to the ceiling as well as the walls, extending the coverage beyond that of exterior air barriers. The disadvantage to both, as compared to strategies at the outside of the framing, is that the materials are not typically continuous. Their surfaces are interrupted by walls and floors that come in contact with the exterior wall of the building. Also, they are penetrated by many services (mostly electrical boxes) that are not present on the exterior surface of the building. Nevertheless, effectual air barrier systems that combine these materials with caulking and gaskets have been developed.

The system that uses drywall as an air barrier is called the *airtight drywall approach (ADA)*. In this approach, meticulous attention is given to sealing all the joints in the drywall panels that are used to finish the interior walls and ceilings. Compressible foam gaskets or sealants are used to eliminate air leakage around the edges of the floor platforms (Figure 15.32). Gypsum board is applied to all the interior surfaces of the outside walls before the interior partitions are framed, thus eliminating potential air leaks where partitions join the outside walls. Gaskets, sealants, or special airtight boxes are used to seal air leaks around electrical fixtures.

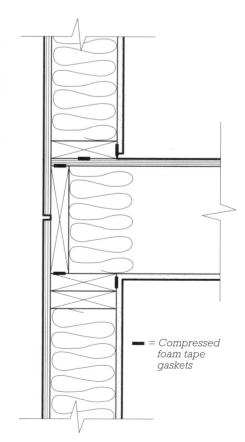

= Compressed foam tape gaskets

FIGURE 15.32
In the airtight drywall approach, foam tape gaskets or beads of sealant are applied to joints before assembly, greatly reducing air leakage through the exterior walls.

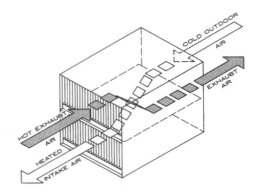

The use of the polyethylene vapor retarder as an air barrier is more practical than ADA when considering the overall residential construction process, because it enables drywall to be applied and finished all in one operation. Special care must be taken, however, where interior partitions would interrupt the vapor retarder, now called an *air/vapor barrier (AVB)*. The usual way to accomplish continuity of the AVB in these locations is to apply a strip of polyethylene to the ends and tops of the interior partitions as they are being installed. These strips are subsequently sealed with tape to the larger AVB sheet.

Hybrid Air Barriers

It is frequently the case that one air barrier system is best for the walls, another for the ceiling, and yet another for the floors or foundation. A different contractor may be responsible for each of these, and still another for *weatherstripping*, which seals the gaps around windows and doors (see Chapter 11) and is an integral part of any air barrier system. For these reasons, air barriers are often ineffective unless the general contractor takes responsibility for their completeness and continuity.

The Need for Ventilation

The trend toward tighter envelopes has sometimes resulted in houses and apartments that exchange so little air with the outdoors that indoor moisture, odors, and chemical pollutants build up to intolerable and often unhealthy levels. Opening a window to ventilate the dwelling in cold weather, of course, wastes heating fuel, so many designers and builders install forced-ventilation systems that employ *air-to-air heat exchangers* (Figure 15.33). These devices recover most of the heat from the air exhausted from the building and add it to the outside air that is drawn in.

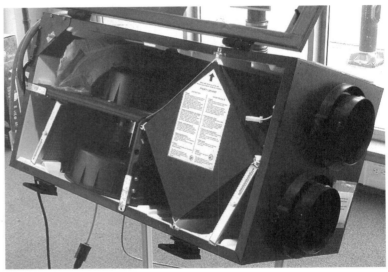

FIGURE 15.33
The diagram shows how ventilation air entering the house is heated by warm exhausted air as the two streams pass in a heat exchanger. The process produces condensate that is drained to the exterior. The photo shows a heat exchanger with fittings for 6-inch intake and exhaust ducts located at the end of the apparatus. The diagonally placed box is where the passing streams of air exchange their heat. The left of the apparatus contains fans, and there is a filter for both entering and exhausted air.
(*Diagram reprinted by permission of John Wiley & Sons, Inc. from Ramsey/Sleeper,* Architectural Graphic Standards, *10th ed., John Ray Hoke Jr., A.I.A., Editor © 2000 by John Wiley & Sons, Inc.; Photo by Greg Thomson.*)

C.S.I./C.S.C MasterFormat Section Numbers for Thermal Insulation	
07200	**THERMAL PROTECTION**
07210	**Building Insulation**
07220	**Roof and Deck Insulation**
07240	**Exterior Insulation and Finish System (EIFS)**
07260	**Vapor Retarder**
07270	**Air Barrier**

SELECTED REFERENCES

1. Lstiburek, Joseph. *Builder's Guide.* Westford, MA: Building Science Corporation, 1997, 1998.

A set of four guides that outline a comprehensive approach to thermal comfort. The guides are based on the logic of conservation and are organized according to climatic zones—one guide each for cold, mixed, hot-humid, and hot-dry. Each guide is profusely and clearly illustrated with chapters covering topics from foundations to painting.

KEY TERMS AND CONCEPTS

thermal insulation
thermal resistance
R-value
thermal envelope
perimeter insulation
glass fiber batt
loose-fill insulation
rigid board insulation
foamed-in-place insulation
radiant barrier

relative humidity
dew point
condensation
vapor retarder
perm
net ventilation area
soffit vent
frieze vent
vent spacer
ridge vent

gable vent
through-roof vent
ice dam
infiltration
air change per hour
air barrier
airtight drywall approach (ADA)
air/vapor barrier (AVB)
weatherstripping
air-to-air heat exchanger

REVIEW QUESTIONS

1. Make a list of four basic types of thermal insulation and describe their most common uses. Can you think of disadvantages for each type?

2. What are some alternative ways of insulating the walls of a wood light frame building to R-values beyond the range normally possible with ordinary 2 × 4 studs?

3. Discuss the use of perimeter insulation in very cold climates.

4. Compare the installation of attic, floor, and wall insulation.

5. Explain the need for a vapor retarder. Where in a wall or ceiling should the vapor retarder be located?

6. Why does an attic need to be ventilated? Where are the vents best located?

7. What is the purpose of an air barrier? Explain the advantages and disadvantages of three types of air barriers.

EXERCISES

1. Visit a local building supply outlet and make a list of the various types of wall insulation available. What is the R-value of each? What would be an appropriate use for each type?

2. Make a section drawing of the meeting of a wood-framed wall and vaulted wood-framed roof. Detail the section showing the type and location of all insulation, vapor barrier, air barrier, and ventilation. Have a local expert such as a building inspector, architect, or building designer review your drawing.

3. Procure from the local building department a list of insulation requirements for your area. Make a cross-sectional drawing of an entire house with notes sufficient to direct a contractor to comply with the requirements of the code.

4. Make a detailed design for the same house that would increase the code-required insulation values by 50 percent. How, exactly, would you propose to increase the insulation for the floor, walls, and roof? Would it be more cost effective to increase the thickness of the framing or to add insulation to the standard framing?

16

FIREPLACES AND STOVES

In earlier times, every house was built to include at least one fireplace. The fireplace served several useful purposes at once: primary heat source, cooking, and wintertime clothes dryer. Houses for the wealthy included a fireplace in almost every room.

With the advent of modern heating systems and appliances, the functional need for the fireplace disappeared. However, the percentage of houses with fireplaces is still quite high and is actually increasing. In 1975, only 52 percent of new single-family houses had fireplaces, and the number rose to 61 percent by 1998. We seem to embrace the fireplace for nostalgic reasons. There is a primordial attraction to the dancing flames, the smell, and the sound of a crackling fire.

FIGURE 16.1
The visual differences between a site-built masonry fireplace and a factory-built unit are chiefly apparent at the firebox. A masonry fireplace such as this can be made in a wide range of sizes and shapes. (*Photo by Rob Thallon*)

FIGURE 16.2
When building with masonry, the opportunity exists to express solidity and craftsmanship at the exterior as well as the interior. This chimney is made of long pieces of basalt oriented vertically except at the top, where horizontal pieces are corbelled to make the transition to a rain cap. (*Photo by Rob Thallon*)

SELECTING A FIREPLACE

There is such a wide variety of fireplaces available that their only shared characteristic is the *firebox,* the fireproof enclosure around the fire itself. Within this vast range, the most fundamental choice to be made when selecting a fireplace is the choice between a traditional *masonry fireplace* and a *factory-built* metal one. The differences between the two are significant. Masonry fireplaces are heavier than factory-built ones, so they need their own foundation. They are less efficient as a source of heat than factory-built fireplaces, and they generally are more expensive. For these reasons, in 1999, factory-built fireplaces accounted for more than 80 percent of fireplaces in new construction, up from 56 percent in 1960. However, there are still plenty of traditional masonry fireplaces constructed for people who prefer the solidity, simplicity, genuine appearance, and durability of bricks and mortar.

Many people choose between masonry and factory-built fireplaces on the basis of the firebox. The firebox of a masonry fireplace, because it is made of bricks on site, can be designed to a wide range of shapes and sizes. It can be very large or very small. It can be tall, shallow, arched, curved, or decorated inside (Figure 16.1). In addition, masonry chimneys can express a solidity and permanence that cannot be achieved with the lightweight wood-clad metal flues used with factory-built fireplaces (Figure 16.2).

Although firebox design is limited, the factory-built fireplace has an advantage over the masonry fireplace in almost every other category (Figure 16.3). It is considerably less expensive than a masonry fireplace. It is a more efficient source of heat than the masonry fireplace. In addition, because all of the components are relatively lightweight, it can be installed on a standard wood floor structure without requiring additional support (Figure 16.4). The one major drawback of factory-built fireplaces is that the metallic components are subject to corrosion and premature failure.

FIGURE 16.3
Factory-built fireplaces are available in a limited number of designs, usually incorporating glass doors. (*Courtesy of Midgley's, Eugene, Oregon*)

FIGURE 16.4
A factory-built wood-burning fireplace bearing on a wood floor in its approximate final location. Note the flange on top that will receive the round metal flue. (*Photo by Rob Thallon*)

PLANNING FOR FIREPLACE CONSTRUCTION

There are many planning decisions involved in the construction of a fireplace. If the fireplace is to be built of masonry, a special foundation to support its weight must be integrated with the foundation of the house, and the floor framing must be significantly altered to provide an opening for the fireplace and hearth. For both masonry and factory-built fireplaces, upper floor, ceiling, and roof framing need to be adjusted to allow for the passage of the chimney. Code-mandated clearances to combustible material from both fireplace and chimney need to be incorporated into the design of the framing of the house.

The location of the fireplace within the building deserves careful consideration. A central location is beneficial for solid-fuel-burning fireplaces because it promotes tall, weather-protected chimneys that improve draft. A central location also minimizes heat loss from the fire to the outdoors. For gas-burning fireplaces, however, locations at or near the outer wall that allow short flues are advantageous. Most fireplaces require a hearth that consumes space in a room, so room sizes need to be increased when a fireplace is included.

THE MASONRY FIREPLACE

The masonry fireplace can be built in an unlimited range of shapes and sizes. It can be fitted with gas burners to ignite or maintain the fire and with a door or heat-circulating device to increase efficiency. This flexibility is part of the attraction of the masonry fireplace, but also makes a concise description of its construction troublesome. For this reason, the following account will chronicle the most conventional masonry fireplace (Figure 16.5).

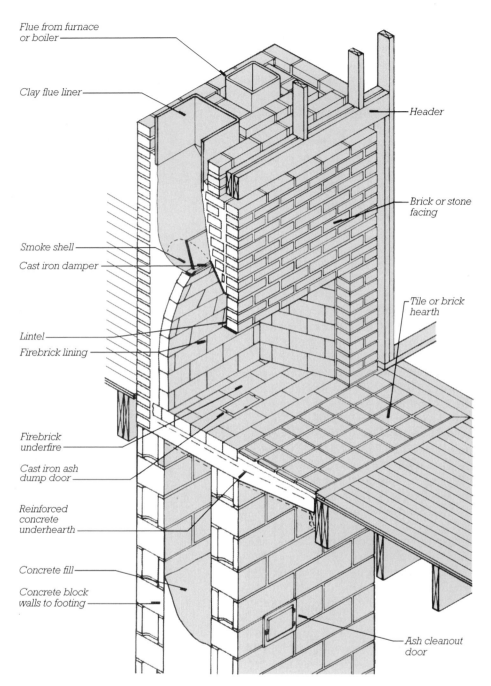

Flue from furnace or boiler

Clay flue liner

Smoke shell

Cast iron damper

Lintel

Firebrick lining

Firebrick underfire

Cast iron ash dump door

Reinforced concrete underhearth

Concrete fill

Concrete block walls to footing

Header

Brick or stone facing

Tile or brick hearth

Ash cleanout door

FIGURE 16.5

A cutaway view of a conventional masonry fireplace. Concrete block masonry is used wherever it will not show, to reduce labor and material costs. The damper, ash doors, and combustion air doors are prefabricated units of cast iron or steel. The flue liners are made of fired clay or cast pumice concrete and are highly resistant to heat. The flue from the furnace or boiler in the basement slopes as it passes the firebox until it adjoins the fireplace flue, to keep the chimney as small as possible.

Design of Masonry Fireplaces

Ever since fireplaces were first built in the Middle Ages, people have sought formulas for their construction that would ensure that the smoke from the fire would go up the chimney and the heat into the room, rather than the other way around. Surprisingly, to this day, there is still disagreement about how fireplaces actually work and how best to design them. A number of people over the ages have attempted to approach the problem scientifically, while others have taken note of the dimensions of fireplaces that seem to work reasonably well. From this, several general principles can be derived.

The firebox can be designed to emphasize the radiation of heat into the room or the prevention of smoke in the room. As heat production has become a less important feature of fireplaces, there has been a tendency for designers and masons to focus on the elimination of smoke. The tables of dimensions compiled over the past 150 years have resulted in recommendations for the proportions of fireboxes that have become increasingly shorter and deeper than their predecessors—resulting in a fire that is smaller and farther from the room but less likely to smoke (Figures 16.6 and 16.7).

FIGURE 16.6
The critical dimensions of a conventional masonry fireplace, keyed to the table in Figure 16.7. Dimension *D*, the depth of the hearth, is commonly required to be 16 inches (405 mm) for fireplaces with openings up to 6 square feet (0.56 m2) and 20 inches (510 mm) if the opening is larger. The side extension of the hearth, *E*, is usually fixed at 8 inches (203 mm) for smaller fireplace openings, and 12 inches (305 mm) for larger ones.

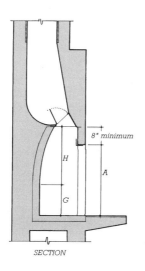
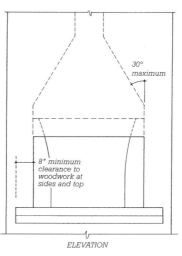
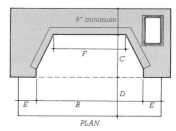

	Fireplace Opening		Minimum Backwall Width *(F)*	Vertical Backwall Height *(G)*	Inclined Backwall Height *(H)*	Flue Lining	
						Rectangular (outside dimensions)	Round (inside diameter)
Height *(A)*	Width *(B)*						
24" (610 mm)	28" (710 mm)		16 to 18" (405 to 455 mm)	14" (355 mm)	14" (355 mm)	8½ × 13" (216 × 330 mm)	10" (254 mm)
28 to 30" (710 to 760 mm)	30" (760 mm)		16 to 18" (405 to 455 mm)	16" (405 mm)	14" (355 mm)	8½ × 13" (216 × 330 mm)	10" (254 mm)
28 to 30" (710 to 760 mm)	36" (915 mm)		16 to 18" (405 to 455 mm)	22" (560 mm)	14" (355 mm)	8½ × 13" (216 × 330 mm)	12" (305 mm)
28 to 30" (710 to 760 mm)	42" (1065 mm)		16 to 18" (405 to 455 mm)	28" (710 mm)	14" (355 mm)	13 × 13" (330 × 330 mm)	12" (305 mm)
32" (815 mm)	48" (1220 mm)		18 to 20" (455 to 510 mm)	32" (315 mm)	14" (355 mm)	13 × 13" (330 × 330 mm)	15" (381 mm)

FIGURE 16.7
Recommended proportions for conventional masonry fireplaces, based largely on figures given in Ramsey/Sleeper, *Architectural Graphic Standards*, 9th ed., John Ray Hoke, Jr., F.A.I.A., Editor, John Wiley & Sons, New York, 1994.

Recent reexamination of these assumptions has brought to light the work of Count Rumford, an American expatriate who published formulas in the 1790s for fireplaces that were designed to radiate as much heat as possible into the room. The principles of the Rumford design are to make the firebox tall and shallow, which encourages a large fire and brings it closer to the room. In addition, the side walls of the firebox are splayed very broadly toward the room so that they radiate more heat into the room rather than back into the fire (Figure 16.8). Many *Rumford-style fireplaces* have been built in the past 30 years—for their advantages of greater heat production and the stately appearance of their more vertical and statuesque firebox opening (Figure 16.9). These experiences have shown that the Rumford principles create successful fireplaces.

The design of the flue to create a strong *draft* is an extremely important factor in assuring that the fireplace does not smoke. Several general principles are clear: *(1)* The cross-sectional area of the flue should be about one-tenth to one-twelfth the area of the front opening of the fireplace, depending on flue height—larger for a shorter flue and smaller for a taller one. *(2)* The chimney should be as tall as is practical so as to produce the largest possible convective draft. *(3)* The flue should be located in the central portion of the building, if possible, so as to avoid the cooling effect on the flue of outside air because the hotter the flue gases can be maintained, the more prone they will be to rise.

Constructing a Masonry Fireplace

The construction of a masonry fireplace is performed by a masonry subcontractor except for the footing (and foundation in some cases), which is done by the foundation subcontractor. The trim around the firebox (Figures 16.23 to 16.25) is completed by a finish carpenter and/or a tilesetter.

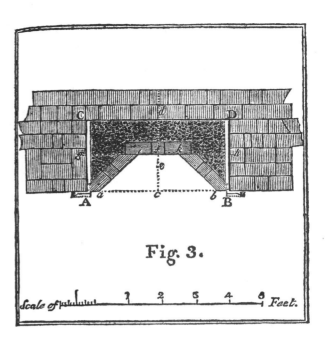

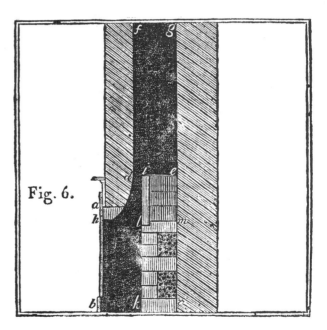

FIGURE 16.8
Plans and sections of the fireplace design developed by Count Rumford in 1790. The proportions are based on thermodynamic principles known at the time. Compared to Figure 16.6, the Rumford firebox is taller, shallower, and more closely aligned with the flue above.

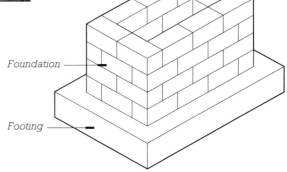

FIGURE 16.10
The footing for a masonry fireplace is required to be 12 inches (305 mm) deep and 6 inches (152 mm) larger on each side than the fireplace it supports. Reinforcing steel in the footing provides tensile strength. The fireplace foundation, which rests on the footing, is necessary if the fireplace is located in a room above a crawlspace or basement.

Foundation

The construction of a durable masonry fireplace starts with a strong footing on solid soil (Figure 16.10). In many cases, the footing is continuous with the footing for the house and only gets wider and deeper for the fireplace. Most codes call for the footing to be 12 inches deep and to extend 6 inches beyond the fireplace masonry on all sides. Prudent contractors will place #4 rebar in the fireplace footing at 12 inches (305 mm) on center in both directions whether it is required or not so as to provide tensile strength in this footing that bears the enormous weight of the masonry fireplace and chimney above.

Next, a foundation wall to support the fireplace is built up from this footing to a level just below the firebox floor (Figure 16.10). This wall, which passes vertically through the crawlspace or the basement, may be made either of concrete block masonry or poured-in-place concrete. It is tied to the footing and reinforced in the same manner as the house foundation wall. In the case where the house is constructed on a slab on grade, there is no foundation wall, as the fireplace will bear directly on its footing.

Firebox and Hearth

At the level of the firebox, a concrete slab to support both the firebox floor and the *hearth* (called "hearth extension" in the building code) must be cast (Figure 16.11). This slab must cantilever to support the hearth and must be set at the right level to accept the firebrick of the firebox floor and the noncombustible finish (such as tile) of the hearth.

The firebox level consists principally of an outer masonry loadbearing structure and an inner receptacle, the *firebox,* to contain and radiate the heat of the fire. The firebox (Figure 16.11) is lined with *refractory firebrick* made of special clays capable of withstanding the high temperatures of a wood fire. The firebricks are laid with very thin joints of a specially formulated *refractory mortar.* The outer masonry at the firebox level may be made of any structural masonry material and aligns with the foundation walls below.

The masonry fireplace *facing* is a special part of the outer masonry structure because, along with the firebox, it is exposed to the room when the fireplace is complete. The facing is constructed simultaneously with the firebox and may be made of exposed brick, stone, or concrete masonry to be covered later with tile or plaster. The opening in this facing must be spanned with a lintel or arch to support the masonry above. The most conventional approach is to use a steel angle as a lintel, but an arch will also span the opening, provided there is sufficient masonry at the sides of the opening to contain the thrust of the arch (Figure 16.12).

Incorporated into the firebox are the *ash dump* and the *combustion air* door. The ash dump, located in the floor of the firebox, is a cast iron door factory-built to the size of a firebrick. The ash dump connects to an ash collection chamber below the firebox via an opening cast into the underhearth slab. Much more important to the operation of the fireplace is the combustion air door, a recent innovation that is usually located near the base of the firebox wall (Figure 16.13). Now mandated by most codes, this operable door brings combustion air to the fire from outside the building. The external combustion air relieves the negative pressure in the house caused by the flow of flue gases up the chimney. The draft is improved because the fire can draw air directly from the outside without having to pull it through cracks in the building envelope. The overall efficiency of the fireplace is increased because the heated air in the living space is not sent up the flue to be replaced by cold outside air seeping into the house through cracks.

Damper and Smoke Chamber

At the top of the firebox is the *damper,* a metal flap that seals off the flue when the fireplace is not in use (Figure 16.14). The damper assembly is made of steel or, better, cast iron or stainless steel, and is sealed to the top of the firebox (called the *throat*) with mortar. The flap is controlled by means of a handle or rotating knob.

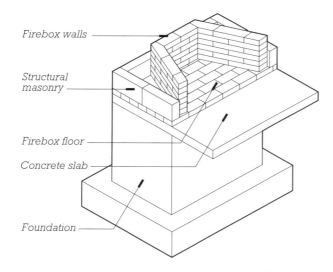

FIGURE 16.11
A concrete slab at the level of the hearth supports the firebox floor and the hearth. The firebox floor is lined with firebricks, and the hearth is usually covered with a noncombustible material such as tile or brick. The firebox walls are made of firebrick and are independent of the structural masonry wall that surrounds them. Total wall thickness at the firebox is required by code to be a minimum of 8 inches (203 mm).

FIGURE 16.12
The masonry facing of this fireplace has been formed into an arch to span the fireplace opening. This rough work will be covered with tile. (*Photo by Rob Thallon*)

FIGURE 16.13
The combustion air doors at the base of the firebox wall will supply air to the fire through an insulated duct from the outside. Most fireplaces have only one combustion air supply. (*Photo by Rob Thallon*)

FIGURE 16.14
The fireplace damper is installed at the base of the smoke chamber. Glass fiber insulation is placed around the edges of the metal damper to allow it to expand without cracking the masonry. (*Photo by Donald Corner*)

(a)

(b)

(c)

(d)

(e)

FIGURE 16.15

In this construction series of a Rumford masonry fireplace, the masons were trying to make a point about how rapidly the work could be accomplished. (a) At 9:00 A.M. on the first day, precast concrete slabs are set on the fireplace foundation to make a base for the fireplace and hearth. (b) At 9:50 A.M., the firebox floor has been made of firebrick, and the firebox walls and outer CMU structural walls are being laid. (c) At 10:40 A.M., a single-piece precast fireclay throat is mortared to the top of the firebox. (d) At 11:30 A.M., a two-piece precast fireclay smoke chamber is set over the damper. (e) At 3:00 P.M., the chimney emerges through the roof. It took about four more hours to complete the job including building a scaffold on the roof, tooling joints, and cleaning up. (*Photos courtesy of Buckley Rumford Co.*)

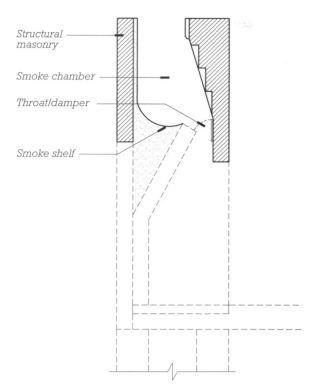

Structural masonry

Smoke chamber

Throat/damper

Smoke shelf

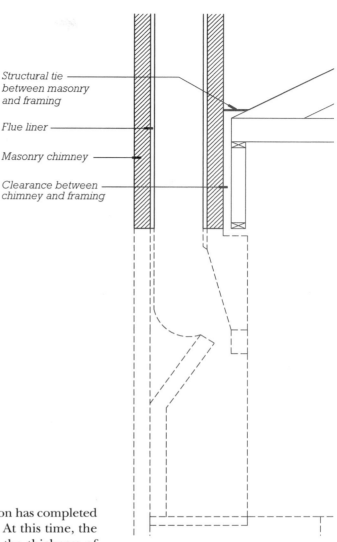

Structural tie between masonry and framing

Flue liner

Masonry chimney

Clearance between chimney and framing

FIGURE 16.16
The damper at the base of the smoke chamber is closed to avoid loss of air from the room when there is no fire. The smoke chamber has traditionally been made of bricks, corbelled in to meet the flue. Precast fireclay smoke chambers now save considerable labor.

Above the damper is the *smoke chamber* (Figure 16.16). Its purpose is to make a transition from the shape of the damper to the shape of the flue and to absorb and redirect the downdrafts that occur in the flue above—preventing them from reaching the firebox, where they would force smoke into the room. The chamber is shaped like an inverted hopper. It traditionally has been made of corbelled bricks that are parged with plaster to make them smooth. Precast fireclay smoke chambers (Figure 16.15*d*) are now available and are becoming popular, despite their high purchase cost, because they are so much faster to install and are therefore less expensive overall. The top of the smoke chamber meets the base of the flue and provides its structural support.

The main inspection of the fireplace for building code compliance occurs when the mason has completed the smoke chamber. At this time, the inspector will check the thickness of materials, the structure of the lintel or arch support over the firebox opening, the installation of the damper, and the combustion air supply.

Flue

Above the smoke chamber, the *flue* and chimney begin (Figure 16.17). The flue is the passageway that conveys the combustion gases from the fire up to the top of the chimney where they are released into the atmosphere. The *chimney* is the masonry shaft that contains the flue. Chimneys used to be made entirely of bricks *parged* (plastered with mortar) on the inside to make a smooth flue. Today, chimneys are made with a *flue liner* surrounded by 4 inches of solid masonry. The flue

FIGURE 16.17
The flue is sized to match the firebox opening for maximum draft. The chimney supports the flue and must be reinforced and tied to the structure in high-risk seismic zones.

liner is either round or rectangular and composed of terra cotta or pumice. Round shapes are more efficient in conveying flue gases but do not fit as efficiently within the rectangular format of masonry units.

Where the flue and chimney pass through the floor, ceiling, or roof, the code prescribes clearances between the masonry and wood framing to minimize the risk of fire. In high-risk seismic zones, the code also requires that the masonry be tied to the framing. This is accomplished with metal straps embedded in the masonry and screwed to the framing.

Chimney

Where the chimney emerges from the building, it becomes exposed to the weather and must be detailed to withstand the forces of nature (Figure 16.18). The surface of the chimney, which can be unfinished concrete block when located within the building, must be impervious to moisture when located at the exterior. Common materials for exterior chimney surfaces include stucco, tile, brick, and stone masonry. Sealers applied to the surface of any of these materials can increase moisture resistance.

The top of the chimney must be protected from the weather with a cast-in-place concrete *chimney cap* that surrounds the flue and slopes away from it in all directions for drainage (Figure 16.19). The chimney cap should be reinforced with a loop of rebar to help prevent it from cracking. It should be at least 4 inches thick at its thinnest point. It is often advisable to shelter the flue opening with a *rain cap,* which is a small roof that is built above the flue to keep out rain and snow. These rain caps, usually constructed of masonry or sheet metal, can extend the life of a flue by minimizing the moisture that enters it (Figure 16.20). In areas with high fire risk, *spark arresters* that minimize the dispersal of burning cinders are a part of the typical rain cap assembly. Care should be taken when designing rain caps to allow easy removal for periodic flue cleaning.

The intersection between roof and chimney must be flashed, and this is best achieved with standard roof flashing plus a counterflashing embedded into the masonry (Figure 16.21). At the upslope side of a wide chimney, a *cricket* is a required flashing to shed water around the chimney (Figure 16.22). If the chimney intersects an exterior wall, the joint between wall and chimney must also be sealed against the penetration of the weather.

Fireplace Finish

The fireplace, the place of fire, is the last part of the assembly to be finished, but aesthetically the most important (Figure 16.23). The noncombustible facing around the fireplace is required by code to be at least 6 inches (152 mm) wide. Common materials include brick, ceramic tile, stone tile, and stone slabs. A facing material that can be finished in a rather smooth plane

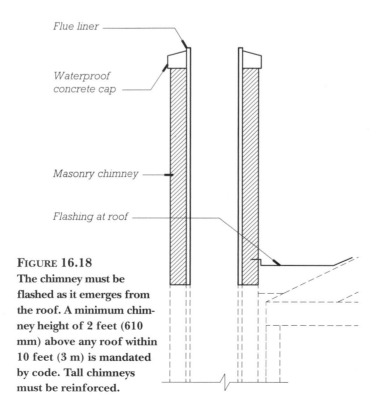

Flue liner

Waterproof concrete cap

Masonry chimney

Flashing at roof

FIGURE 16.18
The chimney must be flashed as it emerges from the roof. A minimum chimney height of 2 feet (610 mm) above any roof within 10 feet (3 m) is mandated by code. Tall chimneys must be reinforced.

FIGURE 16.19
A concrete cap at the top of a masonry chimney seals it against the weather. (*Photo by Rob Thallon*)

FIGURE 16.20
The principal function of a rain cap is to keep rain out of the flue. The rain cap also affords the opportunity to decorate the top of the chimney as does this 1985 example from Portland, Oregon, made of fired clay. Rain caps are usually made of masonry or metal and are most common in regions with heavy annual rainfall. Caps are also designed to increase the draft of flue gases and to prevent sparks from escaping into the atmosphere. (*Photo by Rob Thallon*)

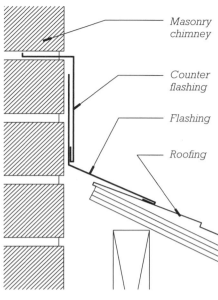

FIGURE 16.21
Counterflashing is embedded into the masonry so that moisture running down the surface of the chimney does not get behind it. The counterflashing laps over flashing that is integrated with the roofing. It is quite difficult to replace counterflashing, so very durable metals such as copper and stainless steel are often used.

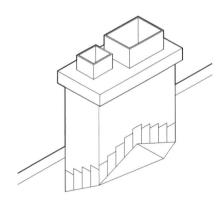

FIGURE 16.22
A cricket made of 2x's and plywood covered with flashing is required above wide chimneys in order to direct rainwater runoff around the chimney.

FIGURE 16.23
This masonry fireplace is faced with marble and trimmed with a traditional wooden fireplace surround, painted white. The hearth, also made of marble, is flush with the wood floor and almost disappears because the wood and marble are similar colors. The dark circle above the firebox contains a chain device that controls the damper. (*Photo by Rob Thallon*)

FIGURE 16.24
Fireplace surrounds and mantels are available from millwork suppliers in a number of designs. Each is furnished largely assembled but is detailed in such a way that it can easily be adjusted to fit any fireplace within a wide range of sizes. (*Courtesy of Morgan Products, Ltd.*)

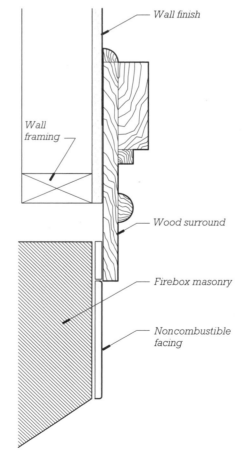

FIGURE 16.25
This plan drawing shows how the fireplace surround trims the gap between the firebox masonry and the framing. If the noncombustible fireplace facing is on the same plane as the finish surface of the surrounding wall as shown here, the installation is easiest. A section drawing at the top of the firebox opening would show similar details but would likely include a projecting mantel.

Wall finish

Wall framing

Wood surround

Firebox masonry

Noncombustible facing

is easier to join to trim materials at the outer edge and to the firebox at the inner edge. If the facing is brick or another masonry material, the mason usually constructs it at the same time as the firebox. If the facing is tile or plaster, a subcontractor usually installs it during the interior finish phase of construction. The hearth is often made of the same material as the facing. The mason needs to know what the facing and hearth finishes will be as the rough hearth and firebox are being built in order to allow for the thickness of these materials.

The noncombustible facing on a traditional fireplace is encircled by a *fireplace surround* capped by a mantel. Usually made of wood, the surround and mantel may be site built, shop built, or selected from a catalog (Figure 16.24). The fireplace surround typically laps over the facing material and trims to the surrounding wall, covering the gap between masonry and framing (Figure 16.25). The minimum clearance of the surround from the firebox is mandated by code. The projection (depth) of a combustible surround and mantel is also limited by code if it is within 12 inches (305 mm) of the fireplace opening.

FACTORY-BUILT FIREPLACES

Factory-built metal fireplaces are now much more common than those made in the traditional way with bricks and mortar. Typically called zero clearance because there is virtually no clearance required from the surrounding framing, these fireplaces are self-contained, fully insulated units that require no masonry whatsoever. The vast majority of them are made to burn wood or gas, but models that burn wood pellets, coal, or oil are also available. Factory-built fireplaces are typically fitted with glass doors to control combustion for efficiency.

Wood-burning types are most similar to the masonry fireplace and consist of an insulated metal firebox with

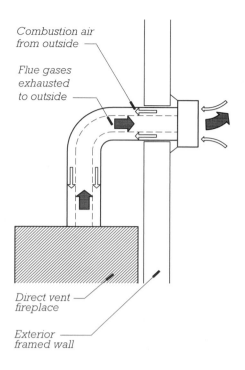

Combustion air from outside

Flue gases exhausted to outside

Direct vent fireplace

Exterior framed wall

FIGURE 16.26
Section of a direct-vent pipe, which exhausts combustion products through an inner chamber of the pipe and draws combustion air into the fireplace or stove through an outer, concentric chamber of the pipe. This concentric pipe arrangement has several advantages. (1) Only one wall penetration is required for both incoming combustion air and outgoing flue gases. (2) The incoming combustion air is preheated for greater combustion efficiency. (3) The incoming air reduces the surface temperature of the assembly where it joins wood construction.

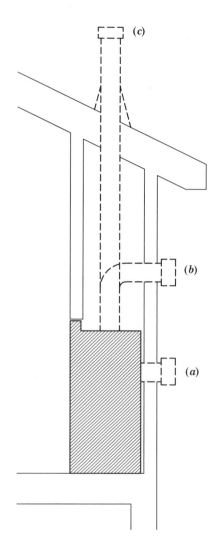

FIGURE 16.27
A direct vent can be routed (a) horizontally, (b) up through the wall, or (c) up through the roof.

FIGURE 16.28
A horizontally vented direct-vent fireplace as viewed from the outside. (*Photo by Rob Thallon*)

a refractory lining (Figure 16.4). Many wood-burning models have provisions to add a gas lighter or gas log sets. Some are equipped with a circulating fan that passes air through chambers surrounding the firebox and forces heated air into the room. Because they are designed to burn wood, which burns less predictably than gas, these fireplaces must generally meet the code provisions required of open masonry fireplaces. This includes a *class A flue*, an insulated stainless steel flue rated for solid-fuel combustion, and a noncombustible hearth.

Gas fireplaces are designed to look like a traditional wood-burning fireplace with an arrangement of noncombustible, artificial "logs" and a gas flame positioned to make it appear that the logs are burning. In reality, a gas fireplace is more like a furnace or radiant heater than a traditional fireplace. Because the fuel—either natural gas or propane—burns very cleanly, the required flue is considerably smaller than that for a comparable sized wood-burning appliance. Most gas fireplaces use either a *natural venting* system (also called a B vent,

which is similar to a class A flue but smaller and thinner) or a *direct venting* system, which has an inner exhaust pipe surrounded by an outer sleeve that conducts combustion air from the outside to the appliance (Figure 16.26). A direct vent can pass from the rear of the firebox horizontally through the wall of the building to the exterior, which significantly reduces the cost of venting as compared to venting through the roof (Figures 16.27 and 16.28). Some gas fireplaces are designed primarily for decorative effect and burn a minimal amount of fuel efficiently enough to be rated as *vent-free*, requiring no exhaust to the exterior. These units are equipped with oxygen depletion sensors and are limited by code to installation only in rooms with sufficient air volume to absorb the byproducts of combustion.

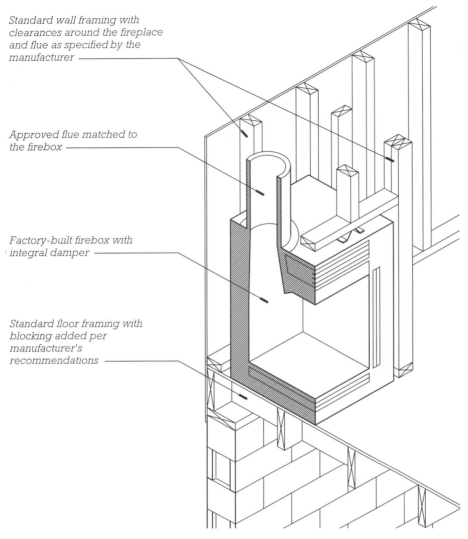

Standard wall framing with clearances around the fireplace and flue as specified by the manufacturer

Approved flue matched to the firebox

Factory-built firebox with integral damper

Standard floor framing with blocking added per manufacturer's recommendations

FIGURE 16.29
A factory-built fireplace supported by standard floor framing. The installation must comply with the specifications furnished by the manufacturer or warranties are not valid. The firebox must be anchored, clearances must be recognized, and factory-supplied firestops must be installed where flues pass through framing. Finish trim is similar to that of masonry fireplaces.

FIGURE 16.30
A gas-fueled factory-built fireplace with a direct-vent flue routed up through the middle of the house. The instructions and accessories are left inside the unit behind the glass door for later use during the finish phase of construction. Carpenters have installed the fireplace and flue, but the gas will be connected by a licensed plumber. (*Photo by Rob Thallon*)

Installing Factory-Built Fireplaces

The most important consideration when installing a factory-built fireplace is to follow the specifications of the manufacturer. These specifications are developed through extensive testing by Underwriters Laboratories or other testing agency and must be followed to the letter in order to ensure safety and not to void the warranty. Units must be anchored to the framing, clearances to combustible materials must be maintained, and factory-supplied firestops must be installed where flues pass through framing. The framers are responsible for leaving a rough opening corresponding to the size of the specific unit and for allowing space for the flue or vent. The fireplace and its flue or vent are typically installed by a subcontractor specializing in this work.

Most units can be supported on standard floor framing with no modifications (Figure 16.29). Framing for the flue needs to allow for unobstructed passage of the flue itself plus some clearance between flue and framing. In the case of wood-burning units located on an exterior wall, a wooden chimney is often framed around the metal flue (Figure 16.31).

Interior finish around a factory-built fireplace is essentially the same as for a masonry fireplace (Figures 16.3 and 16.32). However, some gas- and oil-burning fireplaces—because there is no chance that burning embers will pop out of the fire—are not required to have a noncombustible hearth (Figure 16.30).

FIGURE 16.31
A wood-framed enclosure of a metal flue. (*Photo by Greg Thomson*)

FIGURE 16.32
A factory-built fireplace, trimmed in a traditional style. (*Photo by Rob Thallon*)

FREESTANDING STOVES

Freestanding metal and ceramic stoves are somewhat similar to factory-built fireplaces except that they are not built into the structure (Figure 16.33). Most require a noncombustible surface upon which to stand. Freestanding stoves require rather large clearances to surrounding walls and floors, for which reason they occupy large amounts of space in a house, and need to be incorporated into the planning of a house at a very early date. These stoves are made to burn the same fuels as factory-built fireplaces, and the flue or venting requirements are similar to those of factory-built fireplaces.

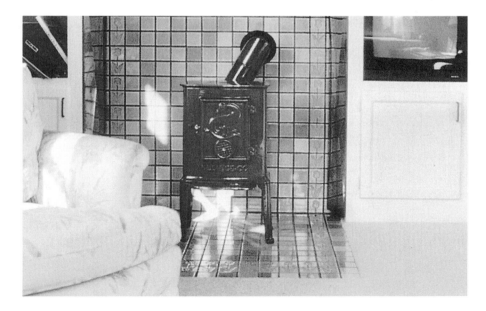

FIGURE 16.33
A freestanding wood-burning stove. Clearances from the stove to combustible materials are specified by the manufacturer and mandated by the code. (*Photo by Rob Thallon*)

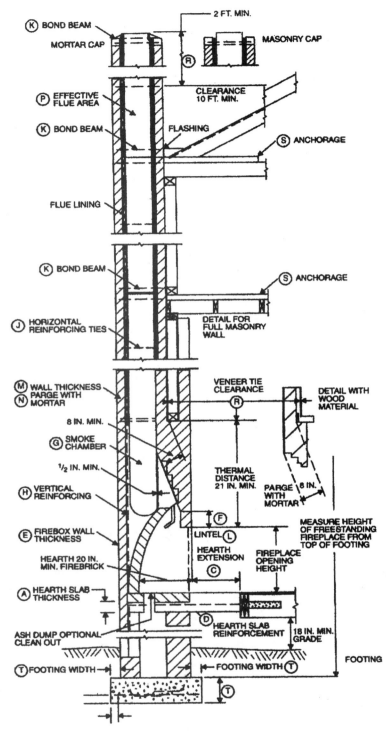

BRICK FIREBOX AND CHIMNEY—
SECTIONAL SIDE VIEW ON WOOD FLOOR

For SI: 1 inch = 25.4 mm, 1 foot = 304.8 mm.

FIREPLACES AND THE BUILDING CODES

Site-built masonry fireplaces are strictly regulated by building codes. Typically, the codes call for a 2-inch (51-mm) clearance between the wood framing and the masonry of a chimney or fireplace, and a 6-inch (152-mm) clearance to combustible materials around the opening of the fireplace. The minimum thickness of masonry around the firebox and flue, the minimum size of flue, and the minimum extension of the chimney above the roof are also specified (Figure 16.34 and 16.35). In high-risk seismic zones, masonry chimneys must be reinforced and structurally anchored at each floor, ceiling, and roof. Hearth extensions must be self-supporting and noncombustible.

FIGURE 16.34
This drawing from the International Residential Code shows the critical dimensions, elements, and connections for a masonry fireplace. The circled letters on the drawing refer to the table shown in Figure 16.35. (*Courtesy of International Code Council, Inc.*)

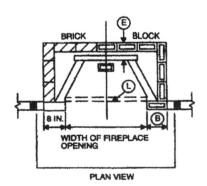

WIDTH OF FIREPLACE
OPENING

PLAN VIEW

Summary of Requirements for Masonry Fireplaces and Chimneys

Item	Letter[a]	Requirements
Hearth slab thickness	A	4"
Hearth extension (each side of opening)	B	8" fireplace opening < 6 ft² 12" fireplace opening ≥ 6 ft²
Hearth extension (front of opening)	C	16" fireplace opening < 6 ft² 20" fireplace opening ≥ 6 ft²
Hearth slab reinforcing	D	Reinforced to carry its own weight and all imposed loads.
Thickness of wall of firebox	E	10" solid brick or 8" where a firebrick lining is used. Joints in firebrick ¼" max.
Distance from top of opening to throat	F	8"
Smoke chamber wall thickness unlined walls	G	6"
Chimney Vertical reinforcing[b]	H	8"
Horizontal reinforcing	J	Four #4 full-length bars for chimney up to 40" wide. Add two #4 bars for each additional 40" or fraction of width or each additional flue.
Bond beams	K	¼" ties at 18" and two ties at each bend in vertical steel.
Fireplace lintel	L	Noncombustible material.
Chimney walls with flue lining	M	Solid masonry units or hollow masonry units grouted solid with at least 4 inch nominal thickness
Walls with unlined flue	N	8" solid masonry.
Distances between adjacent flues	—	See Section R1001.10.
Effective flue area (based on area of fireplace opening)	P	See Section R1001.12.
Clearances:	R	
Combustible material		See Sections R1001.15 and R1003.12.
Mantel and trim		See Section R1001.13.
Above roof		3' at roofline and 2' at 10'.
Anchorage[b]	S	
Strap		³⁄₁₆" × 1"
Number		Two
Embedment into chimney		12" hooked around outer bar with 6" extension
Fasten to		4 joists
Bolts		Two ½" diameter.
Footing	T	
Thickness		12" min.
Width		6" each side of fireplace wall

For SI: 1 inch = 25.4 mm, 1 foot = 304.8 mm, 1 square foot = 0.0929 m².

NOTE: This table provides a summary of major requirements for the construction of masonry chimneys and fireplaces. Letter references are to Figure R1003.1, which shows examples of typical construction. This table does not cover all requirements, nor does it cover all aspects of the indicated requirements. For the actual mandatory requirements of the code, see the indicated section of text.

[a]The letters refer to Figure R1003.1.

[b]Not required in Seismic Design Category A, B, or C.

FIGURE 16.35
Table R1003.1 from the International Residential Code. This table corresponds with the drawing shown in Figure 16.34.
(*Courtesy of International Code Council, Inc.*)

Regulations for factory-built fireplaces and flues are largely left to government-authorized testing agencies that are paid by the manufacturer to test their units for safety and emissions control. Once approval is granted, the building codes rely on the testing agency to specify clearances, connections, and methods of installation. The code does require that flues passing through a floor or ceiling have a factory-supplied firestop and that those passing through living spaces be enclosed to avoid contact with persons or combustibles. There are requirements for hearth extensions, which are mandated for all factory-built wood-burning fireplaces but only some gas-burning fireplaces.

A supply of combustion air to all fireplaces, masonry and factory built, is now required and regulated by code. The location and design of the air intake, the minimum cross-sectional area of the air passageway, and the location and design of the inlet at the firebox are all regulated by code.

C.S.I./C.S.C
MasterFormat Section Numbers for Fireplaces

04800	**MASONRY ASSEMBLIES**
04880	**Masonry Fireplace**
15500	**HEAT GENERATION EQUIPMENT**
15540	**Fuel-Fired Heater**
15550	**Breechings, Chimneys, and Stack**

SELECTED REFERENCES

1. Amrhein, James E. *Residential Masonry Fireplace and Chimney Handbook*, 2nd ed. Los Angeles: Masonry Institute of America, 1995.

2. Orton, Vrest. *The Forgotten Art of Building a Good Fireplace.* Dublin, NH: Yankee, Inc., 1969.

This was the original book that revived the Rumford fireplace. It is an interesting short book with some very long sentences.

3. The building regulations regarding fireplaces are extremely precise because the dangers are so great. The residential code listed in reference 4 at Chapter 1 is therefore a logical starting place for anyone designing a masonry (or manufactured) fireplace and an excellent resource for the study of fireplace construction.

KEY TERMS AND CONCEPTS

masonry fireplace
Count Rumford
Rumford fireplace
draft
hearth
firebox
refractory firebrick
refractory mortar
facing
ash dump
combustion air

damper
throat
smoke chamber
flue
chimney
parge
flue liner
chimney cap
rain cap
spark arrester

cricket
fireplace surround
mantel
factory-built fireplace
zero-clearance fireplace
class "A" flue
natural venting
direct venting
vent-free
freestanding stove

REVIEW QUESTIONS

1. What are the advantages and disadvantages of masonry fireplaces and factory-built fireplaces?

2. What are the factors that contribute to the decision about where in a house to locate a fireplace?

3. Discuss the differences between a standard firebox and a Rumford-style firebox.

4. Describe the components of a masonry fireplace and discuss the important aspects of the construction of each.

5. Compare the qualities of fireplaces that require natural-vent, direct-vent, and vent-free systems.

EXERCISES

1. Design and detail a masonry fireplace for a building that you are designing, using the information provided on page 385 to work out the exact dimensions, as well as the information in Chapter 5 to help in detailing the masonry.

2. Repeat the preceding exercise but with a factory-built firebox and flue. How does the appearance of the fireplace and chimney differ from the masonry design?

3. Call a local mason and arrange to document the construction of a masonry fireplace. Take photographs of each phase of construction and make a presentation to your class. Does the mason have any recommendations for construction that differ from standard practice?

4. Visit a local fireplace and stove retailer to explore the range of available factory-built appliances. Collect brochures and share these with your class. Does the retailer have information on the efficiency of models designed to produce heat?

INTERIOR SURFACES

Finishing the interior of a house is a complex and time-consuming phase of construction that requires careful attention to detail. Finishing operations typically begin with subcontractors who install drywall, hardwood flooring, and tile. They are followed by finish carpenters who install doors, set cabinets, and apply interior trim. As completion draws near, carpet and vinyl flooring are installed, and the plumber, electrician, and heating/cooling contractor make final visits. Painters come and go throughout the process as necessary. All of these workers must be scheduled in a carefully ordered sequence that varies somewhat from one building to another, depending primarily on the materials used, the finish details specified, and the availability of the workers and supplies (Figure 1.27).

Three-coat plaster applied to wood strip lath was the prevalent wall and ceiling finish system until World War II, when *gypsum board* came into widespread use because of its lower cost, more rapid installation, and utilization of less skilled labor. Currently, there are two popular systems that use gypsum board panels as a base. The most common is colloquially called *drywall*, in which the panels are finished only at the edges with a plaster-like joint compound. The other system is called *veneer plaster* or *thincoat*, in which the entire panel is covered with a thin coat of plaster.

WALL AND CEILING FINISH

The first finish operation, the application of a smooth, continuous surface to the inside of the wall studs and ceiling joists, is one of the most dramatic and satisfying stages in the construction a house. The planes that define the rooms suddenly emerge from the confusing complexity of the raw frame, and the spaces of the house are revealed for the first time. The most popular finishes by far for walls and ceilings are the *gypsum wallboard systems,* including gypsum drywall and veneer plaster. Indeed, over 90 percent of residential interior surfaces in North America are finished in this way. Other prevalent finish systems include *sheet paneling, board paneling,* and tile.

The wall and ceiling finish system is applied to the interior plane of the framing and creates a new plane to which molding and other trim is applied (Figure 17.1). The designer will have specified window and door casings to align with this new plane, and electrical boxes will have been installed to be flush with this finish surface. If more than one wall finish system is used in a house (such as a wainscot at the base of a gypsum board wall), it is more convenient if the thickness of both materials is the same so that moldings straddling the two materials can be simple.

Gypsum Board Systems

Plaster-type finishes have always been the most popular for walls and ceilings in wood frame buildings. Their advantages include a substantially lower installed cost than any other type of finish, a unique ability to provide a seamless surface that can be either painted or wallpapered, and a degree of fire resistance that offers considerable protection to a combustible frame (Figure 17.2).

FIGURE 17.1
The interior surface of the drywall or other finish wall surface creates a plane onto which interior casing and other trim is applied. The depths of door and window jambs are made to equal the total thickness of the wall, including sheathing, framing, and interior finish, so that interior casing can lap over the inner edge of the jamb and the wall surface.

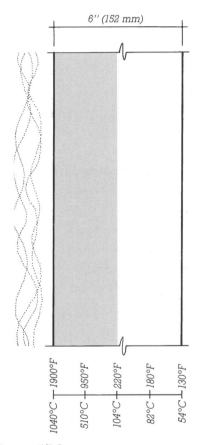

FIGURE 17.2
The effect of fire on gypsum, based on data from Underwriters Laboratories. After a 2-hour exposure to heat following the ASTM E119 time–temperature curve, less than half the gypsum on the side toward the fire, shown here by shading, has calcined (dehydrated). The portions of the gypsum to the right of the line of calcination remain at temperatures below the boiling point of water.

Gypsum Board

Gypsum board is a prefabricated plaster sheet material that is manufactured in widths of 4 feet (1220 mm) and lengths of 8 to 14 feet (2440 to 4270 mm). It is also known as gypsum wallboard (GWB), plasterboard, and drywall. (The term "sheetrock" is a registered trademark of one manufacturer of gypsum board and should not be used in a generic sense.) Gypsum board is the least expensive of all interior wall and ceiling finishing materials. For this reason alone, it has found wide acceptance throughout North America in buildings of every type. Among its other advantages are that it is durable and light in weight as compared to many other materials, it resists the passage of sound better than most materials, and it is highly resistant to the passage of fire. It is installed with minimal labor by semiskilled workers and can be fashioned into surfaces that range from smooth to heavily textured. Finally, because it is installed largely in the form of dry materials, it eliminates some of the waiting that is associated with the curing and drying of traditional plaster on lath.

The core of gypsum board is formulated as a slurry of calcined gypsum, starch, water, pregenerated foam to reduce the density of the mixture, and various admixtures. This slurry is sandwiched between special paper faces and passed between sets of rollers that reduce it to the desired thickness. Within 2 or 3 minutes, the core material has hardened and bonded to the paper faces. The board is cut to length and heated to drive off residual moisture, then bundled for shipping (Figure 17.3). In recent years, a gypsum board that has no paper faces has come into production. The gypsum slurry from which the board is made contains an admixture of cellulose fibers that greatly strengthen the hardened material, making paper faces unnecessary.

FIGURE 17.3
Sheets of gypsum board roll off the manufacturing line, trimmed and ready for packaging. (*Courtesy of United States Gypsum Company*)

Types of Gypsum Board

Gypsum board is manufactured in a number of different types in accordance with ASTM C36.

• *Regular gypsum board* is used for the majority of applications.

• *Water-resistant gypsum board,* a special board with a water-repellent paper facing and a moisture-resistant core, is used in locations exposed to moderate amounts of moisture such as bathrooms and kitchens.

• *Type X gypsum board* is required by the building code for many types of fire-rated assemblies. The core material of Type X board is reinforced with short glass fibers. In a severe fire, the fibers hold the calcined gypsum in place to continue to act as a barrier to fire, rather than permitting it to erode or fall out.

• *Ceiling board* is a high-strength board developed for use in ceilings with widely spaced joists where regular gypsum board would sag over time. Ceiling board ½ inch thick has strength roughly equivalent to ⅝ inch regular board.

• *Foil-backed gypsum board* can be used to eliminate the need for a separate vapor retarder in outside wall assemblies. If the foil faces an airspace, it also adds R-value.

• *Veneer-plaster base board,* also called "blueboard" because of its color, has a high-density gypsum core and an absorbent paper face that is treated with a catalyst to speed the curing of veneer plaster applied to it. Veneer plaster is discussed later in this chapter.

Gypsum board is manufactured with a variety of edge profiles, but the most common by far is the tapered edge, which permits sheets to be joined with a flush, invisible seam by means of subsequent joint finishing operations (Figure 17.4). A number of different thicknesses of board are produced, ranging from ⅛ inch (3 mm) to ¾ inch (20 mm). The most common thickness for residential work is ½ inch (13 mm).

Installing Gypsum Board

Gypsum board may be installed using either screws or nails to fasten it to the framing. If the framing lumber is not dry, it will usually shrink somewhat after the board is installed, which can cause nails to loosen slightly and "pop" through the finished surface of the board (Figure 17.5). Nail popping can be minimized by using only fully dried framing lumber, using the shortest nail that will do the job, and using ring-shank nails that have extra gripping power in the wood. Screws have less of a tendency to pop than nails. When screws or nails are driven into gypsum board, their heads are driven to a level slightly below the surface of the board, but not far enough to tear the paper surface. Adhesive is sometimes applied to the edges of the studs before the gypsum board is applied and fastened, in order to make a stronger joint. To minimize the length of joints that must be finished, and to create the stiffest wall possible, gypsum board is usually installed with the long dimension of the boards horizontal. The longest possible boards are used, to eliminate or at least minimize end joints between boards, which are difficult to finish because the ends of boards are not tapered. Gypsum board is cut rapidly and easily by scoring one paper face with a sharp knife, snapping the brittle core along the score line with a blow from the heel of the hand, and cutting the other paper face along the fold created by the snapped core (Figure 17.6). Notches, irregular cuts, and holes for electric boxes are made with a small saw.

Trim accessories of metal or plastic are required at exposed edges and external corners to protect the brittle board and present a neat edge (Figure 17.7). These are installed by the *wallboard hanger* (the worker who attaches the wallboard to the framing) at the same time as the wallboard. Trim accessories are also used with veneer plaster, in which case they are designed to act as lines that gauge the proper thickness and plane of plaster surface. A straightedge may be run across them to level the wet plaster. In this role, the trim accessories are known collectively as *grounds*.

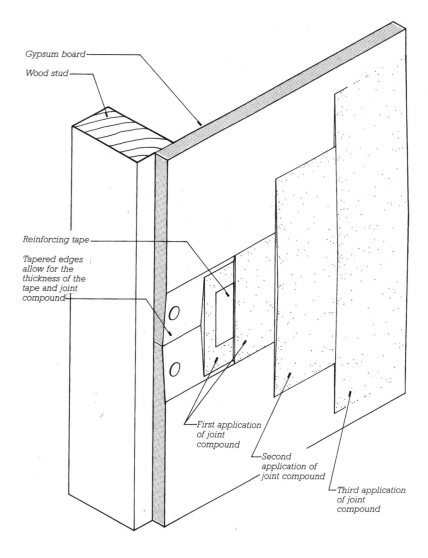

Gypsum board
Wood stud
Reinforcing tape
Tapered edges allow for the thickness of the tape and joint compound
First application of joint compound
Second application of joint compound
Third application of joint compound

FIGURE 17.4
Finishing a joint between panels of gypsum board.

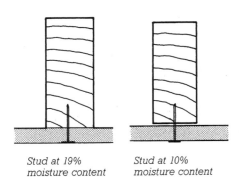

Stud at 19% moisture content

Stud at 10% moisture content

FIGURE 17.5
When wood studs dry and shrink during a building's first heating season, nail heads may pop through the surface of gypsum board walls.

(a)

(b)

FIGURE 17.6

Cutting gypsum board: (*a*) A sharp knife and metal T-square are used to score a straight line through one paper face of the panel. (*b*) The scored board is easily "snapped," and the knife used a second time to slit the second paper face. (*Courtesy of United States Gypsum Company*)

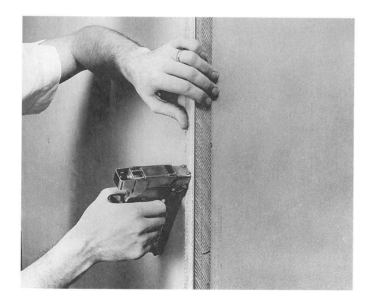

FIGURE 17.7

Stapling a metal corner bead to an external corner to create a straight, durable finished edge. (*Courtesy of United States Gypsum Company*)

FIGURE 17.8
Gypsum board can be curved to a large radius simply by bending it around a curving line of studs. (*Courtesy of United States Gypsum Company*)

the finishing process, a smooth, sticky plaster called *joint compound* (and fondly referred to by workers as "mud") is used. Ordinary joint compound dries by evaporation. It is a mixture of marble dust, binder, and admixtures, furnished either as a dry powder to be mixed with water or as a premixed paste. For most purposes, this all-purpose joint compound is adequate, but two-component, fast-setting joint compounds are available for high-production operations.

The finishing of a joint between panels of gypsum board begins with the troweling of a layer of joint compound into the tapered edge joint and the bedding of a paper *reinforcing tape* in the compound (Figures 17.10 and 17.11). Compound is also troweled over the indentations made by nails or screws. After overnight drying, a second layer of compound is applied to the joint, to bring it level with the face of the board and to fill the space left by the slight drying shrinkage of the joint compound. When this second coat is dry, the joints are lightly sanded before a very thin final coat is

Gypsum board can be curved when a design requires it. For gentle curves, the board can be bent into place dry (Figure 17.8). For somewhat sharper curves, the paper faces are moistened to decrease the stiffness of the board before it is installed. When the paper dries, the board is once again as stiff as before. Special high-flex ¼-inch (6.5-mm) board can be bent dry to a relatively small (5 foot) radius, and bent wet to an even smaller radius. It is usually installed in two layers so as to achieve a total thickness of ½ inch.

Drywall Finishing System

In the drywall system, the applied gypsum board remains exposed, but joints and fasteners are finished to create the appearance of a mono-lithic surface, indistinguishable from traditional plaster (Figure 17.9). In

FIGURE 17.9
Drywall after the nails and joints have been finished. The narrow vertical patches are finishing of nails at studs. The horizontal bands are finishing of the long tapered edges of gypsum board. By using 12-foot-long boards, very few square-cut short ends of gypsum board need to be finished. (*Photo by Rob Thallon*)

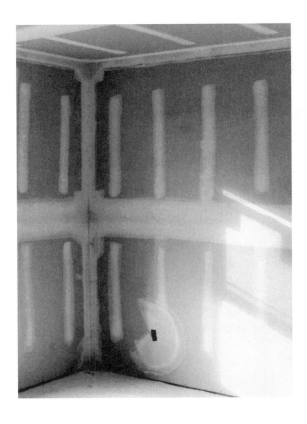

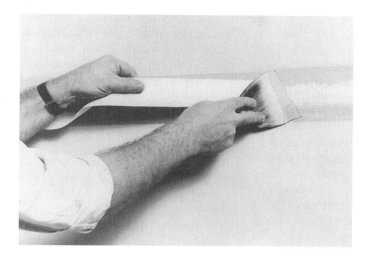

FIGURE 17.10
Applying paper joint reinforcing tape to gypsum board.
(*Courtesy of United States Gypsum Company*)

FIGURE 17.11
An automatic taper simultaneously applies tape and joint
compound to gypsum board joints. (*Courtesy of United States
Gypsum Company*)

FIGURE 17.12
To obtain a smooth surface, three applications of joint compound
are required, each of which is sanded after it has dried. Stilts
allow workers to reach the ceiling without ladders or scaffolding.
(*Photo by Rob Thallon*)

applied to fill any remaining voids and feather out to an invisible edge. Before painting, the wall is again sanded lightly to remove any roughness or ridges. If the finishing is done properly, the painted or papered wall will show no visible signs that it is made of discrete panels of material.

Gypsum board has a smooth surface finish, but a number of spray-on textures and textured paints can be applied to give a rougher surface. Most gypsum board contractors prefer to apply a *texture* to mask the minor irregularities in workmanship that are likely to occur. Texturing also hides the slight surface differences between the paper of the gypsum board and the smoothly sanded joint compound (Figure 17.13).

Veneer Plaster Finishing System

Veneer plaster (also called *thincoat*) is a one-coat system that uses gypsum board as a base, but produces a much harder, more durable surface. The gypsum board base, the fasteners, and the edge trims are essentially the same as drywall, and the tape between panels is a *fiberglass mesh* instead of paper. The similarities stop here, however, because, instead of covering just the fasteners and the joints between panels with joint compound, as drywall does, veneer plaster covers the entire wall or ceiling with a coat of plaster approximately ⅛ inch (3 mm) thick. The plaster cures by hydration rather than evaporation and forms surfaces whose quality and durability are superior to those of drywall, at comparable prices.

The application of veneer plaster is more difficult than finish drywall joints, but in geographic areas where there are plenty of skilled plasterers, veneer plaster captures a substantial share of the market. Application of plaster is done by hand with two very simple tools, a *hawk* in one hand to hold a small quantity of plaster ready for use and a *trowel* in the other hand to lift the plaster from the hawk, apply it to the surface, and smooth it into place (Figure 17.14). Plaster is transferred from the hawk to the trowel with a quick, practiced motion of both hands, and the trowel is moved up the wall or across the ceiling to spread the plaster, much as one uses a table knife to spread soft butter. After a surface is covered with plaster, it is leveled by drawing a straightedge, called a *darby*, across it, after which the trowel is used again to smooth the surface.

Board Paneling

Commonly used when wood was plentiful and plastering more involved, board paneling has declined in popularity but is still employed to give visual relief from relentless plastered surfaces. Boards can be installed quickly by carpenters, do not intro-

FIGURE 17.13
Drywall texture can be sprayed or hand applied. The texture helps to hide imperfections in the work. (*Courtesy of United States Gypsum Company*)

FIGURE 17.14
Applying veneer plaster with a hawk and trowel. The plaster covers the entire surface of the gypsum board with ⅛ inch (3 mm) of plaster. The process takes less time than the finishing of drywall, but introduces much more moisture into the interior finish environment. (*Courtesy of United States Gypsum Company*)

FIGURE 17.15
Board paneling contributes a pattern to walls and ceiling. In this example, tongue-and-groove boards are used as a wainscot. Panel products can also be used for this purpose, but the joints between panels are usually difficult to conceal. (*Photo by Rob Thallon*)

duce moisture into the building, and are ready for paint or stain as soon as they are in place. The most popular paneling consists of softwood boards (usually pine, fir, or cedar) with a tongue-and-groove edge detail. Individual boards are blind nailed to the framing, resulting in a continuous surface free of exposed fasteners (Figure 17.15). When oriented parallel to framing, blocking or strapping must be used to support the paneling.

Board paneling is usually 1-inch nominal (19 mm) material, but some thin veneer paneling (about ¼ inch, or 6 mm, thick) that must be applied over a gypsum board backing is also common. Whatever its thickness, board paneling is seldom the same thickness as standard drywall finish and needs to be coordinated with any other finish surface adjacent to it.

Sheet Paneling

There is a wide range of prefinished sheet paneling available for the finishing of walls and ceilings. It usually consists of a structural core of plywood, OSB, hardboard, or gypsum board that is covered with a veneer of wood, paper, fabric, or vinyl. Thicknesses range from ⅛ inch (3 mm) to ¾ inch (20 mm). The thicker panels may be applied directly to framing, but thinner panels require a structural backer such as OSB or gypsum board. Because the panels are prefinished, fasteners will be visible, so many panel designs incorporate grooves that will camouflage the existence of the small painted fasteners. The edges of panels are shiplapped, butted, or covered with trim strips. Like board paneling, sheet paneling is generally installed by the finish carpenters.

FINISH FLOORING

Floors have a lot to do with one's visual and tactile appreciation of a building. People sense their colors, patterns, and textures, their "feel" underfoot, and the noises they make in response to footsteps. Floors affect the acoustics of a room, contributing to a noisy quality or a hushed quality, depending on whether a hard or soft flooring material is used. Floors also interact in various ways with light. Some give mirrorlike reflections, some diffuse reflections, and some give no reflection at all. Dark flooring materials absorb light and contribute to the creation of a darker room, while light materials reflect most incident light and help create a brighter room (Figure 17.22).

Floors are also a major functional component of a building. They are its primary wearing surfaces, subject to water, grit, dust, and the abrasive and penetrating actions of feet and furniture. They require more cleaning and maintenance effort than any other component of a building. They must be designed to deal with problems of sanitation, skid resistance, and noise reduction between floors. And, like other interior finish components, floors must be selected with an eye to fire resistance, as well as the structural loads that they will place on the frame of the building.

The installation of finish flooring materials may occur at any time after the finish materials for the walls and ceilings have been applied and painted. Scheduling depends on the type of flooring and needs to be coordinated with other finish operations. There is an impulse to delay flooring installation as long as possible, to let the other trades complete their work and get out of the building to avoid damage caused by dropped tools, spilled paint, coffee stains, and construction debris ground underfoot. However, some flooring materials are best installed early, with trim and cabinets installed over them. When floors are installed early, they are protected by taping sheets of durable paper, plastic, or cardboard to their surface.

The thicknesses of floor finishes vary from the ⅛ inch (3 mm) or less for resilient flooring to 3 inches (76 mm) or more for mud-set ceramic pavers. Frequently, several different types of flooring are used in different rooms on the same floor of a house. If the differences in thickness of the flooring materials are not great, they can be resolved by using tapered edgings or thresholds at changes of material. Otherwise, the level of the subfloor must be adjusted from one part of the house to the next, to bring the finish floor surfaces to the same elevation. In many cases, special structural details must be drawn by the designers to indicate how the level changes should be made. In wood framing, they can usually be accomplished either by notching the ends of the floor joists to lower the subfloor in parts of the building with thicker floor materials or by adding sheets of *underlayment* material of the proper thickness to areas of thinner flooring (Figure 17.16).

Floor finishing operations require cleanliness and freedom from traffic, so other trades are banished from the area when the flooring materials are applied. Before the flooring can be installed, the subfloor must be scraped free of plaster droppings and swept thoroughly. Resilient-flooring installers vacuum the underlayment meticulously so that particles of dirt will not become trapped beneath the thin flooring and cause bumps in the surface.

Wood Flooring

Wood is used in several different forms as a finish flooring material. The most common type is wood *strip flooring*, which is typically made of white oak, red oak, pecan, or maple (Figure 17.17). The strips are held tightly together and blind nailed by driving nails diagonally through the upper interior corners of the tongues, where they are entirely concealed from view

(Figure 17.18). The entire floor is then sanded smooth, stained if desired, and finished with two or more coats of a varnish or other clear coating. The finish coatings are applied in as dust-free an atmosphere as possible to avoid embedded specks. When its surface becomes worn, the flooring can be restored to a new appearance by sanding and refinishing.

A less costly form of wood flooring consists of factory-made, *prefinished wood planks* and *parquet tiles*. These are furnished in many different woods and patterns, and are usually fastened to the subfloor with a mastic adhesive. Most are too thin to be able to withstand subsequent sanding and refinishing.

Some wood floors are not nailed or glued to the subfloor but instead "float" above it. These so-called *floating floors* are made with a tongue-and-groove detail at the sides and ends that are glued together to make one continuous piece as large as the room in which the flooring is laid. The flooring can expand and contract with changes in humidity or temperature, so a gap is left around the edges at the walls, and this gap is covered with base trim. Because of their tendency to move, most floating floors are made of laminated material, which is more dimensionally stable than solid wood. The laminations are thin in relation to strip flooring, so some manufacturers of floating floors impregnate the finish layer of veneer with acrylic resin or make it of plastic laminate to increase durability.

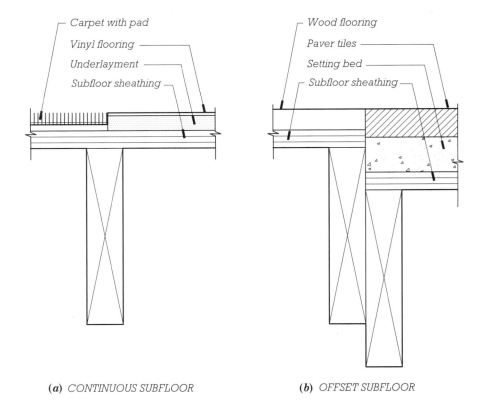

(a) CONTINUOUS SUBFLOOR   (b) OFFSET SUBFLOOR

FIGURE 17.16
When dissimilar flooring materials are adjacent, their level may need to be adjusted: (*a*) **When the thickness of finish flooring is similar, levels may be adjusted with underlayment below the thinner material.** (*b*) **When thickness varies considerably, the subfloor may need to be lowered to accommodate the thicker material.**

FIGURE 17.17
Oak strip flooring being installed in a residence. The flooring subcontractor blind nails the tongue edge of each board at closely spaced intervals with a pneumatic nailer designed for the purpose. Notice the 15-lb building paper laid under the flooring to cushion it from the subfloor and minimize squeaks. (*Photo by Rob Thallon*)

FIGURE 17.18
Details of hardwood strip flooring installation. At left, the flooring is applied to a wood joist floor, and at right, to wood sleepers over a concrete slab. The blind nailing of the flooring is shown only for the first several strips of flooring at the left. The baseboard makes a neat junction between the floor and the wall, covering up the rough edges of the wall material and the flooring. The three-piece baseboard shown at the left does this job somewhat better than the one-piece baseboard, but it is more expensive and elaborate.

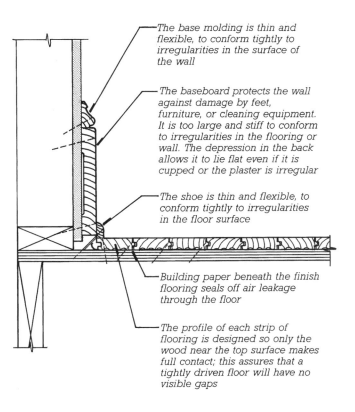

The base molding is thin and flexible, to conform tightly to irregularities in the surface of the wall

The baseboard protects the wall against damage by feet, furniture, or cleaning equipment. It is too large and stiff to conform to irregularities in the flooring or wall. The depression in the back allows it to lie flat even if it is cupped or the plaster is irregular

The shoe is thin and flexible, to conform tightly to irregularities in the floor surface

Building paper beneath the finish flooring seals off air leakage through the floor

The profile of each strip of flooring is designed so only the wood near the top surface makes full contact; this assures that a tightly driven floor will have no visible gaps

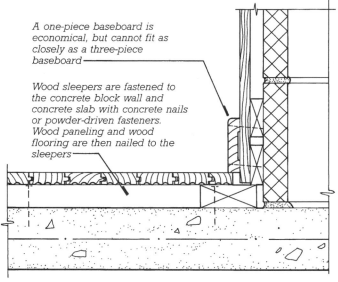

A one-piece baseboard is economical, but cannot fit as closely as a three-piece baseboard

Wood sleepers are fastened to the concrete block wall and concrete slab with concrete nails or powder-driven fasteners. Wood paneling and wood flooring are then nailed to the sleepers

If the concrete slab lies directly on grade, a sheet of polyethylene is laid beneath the sleepers to prevent moisture from entering the building

Resilient Flooring

The oldest *resilient flooring* material is *linoleum,* a sheet material made of ground cork in a linseed oil binder over a burlap backing. Asphalt tiles were later developed as an alternative to linoleum, but most of today's resilient sheet floorings and tiles are made of vinyl (polyvinyl chloride), often in combination with reinforcing fibers. Thicknesses of these *vinyl composition floorings* typically measure around ⅛ inch (3 mm), slightly thinner for lighter duty floorings, and slightly thicker if a cushioned back is added to the product. Most resilient flooring materials are glued to the subfloor or underlayment (Figure 17.19). The primary advantages of resilient floorings are the wide range of available colors and patterns, a moderately high degree of durability, and low initial cost.

Vinyl composition tile has the lowest installed cost of any flooring material except concrete and is used in vast quantities on the floors of residences. The tiles are usually 12 inches square. Sheet flooring is furnished in rolls 6 to 12 feet (1.83 to 3.66 m) wide. If skillfully made, the seams between strips of sheet flooring are virtually invisible.

Resilient floorings of rubber, cork, and other materials are also available. Each offers particular advantages of durability or appearance.

Most resilient flooring materials are so thin that they show even the slightest irregularities in the floor deck beneath. Wood subfloors are covered with a layer of smooth underlayment panels, usually of hardboard, particleboard, or sanded plywood, to prepare them for resilient flooring materials. Joints between underlayment panels are offset from joints in the subfloor to eliminate soft spots. The thickness of the underlayment is chosen to make the surface of the resilient flooring level with the surfaces of flooring materials such as hardwood and ceramic tile that are used in adjacent areas of the building.

Carpet

Carpet is manufactured in fibers, styles, and patterns to meet almost any flooring requirement, indoors or out, except for rooms such as bathrooms and kitchens that need thorough sanitation. Some carpets are tough enough to wear for years in heavily used corridors, yet others are soft enough for intimate residential interiors. The costs of carpeting are often competitive with those of other flooring materials of similar quality, whether measured on an installed-cost or life-cycle-cost basis.

Carpet is usually stretched over a *carpet pad* and attached around the perimeter of the room by means of a *tackless strip,* which is a continuous length of wood, fastened to the floor, that has protruding spikes along the top to catch the backing of the carpet and hold it taut (Figure 17.20). If carpet is laid directly over a wood panel subfloor such as plywood, the panel joints perpendicular to the floor joists should be blocked beneath to prevent movement between sheets. Tongue-and-groove plywood subflooring accomplishes the same result without blocking. Alternatively, a layer of underlayment panels may be nailed over the subfloor with its joints offset from those in the subfloor.

Hard Flooring Materials

Hard finish flooring materials (concrete, tile, and stone) are often chosen for their resistance to wear and moisture. Being rigid and unyielding, they are not comfortable to stand upon for extended periods of time, and they contribute to a live, noisy acoustic environment. However, many of these materials are so beautiful in their colors and patterns, and so durable, that they are considered among the most desirable types of flooring by designers and building owners alike.

With a steel trowel finish, the concrete surface of a slab on grade finds its way into many dwellings as a finish floor (Figure 17.21). Color can be added with a colorant admixture, a concrete stain, or a couple of coats of floor paint. Concrete's chief advantages as a finish flooring material are its low initial cost and its durability. On the minus side, extremely good workmanship is required to make an acceptable floor finish, and even the most carefully protected of concrete surfaces is likely to sustain some damage and staining during construction.

Thin, durable, decorative *floor tile* may be applied to a concrete slab (Figure 17.22). These materials in the form of brick pavers, quarry tiles, ceramic tiles, or stone add texture, pattern, and color, and are usually even more durable than concrete. Because they are relatively thin and thus do not add much dead weight to a floor, these materials also lend themselves to use as finish flooring over wood framing. When laying tile on wood framing, the floor structure must be very stiff so that it does not flex and crack the tiles or mortar joints. This may require the use of deeper joists, reduced joist spacing, and/or thicker subfloor sheathing.

Some tiles may be applied directly to a wood subfloor, but most require a *setting bed,* a cement-based surface that provides bonding for the *setting material,* which adheres the tile to the building. The most common setting bed is a *thinbed,* a cementitious backer board about ½ inch (13 mm) thick, securely fastened (with adhesive and nails or screws) over the subfloor (Figure 17.23). Thicker setting beds, required in some cases such as a sloping shower floor, can be built up with reinforced mortar and may require extra framing because of their weight.

Collectively, the term *ceramic tile* refers to a wide range of relatively small, flat, geometrically shaped materials made of fired clay. The most common types for floors are quarry tile, paver tile, and glazed wall tile. *Quarry tile* is a hard unglazed tile, usually red or brown brick color, about 4 inches (100 mm) to 6 inches (150 mm) square and ⅜ inch (9 mm) to ½

FIGURE 17.19
A worker spreads adhesive on underlayment in front of a long roll of sheet vinyl flooring. The flooring will be cut to fit at the edges as it is unrolled. (*Courtesy of Imperial Floors, Eugene, Oregon*)

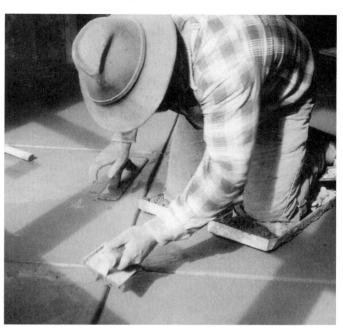

FIGURE 17.21
Finishing a concrete slab in the living room of a new residence. Added color and a diamond pattern of control joints add visual interest to this very economical floor finish. (*Photo by Rob Thallon*)

FIGURE 17.20
Residential carpet being laid. The tool in the worker's left hand has hooks on the bottom that grab the carpet. As the tool is hit with the worker's knee, it stretches the carpet over tacks projecting up from strips nailed to the subfloor along each wall. The carpet is trimmed to fit as it is installed. (*Photo by Rob Thallon*)

FIGURE 17.22
The waxed tile floor in the hallway of this residence contrasts with the hardwood floors in the room to the right. The distinct floor finishes help to define the spaces. The framing under the tile floor is lower than that under the adjacent wood floor in order to account for the extra thickness of the tile setting bed. (*Photo by Rob Thallon*)

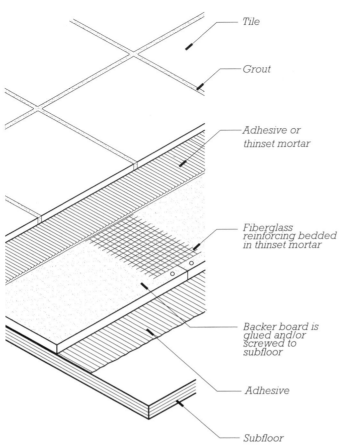

FIGURE 17.23
A setting bed made of cementitious backer board provides stiffness, moisture resistance, and improved bonding over a wood subfloor. The backer board is attached to the subfloor with mastic and nails or screws.

FIGURE 17.24
A diamond-blade mason's saw is used to cut masonry products, including stone, quickly and precisely. Water circulated over the blade helps to cool it and to eliminate dust. (*Photo by Rob Thallon*)

inch (13 mm) thick. *Brick pavers* are thin bricks that look the same as regular bricks after installation, but whose installation is similar to that of quarry tiles. *Glazed tiles* are generally smaller than quarry tiles and have a protective fired glaze on the exposed surface to give the tile color and texture. The most common shape is square, but rectangles, hexagons, circles, and more elaborate shapes are also available (Figure 17.25). Sizes range from ½ inch (13 mm) to 4 inches (100 mm) and more. The smaller sizes of tiles are shipped from the factory adhered to large backing sheets of plastic mesh or perforated paper. The tilesetter is thus able to lay a sheet of a hundred or more tiles together in a single operation, rather than as individual units.

Nonceramic tiles made of natural stone, cement-based materials, glass, and plastic resins are also available, and many are appropriate for use on floors. Many types of building stones are manufactured into floor tiles, in surface textures ranging from mirror-polished marble and granite to split-face slate and sandstone (Figure 17.26). Installation of any of these materials is similar to that of ceramic tile.

A specialized subcontractor does the preparation for and setting of tile. First, the setting bed is installed. Next, the tiles are laid out dry to determine how best to fit them to the available space. Then the tiles are adhered to the setting bed using an organic adhesive or a thin-set mortar, which generally requires some time to

cure. Once the tile is firmly adhered, *grout*, a dense, cement-based paste, is pressed into the spaces between the tiles. Grout color has a strong influence on the appearance of tile surfaces, and many different premixed colors are available, or the tilesetter may color a grout with pigments. Finally, the excess grout is cleaned from the surface of the tiles with a damp sponge and the entire surface is sealed if necessary.

The principles relating to the use and installation of tile on floors discussed previously apply equally to countertops and walls where tile use is also common. The application of tile to countertops is discussed in Chapter 18. The application of ceramic tile to a shower stall is illustrated in Figure 17.27.

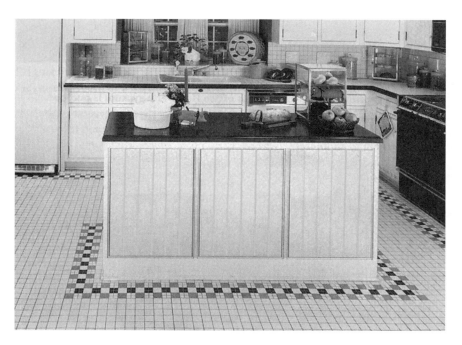

FIGURE 17.25
Ceramic tile is used for the floor, countertops, and backsplash in this kitchen. The border was made by selectively substituting tiles of four different colors for the white tiles used for the field of the floor. (*Designer: Kevin Cordes. Courtesy of American Olean Tile*)

FIGURE 17.26
Slate tiles being laid on a concrete backer board setting bed. Tiles are generally laid out from the center of a space and worked toward the walls in order to minimize the number of joints from the measured starting point. (*Photo by Rob Thallon*)

FIGURE 17.27
Installing sheets of ceramic tile in a shower stall. The base coat of portland cement plaster over metal lath has already been installed. Now the tilesetter applies a thin coat of tile adhesive with a trowel and presses a sheet of tiles into it, taking care to align the tiles individually around the edges. A day or two later, after the adhesive has hardened sufficiently, the joints will be grouted to complete the installation. (*Photos by Joseph Iano*)

C.S.I./C.S.C
MasterFormat Section Numbers for Interior Surfaces

06250	**Prefinished Paneling**
06260	**Board Paneling**
09200	**PLASTER AND GYPSUM BOARD**
09205	**Furring and Lathing**
09210	**Gypsum Plaster**
09250	**Gypsum Board**
09300	**TILE**
09310	**Ceramic Tile**
09330	**Quarry Tile**
09340	**Paver Tile**
09380	**Cut Natural Stone Tile**
09600	**FLOORING**
09620	**Specialty Flooring**
09630	**Masonry Flooring**
09640	**Wood Flooring**
09650	**Resilient Flooring**
09680	**Carpet**

SELECTED REFERENCES

1. Thallon, Rob. Graphic *Guide to Interior Details*. Newtown, C:, Taunton Press, 1996.

The materials and methods used to finish walls, ceilings, and floors are described extensively and comprehensively with drawings and text in the first half of this book.

2. Ferguson, Myron. *Drywall: Professional Techniques for Walls and Ceilings*. Newtown, CT: Taunton Press, 1997.

A how-to manual written by a drywall contractor who discusses materials, tools, installation, and texturing of drywall.

3. Bollinger, Don. *Hardwood Floors: Laying, Sanding, and Finishing*. Newtown, CT: Taunton Press, 1990.

A well-illustrated how-to manual written by a flooring contractor in which materials, tools, installation, and finishing of hardwood floors are discussed.

4. Byrne, Michael. *Setting Tile*. Newtown, CT: Taunton Press, 1995.

A well-illustrated how-to manual written by a tile contractor who discusses materials, tools, design, and installation of tile for floors, walls, and ceilings.

KEY TERMS AND CONCEPTS

gypsum wallboard systems
gypsum board
drywall
veneer plaster
thincoat
regular gypsum board
water-resistant gypsum board
Type X gypsum board
ceiling board
foil-backed gypsum board
veneer plaster base
wallboard hanger
ground
joint compound

reinforcing tape
fiberglass mesh tape
hydration
hawk
trowel
darby
board paneling
sheet paneling
underlayment
strip flooring
prefinished wood plank
parquet tile
floating floor
resilient flooring

linoleum
vinyl composition flooring
carpet
carpet pad
tackless strip
tile flooring
setting bed
setting material
thinbed
ceramic tile
quarry tile
brick paver
glazed tile
grout

REVIEW QUESTIONS

1. Why is gypsum used so much in interior finishes? List as many reasons as you can.

2. Describe step by step how the joints between sheets of gypsum board are made invisible.

3. Explain the differences between the drywall system and veneer plaster. For what reasons would veneer plaster be preferable to drywall?

4. Discuss three ways to accommodate thickness differences of finish floor materials installed adjacent to one another in new construction.

5. Compare the advantages and disadvantages of wood strip flooring, resilient flooring, and carpet.

6. Discuss the importance of stiffness in wood framing over which a tile floor is to be installed.

EXERCISES

1. Visit a residential construction site where drywall is being installed. Discuss with the contractor the steps in the process and the scheduling of each step. Report your findings to your class.

2. Sketch typical details showing how the various trim accessories are used in a gypsum board wall.

3. Take a survey in your class of the types of flooring in each student's house according to the rooms in which the flooring is located. Discuss the reasons for the results of your survey.

18

FINISHING THE INTERIOR

- **Interior Doors**
- **Cabinets**
  - Cabinet Types
  - Cabinet Hardware
  - Planning for Appliances and Fixtures
  - Cabinet Installation
  - Countertops
  - Sinks

- **Finish Carpentry and Trim**
  - Window and Door Casings and Baseboards
  - Finish Stairs
  - Miscellaneous Finish Carpentry
- **Paints and Coatings**
  - Finishing Touches

Once the interior surfaces of walls, ceilings, and floors have been installed, the interior doors, cabinets, plumbing fixtures, lighting fixtures, and appliances are added to make the interior functional. Trim is nailed on to cover the gaps between interior elements, and paint and other coatings are applied to protect the surfaces and add color. Cleaning, touch-up, and adjustments complete the construction process. The sequence for incorporating all of these interior components varies slightly from project to project, but generally follows a strategy of layering one material over another (Figure 18.1).

Figure 18.1
The walls and ceilings of this new house have been finished and painted. Next, the interior doors, cabinets, and trim will be installed. The flooring in this house will be carpet, which will be one of the last items to be installed. (*Photo by Rob Thallon*)

INTERIOR DOORS

Interior doors fall into three general types of construction: panel, glazed, and flush (Figure 18.2). *Panel doors* were developed centuries ago to minimize dimensional changes and distortions caused by seasonal changes in the moisture content of the wood. They are now manufactured primarily with laminated and composite materials and are widely available in natural wood or paint-grade surfaces from millwork dealers. *Glazed doors* are derived from panel doors, essentially substituting tempered glass for one or more of the panels.

Flush doors may be either solid core or hollow core. Solid core doors consist of two veneered faces glued to a core of wood blocks or bonded wood chips (Figure 18.3). They are much heavier, stronger, and more resistant to the passage of sound than hollow core doors, and are also more expensive. In residential buildings, their use is usually confined to entrance doors. Hollow core doors have an interior grid of wood or paperboard spacers to which the face veneers are bonded. Flush doors of either type are available in a variety of veneer species, the least expensive of which are intended to be painted.

The standard height of residential interior doors is 6⅔ feet (2.025 m), and widths start at 2 feet (610 mm) and increase by 2-inch (50.8-mm) increments to 2⅔ feet. The most common width is 2½ feet (76.2 cm). The standard thickness of interior doors is 1⅜ inches (35 mm), which is ⅜ inch (9.5 mm) thinner than exterior doors.

Most doors are *hinged* and can swing either right-handed or left-handed like exterior hinged doors. Because interior doors do not need to seal against the weather like exterior doors, a number of loose-fitting alternatives to the hinged door are available for use in closets and at other locations where privacy is not an issue. *Bypass doors* and *bifold doors* are most commonly used for closets (Figure 18.4), and *pocket doors* are used where an open hinged door would be in the way. These common doors all have tracks at the top of the

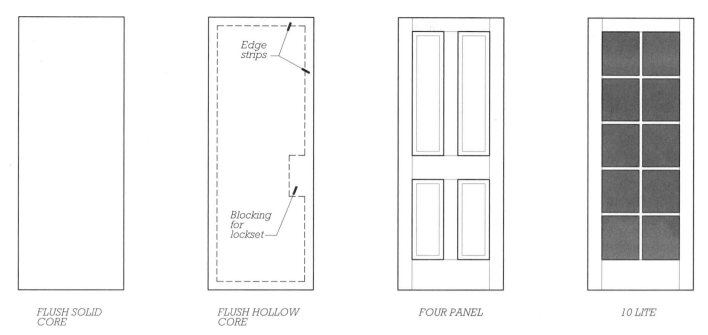

FLUSH SOLID
CORE

FLUSH HOLLOW
CORE

*Edge
strips*

*Blocking
for
lockset*

FOUR PANEL

10 LITE

FIGURE 18.2
Types of wood doors. Panel doors are commonly available with one, two, four, and six panels. Glazed doors are available in a wide variety of configurations. All doors are available in paint grade or stain grade.

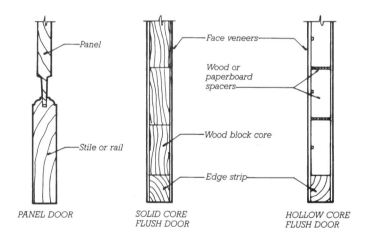

Panel

Stile or rail

PANEL DOOR

Face veneers

*Wood or
paperboard
spacers*

Wood block core

Edge strip

SOLID CORE
FLUSH DOOR

HOLLOW CORE
FLUSH DOOR

FIGURE 18.3
Edge details of three types of wood doors. The panel is loosely fitted to the stiles and rails in a panel door to allow for moisture expansion of the wood. The spacers and edge strips in hollow core doors have ventilation holes to equalize air pressures inside and outside the door.

FIGURE 18.4
Bifold doors such as these and bypass doors that slide past one another are typically used for closets. (*Photo by Greg Thomson*)

door to support them and keep them in line, and the casings to conceal this track are usually set lower than those of a typical hinged door.

For speed and economy of installation, most hinged interior doors are furnished *prehung,* meaning that they have been hinged and fitted to their jambs at the mill (Figure 18.5). The carpenter on the site merely tilts the prehung door and frame unit up into the rough opening, plumbs it carefully with a spirit level, shims it with pairs of wood shingle wedges between the finish and rough jambs, and nails

it to the studs with finish nails through the jambs (Figure 18.6). Casings are then nailed around the frame on both sides of the partition to close the ragged gap between the door frame and the wall finish (Figure 18.7). Bifold, bypass, and pocket doors are not usually furnished prehung, but their jambs are installed similarly.

The finish carpenter generally applies doorknobs at the site after the doors are hung. If doorknobs have been specified at the time the doors are ordered, hinged doors are often predrilled for this hardware at the

factory. Bypass doors usually have recessed pulls instead of knobs, bifold doors have knobs mounted to one side of the door, and pocket doors have a double-sided catch mechanism cut into the stile of the door.

To save the labor of applying casings, door units can be purchased with split jambs that enable the door to be cased at the mill. At the time of installation, each door unit is separated into halves, and the halves are installed from opposite sides of the partition to telescope snugly together before being nailed in place (Figure 18.8).

FIGURE 18.5
A prehung single-panel interior door. The door and frame are in the basement of a house under construction where the assembly will be varnished before it is installed. (*Photo by Rob Thallon*)

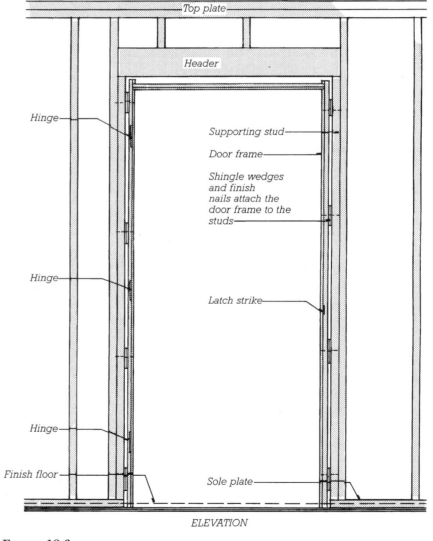

ELEVATION

FIGURE 18.6
Installing a door frame in a rough opening. The wood shingle wedges at each nailing point are paired in opposing directions to create a flat, precisely adjustable shim to support the frame.

(*a*)

(*b*)

(*c*)

(*d*)

FIGURE 18.7
**Casing a door frame: (*a*) The finish nails in the frame are set with a steel nail set. (*b*)
The top piece of casing, mitered to join the vertical casings, is ready to install, and
glue is spread on the edge of the frame. (*c*) The top casing is nailed into place. (*d*) The
nails are set below the surface of the wood, ready for filling. (*Photos by Joseph Iano*)**

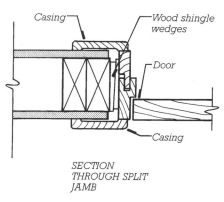

SECTION
THROUGH SPLIT
JAMB

FIGURE 18.8
**A split-jamb interior door arrives on the
construction site prehung and precased.
The halves of the frame are separated
and installed from opposite sides of
the partition.**

CABINETS

In the 18th century, cabinets were free-standing pieces of furniture that could be moved from room to room. With the advent of indoor plumbing, cabinets became stationary and integral with the house. Early built-in cabinets were made entirely of wood like their furniture predecessors. Wood is still an important material in modern cabinets, although composite wood products such as plywood and fiberboard predominate today, and most countertops are faced with plastic laminates.

Residential cabinets are either custom made (Figure 18.9) or mass produced (Figure 18.10). *Custom cabinets* are typically made in a local cabinet shop and are trucked to the building site in sections that are as large as is practical. Mass-produced cabinets are manufactured in large centralized factories and are composed of small modules that are joined together at the building site to make a large ensemble. These mass-produced cabinets are thus known as *modular cabinets.*

The quality of both custom and modular cabinets is regulated by the Architectural Woodwork Institute (AWI). This organization publishes the AWI Quality Standards, which describe properties and tolerances for materials and joinery in all cabinets. There are three grades of custom-made cabinets: Premium, Custom, and Economy. Premium grade is reserved for the very finest cabinets—

FIGURE 18.9
Custom-designed cabinets brighten a remodeled kitchen in an older house. A large percentage of the cabinets in a house are installed in the kitchen, where they must be carefully designed to fit a number of appliances including: refrigerator, cooktop, ovens, sink, and dishwasher. (*Architects: Dirigo Design, Belmont, Massachusetts. Photo © 1989 by Lucy Chen*)

very expensive and rare in residential work. Custom grade requires very good materials and craftsmanship and is the most widespread grade of residential cabinet. Economy is the lowest and least expensive grade. For mass-produced cabinets, there is only one grade, Modular, which is equivalent to the Custom grade.

The cross-sectional dimensions of cabinets are standardized within the industry (Figure 18.11). Unless specified otherwise, one can assume that cabinets will conform to these ergonomically based dimensions. Special dimensions for disabled persons are legislated.

Cabinet Types

In both custom and modular cabinets, there is a basic distinction between face-frame and frameless (European style) cabinets (Figure 18.12). These two cabinet types are constructed differently and have distinct appearances. The *face-frame cabinet* is constructed with a frame that attaches to the front of the case and surrounds the doors and drawers that are set within it. The

FIGURE 18.10
Inexpensive modular cabinets are used in this small kitchen, and the space is expanded with the addition of a half-wall and eating bar on the living room side. (*Photo by Rob Thallon*)

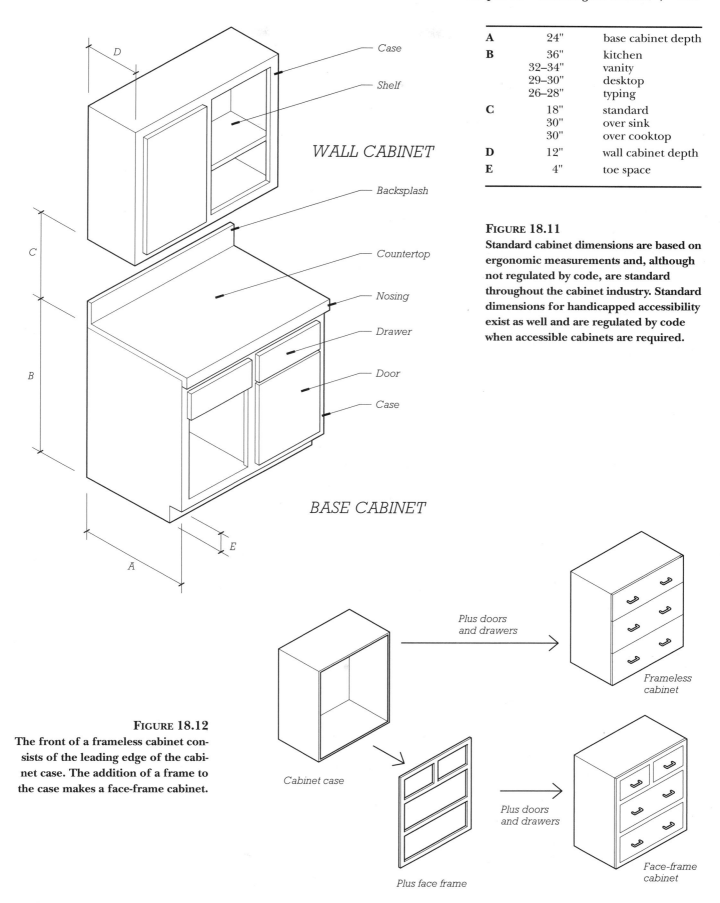

A	24"	base cabinet depth
B	36"	kitchen
	32–34"	vanity
	29–30"	desktop
	26–28"	typing
C	18"	standard
	30"	over sink
	30"	over cooktop
D	12"	wall cabinet depth
E	4"	toe space

Case

Shelf

WALL CABINET

Backsplash

Countertop

Nosing

Drawer

Door

Case

BASE CABINET

D

C

B

A

E

FIGURE 18.11
Standard cabinet dimensions are based on
ergonomic measurements and, although
not regulated by code, are standard
throughout the cabinet industry. Standard
dimensions for handicapped accessibility
exist as well and are regulated by code
when accessible cabinets are required.

FIGURE 18.12
The front of a frameless cabinet consists of the leading edge of the cabinet case. The addition of a frame to
the case makes a face-frame cabinet.

Plus doors
and drawers

Frameless
cabinet

Cabinet case

Plus doors
and drawers

Plus face frame

Face-frame
cabinet

frame allows design flexibility in that it can support doors or drawers independently of the cabinet (Figure 18.13). The *frameless cabinet* has no face frame; only the doors and drawers are visible at the front of the cabinet. The frameless cabinet has less design flexibility but many homeowners prefer its clean, minimal appearance, and it is more suited for use in modular systems because joints between the modular sections do not show in the completed assembly.

Cabinet doors may be either single piece or frame and panel (Figure 18.14). *Single-piece doors* are made of a simple piece of panel product such as plywood or fiberboard—often *banded* (covered with a veneer) at the edges. *Frame-and-panel doors* are constructed of several pieces in the traditional fashion that was standard before panel products made single-piece doors possible. Drawers usually have single-piece fronts because their size makes a frame-and-panel system impractical.

In the frameless cabinet, door and drawer fronts are invariably flush, with a ⅛-inch gap between them (Figure 18.15*A*). In the face-frame cabinet, on the other hand, there are several choices for how the doors and drawers can meet the frame (Figure 18.15*B*). Simplest and most economical is the *full-overlay* relationship where the doors and drawers fully lap over the face frame. The *inset* door (and drawer) is the most complicated and expensive type, because the gap between door and frame must be precisely controlled by the cabinet-maker. Between overlay and inset is the *lip* door, which has a rabbet (a groove along the edge) on the rear of the door or drawer front so that it overlays the frame but only projects one-half the thickness of the door.

Cabinet Hardware

The selection of hardware for cabinets is important because of its impact on the function, aesthetic appearance, and overall quality of the cabinet. Functionally, the most important decisions are *drawer guides* and door hinges. The main selection criteria for drawer guides are weight capacity and extension. Drawer guides are designed in a range of capacities from 25 to 250 pounds and should be matched to their intended use. A reg-

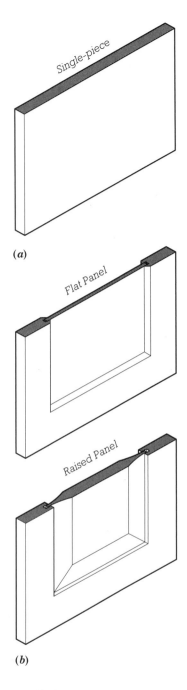

(a)

(b)

FIGURE 18.14
The two basic door styles: (*a*) single-piece door and (*b*) frame-and-panel door. Both types can have any edge detail. The panels in frame-and-panel doors can be made of a variety of materials and are usually either flat or raised.

FIGURE 18.13
Prepainted face-frame kitchen cabinets installed, but lacking shelves, drawers, doors, and countertops. Notice the variety of door widths in these custom cabinets. (*Photo by Edward Allen*)

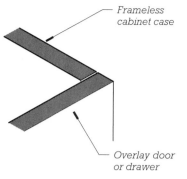

(a) FRAMELESS CABINET

- Frameless cabinet case
- Overlay door or drawer

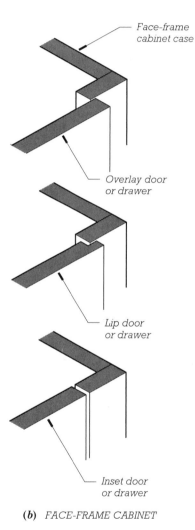

- Face-frame cabinet case
- Overlay door or drawer
- Lip door or drawer
- Inset door or drawer

(b) FACE-FRAME CABINET

FIGURE 18.15
Isometric section drawings showing the relationship of cabinet doors and drawers to cabinet case. In the frameless cabinet, the door or drawer invariably overlays the front edge of the case. In the face-frame cabinet, the doors and drawers can be in one of three different positions.

FIGURE 18.16
Workers are moving cabinets into this house that has not yet had the siding applied, illustrating that interior and exterior finishing can proceed simultaneously. (*Photo by Rob Thallon*)

ular drawer guide allows a drawer to open to approximately 80 percent of its depth, but *full-extension* guides are available that reveal 100 percent of the drawer when it is open. Cabinet door hinges are a much more complicated matter than drawer guides. Hinges must be matched to the door type (overlay, lip, inset) and are quite variable in terms of their ability to be adjusted, their durability, and the angle to which they allow the door to open. There are many types of hinges from which to choose, ranging from simple *butt hinges* to *fully concealed* (and fully adjustable) hinges. Many hinges are *self-closing* in that they contain a spring that engages and pushes the door closed when it is within a certain range of positions. Handles for cabinet doors and drawers are called *pulls* and are available in a very wide selection. Overlay doors and drawers do not require pull hardware because the edge of the door or drawer can be shaped so that it can be grasped easily.

Planning for Appliances and Fixtures

Most residential cabinets are located in the kitchen, bathroom, and laundry. In the kitchen, they must be integrated with a number of appliances, usually including at least a range, a range hood, a refrigerator, and a dishwasher; and with the kitchen sink. Some appliances, like the refrigerator, can be accommodated by merely leaving adequate space. Others, like the range, are often built into the cabinets. In all cases, the size (if not the particular model) of the appliance must be known before the cabinets are constructed in the shop. Appliance and fixture manufacturers typically supply installation dimensions and/or templates to the cabinetmaker.

Cabinet Installation

Cabinets are typically fully finished with paint or lacquer before they are

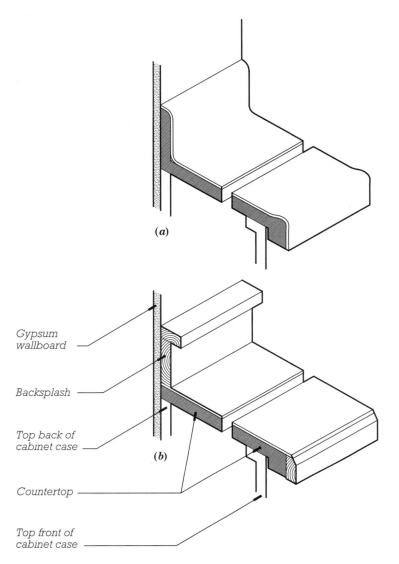

(a)

Gypsum wallboard

Backsplash

Top back of cabinet case

(b)

Countertop

Top front of cabinet case

FIGURE 18.17

Two common ways in which countertops are finished. (*a*) A shop-applied plastic laminate makes a continuous surface from the top of the backsplash to the base of the nose. (*b*) A wooden backsplash and wooden edge band trim a field-applied plastic laminate countertop.

Countertops

After the cabinets are anchored in place, the *countertops* can be installed. Although they appear to be integral with the cabinet, countertops are usually made of an entirely different material, often by someone other than the cabinetmaker (Figure 18.17). Countertops take much more abuse than cabinets as they are frequently exposed to moisture, heat, scratching, and staining. They must be made of material that is easily cleaned to keep them sanitary and they must be able to accept the watertight installation of sinks.

The most common materials for countertops include *plastic laminate*, ceramic tile, stone tile, stone slabs, stainless steel, wood, concrete, and *solid surface* materials composed of powdered stone blended with acrylic resin. Those facing materials too thin or weak to span between cabinet case partitions, such as plastic laminate or tile, must be applied over a stiff substrate. The most common substrates are ¾-inch (19-mm) plywood or medium-density fiberboard (MDF). For tile, a cementitious backer board is often applied over the substrate.

Every countertop must be finished at its edges—both the exposed front edge, the *nosing*, and the rear edge against the wall, the *backsplash* (Figure 18.17). The detailing of these edges depends on the countertop material. Most materials lend themselves to making the nosing and the backsplash of the same material as the countertop. Often, however, the nosing will be made of a contrasting material. It is common, for example, to have a hardwood nosing (called an *edge band*) on a plastic laminate countertop.

Most countertops with shop-applied facings are attached to the cabinet case with screws driven invisibly from the underside, through bracing flanges at the top of the cabinet cases. Field-applied countertop facings such as tile, stone slab, and concrete are usually adhered to a substrate that is fastened to the cabinet from above.

brought to the job site. The finish carpenters or cabinetmakers install them after the walls have been painted. Usually, it is easiest to install the flooring first and install the cabinets over the finish floor. Fastening of cabinets to the building is usually done with screws through the case to the wall(s) and floor. Before being fastened, the cabinets must be leveled and shimmed because walls and floors are never perfectly plumb,

level, or planar. The best way to take up irregularities between cabinet edges and walls is to *scribe* the cabinet to the wall. Scribing consists of shaping the cabinet edge to the contour of the wall. To facilitate scribing, the cabinetmaker makes the cabinet edges thin by cutting away wood at the rear along the length of the edge. Cabinet edges are caulked or covered with slender trim in work of lower quality.

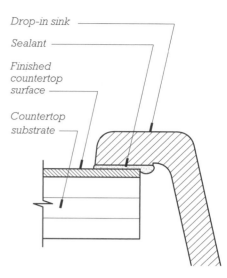

FIGURE 18.18
A drop-in cast iron sink in a tile countertop. The tile is laid first, then the sink is installed from the top, and the rim laps over the raw tile edge and conceals it. The heavy sink is adhered to the tile with adhesive that also acts to seal the gap between tile and sink. (*Photo by Rob Thallon*)

FIGURE 18.19
There are a variety of ways that a sink can seal to a countertop. The popular drop-in method shown here in section requires a quality sealant between the sink and the finished countertop to prevent moisture from reaching the wooden cabinet assembly.

Sinks

There are several ways in which a sink can meet a countertop. The most popular and economical method, the *drop-in* sink, is installed from the top through a hole cut into the countertop, and a perimeter flange on the sink rests on the countertop surface (Figures 18.18 and 18.19). A sealant at the joint between sink and countertop prevents the penetration of water. Other sink installation methods include *flush-mount,* where the top of the sink is flush with the countertop surface; *undermount,* where the sink is attached to the underside of the countertop; and *integral,* where the sink and countertop are made of the same material. The undermount sink requires finishing of the edge of the countertop around the sink opening (Figure 18.20).

FIGURE 18.20
A sink mounted from below requires finishing of the countertop edge where it meets the sink. (*Courtesy of Patrick Morgan Design, Eugene, Oregon*)

FINISH CARPENTRY AND TRIM

Millwork (so named because it is manufactured in a planing and molding mill) includes all the wood finish components of a building—the doors, windows, cabinets, trim, and molding (Figure 18.21). Millwork is generally produced from much higher quality wood than that used for framing. The softwoods used are those with fine, uniform grain structure and few defects, such as sugar pine, ponderosa pine, and Douglas fir. Flooring, stair treads, and millwork intended for transparent finish coatings such as varnish or shellac are customarily made of hardwoods and hardwood plywoods: red and white oak, cherry, mahogany, and walnut. *Moldings* are also produced by specialized manufacturers from high-density plastic foams, parallel strand lumber, and fiberboard as a way of reducing costs and minimizing moisture expansion and contraction; all of these materials must be painted. Doors and cabinets are commonly made of composite wood products such as plywood, particleboard, and fiberboard with facings and edge bands of wood veneer or plastic laminate.

Moldings and trim are manufactured and delivered to the building site at a very low moisture content, typically in the range of 10 percent, so it is important to protect them from moisture and high humidity before and during installation to avoid swelling and distortions. The humidity within the building is high at the conclusion of plastering or gypsum board work. The framing lumber, concrete work, masonry mortar, plaster, drywall joint compound, and paint are still diffusing large amounts of excess moisture into the interior air. As much of this moisture as possible should be ventilated to the outdoors before finish carpentry (the installation of millwork) commences. Windows

should be left open for a few days, and in cool or damp weather the building's heating system should be turned on to raise the interior air temperature and help drive off excess water. In hot, humid weather, the air-conditioning system should be activated to dry the air.

Interior finish carpentry includes the installation of doors and cabinets as well as the application of various moldings that cover the gaps between materials and provide interest and emphasis to the rooms in a house. The moldings, also known collectively as *trim*, need to be sufficiently flexible to conform to the minor irregularities in the wall and floor surfaces, while at the same time rigid enough to hold themselves in place once fastened (Figure 17.18).

The application of trim has become much easier in recent years with the advent of the *pneumatic finish nailer* (Figure 18.22). Not many years ago, trim would be fastened with finish nails driven home with several blows of a hammer and then set below the surface of the wood using a separate tool called a *nail set*. Now the entire operation is performed better and more efficiently with one pull of a trigger. The nails in casings and baseboards must reach through the plaster or gypsum board to penetrate the framing members beneath in order to make a secure attachment. Later, the painters will fill the nail holes with a paste filler and sand the surface smooth after the filler has dried so that the holes will be invisible in the painted woodwork. For transparent wood finishes, either the nail holes may be left unfilled or a filler must be carefully selected to match the color of the wood, which is very difficult since the wood changes color over time. In high-quality work, nail holes in transparent finishes are filled after all finishing has been done, using pigmented wax fillers. For this type of filling, it is best to wait a year or so until the color of the wood has stabilized.

FIGURE 18.21

Some common molding patterns for wood interior trim. *A* and *B* are crowns, *C* is a bed, *D* and *E* are coves. All are used to trim the junction of a ceiling and a wall. *F* is a quarter-round, for general-purpose trimming of inside corners. Moldings *G* through *J* are used on walls—*G* is a picture molding, applied near the top of a wall so that framed pictures can be hung from it at any point on special metal hooks that fit over the rounded portion of the molding. *H* is a chair rail, installed around dining rooms to protect the plaster from damage by the backs of chairs. *I* is a panel molding and *J* a batten, both used in traditional paneled wainscoting. Baseboards (*K* through *N*) include three single-piece designs and one traditional design (*K*) using a separate cap molding and shoe in addition to a piece of S4S stock that is the baseboard itself (see also Figure 17.18). Notice the shallow groove, also called a relieved back, in the single-piece baseboards and many other flat moldings on this page—this serves to reduce cupping forces on the piece and makes it easier to install even if it is slightly cupped. Designs *AA* through *FF* are standard casings for doors and windows. *GG*, *HH*, and *II* are handrail stock. *MM* is representative of a number of sizes of S4S material available to the finish carpenter for miscellaneous uses. *NN* is lattice stock, also used occasionally for flat trim. *OO* is square stock, used primarily for balusters. *PP* represents several available sizes of round stock for balusters, handrails, and closet poles. Wood moldings are furnished in either of two grades: *N* Grade, for transparent finishes, must be of a single piece. *P* Grade, for painting, may be finger jointed or edge glued from smaller pieces of wood. *P* Grade is less expensive because it can be made up of short sections of lower-grade lumber with the defects cut out. Once painted, it is indistinguishable from *N* Grade. The shapes shown here represent a fraction of the moldings that are generally available from stock. Custom molding patterns can be easily produced because the molding cutters used to produce them can be ground quickly to the desired profiles, working from the architect's drawings.

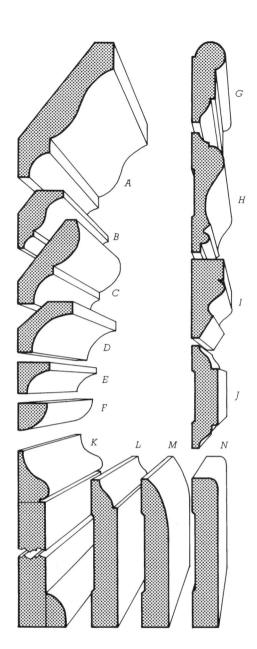

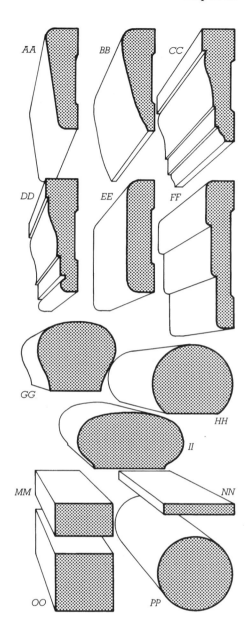

FIGURE 18.22
Almost all trim is applied with a pneumatic finish nailer. The nailer allows the finish carpenter to hold the trim in place with one hand while the other hand positions and drives the nail with one squeeze of the finger. The nail is automatically set below the surface of the trim. (*Photo by Edward Allen*)

FIGURE 18.23

Casing a window: (*a*) Marking the length of a casing. (*b*) Cutting the casing to length with a power miter saw. (*c*) Nailing the casing because the pneumatic finish nailer is in the shop. (*d*) Coping the end of a molded edge of an apron with a coping saw, so that the molding profile will terminate neatly at the end of the apron. (*e*) Planing the edge of the apron, which has been ripped (sawn lengthwise, parallel to the grain of the wood) from wider casing stock. (*f*) Applying glue to the apron. (*g*) Nailing the apron, which has been wedged temporarily in position with a stick. (*h*) The coped end of the apron, in place. (*i*) The cased window, ready for filling, sanding, and painting. (*Photos by Joseph Iano*)

Window and Door Casings and Baseboards

Windows are cased in much the same manner as doors (Figures 18.23 and 18.24). After the finish flooring is in place, *baseboards* are installed at the junction of the floor and the wall to close the ragged gap between the flooring and the wall finish and to protect the wall finish against damage by feet, furniture legs, and cleaning equipment (Figure 17.18).

When a tall baseboard is desired, a slender, flexible *base shoe* is often used between the baseboard and floor, because the baseboard itself will not bend to conform to the contours of the flooring.

Finish Stairs

Finish stairs are either site built in place (Figures 18.25 to 18.28) or shop built (Figure 18.29). *Shop-built stairs* tend to be more tightly constructed and to squeak less in use, but *site-built stairs* can be fitted more closely to the walls and are more adaptable to special situations and framing irregularities. Site-built stairs are much more common in residential construction and are usually rough framed early in the construction process so that workers can use them. When it comes time to finish site-built stairs, the rough framing is covered with finish material and the trim and/or railings are added at the sides.

The most common finish materials for residential stairs are hardwood and carpet. Wood stair treads are usually made of wear-resistant hardwoods such as oak and maple. Risers and stringers may be made of any reasonably hard wood such as oak, maple, or Douglas fir. Wood species for stair treads are often matched to those of the flooring above and/or below the stair and are installed at the same time as the floor by the hardwood flooring subcontractor.

When stairs are finished with carpet, this work is typically done toward the end of the finish schedule, when the carpeting is laid in the rest of the house. Sometimes carpet is laid over wooden stairs, as a runner. Carpet is not nearly as durable as hardwood for stairs, but its initial cost is much lower, and many people prefer the soft, quiet feeling of carpeted stairs.

FIGURE 18.24
Simple but carefully detailed and skillfully crafted window and door casings. The door casings are continuous with the window casings in order to unite these elements visually. (*Photo by Rob Thallon*)

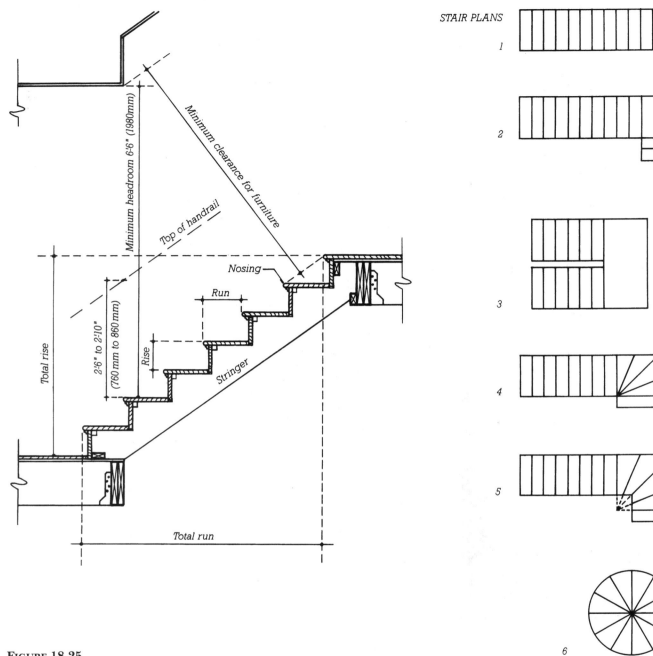

STAIR PLANS

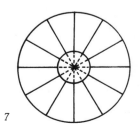

FIGURE 18.25

Left: Stair terminology and clearances for wood frame residential construction. Right: Types of stairs. (1) Straight run. (2) L-shaped stair with landing. (3) 180 degree turn with landing. (4) L-shaped stair with winders (triangular treads). Winders are helpful in compressing a stair into a much smaller space but are perilously steep where they converge, and their treads become much too shallow for comfort and safety. Many codes do not permit winders. (5) L-shaped stair whose winders have an offset center. The offset center can increase the minimum tread dimension to within legal limits. (6) A spiral stair (in reality a helix, not a spiral) consists entirely of winders and is generally illegal for any use but a secondary stair in a single-family residence. (7) A spiral stair with an open center of sufficient diameter can have its treads dimensioned to legal standards. (*Adapted by permission from Albert G. H. Dietz,* Dwelling House Construction, *4th ed., M.I.T. Press, Cambridge, Massachusetts.* © *1974 Massachusetts Institute of Technology*)

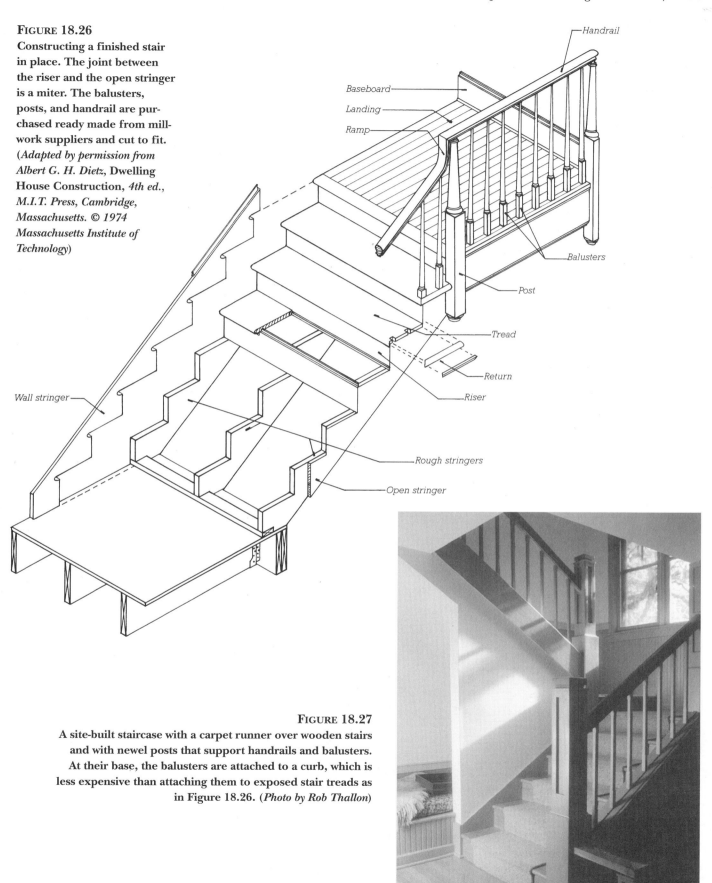

FIGURE 18.26
Constructing a finished stair in place. The joint between the riser and the open stringer is a miter. The balusters, posts, and handrail are purchased ready made from millwork suppliers and cut to fit. (*Adapted by permission from Albert G. H. Dietz,* Dwelling House Construction, *4th ed., M.I.T. Press, Cambridge, Massachusetts. © 1974 Massachusetts Institute of Technology*)

Handrail

Baseboard

Landing

Ramp

Balusters

Post

Tread

Return

Riser

Wall stringer

Rough stringers

Open stringer

FIGURE 18.27
A site-built staircase with a carpet runner over wooden stairs and with newel posts that support handrails and balusters. At their base, the balusters are attached to a curb, which is less expensive than attaching them to exposed stair treads as in Figure 18.26. (*Photo by Rob Thallon*)

FIGURE 18.28
Millwork suppliers offer several different stair "packages" of coordinated parts in traditional designs. (*Courtesy of Morgan Products, Ltd.*)

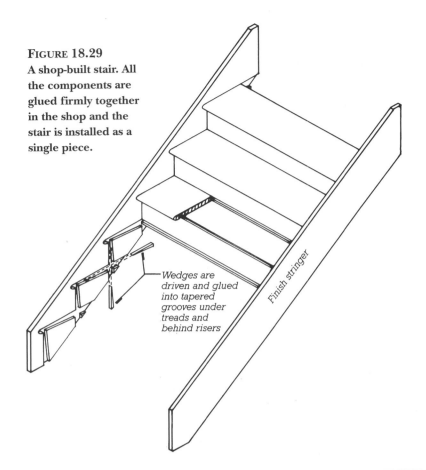

FIGURE 18.29
A shop-built stair. All the components are glued firmly together in the shop and the stair is installed as a single piece.

Wedges are driven and glued into tapered grooves under treads and behind risers

Finish stringer

Miscellaneous Finish Carpentry

Finish carpenters install dozens of miscellaneous items in the average dwelling—closet shelves and poles, pantry shelving, bookshelves, wood paneling, chair rails, picture rails, ceiling moldings, mantelpieces, laundry chutes, door hardware, weatherstripping, doorstops, and bath accessories (towel bars, paper holders, and so on). Many of these items are available ready made from millwork and hardware suppliers (Figure 18.30), but others have to be crafted by the carpenter.

FIGURE 18.30
Fireplace mantels are available from millwork suppliers in a number of designs. Each is furnished largely assembled but is detailed in such a way that it can easily be adjusted to fit any fireplace within a wide range of sizes. (*Courtesy of Morgan Products, Ltd.*)

PAINTS AND COATINGS

Painting the interior of a residence essentially occurs in two phases. The first phase of the painter's work is to coat the extensive areas of walls and ceilings (usually drywall) with latex paint. In the second phase, the painter coats the woodwork, molding, and cabinetry that are installed over the painted walls. A painting subcontractor generally performs the work, but some general contractors like to do the painting themselves.

The walls and ceilings are painted as soon as possible after the plasterwork has dried. With veneer plaster or drywall, this can be as soon as a day or two after the plasterers or joint finishers have finished, depending on interior atmospheric conditions. Most painters spray the paint during this phase, but some apply it with a roller. When spraying, the work begins by masking the windows, doors, and other items that might be damaged by paint overspray. The walls and ceilings are then sprayed with a coat of primer/sealer to provide a good base for the latex paint, which is sprayed next. Roller application of the primer and paint generally requires less time for preparation but takes longer overall (Figure 18.31). The primer may be formulated to furnish a vapor retarder. The latex is sprayed or rolled in one or two coats and provides the color. In bathrooms and kitchens, where extremely humid conditions often occur, a latex enamel or oil-based paint is usually used to better seal the wall against moisture.

The painter returns to the site to paint the woodwork after the doors have been hung, the cabinets installed, and the trim applied (Figure 18.32). This work is usually done with a brush because the surfaces to be painted are relatively small and narrow. Cabinets and doors, given their more extensive surface area, are usually sprayed, and this can be done either on site or off site. The doors, windows, and trim are filled, if necessary, and primed or coated with a transparent sealer to seal the pores in preparation for the paint, stain, or transparent coating that will ultimately

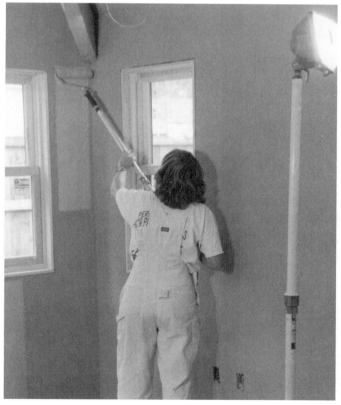

FIGURE 18.31
A painter rolling the finish coat onto drywall. Paint is usually sprayed onto walls and ceilings, but in this case, the owner desired a texture that was most easily achieved with a roller. Walls and ceilings are typically painted before floors, cabinets, or trim are installed. (*Photo by Rob Thallon*)

FIGURE 18.32
Painters brushing cabinet work. Trim is also commonly painted by hand with a brush. (*Photo by Rob Thallon*)

protect the surface. Nail holes are filled, and the primed and filled surface is sanded to remove dust, raised grain, and excess filler, and to create a very smooth surface for the finish coat(s). The inevitable gaps between trim and wallboard and between trim pieces themselves are filled with latex painter's caulk before the final coat is applied. The final coats of paint are applied with a brush in as dust-free an environment as possible, which often means that the painter must seal off areas of the house and schedule the painting work around other trades working at the site. Paint brand and type are often specified by the designer, but good painters will recommend quality paints, either latex or oil-based, whether specified or not. Hard oil-based or industrial finishes should be specified at locations such as doors, newel posts, and cabinets that will receive consistent abrasion from the users of the building.

Finishing Touches

When flooring and painting are finished, the plumbers install and activate the lavatories, water closets, tubs, sinks, and shower fixtures. Gas lines are connected to appliances and the main gas valve is opened. The electricians connect the wiring for electrical equipment, including heating and air-conditioning equipment; mount the receptacles, switches, and lighting fixtures; and put metal or plastic cover plates on the switches and receptacles. The electrical circuits are energized and checked to be sure that they operate properly. The smoke alarms and heat alarms, required by most codes in residential structures, are also connected and tested by the electricians, along with any communications, entertainment, computer network, and security system wiring. The heating and air-conditioning system is completed with the installation of air grills and registers or with the mounting of metal convector covers, then turned on and tested.

Any last-minute problems are identified and corrected through a cooperative effort by the contractors, the owner, and the architect. The building inspector is called in for a final inspection and the issuance of an occupancy permit. After a thorough cleaning, the building is ready for use.

C.S.I./C.S.C
MasterFormat Section Numbers for Finishing the Interior

06200	FINISH CARPENTRY	
06220	Millwork	
06270	Closet and Utility Wood Shelving	
06400	ARCHITECTURAL WOODWORK	
06410	Custom Cabinet	
06415	Countertop	
06430	Wood Stairs and Railing	
06440	Wood Ornament	
06450	Standing and Running Trim	
06460	Wood Frames	
09900	PAINTS AND COATINGS	
09910	Paint	
09930	Stains and Transparent Finishes	

SELECTED REFERENCES

1. Thallon, Rob. *Graphic Guide to Interior Details.* Newtown, CT: Taunton Press, 1996.

The second half of this book, first referenced in Chapter 17, describes the design, construction, and installation of cabinets, countertops, trim, and stairways.

2. Architectural Woodworking Institute. *Architectural Woodwork Quality Standards: Illustrated* (7th ed.). Reston, VA: Architectural Woodworking Institute, 1999.

A collection of standards that specify construction tolerances, joinery, and finish surfaces for lumber, panel products, doors, windows, cabinets, trim, and ornamental woodwork. This book is referenced whenever the quality of woodwork is specified.

3. Savage, Craig. *Trim Carpentry Techniques.* Newtown, CT: Taunton Press, 1989.

An illustrated how-to manual written by a finish carpenter who focuses on the basics of hanging windows and doors and applying molding. Tools and material selection are also covered.

KEY TERMS AND CONCEPTS

panel door
glazed door
flush door
hinged door
bypass door
bifold door
pocket door
prehung door
custom cabinet
modular cabinet
face-frame cabinet
frameless cabinet
single-piece cabinet door
banded
frame-and-panel cabinet door
full-overlay cabinet door

inset cabinet door
lip cabinet door
drawer guides
full-extension guides
butt hinge
fully concealed hinge
self-closing hinge
cabinet pull
scribe
countertop
solid surface countertop
plastic laminate countertop
tile countertop
nosing
backsplash
edge band

drop-in sink
flush-mounted sink
undermount sink
integral sink
millwork
molding
trim
nail set
window casing
baseboard
base shoe
finish stairs
shop-built stairs
site-built stairs
first-phase interior painting
second-phase interior painting

REVIEW QUESTIONS

1. List the sequence of operations required to complete the interior of a house and explain the logic of the order in which these operations occur.

2. Compare the advantages and disadvantages of hinged, pocket, bifold, and bypass doors.

3. Discuss the differences between custom cabinets and modular cabinets.

4. Explain the practical differences between a face-frame cabinet and a frameless cabinet.

5. What is the relationship among countertop, backsplash, and nosing? Why do some styles of sinks work with some countertops and not with others?

6. What is the primary function of trim? What other functions does it perform?

7. Summarize the most important things to keep in mind when designing a stair.

8. Explain the reasons for the phasing of the painter's work. Why are the painting techniques usually different during the different phases of the work?

EXERCISES

1. Design and detail a cabinet for a building that you are designing, using the information provided on page 427 to work out the dimensions.

2. Design and detail a stairway for a building that you are designing, using the information provided in Figures 18.25 and 18.26 to help you figure out the relationship of the parts.

3. Visit a wood frame building that you admire. Make a list of the interior finish materials and components, including species of wood, where possible. How does each material and component contribute to the overall feeling of the building? How do they relate to one another?

4. Make measured drawings of millwork details in an older building that you admire.

Analyze each detail to discover its logic. What woods were used and how were they sawn? How were they finished?

5. Obtain drawings of moldings available from a local lumberyard or home improvement center. Compare these to the drawings of the historic moldings you did in the previous exercise. Can you find other sources of moldings?

19

Finish Sitework

- **Paving Systems**
  - Flatwork
  - Asphalt Paving
  - Unit Pavers
  - Stone Paving
- **Level Changes**
  - Retaining Walls
  - Outdoor Steps
- **Porches and Decks**
  - Structure for Porches and Decks
  - Decking Boards
  - Deck Railing

- **Finish Grading**
- **Fencing**
- **Outdoor Lighting**
- **Irrigation**
- **Planting**

During pleasant weather, a well-planned and well-constructed yard extends the living space of a house in innumerable ways. Usable outdoor spaces, such as lawns, porches, decks, and terraces, are prized by home-owners and serve to increase the value of a residence. In addition to spaces used for social activities or recreation, there are a number of practical exterior features to be designed and built that make a residence complete. Virtually all houses have a driveway to the garage and a walkway to the front door. Most have plantings of lawn, perennials, shrubs, and trees. Many have a fence around the property for enclosure or at one or more edges for privacy. Irrigation systems allow plantings to flourish during dry spells, and outdoor lighting provides safety and nighttime use of outdoor spaces.

The finish sitework must generally wait until the shell of the house has been completed. Only then can the ground be cleared of all construction equipment and materials, and the finishing of the site begin. Often, parts of this work (such as planting, irrigation, and fencing) are deferred—leaving them for the owners to complete after the house has been occupied. Other parts of the finish sitework (such as the driveway, walk, and finish grading), however, are almost always included within the construction contract and are completed under the direction of the contractor.

PAVING SYSTEMS

Flatwork

Nearly every residential construction project includes one or more outdoor concrete slabs—referred to collectively as *flatwork* because of their horizontal orientation. The most common components of flatwork are the driveway, concrete walkways, and backyard patios or terraces. This work is typically performed by a specialized subcontractor.

Because concrete is brittle, it cannot withstand differential settlement. Therefore, the *subgrade,* the soil beneath the flatwork that will support its design load, must be firm and stable. The subgrade may be original soil free of organic material or it may be a structural fill compacted in lifts. Over the subgrade, a compacted or self-compacting *gravel base* helps to distribute the load of the concrete flatwork evenly and isolates it from the moisture that migrates upward through the soil. The thickness of the base depends primarily on the characteristics of the underlying soil and the intended use of the slab above. Weaker soils and heavy-duty use call for a thicker base. Driveways, which carry extreme loads, often have a base of 5 to 6 inches, while a sidewalk may have a base layer of only 2 to 4 inches.

The edges of concrete slabs are formed with 2×4 or 2×6 boards where they do not come up against a building. These forms are staked in place well within the boundaries of the base and serve as guides for the straight-edge used to level the surface of the concrete as it is placed. The forms are removed after the concrete has cured because they will rot over time (Figure 19.1). *Control joints,* made of asphalt-impregnated fiberboard, divide large expanses of slab into smaller rectangles that are less likely to crack as the concrete shrinks after curing. These joints are also located at the edges of slab where it abuts another fixed concrete object such as a foundation, street curb, or another slab, allowing the slab to expand and contract with a change in temperature. In driveways or garage floors, flatwork is reinforced with welded wire mesh or slender rebar reinforcement to reduce shrinkage cracking and provide strength to resist bending of the slab under the wheel loads of vehicles.

A variety of finishes may be applied to concrete at the time it is placed. A smoothly troweled finish is common in garages and is sometimes used for patios or porches. The *smooth finish* is the easiest of finishes to clean, but it can be slippery when wet. A *broom finish,* made by dragging a broom over the fresh, smoothly troweled concrete, has a rough texture to keep people from slipping. A *washed aggregate* surface has more texture yet. This finish is made by spraying the fresh concrete with water to expose the aggregate that lies just below the surface. *Stamped concrete* is made to resemble brick or stonework by forcing a mold down into the surface of the freshly placed concrete (Figure 19.2). Coloring agents can be mixed with the concrete batch or added to the surface of the freshly placed slab.

Asphalt Paving

Asphalt paving is a mixture of bitumens (hydrocarbons from petroleum or coal), gravel, and sand that is applied hot and hardens as it cools. It is often used for driveways and occasionally for paths in residential work (Figure 19.3). Like concrete flatwork, asphalt paving requires a firm and stable subgrade and an aggregate base. However, because asphalt is a more flexible covering than concrete, the base must be deeper. Aggregate bases for driveways range in thickness from 8 to 12 inches on most soils.

A variety of machines may be employed to distribute the hot asphalt mixture evenly across the base, and hand tools are used to smooth out irregularities in the surface. A heavy roller then compacts the asphalt while it is still hot, removing air pockets and forcing the bitumen into close contact with the sand and gravel. When the mixture cools, the bitumen stiffens and gives the surface strength.

FIGURE 19.1
Formwork for a concrete sidewalk and driveway. The gravel base will be smoothed and compacted before the concrete is poured. Notice the storm drain to the left of the driveway, which will be routed under the sidewalk and through the new curb to the gutter. (*Photo by Rob Thallon*)

FIGURE 19.2
Concrete can be stamped and colored before it has set to imitate stone or brick pavers. The difference in cost between stamped concrete and real stone or brick-work is substantial. (*Photo by Rob Thallon*)

FIGURE 19.3
Installing an asphalt driveway usually involves a large number of workers and lots of equipment. This 40-foot driveway took two dump truck loads of hot asphalt which was spread by the machine in the foreground, smoothed with hand tools, and compacted by the small roller and hand tampers. About ten people worked at the site for approximately two hours to complete the job. (*Photo by Rob Thallon*)

Asphalt is durable but is susceptible to damage in very hot weather when it becomes soft.

Asphalt paving is less expensive per unit area than concrete—even considering the deeper aggregate base. For small areas, however, concrete is often competitive in price because it can be included with other flatwork projects. For this reason, asphalt driveways are generally longer and concrete ones shorter.

Unit Pavers

Often employed in patios, terraces, and walks, *unit paving* consists of small units of brick, concrete masonry, or stone that are laid in a sand setting bed (Figure 19.4). The edges of areas paved with unit pavers are restrained with walls or permanent decay-resistant wooden forms so that the units at the periphery do not creep. When all of the paving units are in place, sand is brushed over the surface, falling between the units and locking the system together. Because the system employs many small independent units, the surface is flexible like asphalt and requires a deeper aggregate base than would concrete flatwork on the same soil. A sand setting bed placed over the base provides a smooth surface for setting the pavers. The units may be butted together edge to edge or spaced slightly. In either case, the surface allows water to pass through its joints to the soil beneath. Unit paving work is generally performed by a specialized subcontractor.

Simple, elegant patterns can be made with unit pavers. Concrete pavers are manufactured in a variety of shapes, and some are designed to interlock to form patterns. Simple rectangular or square shapes can be arranged to form patterns as well (Figure 19.5). Bricks are a durable and colorful unit paving material and are the material of many traditional patterns. Stone paving made of flagstones, granite setts, and cobblestones are also common in the East. When designing unit paving, one of the most important considerations is the edge of the installation, where pavers will have to be cut (usually with a diamond-bladed masonry saw) for all but the very simplest of patterns.

Stone Paving

Natural stone is the most expensive paving option but is often selected for its aesthetic appeal despite its high cost. Flagstones, cobblestones, and other quarried local stones such as granite, basalt, and limestone are commonly employed for residential terraces, paths, and steps. Large stones are inherently stable and thus may be set directly onto a bed of sand over a gravel base, while smaller stones must be laid in mortar over a concrete base. As with most forms of paving, stability of the edge of the paved area is a major concern for the overall durability of the installation.

FIGURE 19.4
Brick pavers being laid on a setting bed of compacted gravel topped with a thin layer of sand. (*Photo by Rob Thallon*)

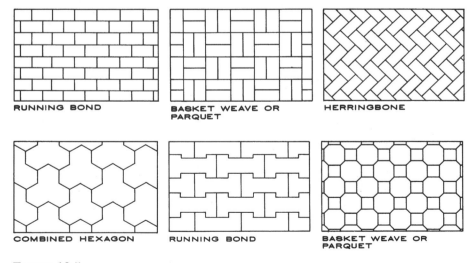

FIGURE 19.5
Common patterns of various unit pavers. (*Reprinted by permission from Ramsey/Sleeper, Architectural Graphic Standards, 10th ed., John Ray Hoke, Jr., A.I.A., Editor, John Wiley & Sons, 2000. © 2000 by John Wiley & Sons*)

LEVEL CHANGES

On sloping sites, slope stabilization and the creation of reasonably level areas for human use are often important. For gently sloping sites, the simplest approach is to allow landscape features such as lawns, planting beds, and paths to follow the slope of the terrain. For steeper sites, however, more dramatic measures such as retaining walls and steps must often be constructed (Figure 19.6). In all cases, runoff needs to be controlled to prevent erosion.

Retaining Walls

Retaining walls are freestanding structural walls that resist the horizontal force of soil at major changes of level (Figure 19.7). The engineering design of retaining walls depends primarily on the type and slope of the soil being retained, the soil upon which the wall bears, and the height of the wall.

FIGURE 19.6
Flagstone steps in this entry path take up the difference in grade between the street and the front porch. Dry stacked stone retaining walls create terraces for planting. (*Photo by Rob Thallon*)

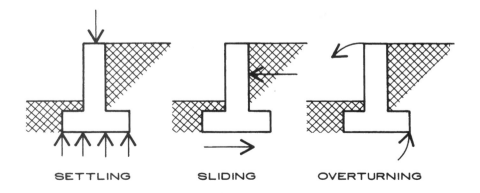

SETTLING   SLIDING   OVERTURNING

FIGURE 19.7
Possible types of failure of a retaining wall. (*Reprinted by permission from Ramsey/Sleeper,* **Architectural Graphic Standards,** *10th ed., John Ray Hoke, Jr., A.I.A., Editor, John Wiley & Sons, 2000. © 2000 by John Wiley & Sons*)

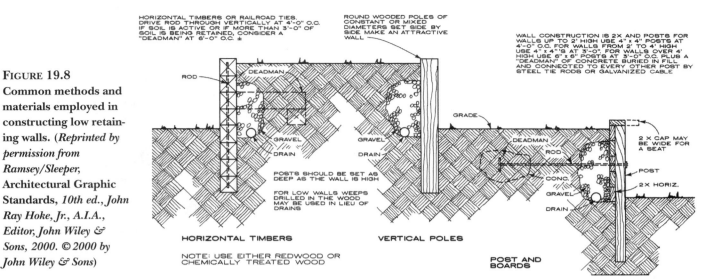

FIGURE 19.8
Common methods and materials employed in constructing low retaining walls. (*Reprinted by permission from Ramsey/Sleeper, Architectural Graphic Standards, 10th ed., John Ray Hoke, Jr., A.I.A., Editor, John Wiley & Sons, 2000. © 2000 by John Wiley & Sons*)

FIGURE 19.9
Sections of typical reinforced concrete retaining walls. Similar walls are often made with concrete block. Reinforcing steel in these walls and footings can be reduced if the retaining wall is braced at regular intervals with perpendicular concrete or masonry walls. (*Reprinted by permission from Ramsey/Sleeper, Architectural Graphic Standards, 10th ed., John Ray Hoke, Jr., A.I.A., Editor, John Wiley & Sons, 2000. © 2000 by John Wiley & Sons*)

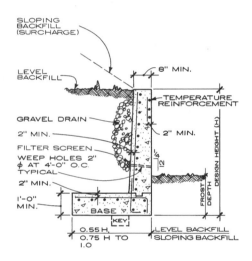

DIMENSIONS AND REINFORCEMENT

WALL	H	B	T	A	"V" BARS	"F" BARS
8"	3' 4"	2' 4"	9"	8"	#3 @ 32"	#3 @ 27"
	4' 0"	2' 9"	9"	10"	#4 @ 32"	#3 @ 27"
	4' 8"	3' 4"	10"	12"	#5 @ 32"	#3 @ 27"
	5' 4"	3' 8"	10"	14"	#4 @ 16"	#4 @ 30"
	6' 0"	4' 2"	12"	16"	#6 @ 24"	#4 @ 25"
12"	5' 4"	3' 8"	10"	14"	#4 @ 24"	#3 @ 25"
	6' 0"	4' 2"	12"	15"	#4 @ 16"	#4 @ 30"
	6' 8"	4' 6"	12"	16"	#6 @ 24"	#4 @ 22"
	7' 4"	4' 10"	12"	18"	#5 @ 16"	#5 @ 26"
	8' 0"	5' 4"	12"	20"	#7 @ 24"	#5 @ 21"
	8' 8"	5' 10"	14"	22"	#6 @ 8"	#6 @ 26"
	9' 4"	6' 2"	14"	24"	#8 @ 8"	#6 @ 21"

NOTE: See General Notes for design parameters.

FIGURE 19.10
Typical cantilever retaining wall showing steel reinforcing. (*Reprinted by permission from Ramsey/Sleeper, Architectural Graphic Standards, 10th ed., John Ray Hoke, Jr., A.I.A., Editor, John Wiley & Sons, 2000. © 2000 by John Wiley & Sons*)

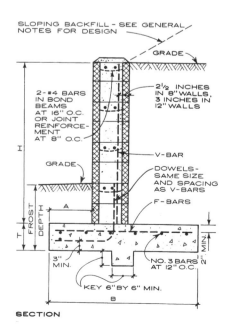

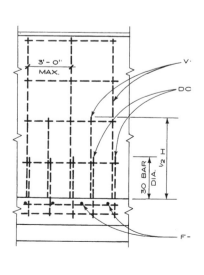

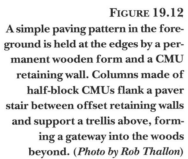

FIGURE 19.11
Dry-stack interlocking masonry retaining systems have become popular in recent years. The masonry units interlock but use no mortar so slight movement of the wall is not detectable or harmful. The serpentine shape of these walls contributes to their stability. (*Photo by Rob Thallon*)

FIGURE 19.12
A simple paving pattern in the foreground is held at the edges by a permanent wooden form and a CMU retaining wall. Columns made of half-block CMUs flank a paver stair between offset retaining walls and support a trellis above, forming a gateway into the woods beyond. (*Photo by Rob Thallon*)

Most codes allow low retaining walls (4 feet or less between the two soil levels) to be designed without the involvement of an engineer. These walls may be made of concrete, concrete block, stone, cast masonry retaining units, preservative-treated wood, and other materials. They may be built with or without a foundation and are sometimes cantilevered from the ground with piers analogous to wooden fence posts (Figure 19.8).

The design review of a registered professional engineer is required for retaining walls over 4 feet, or 1.2 m, in height. These walls are usually made of either cast-in-place reinforced concrete or reinforced con-

crete block, both of which require concrete footings (Figure 19.9). The footing is similar to a foundation wall strip footing but is often wider to help resist the sliding and overturning forces of the retained soil. When the footing is extended on the uphill side of the wall, the soil above the footing contributes to the weight of the construction to help prevent rotation and sliding. The connection between footing and retaining wall is the location of a great deal of stress and is typically heavily reinforced (Figure 19.10). Tall retaining walls made of concrete or concrete block are frequently covered with a veneer of stone or brick for better appearance (Figure 5.42).

Outdoor Steps

Outdoor steps may be made of concrete, wood, stone, bricks, and combinations of these and other materials (Figure 19.13). They can be very wide, making a grand connection between terraces at different levels, or quite narrow, providing utilitarian access to different parts of the site. Regardless of their width, outdoor steps are generally less steep than their indoor equivalents because the human stride is longer when taken outdoors. The typical proportions for outdoor steps include a rise of about 6 inches (150 mm) and a run of 14 inches (350 mm). Steps with a rise as low as 4 inches (100 mm) are not uncommon.

FIGURE 19.13
New wooden steps under construction. Preservative-treated stringers notched for the treads and risers are bolted to preservative-treated 4 × 4 posts set into concrete. These treads and risers are made of cedar. Other decay-resistant species and synthetic decking are often used as well. (*Photo by Cynthia Girling*)

FIGURE 19.14
Concrete steps to a new residence. The upper steps have been backfilled. (*Photo by Rob Thallon*)

Concrete steps are formed much like a sidewalk (Figure 19.14). The work is done by concrete flatwork contractors who place and compact the aggregate base, construct the forms, and pour and finish the concrete. Concrete steps may also be finished by adhering other materials such as tiles, pavers, or stones to them.

Durable and inexpensive steps may also be made of decay-resistant timbers backfilled with gravel. When bricks or other unit pavers are used as treads in such steps, the width of the steps should be carefully calculated to correspond to paver modules in order to avoid undue cutting of masonry units.

PORCHES AND DECKS

Outdoor porches, decks, and stairs that are above grade are usually made of wood (Figure 19.16). The materials and construction methods are quite similar to those used to construct the house, so the work is often done by the framing and/or finish crew.

Because the work is exposed to the weather, decay-resistant heartwoods, wood that is pressure treated with preservatives, or wood/plastic composite planks are used. If nondurable woods are used, they will decay at the joints, where water is trapped and held by capillary action. Even with durable woods, joints should be minimized, drainage space should be provided wherever possible, and gaps should be left between adjacent planks to allow for expansion, contraction, and drainage.

Structure for Porches and Decks

A post-and-beam structure generally supports the level surface of the deck (Figure 19.17). The posts bear on precast or cast-in-place concrete piers located on solid bearing soil. Beams sit on top of the posts and carry joists that run perpendicular to them. The joists carry the decking or, alternatively, the decking is fastened directly to the beams.

This structural system of posts, beams, and joists must be braced against lateral movement, which is usually accomplished by diagonal bracing between the posts. If one or more edges of the deck are against the house, as is often the case, this edge can be braced by bolting it to the house using a detail that has been carefully thought out to avoid trapping water. Diagonal braces in a horizontal plane beneath the joists then impart stability to the entire deck. If it is exposed to view, the bracing structure may be carefully detailed and crafted. Alternatively, vertical planes of bracing may be enclosed with screening that allows ventilation (but not pets or wildlife) into the space below the deck.

FIGURE 19.15
The transition from a brick terrace to a wooden porch is made in this case with a stairway constructed partly of brick and partly of wood. Notice the step light mounted in the second wood riser. (*Photo by Brad Stangeland*)

FIGURE 19.16
An exterior deck of redwood. The decking boards have spaces between to allow water to drain through. (*Designer: John Matthias. Photo by Ernest Braun. Courtesy of California Redwood Association*)

Decking Boards

The decking boards are the most visible part of a deck, so they should be carefully selected and applied. Clear and tight-knot grades of decay-resistant species of wood make the most durable and beautiful decks. A 2×6 or $5/4 \times 6$ board is the most common size for decking because its width minimizes expansion, contraction, and cupping. Decking boards should always have open, spaced joints to allow for drainage of water through the deck and for expansion and contraction of the decking. Fastening the boards with specialized deck screws or with hot-dipped galvanized casing nails will minimize the withdrawal of fasteners due to expansion and contraction of the material.

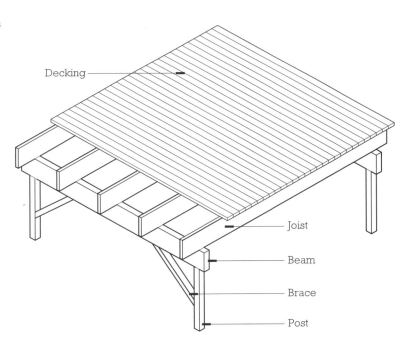

FIGURE 19.17
Typical structural organization of a deck. Decking, joists, and beams must all be designed to have sufficient structural strength and stiffness. Diagonal bracing must usually be added to the system for lateral stability.

Synthetic decking boards are manufactured of reclaimed wood fiber and recycled plastic. They are not susceptible to decay and have an integral nonskid surface. They are fastened to joists or beams with nails or deck screws in accordance with the recommendations of the decking manufacturer. Synthetic boards are not as stiff as wood, for which reason the joists that support them must be spaced closer together than for wood decking.

Deck Railing

If a deck or porch is 30 inches or more above finish grade, a 36-inch-high (900 mm) guardrail is required by code to prevent people from falling off. This rail must be firmly connected to the framing of the deck, and often is connected to the same posts that support the deck as they extend above deck level. The building code has stringent requirements for the resistance of guardrails to lateral pushing actions. The most common rail is formed with 2×4 or 2×6 horizontal members at top and bottom with vertical 1-inch pieces between (Figure 19.18). More imaginative designs can incorporate seating and planters with the railing.

FINISH GRADING

Finish grading is the final placement and smoothing of the soil. The soil must follow contours specified on the drawings, must be backfilled as specified against the building, and must be flush with site features such as driveway, walks, and terraces. If topsoil was removed and stockpiled earlier during excavation, this soil is now distributed across the site for finish grading. If topsoil must be imported, it should be spread to a minimum thickness of 6 inches. Where soil is highly cohesive, the contract may specify a maximum size of clod that is allowed on the finish surface of the ground.

FIGURE 19.18
A simple deck railing made with 1×8 boards spaced 4 inches apart. A 2×6 cap resists outward thrust on the rail. (*Photo by Rob Thallon*)

FIGURE 19.19
Finish grading is commonly done with a small tractor. (*Photo by Rob Thallon*)

Finish grading may be done with a variety of small earth-moving equipment and is usually the responsibility of a landscape contractor. The first phase of placing the topsoil is necessarily carried out after the paving is in place, but before the fencing, lighting, or irrigation systems are installed (Figure 19.19). This timing allows the equipment to move freely around the site and to backfill against the paving without damaging buried pipes or wires. After the fencing, lighting, and irrigation are installed, hand raking is employed to make fine adjustments in the level and smoothness of the soil.

FENCING

Fences, screens, and walls may be employed for a variety of purposes, including privacy, security, pet containment, deer control, and aesthetics. The barriers used for these purposes are made of an assortment of materials (wood, masonry, metal) and in a wide range of heights. Details that must be considered in their design include not only the basic section, but also transitions at corners, ends, slopes, and gates. If there is a fence or wall to be constructed as part of the general contract, it will be detailed and specified by the designer and included in the basic bid for construction. The work is usually performed by a specialized subcontractor.

By far, the most common fencing is the wooden fence (Figures 19.20 and 19.21). Even with just the simple components of post, (horizontal) rail, and (vertical) slat, there are literally hundreds of ways to build the "typical" fence. This is because there

FIGURE 19.20
Two of the most common fence types: the double-sided privacy fence and the picket fence. There are many ways to make a simple wood fence. Great variety is possible by selecting different materials and by spacing, orienting, and layering them in different ways.

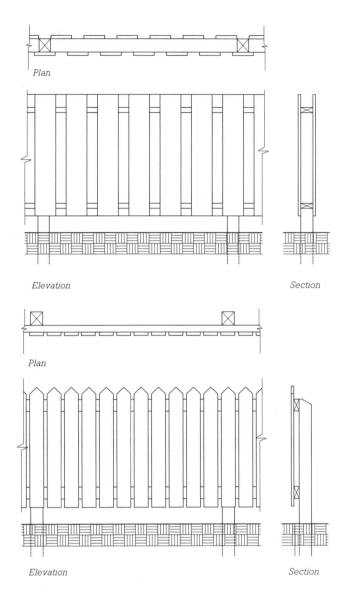

FIGURE 19.21
An elegant fence made of simple materials provides privacy and contributes to the aesthetics of the street. (*Photo by Rob Thallon*)

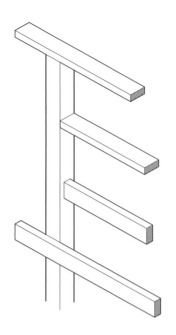

FIGURE 19.22
There are four possible basic relationships of simple horizontal fence rails to a fence post.

are numerous ways to make each connection and a wide variety of materials and material sizes (Figures 19.22 and 19.23). Like decks and other outdoor structures, wooden fences, in order to last, must be constructed of weather-resistant materials and detailed to minimize the accumulation and retention of moisture in the connections. Preservative-treated lumber is used for direct-burial posts and posts embedded in concrete, and lumber made of naturally decay-resistant species of wood such as cedar is usually used for other parts of the fence. Details should encourage the drainage of water from horizontal surfaces and avoid the trapping of moisture in joints.

OUTDOOR LIGHTING

Landscape lighting is used to illuminate pathways and gathering areas and to highlight landscape features. Electrical power may also be required at outdoor cooking areas. If there is outdoor lighting or power specified in the landscape plan, it is best installed before planting trees, shrubs, and flowers, because the wires run underground and trenching for wires is very disruptive to plants.

There are two basic types of outdoor lighting (Figure 19.24). One type uses line voltage (110 volts) transmitted through direct burial cable that must be buried at least 12 inches below the surface. *Line voltage lighting* is wired directly into the house electrical system and is most commonly used to illuminate garden structures. The other type of lighting is a *low-voltage system*, which requires a transformer to reduce the voltage. Low-voltage wires do not have to be buried deep in the ground for safety because they are incapable of giving a fatal electric shock or causing a fire. Low-voltage systems are most commonly used for pathway and step illumination. Installation of a line voltage system requires a licensed electrician, whereas that of a low-voltage system does not.

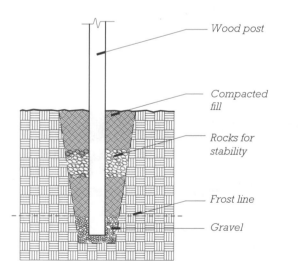

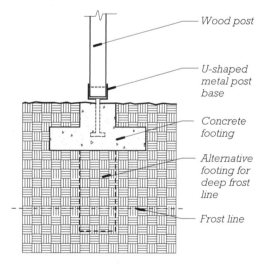

FIGURE 19.23
Fence posts must be stabilized by the ground. The most common method is to "plant" a preservative-treated post directly into the soil. An alternative method is to support the post with a metal bracket anchored to a concrete footing. The metal bracket must be bolted to the post and be sufficiently tall and strong to provide lateral stability.

FIGURE 19.24
Exterior lighting is important for safety and can highlight the features of a house and its landscape. At this residence, line voltage is used for path lights and lights on the building. Low-voltage lights are located in the trees. (*Photo by Kenneth Rice. From Jan Moyer*, The Landscape Lighting Book, *copyright 1992, John Wiley & Sons, New York. Reprinted by permission of John Wiley & Sons, Inc.*)

IRRIGATION

An irrigation system, when specified, is usually installed after finish grading because *sprinkler heads* must be set flush with finish grade. In some cases, the distribution pipes are set prior to finish grading, as this sequence saves some trenching, but great care must be taken not to crush the pipes with grading equipment. Irrigation systems are typically installed by a specialized subcontractor.

Sprinkler systems employ plastic pipe with glued or clamped connections (Figure 19.25). The distribution pipes do not usually need to be set below the frost line because the system is drained in the cold months when not in use. The most common sprinkler head is the pop-up type made

FIGURE 19.25
A new sprinkler system being installed. The rigid white PVC line conducts water under pressure from a zone controller at the water supply. The sprinkler head is mounted on a flexible line, and its position can be adjusted as finish grading is refined. (*Photo by Rob Thallon*)

of plastic or plastic and metal. These heads retract automatically when not in use so they are not likely to be damaged by yard equipment. There are a variety of head types, ranging from a 360-degree spray to a narrow-angle, adjustable spray to a fine mist.

Sprinkler systems are organized into zones of about four to six heads each so that adequate pressure can be delivered to each head. Zoning also allows different amounts of water to be delivered to different kinds of plantings. The zones are controlled by an electric timer that turns them on sequentially by means of solenoid valves, one at a time, according to a set schedule.

PLANTING

Planting around newly constructed houses is typically limited to trees, shrubs, and lawn grass. Trees can provide shade, texture, color, and scale to a house, and shrubs have similar qualities, sometimes adding fragrance and flowers as well. Lawns are appreciated for providing verdant open space that is soft underfoot, resistant to erosion, and relatively easy to maintain. Whatever ground is not planted in these materials is usually *mulched* with shredded bark to deter weed growth. Mulched areas are either designed to be permanent (in which case gravel is sometimes used) or with the expectation that the occupants of the new house will continue with the planting by adding perennials, annuals, and/or vegetables. The initial installation of trees, shrubs, lawn, and mulch is typically done by a landscape contractor.

The planting of trees and shrubs requires the digging of large holes that displace considerable soil (Figure 19.26). This work is therefore generally undertaken before final grading occurs. Large holes are dug with heavy machinery when possible, and small holes for shrubs are often dug out by hand. It is important to make each hole large enough to allow for root growth while a tree or shrub becomes established, especially if the underlying soil is poor. A hole diameter 1½ to 2 times the diameter of the root ball is a rule of thumb used by landscape contractors. The hole should be backfilled with native soil *amended* with components that promote aeration, drainage, and root growth.

A lawn for a new residence is either grown from seed or laid as strips of sod. *Seeded lawns* are less expensive but can only be planted during certain seasons of the year (typically spring and fall) and must endure a long period of vulnerability while the seed germinates and the grass matures. During this period, seeded lawns must be kept moist and protected from foot traffic of all types. *Sod lawns* are installed like a carpet of established grass and thus are initially less susceptible to damage than

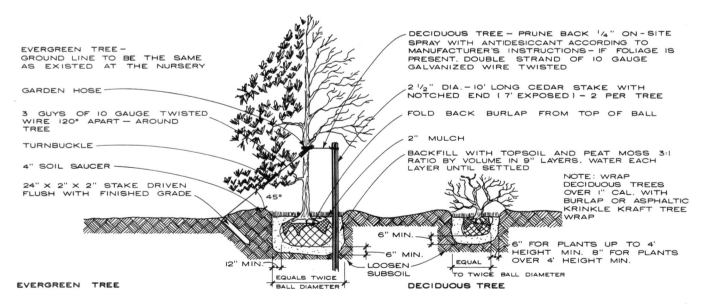

FIGURE 19.26
Typical planting details for trees and shrubs. (*Reprinted by permission from Ramsey/Sleeper*, Architectural Graphic Standards, *10th ed., John Ray Hoke, Jr., A.I.A., Editor, John Wiley & Sons, 2000. © 2000 by John Wiley & Sons*)

FIGURE 19.27
These workers are installing a sod lawn on an irregularly shaped, sloping site. The worker at the left is unrolling a sod strip that will be trimmed to fit against the strips already laid. The worker at the right is resting after having transported the roll of sod to the work site in a wheelbarrow and lifted it into position. (*Photo by Helen Etzwiler*)

seeded lawns. However, sod should not be walked upon for a couple of weeks while the root system knits itself to the ground. Although sod lawns look good immediately and are usable relatively soon after installation, they are generally not as hardy as seeded lawns in the long run. The economics of lawn installation are such that large lawn areas are typically seeded while small lawn areas commonly employ sod (Figure 19.27).

C.S.I./C.S.C MasterFormat Section Numbers for Finish Sitework	
02700	**BASES, BALLASTS, PAVEMENTS, AND APPURTENANCES**
02720	**Unbound Base Courses and Ballast**
02740	**Flexible Pavement**
02750	**Rigid Pavement**
02775	**Sidewalk**
02780	**Unit Paver**
02800	**SITE IMPROVEMENTS AND AMENITIES**
02810	**Irrigation System**
02820	**Fences and Gate**
02830	**Retaining Wall**
02900	**PLANTING**
02910	**Plant Preparation**
02920	**Lawns and Grasses**
02930	**Exterior Plants**

SELECTED REFERENCES

1. Weinberg, Scott, and Coyle, Gregg A., Eds. *Handbook of Landscape Architecture Construction, Volume 4: Materials for Landscape Architecture Construction.* Washington, DC: Landscape Architecture Foundation, 1992.

A reference book for landscape architects and designers that covers the topics of soils, paving, retaining walls, and wood structures.

2. Thompson, J. William, and Sorvig, Kim. *Sustainable Landscape Construction.* Washington, DC: Island Press, 2000.

A book that treats environmentally responsible landscape construction at all scales by laying out a set of 10 principles. The principles such as "keep healthy sites healthy," "heal injured sites," and "pave less" are discussed and illustrated with examples.

KEY TERMS AND CONCEPTS

flatwork
subgrade
gravel base
control joint
smooth finish
broom finish
washed aggregate

stamped concrete
asphalt paving
unit paving
stone paving
retaining wall
outdoor step
wood porch

wood deck
wood decking board
synthetic decking board
deck railing
finish grading
fencing
outdoor lighting

line voltage lighting
low-voltage lighting
irrigation system
sprinkler head
mulch
amended soil
seeded lawn
sod lawn

REVIEW QUESTIONS

1. Explain the logic of the gravel base. What variables does the design of a base depend on?

2. Discuss the advantages of the various finishes that can be applied to flatwork.

3. What are the principal design considerations for a unit paver surface?

4. Describe a logical schedule for the construction of a full complement of residential site development, including planting.

What components of the construction may overlap? What components are required to overlap?

EXERCISES

1. Design a deck with railing and stairs for an existing house. How is the deck braced? What is the structure of the railing? Make a wooden model of the junction between deck, rail, and stair.

2. Take a tour of a residential neighborhood and photograph as many types of fences as possible. Make a presentation to your class and discuss the functions, advantages, and disadvantages of each type.

3. Obtain a site development plan from a contractor or landscape architect and discuss the implementation of the plan in class. How would the work be scheduled?

How many subcontractors would likely be involved? What would be the detailed step-by-step process employed by each subcontractor?

PART FOUR

ALTERNATIVE CONSTRUCTION SYSTEMS

LOW-TECH, LOW-ENERGY CONSTRUCTION

A number of construction systems employ naturally occurring materials that are subjected to a minimum of processing and can be assembled with simple tools and unskilled labor. Often referred to as "primitive," most of these systems have been used for centuries on a global scale.

The systems include the following:

1. Earthen construction—including adobe bricks and rammed earth. Soil, which is usually excavated from the very site upon which the building is constructed, is the principal ingredient in these systems.

2. Stacked log construction—also known as log cabin construction or just log construction. Tree trunks are processed only minimally before being incorporated into the walls of the building.

3. Straw bale construction—a relatively new system that originated in Nebraska in the 1890s and has been revived recently in response to an interest in green building. Straw bales, essentially a farm waste product, are stacked as running bond masonry to form insulated walls.

Because the materials are inexpensive and a structure can be erected using unskilled labor, all of these systems have been embraced by owner/builders. A number of the systems have also become popular with the general public, and professional designers and contractors are involved in their production. Professionally designed and built systems will be emphasized in this chapter.

EARTHEN CONSTRUCTION

Between one-third and one-half of the world's population lives in houses made of earth—mostly adobe or rammed earth (Figure 20.1). Earth is used as a building material primarily because of its low cost and its availability. In the current climate of declining resources and increased pollution, earth-based construction offers the additional advantage of low environmental impact because the material can come directly from the building site with little or no energy aside from human labor involved in its production. An adobe brick, for example, can be made with 1 percent of the energy required to manufacture a concrete block or a fired clay brick of the same volume.

When an earth building is complete, the solid earth walls afford thermal mass and low sound transmission, contain no toxic substances, and are not susceptible to attack by insects or microorganisms. The construction process, although long and labor intensive, creates virtually no construction waste.

Earthen construction, like masonry, can be used to make vaulted roofs, but this is generally not done in the United States because of the limited distance that can be spanned with mud vaults, the expensive engineering that is required to satisfy building officials, and the need to make the walls thicker in order to resist the thrust of the vault. Modern earthen construction in the United States is therefore typically limited to the construction of walls.

There are four basic types of earthen wall construction:

1. Adobe—air-dried mud bricks are laid up with mud mortar in the same fashion as concrete blocks.

2. Rammed earth—a moistened soil mixture is compacted into a form that is reused repeatedly to make a monolithic wall.

3. Cob—a mud mixture of clay, sand, and straw is piled onto a wall and formed by hand.

4. Contained earth—plastic or textile sacks filled with soil (and sometimes sand) are stacked to form walls.

Adobe and rammed earth are the systems most commonly used in North America.

Adobe

The oldest known adobe structure was built in Israel over 10,000 years ago. In the American Southwest, Pueblo tribes built apartments of adobe as early as 1250 A.D. When the Spanish arrived in the area, they assimilated the Pueblo building techniques for the construction of their own buildings. Later generations of Americans have continued the adobe tradition, mostly in the Southwest, but also in other parts of the country (Figure 20.2). Currently, there are about 50 professional adobe builders concentrated in the states of New Mexico and Arizona, where adobe bricks are commercially produced and building codes do not discourage adobe construction. Most of these builders construct large, expensive, custom residences.

FIGURE 20.1
Earthen houses such as these are
common throughout the world.
(*Photo by Amy Houghton*)

FIGURE 20.2
Adobe mud plaster being applied
to a traditional adobe residence in
New Mexico. (*Courtesy of Southwest
Solar Adobe School, © 2001 SWSA*)

Adobe is an air-dried brick made of mud that is composed of inorganic soil and sand (Figure 20.3). It is important that the soil have approximately 8 to 15 percent clay content. Straw can be added for strength and to prevent cracking. Modern adobe is also often modified with emulsified asphalt that minimizes moisture penetration.

Adobe bricks in the United States generally range from 10 to 12 inches (254 to 305 mm) wide and weigh from 32 to 48 pounds (15 to 22 kg). (Smaller sizes are made for domes and vaults.) Walls are laid up with a single wythe of adobe bricks, making them 10 to 12 inches thick. Modern adobe walls are produced much more efficiently than the historic Pueblo or Spanish walls, which were made with a double wythe and were 24 or more inches thick.

Modern Adobe Construction

Modern adobe structures are built on a concrete footing with a concrete stem wall that extends above grade (Figures 20.4 and 20.5). As a measure to control moisture migration into the adobe from the ground, the first courses are often constructed with asphalt-modified adobe. At the top of the wall, a reinforced concrete or laminated wood bond beam ties the adobe bricks together and provides anchoring for the roof structure (Figure 20.5). Openings for windows and doors are typically spanned with wooden lintels that are as deep as the wall (Figure 20.6).

Once in place, adobe walls are typically insulated on the outside with extruded polystyrene (EPS) boards before being covered with wire and cement plaster or mud plaster (Figures 20.2 and 20.7). Electrical wiring and plumbing can be installed in grooves cut into the walls before the insulation and plaster are installed.

Adobe Construction and the Building Codes

The model codes require adobe walls to be provided with both horizontal

FIGURE 20.3
An adobe field in New Mexico. The adobe bricks in the background have been turned on edge to promote drying from both sides. (*Courtesy of Southwest Solar Adobe School, © 2001 SWSA*)

FIGURE 20.4
A new adobe wall before insulation and stucco have been applied. Ten-inch CMU blocks at the base of the wall elevate the adobe bricks above potential sources of water. Traditional canales to drain the roof project through a wooden parapet wall. (*Photo by Betsy Kramer*)

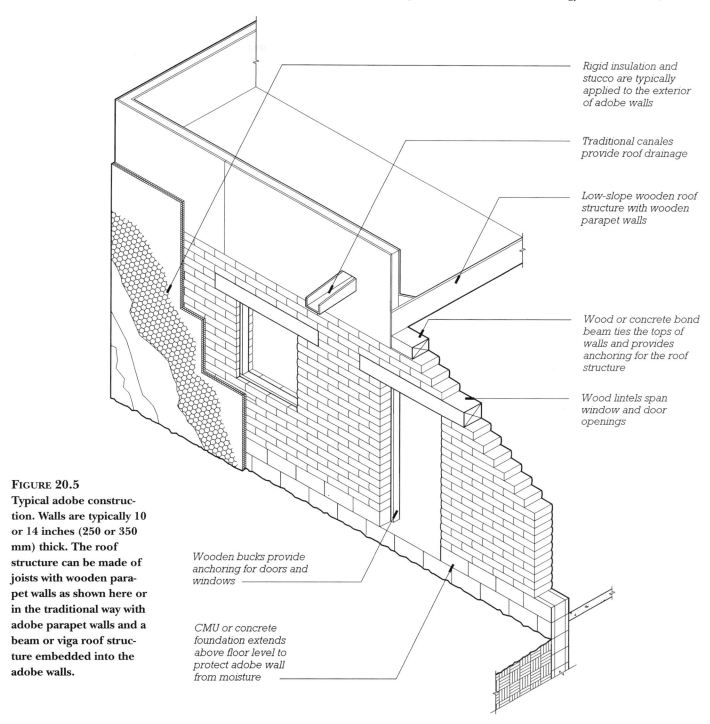

Rigid insulation and stucco are typically applied to the exterior of adobe walls

Traditional canales provide roof drainage

Low-slope wooden roof structure with wooden parapet walls

Wood or concrete bond beam ties the tops of walls and provides anchoring for the roof structure

Wood lintels span window and door openings

Wooden bucks provide anchoring for doors and windows

CMU or concrete foundation extends above floor level to protect adobe wall from moisture

FIGURE 20.5
Typical adobe construction. Walls are typically 10 or 14 inches (250 or 350 mm) thick. The roof structure can be made of joists with wooden parapet walls as shown here or in the traditional way with adobe parapet walls and a beam or viga roof structure embedded into the adobe walls.

and vertical steel reinforcement in zones with seismic risk. Because adobe bricks do not have a void for the insertion of vertical reinforcing, it is difficult to comply with the vertical reinforcement requirement of the model codes. Therefore, adobe construction has all but disappeared in most regions—

including California, which has a rich tradition of adobe construction and a large stock of historic adobe houses and missions.

However, the state of New Mexico, where adobe construction is traditional, has written its own adobe code. The New Mexico code recog-

nizes the inherent structural properties of adobe and does not require vertical reinforcing, so adobe construction can flourish there as well as in parts of some neighboring states where the New Mexico code has been adopted.

FIGURE 20.6
A wooden lintel over a doorway in an adobe wall. A wooden buck at the side of the doorway provides attachment for the door frame. (*Courtesy of Southwest Solar Adobe School, © 2001 SWSA*)

FIGURE 20.7
Rigid insulation and stucco have been applied to the new adobe wall pictured in Figure 20.5. (*Photo by Betsy Kramer*)

FIGURE 20.8
The southern wall of a new passive solar adobe house in Colorado. Generous glazing is combined with adequate roof overhangs to capitalize on the effectiveness of adobe as thermal mass. (*Courtesy of Southwest Solar Adobe School, © 2001 SWSA*)

Rammed Earth

Rammed-earth construction consists of packing slightly moistened soil between rigid forms. The forms are relatively small and lightweight and are removed immediately upon completion of the compaction to form another section of the wall (Figures 20.9 and 20.10).

This type of construction was used and described by the Romans as early as the 1st century A.D. Rammed earth was rediscovered in the mid-19th century by European and North American builders who were attracted by its simplicity, its low cost, and the availability of the materials. Numerous houses were built of rammed earth until the building boom after World War II raised wages and placed an emphasis on construction speed, and the use of rammed earth all but ceased. Interest resumed, however, when the energy crisis of the 1970s prompted a search for more economical, ecologically sound ways to build. Because the method is so labor intensive, rammed earth construction has tended to be limited to owner/builders and to professional builders who use tractors and air-driven compactors to reduce labor costs.

The ideal soil for rammed earth walls is composed of 30 percent clay and 70 percent sand and small gravel. Up to 5 percent by volume of portland cement is sometimes added as a stabilizer. The moisture content of the mixture as it is placed into the forms is important because insufficient moisture leads to a crumbly wall in which the soil particles are not consolidated, and too much moisture causes the mix to stick to the forms and leads to shrinking and cracking of the finished wall. A quality mix that is compacted properly develops approximately half the strength of typical concrete.

The Rammed-Earth Construction Process

A rammed-earth wall should be constructed on a concrete foundation with a footing sufficiently wide to support its weight. A foundation stem wall as thick as the rammed-earth wall (typically 12 to 24 inches) extends

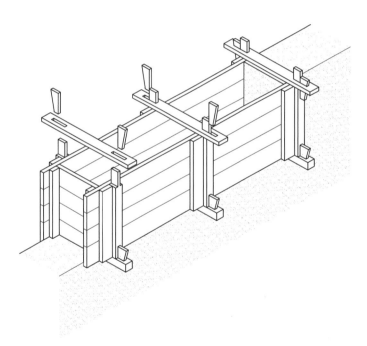

FIGURE 20.9
Traditional rammed-earth formwork made entirely of sawn boards. The boards passing through the wall at the base of the form must be left in place. They are later sawn off flush with the wall surface and covered with stucco or plaster.

FIGURE 20.10
Rammed-earth forms being assembled. Bar clamps pass through plywood side forms and hold wooden stiffeners in place. The end forms are constrained by the bars. (*Photo © Cynthia Wright, Rammed Earth Works*)

well above finish grade to ensure that excessive moisture does not enter the base of the earthen wall after it is built (Figure 20.11).

To begin construction of the rammed-earth wall, forms are set on the top of the concrete foundation, vertical reinforcing steel is put in place, and the soil mixture is dumped into the forms to a depth of about 6 inches (150 mm). It is then compacted with wooden tampers or pneumatic compactors (Figure 20.12). Compaction reduces the earth mixture to about half of its original volume and brings the soil particles into such intimate contact with one another that they bond together to produce a strong, coherent wall. When the form has been filled, it can be removed immediately and relocated to the next position, where the process is repeated (Figure 20.13).

Horizontal reinforcing steel is placed on top of compacted lifts at appropriate levels. This procedure is continued to the top of the wall where, in zones with seismic activity, roof anchoring is provided by a wood or reinforced concrete bond beam. In other zones, the roof may be connected to a simple wooden plate anchored to the top of the wall by anchor bolts embedded into the com-

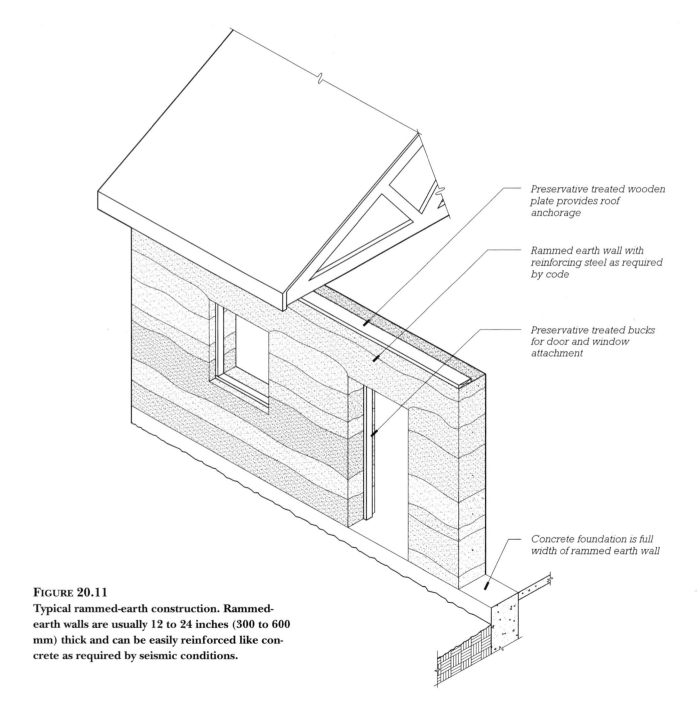

Preservative treated wooden plate provides roof anchorage

Rammed earth wall with reinforcing steel as required by code

Preservative treated bucks for door and window attachment

Concrete foundation is full width of rammed earth wall

FIGURE 20.11
Typical rammed-earth construction. Rammed-earth walls are usually 12 to 24 inches (300 to 600 mm) thick and can be easily reinforced like concrete as required by seismic conditions.

FIGURE 20.12
A pneumatic tamper compacts the earth in the forms. A worker in the background is using a shovel to spread earth that has been stored on a shelf at the top of the formwork. (*Photo © Cynthia Wright, Rammed Earth Works*)

FIGURE 20.13
Forms being removed from a newly constructed rammed-earth wall. The dark top layer of the wall is a bond beam containing cement and reinforcing steel. (*Photo © Cynthia Wright, Rammed Earth Works*)

pacted earth. Openings in the walls are made by blocking out the space with forms that are placed between the wall forms and removed after the wall forms come down.

Rammed Earth and the Building Codes

Numerous engineering tests have been conducted in recent years on rammed-earth walls made with a variety of soils and a range of mix designs and reinforcing steel. These tests generally offer sufficient data for an engineer to design a rammed-earth structure for any region. The walls are designed using the same procedures as are used for concrete design. These are based on an assumed compressive strength of 800 to 1200 psi as opposed to the minimum 2000 psi used for concrete.

FIGURE 20.14
Completed rammed-earth walls. (*Photo © Cynthia Wright, Rammed Earth Works*)

STACKED LOG CONSTRUCTION

Stacked log structures have their origin in the centuries-old cultures of Russia, from where the technique spread to Scandinavia and northern Europe. North American immigrants from these regions found abundant forests on the frontier that needed to be cleared for agriculture, and the downed logs were employed to make simple houses. Early log houses in North America were hastily built and not as sophisticated as their European precedents, and few survive (Figure 20.15). The allure of the log house, however, has prompted people to carry on the tradition long after log construction was the only practical way to provide housing. Indeed, many log dwellings have been built in recent years in locations that were never forested.

The production of log houses for their sentimental association with pioneer life had persisted quietly and mostly unnoticed until the practice enjoyed something of a renaissance for the construction of resorts and park structures in the 1920s and 1930s. During this period, many notable lodges and houses were built in the Adirondack Mountains, the western United States, and Canada. A renewed interest in log building started in the 1970s and has led to a multimillion-dollar industry supported by hundreds of companies that build exclusively with logs (Figure 20.16). Most professionally designed and built log houses are located in mountain resort and rural areas. Approximately 1000 to 2000 of these structures are built by professionals each year in North America, and untold numbers are constructed by owner/builders.

Log houses tend to be made of coniferous tree species because their trunks are available in long, straight sections with minimal taper (Figure 20.18). Preferred species are white, yellow, and lodgepole pine, western red cedar, and Douglas fir. Logs are typically peeled to remove the bark

FIGURE 20.15

A historic early-19th-century log cabin from Missouri. The form is simple, and the fit between logs is relatively crude. (*Photo: Library of Congress, Prints and Photographs Division, Historic American Buildings Survey, Reproduction Number HABS, MO, 92-FEMO.V,1-1***)**

for aesthetic and practical reasons, as bark left on can harbor insects. In the case of logs harvested green, log suppliers often fell the trees at a time in the growing season when the bark virtually falls off of the logs. Standing-dead timber, killed by high elevation, fire, or insects, is a primary source of logs because it is dry. It no longer has bark remaining, but is peeled to renew the surface.

Nine out of ten log houses are now preconstructed in a commercial log yard before being disassembled and shipped to a site for reerection (Figure 20.19). The advantages of preconstruction are many. Difficult joinery can be accomplished under controlled conditions, and on-site construction time can be reduced considerably. For example, a typical 3000-square-foot (280 m²), full-scribed,

handcrafted log house will take approximately 12 weeks to construct in the yard, but can be reassembled at the site in 3 to 5 days.

The basic strategy for constructing a log house is to stack logs one on top of the other and notch them to interlock at corners (Figure 20.20). Logs in perpendicular walls are offset in height by one-half log diameter in order to allow the corner joints to lap. Thus, at the base of a simple four-walled house, two of the walls start with the top halves of logs that have been ripped lengthwise. The logs are left exposed both inside and outside the house in order to gain their full aesthetic benefit on both sides of the wall. This basic system is the same as that used in Lincoln Logs, which were invented (ironically) by Frank Lloyd Wright's son, Jon, in 1916.

FIGURE 20.16
A multimillion dollar log construction industry involves professional designers and builders in the construction of 1000 to 2000 new log houses each year. (*Architect: The Jarvis Group, Ketchum, Idaho, Photo by Fred Lindholm*)

FIGURE 20.17
The loft of this log residence has milled columns and a framed wall that contrast with the log walls and roof structure. (*Architect: The Jarvis Group, Ketchum, Idaho; Photo by Cindy Theide and Jonathan Stoke: Inside Log Homes*)

FIGURE 20.18
The trunks of coniferous trees, long and straight with minimum taper, are best suited for stacked log construction. The bark is typically removed to discourage insect infestations. (*Photo by Sandy Burke*)

FIGURE 20.20
Starting with a half log at the base of parallel walls, logs are stacked on one another to make a wall. Walls perpendicular to the half-log walls are made with a full log at the base. Notches at the corners lock each log in place and allow it to rest on the log below. The top logs in this photo have been positioned but are not yet notched. (*Photo by Rob Thallon*)

FIGURE 20.19
Most new log houses are preconstructed in an industrial yard before being disassembled and shipped to the construction site. (*Photo by Sandy Burke*)

Because of the desire to expose them as primary structure, logs are often used for roof structure as well as for walls. A framework of log beams, trusses, purlins, or log rafters typically supports exposed decking or a conventionally framed roof structure (Figure 20.21).

Chinked Style Versus Full-Scribed Style

There are two principal ways to seal between logs: the chinked style and the full-scribed style. The chinked style is the simpler of the two because it allows the logs to be stacked on one another without any advance treat-ment to make them fit precisely (Figure 20.22).

The space between the logs is later filled with one of many varieties of chinking material. Early North American chinking consisted of mosses, mortar, or other convenient material. Most log walls are now chinked with a flexible synthetic material applied over a split, closed-cell foam backer rod. One benefit of chinking is that electrical wires can be hidden in the chinking. Synthetic chinking has a long performance life because it adheres tightly and flexes with the movement of the logs caused by changes in climatic conditions.

Scribing a log wall involves cutting away the underside of each log so that it is contoured to match the irregularities of the log below (Figure 20.23). The process of scribing requires that each log be put in place in order to mark (scribe) it according to the shape of the log below, and then removed from the wall to cut the wood away to the scribed line. This process slows the construction considerably, but leads to a clean fit without major gaps between the logs. Scribed joints are typically sealed against air infiltration with expanding foam that is injected between the logs from the end after the logs are in place. Scribing originated in Scandinavia and is now more popular in Canada than in the United States. It produces a superior result, but can add 35 percent to the cost of the log package for a new house.

The Unique Characteristics of Log Construction

Because their walls consist entirely of horizontally stacked wood, log houses can exhibit significant settling. Even dry logs (moisture content 15 percent or below) can settle ⅛ inch per foot of wall height (10 mm/m) over the course of the first one or two heating seasons after construction. A green log structure can settle up to eight times

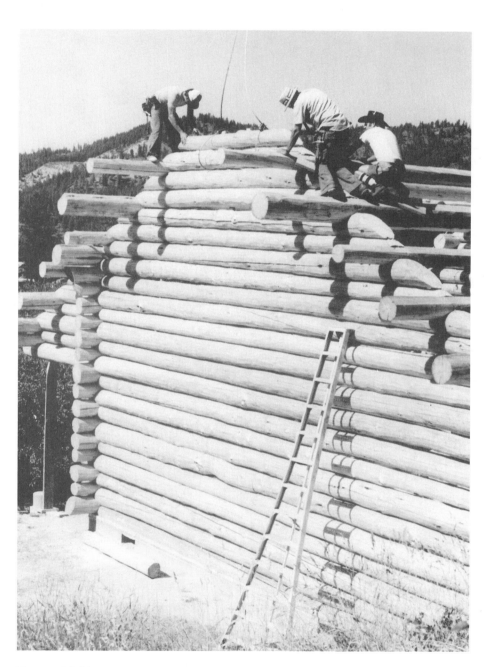

FIGURE 20.21
Log purlins form the principal roof structure of this log house. Following customary practice, openings for windows have not yet been cut into the wall. (*Photo by Sandy Burke*)

FIGURE 20.22
Detail at a corner of logs that will be chinked. Synthetic chinking applied from both sides of the wall will seal the gaps between logs. The "shrink-to-fit" joint is derived from the traditional saddle joint. (*Photo by Sandy Burke*)

FIGURE 20.23
Detail of scribed logs at the corner. A concealed compressible gasket between logs retards air infiltration. (*Photo by Sandy Burke*)

as much. Thus, the walls of a typical single story-log house will shorten vertically from 1 to 8 inches after it has been inhabited. This characteristic calls for special detailing to prevent the stressing of building elements that do not settle such

as windows, doors, columns, and framed walls.

Experienced log builders have a number of strategies to deal with the problems of settling and shrinkage. Loadbearing columns are fitted with screw jacks in the floor and are low-

ered as the logs settle. Windows and doors are not installed into their rough openings until the very end of the construction process when much of the settling can be expected to have occurred. Even then, space is left above installed windows and doors,

PLAN

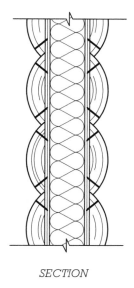

SECTION

FIGURE 20.24
Many people wanting the look of log construction at lower cost are turning to log siding. Log slabs are applied as siding to both the inside and the outside of 2 × 6 framed walls. Corners are made with full log ends.

and trim fitting and installation is delayed until most drying of the logs has taken place.

The exterior side of any wooden structure is exposed to the weather and to destructive insects such as termites, carpenter ants, and powder post beetles. This is especially true for log structures. At the log ends, the soft end grain is exposed and wood fibers are oriented such that they tend to wick moisture into the log. This problem can be minimized by providing generous roof overhangs and by smooth sanding and sealing these ends. As with other wood buildings, care must be taken to keep the wood clear of the ground and insect-inhabited material. Insecticides, mildewcides, and fungicides usually can be applied to the logs at the factory as an option.

Stacked log structures use more wood fiber per area of wall than any other method of construction. While it may be argued that very little energy is consumed in the production of the logs, it is also true that the material is used very inefficiently, especially in green log construction. Standing-dead timber, because it has checked in the drying process, is not suitable as saw logs, and therefore does not represent such a waste of resource. The quadrupling of log prices in recent years, both green and dry, has led to reduced interest in log construction except at the high end of the market. As a result, hybrid log structures that use only a fraction of the wood are becoming popular (Figure 20.24).

The Thermal Performance of Log Construction

Much is said by those promoting log construction about the ease of heating and cooling a log house. Scientific studies do demonstrate the superior capacity of logs to act as thermal mass. But the R-values of log walls do not meet code-mandated standards, and the literally thou-

sands of feet of joints between logs that must be sealed are the Achilles heel of log house construction. Even when great care and skill are used to seal these joints against air infiltration, the constant movement of the logs due to changes in moisture content can lead to deterioration and failure of the seal. Finding the sources of infiltration is difficult due to the extensiveness and hidden nature of the joints. The introduction of flexible chinking systems has mitigated this disadvantage.

Log Houses and the Building Codes

Although building codes do not specifically address the structural values of stacked logs, log houses have a history of acceptance in the culture and tend to be constructed in compact geographical areas where building officials are accustomed to them. In cases where obstreperous local officials need to be convinced that a log residence can satisfy the structural code, an engineer can easily perform this task.

More troublesome than structural concerns is the inability of a log house to meet the energy conservation provisions of the code in the normal manner. Because the commonly required R-19 wall insulation rating cannot be obtained in a typical log wall, calculations must be submitted to the building department demonstrating that insulative values have been boosted in other parts of the building to achieve a satisfactory overall rating. Studies that have shown the thermal mass of log walls to be effective deterrents to heat loss can be used to help convince building officials that log houses comply with building code energy requirements, but the fact is that thermal mass conserves energy only when outdoor temperatures fluctuate widely on a daily basis.

FIGURE 20.25
A straw bale house under construction near Purdum, Nebraska in 1907. (*Courtesy of Out on Bale, (un)Ltd. Tucson, Arizona*)

STRAW BALE CONSTRUCTION

Straw bale construction employs bales that are stacked up as if they were huge masonry units to make walls. The method was first used in the plains of western Nebraska in the 1890s (Figure 20.25). The mechanical baler had been invented several decades earlier, and the pioneers recognized the potential of the bales as an inexpensive alternative to costly imported lumber. In the 50 years between 1890 and 1940, approximately 70 straw bale structures were built in the vicinity of the first experimental house.

Interest in straw bales waned with the onset of World War II but was rekindled in 1973 by a one-page description in the book *Shelter*, which describes low-cost residential construction methods from around the globe. This more recent interest in bales was

FIGURE 20.26
This 3000-square-foot single-story straw bale structure has a simple form with straightforward openings for windows and doors. These features are characteristic of most straw bale buildings. (*Courtesy of Trinity Springs Ltd. Bottling Facility, © 2001 Kirk Anderson*)

FIGURE 20.27
A two-story straw bale house built in California. (*Photo by Kelly Lerner*)

fueled principally by a search for environmentally responsible building materials with high insulation values for energy conservation. A later article in *Fine Homebuilding* magazine further stimulated curiosity, and there has been a groundswell of activity ever since. There are now several thousand straw bale structures in North America with representative buildings in each of the 50 states and all the Canadian territories (Figures 20.26 and 20.27).

The basic building unit is a two-string or a three-string bale of straw (Figure 20.28). Straw is a waste product, the stems of wheat, oats, barley, or other small grains, and is traditionally used only for mulching or for bedding material for farm animals. It is different from hay in that it contains only the shaft of the grain plant from which it is cut, and it has little nutritional content. Straw is therefore not attractive to vermin as food.

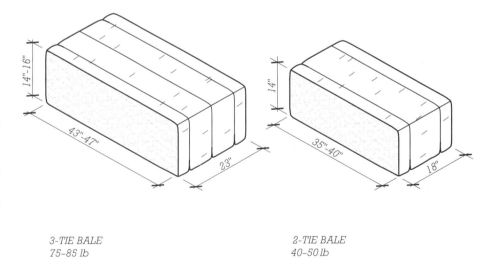

3-TIE BALE
75–85 lb

2-TIE BALE
40–50 lb

FIGURE 20.28
Two-string and three-string bales. The bales can be made shorter in length by threading new strings through the intact bale, wrapping the new strings around the end of the bale to be saved, tightly tying the new strings, and cutting the original strings to release the unused part of the bale. This is a time-consuming process, so experienced straw bale designers and builders do everything possible to use complete bales.

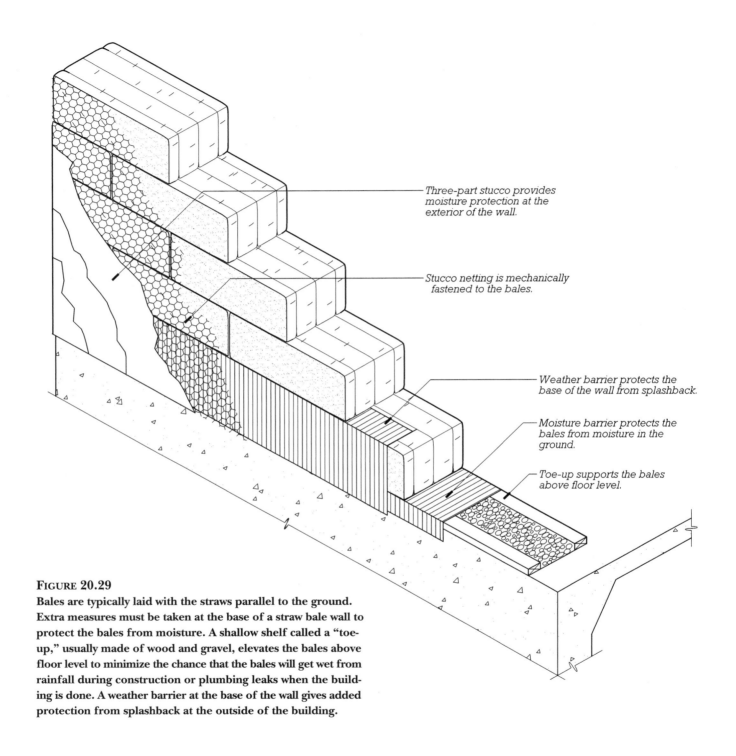

Three-part stucco provides
moisture protection at the
exterior of the wall.

Stucco netting is mechanically
fastened to the bales.

Weather barrier protects the
base of the wall from splashback.

Moisture barrier protects the
bales from moisture in the
ground.

Toe-up supports the bales
above floor level.

FIGURE 20.29
Bales are typically laid with the straws parallel to the ground.
Extra measures must be taken at the base of a straw bale wall to
protect the bales from moisture. A shallow shelf called a "toe-
up," usually made of wood and gravel, elevates the bales above
floor level to minimize the chance that the bales will get wet from
rainfall during construction or plumbing leaks when the build-
ing is done. A weather barrier at the base of the wall gives added
protection from splashback at the outside of the building.

Not all bales are suitable for construction, and straw bale manuals and books expend considerable page space discussing how to find bales that are dry, dense, and of consistent size. The bales are typically laid with the straws parallel to the ground for wall construction because they are stronger in this orientation, the baling strings are protected within the finished wall, and the wall surfaces can be easily notched for utilities without interfering with the strings (Figure 20.29). Interior walls are usually framed with wood because bales take up so much space, and there is no advantage to the insulative capacity of straw at the interior of the house.

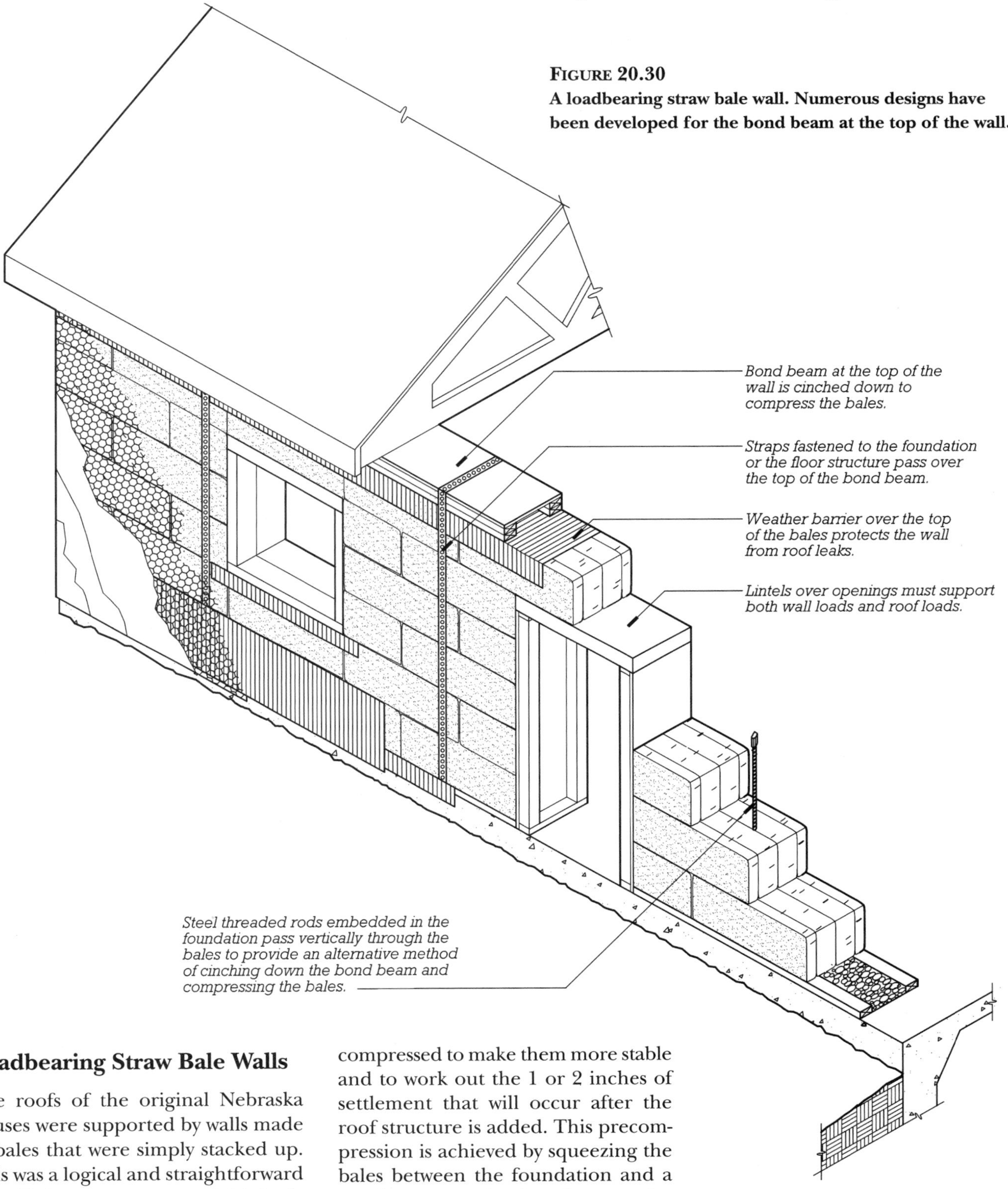

FIGURE 20.30
A loadbearing straw bale wall. Numerous designs have
been developed for the bond beam at the top of the wall.

*Bond beam at the top of the
wall is cinched down to
compress the bales.*

*Straps fastened to the foundation
or the floor structure pass over
the top of the bond beam.*

*Weather barrier over the top
of the bales protects the wall
from roof leaks.*

*Lintels over openings must support
both wall loads and roof loads.*

*Steel threaded rods embedded in the
foundation pass vertically through the
bales to provide an alternative method
of cinching down the bond beam and
compressing the bales.*

Loadbearing Straw Bale Walls

The roofs of the original Nebraska houses were supported by walls made of bales that were simply stacked up. This was a logical and straightforward use of bales as structure, and this basic strategy is still employed in modern straw bale houses. Bales are stacked in a running bond pattern, which ties them together structurally. To increase the continuity and strength of the walls, however, they are generally pre-compressed to make them more stable and to work out the 1 or 2 inches of settlement that will occur after the roof structure is added. This precompression is achieved by squeezing the bales between the foundation and a continuous wooden bondbeam at the top of the wall. The foundation and bondbeam are connected either by threaded rods running through the bales or with straps made of wire mesh or wood at both faces of the bales (Figures 20.30 and 20.31).

FIGURE 20.31 Workers strapping a bond beam to the top of a loadbearing straw bale wall. (*Photo copyright David Eisenberg*)

Infill Straw Bale Walls

Most straw bale structures are now constructed with an independent wooden frame that supports the roof with the bales used merely as infill (Figure 20.32). The benefits of this approach are that code approval is usually simpler, loads can be greater, and greater design flexibility can be achieved, including more wall openings and the addition of second stories. The wooden frame can be made with a post-and-beam structure composed of conventional 4×4 or 6×6 members, but it is more frequently made with flat box columns that are composed of 2×4s and plywood and extend the full thickness of the wall (Figures 20.33 and 20.34). The amount of lumber used to make this vertical structure often does not exceed that which would be used to make the bond beam at the top of a loadbearing bale wall. There are also experiments in progress for infill straw bale walls that use metal, concrete, con-

FIGURE 20.32
An infill straw bale house using a post-and-beam structure to support the roof. The straw bales provide only insulation and enclosure in this type of structure. (*Photo by Kelly Lerner*)

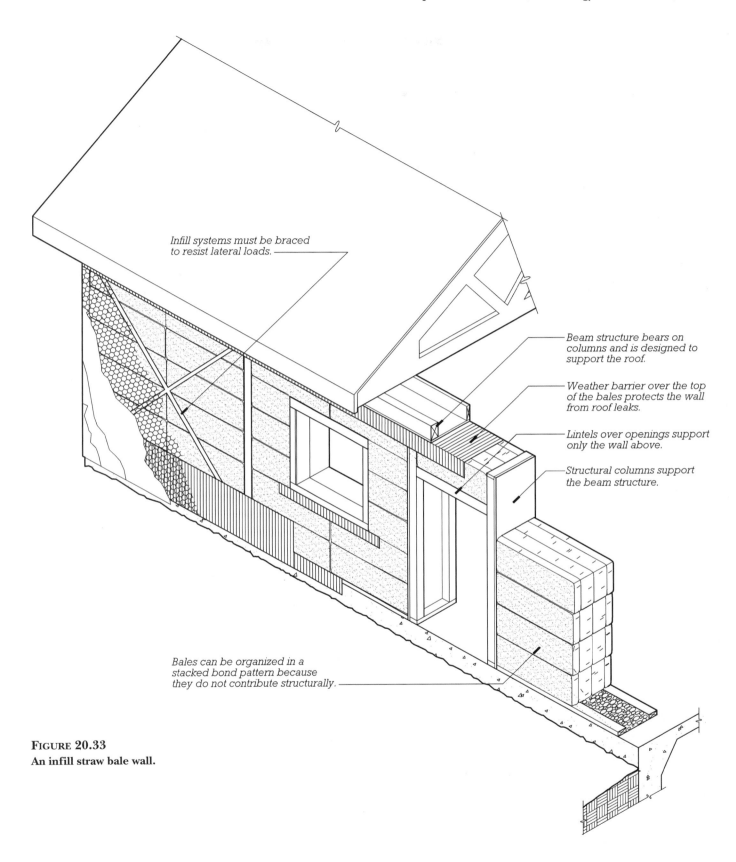

Infill systems must be braced
to resist lateral loads.

Beam structure bears on
columns and is designed to
support the roof.

Weather barrier over the top
of the bales protects the wall
from roof leaks.

Lintels over openings support
only the wall above.

Structural columns support
the beam structure.

Bales can be organized in a
stacked bond pattern because
they do not contribute structurally.

FIGURE 20.33
An infill straw bale wall.

FIGURE 20.34
Framing for an infill straw bale house. The wall at the right has box columns framed around a window. The wall at the left is being framed conventionally with 2 × 6 studs. Compare the amounts of lumber in the two types of wall. (*Photo by James McDonald*)

FIGURE 20.35
An infill straw bale wall is ready for its stucco finish. The steel crossbracing between the door and the window provides resistance to lateral forces along the plane of the wall. (*Photo by Rob Thallon*)

FIGURE 20.36
The window positioned at the outside face of the straw bale wall allows a window seat to be located within the thickness of the wall. (*Photo by Kelly Lerner*)

FIGURE 20.37
Wire lath is pinned to the straw bales
to improve the adherence of stucco.
(*Photo by Kelly Lerner*)

FIGURE 20.38
Stucco work is beginning in the gable of this straw bale house. The weather barriers at
the tops and bottoms of the bale walls provide extra protection against unwanted
moisture. The tarp over the roof adds protection from rain during construction. Straw
bale construction projects tend to attract numerous volunteer workers and onlookers
because of the novelty of the construction system. (*Photo by Kelly Lerner*)

crete block, or wooden trusses for columns. Infill straw bale walls must be braced to resist lateral loads, and diagonal metal straps attached to the structural frame are typically used for this purpose (Figures 20.33 and 20.35).

Window and Door Openings

Openings for windows and doors in straw bale walls present interesting challenges. Because of the thickness of the bale walls (18 to 23 inches), a decision must be made to locate the window or door at the exterior, the interior, or the center of the wall. Most windows are located at the exterior, giving the interior a wide stool at the base of the window and wide reflective surfaces at the sides and top (Figure 20.36). Doors can be installed at or near the interior face of the wall to give them more protection from the weather.

To create a rigid, regular surface for the attachment of windows and doors, rough openings are framed with

wood. In loadbearing walls, a wooden frame is pinned to the bales with dowels. For infill walls, the wooden structural frame provides the rough opening (Figure 20.34). Openings in load-bearing bale walls must have a structural header capable of supporting roof loads, while openings in infill walls need to provide support only for the dead load of the bales directly above them.

Finishing Straw Bale Walls

The protection of straw bale walls from moisture is critical because the straw will decay and collapse if it is wetted for a significant period of time. It is wise to build a broad roof overhang that will keep rainfall from striking the walls directly.

A wall finish carefully designed and constructed to prevent the intrusion of water is essential. The typical exterior finish for a straw bale wall is three-coat portland cement–based stucco. The stucco may be applied

directly to the bales, but is more typically applied over stucco netting that is wired or pinned to the bale walls (Figure 20.37). A weather barrier beneath the stucco netting at the base of the walls can be added for insurance against the intrusion of wind-driven rain or snow, but the inclusion in the wall assembly of any membrane must be measured against the need for the wall to dissipate any moisture that enters it (Figure 20.38).

The interior surfaces of bale walls do not need protection from the weather. They are usually coated with gypsum plaster rather than stucco.

The irregularity of a typical bale wall consumes up to 70 percent more stucco or plaster than a smoothly framed wall, so bale walls are often filled with a mixture of mud and straw before the finish is applied. The stucco and plaster contribute significantly to the strength of the walls. When stucco is shot onto the walls (gunite), strength is increased even more.

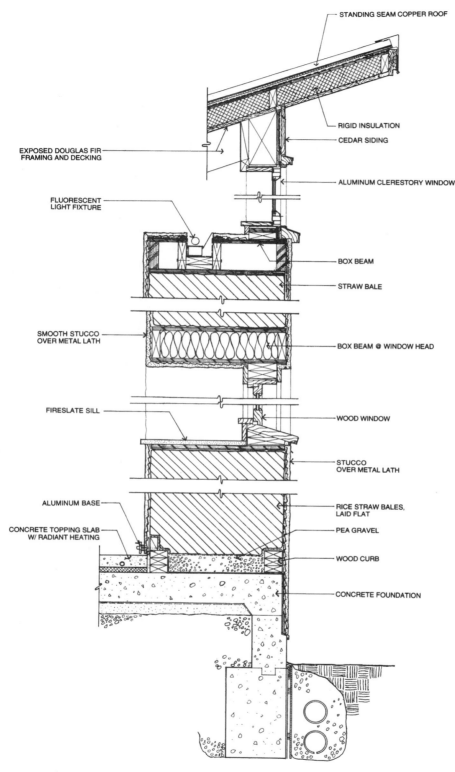

STANDING SEAM COPPER ROOF

RIGID INSULATION

CEDAR SIDING

EXPOSED DOUGLAS FIR FRAMING AND DECKING

ALUMINUM CLERESTORY WINDOW

FLUORESCENT LIGHT FIXTURE

BOX BEAM

STRAW BALE

SMOOTH STUCCO OVER METAL LATH

BOX BEAM @ WINDOW HEAD

FIRESLATE SILL

WOOD WINDOW

STUCCO OVER METAL LATH

ALUMINUM BASE

RICE STRAW BALES, LAID FLAT

CONCRETE TOPPING SLAB W/ RADIANT HEATING

PEA GRAVEL

WOOD CURB

CONCRETE FOUNDATION

FIGURE 20.39
A section detail of a straw bale house. (*Courtesy of Fernau & Hartman Architects, Inc., Berkeley, California*)

Advantages and Disadvantages of Straw Bale Construction

The main advantages of straw bale construction relate to its use as a "green" building material. The use of straw for construction decreases the need to dispose of it by burning or composting, thus reducing the amount of smoke and methane in the atmosphere. The straw bales have minimal embodied energy, especially if produced locally. Their insulative value is high—R-33 or more for a 23-inch wall. In addition to these environmental advantages, straw bale construction produces a richly textured surface and deep window and door reveals, which most people consider to be aesthetic benefits.

The principal disadvantage of straw bale construction is its vulnerability to moisture, which can cause the formation of mold and mildew or, in the worst case, can lead to the decay and complete disintegration of the straw. Concerns about the flammability of bale construction are unfounded: Straw bale structures have survived catastrophic fires, and the results of fire tests have shown straw bale construction to be superior to wood frame construction with regard to combustibility. Termites and rodents appear not to be a problem because the straw is not attractive for either to eat, and there are fewer voids for nesting than in wooden structures.

Straw Bale Construction and the Building Codes

The national model codes do not recognize straw bale construction as an approved building method. Some jurisdictions in the southwestern United States, however, have adopted prescriptive codes for straw bale construction. Pima County, Arizona, was the first to adopt such a code in 1997; the state of New Mexico soon followed with a post-and-beam straw bale code. Various counties in western states have adopted amended versions of these codes. In other locations, approval depends on the acceptance of straw bales as an alternative material.

SELECTED REFERENCES

1. Elizabeth, Lynne, and Adams, Cassandra, Eds. *Alternative Construction: Contemporary Natural Building Methods.* New York: John Wiley & Sons, 2000.

A profusely illustrated book of global low-tech building methods including adobe, rammed earth, straw bale, and many more. The history and logic of each system are explained, and applications are documented in case studies. An excellent starting place for someone wanting an overview of alternative building methods.

2. Kahn, Lloyd. *Shelter.* Bolinas, CA: Shelter Publications, 1973.

This covers the entire spectrum of alternative building methods. A superficial survey, but inspiring!

3. Romero, Orlando, and Larkin, David. *Adobe: Building and Living with Earth.* Boston: Houghton Mifflin, 1994.

A book of handsome color photographs and line drawings that documents the history, technology, and beauty of adobe construction.

4. Easton, David. *The Rammed Earth House.* White River Junction, VT: Chelsea Green, 1996.

A beautiful and inspirational book in which every aspect of rammed-earth construction is discussed—from history to soil identification. Practical step-by-step construction techniques are also included.

5. Phelps, Herman. *The Craft of Log Building.* Ottawa: Lee Valley Tools, 1982.

The author of this book uses historical styles to illustrate the craft of joinery in log construction. Translated from the German and profusely illustrated with three-dimensional drawings.

6. Myhrman, Matts, and MacDonald, S. O. *Build It with Bales,* Version 2. Tucson: Out On Bale, 1997.

A chatty, humorous, and very informative illustrated text covering everything you would want to know (and a little bit more) about straw bale construction.

7. Magwood, Chris, and Mack, Peter. *Straw Bale Building.* Gabriola Island, BC: New Society Publishers, 2001.

One of a half-dozen or so good references for straw bale builders, but the only one written for straw bale builders in cool, moist, northern climates.

KEY TERMS AND CONCEPTS

earthen construction
adobe
adobe brick
asphalt-modified adobe
rammed earth
rammed-earth stem wall
rammed-earth compaction

log construction
preconstruction
chinked style
chinking
scribed style
scribing
settling of log walls

thermal performance of logs
straw bale construction
two-string bale
three-string bale
loadbearing straw bale wall
infill straw bale wall

REVIEW QUESTIONS

1. Compare the advantages and disadvantages of adobe and rammed-earth construction. Is either of these systems prevalent in your area? Why or why not?

2. What are the two basic styles of stacked log construction? Discuss these two styles in terms of their advantages and disadvantages.

3. Explain the difficulty with using a column incorporated into stacked log construction.

4. Compare the two basic structural types of straw bale walls. Do you imagine the differences can be seen once the walls are finished?

5. What measures are typically taken to keep moisture out of a straw bale wall?

EXERCISES

1. Contact a local contractor, designer, or building inspector and inquire as to whether there has been any recent construction activity using the alternative materials outlined in this chapter. If so, try to visit the site and interview the owner about his or her philosophical reasons for choosing one of these alternative building methods. Interview the builder (who is often the owner) about difficulties and delights with the construction process. What would be done differently if the project were repeated?

2. Following the example of the three little pigs, design a simple one-room cottage using first straw bales, then stacked logs, and, finally, adobe bricks. What adjustments do you need to make to the design as it is reinterpreted from one material to the next? Which design do you imagine would withstand the highest wind force?

LOADBEARING MASONRY AND CONCRETE CONSTRUCTION

For the past century, masonry and concrete have been used extensively to construct perimeter foundations for residences in North America. They are uniquely suited for this purpose because they are strong, durable, and virtually unaffected by moisture. When extra strength is required, both masonry and concrete can be reinforced with steel.

Above grade, loadbearing masonry and concrete walls may be used for the exterior walls of a house, with interior partitions made of frame construction (Figure 21.1). The strength, aesthetic effect, and weather resistance of masonry and concrete are exploited in the exterior walls, with thinner, space-saving frame walls inside.

In two-story houses, the designer must keep in mind that the wood framing at the core of the house is subject to settling due to changes in moisture content, whereas the masonry or concrete perimeter walls are not. It is especially important in this situation to ensure that interior supports do not contain significant amounts of cross-grain wood because this can lead to floors that slope. The common solution is to use balloon framing or continuous columns upon which the floor and ceiling framing is hung. Another solution is to use manufactured I-joists, which shrink scarcely at all, for floor framing. Floor and wall framing of light-gauge steel eliminates the shrinkage problem and allows platform framing to be used.

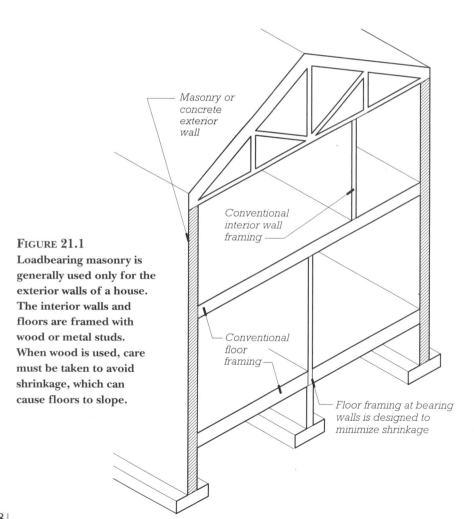

Masonry or concrete exterior wall

Conventional interior wall framing

Conventional floor framing

Floor framing at bearing walls is designed to minimize shrinkage

FIGURE 21.1
Loadbearing masonry is generally used only for the exterior walls of a house. The interior walls and floors are framed with wood or metal studs. When wood is used, care must be taken to avoid shrinkage, which can cause floors to slope.

LOADBEARING MASONRY

Loadbearing Masonry

Loadbearing masonry is one of the most ancient systems of construction. Worldwide, it remains one of the most popular methods of house construction. In North America, houses with stone or brick loadbearing walls were very common until the early 1900s. By the middle of the 20th century, CMU construction had almost entirely replaced brick for structural walls, and the costlier brick wythe came to serve only as a thin veneer over a CMU or wood light frame structural wall.

Today, both brick and CMU are finding increased use as loadbearing walls for residences as people recognize the advantages of masonry over frame construction. Masonry is often chosen for its qualities of permanence, solidity, and fire resistance; for its unique colors, textures, and patterns; and for its association with beautiful buildings of the past (Figures 21.2 and 21.3).

Masonry, like wood light frame construction, is a construction process that is carried out largely at the building site with small, relatively inexpen-

FIGURE 21.2
Loadbearing masonry walls made of manufactured masonry units can simulate traditional stone walls. (*Photographs from National Concrete Masonry Association* (*NCMA*) *publication* **Residential Technology: Volume 2, [*RESTEK 2*])**

FIGURE 21.3
Manufactured masonry units are rectangular and thus are most readily assembled into rectangular forms. An assemblage of simple rectangular forms can produce a functional and pleasing residence. (*Photographs from National Concrete Masonry Association* (*NCMA*) *publication* **Residential Technology: Volume 2, [*RESTEK 2*])**

sive tools and machines. It does not require an extensive period of preparation or fabrication in advance of the beginning of construction. However, construction can be delayed during periods of very hot, very cold, or very wet weather, which interfere with the proper curing of the mortar.

Reinforced Masonry Walls

Loadbearing walls may be built with or without reinforcing. Unreinforced masonry walls cannot carry such high stresses as reinforced walls and are unsuitable for use in regions with high seismic risk, but they are com-

mon in many other regions. Unreinforced walls must be thicker than reinforced ones in order to resist buckling under the lateral forces of wind and other dynamic loads. With steel reinforcing, the wall achieves tensile strength as well as compressive strength, and the minimum required thickness is less than that of an unreinforced wall.

In residential construction, *reinforced brick loadbearing walls* are usually made with a single wythe of bricks that are slightly thicker than normal with an open core to accommodate vertical reinforcing steel (Figures 21.4 and 21.5). Horizontal reinforcement is achieved with welded wire joint reinforcement. Tall brick walls must be temporarily braced during construction but are ultimately stabilized by floor and roof diaphragms that are bolted to them (Figure 21.6).

CMU loadbearing walls are much more prevalent than brick because they are less expensive to construct and easier to insulate (Figure 21.7). Both of these advantages stem primarily from the larger size of a CMU, which vastly reduces the number of units that the mason must handle and which allows for larger hollow cores to receive insulation. The height of a CMU unit allows horizontal reinforcement to be achieved either with wire joint reinforcing or with reinforcing bars grouted into a bond beam (Figure 21.8).

For both brick and CMU walls, the number, locations, and sizes of the steel reinforcing bars are determined by the structural engineer. The engineering design of masonry loadbearing walls is governed by ACI 530/ASCE 5, a standard established jointly by the American Concrete Institute and the American Society of Civil Engineers. This document establishes quality standards for masonry units, reinforcing materials, metal ties and accessories, grout, and masonry construction. It also sets forth the procedures by which the strength and stiffness of masonry structural elements are calculated.

FIGURE 21.4
Section of a 4-inch brick bearing wall. The ground floor can be framed with wood like the upper floor if a crawlspace or basement is desired.

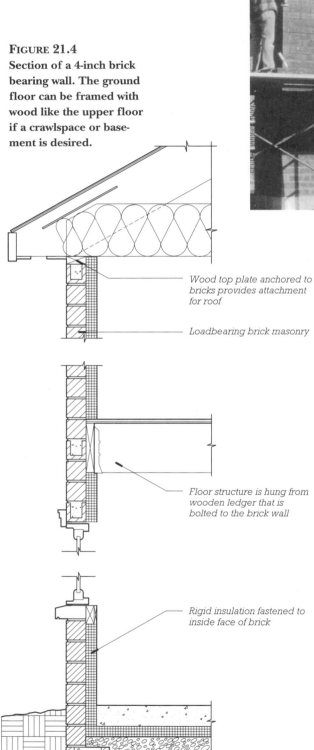

Wood top plate anchored to bricks provides attachment for roof

Loadbearing brick masonry

Floor structure is hung from wooden ledger that is bolted to the brick wall

Rigid insulation fastened to inside face of brick

FIGURE 21.5
A single-wythe brick loadbearing wall with vertical rebar projecting from the top of the wall. The masons are working at the level of the second floor. (*From* Masonry Concept Home *"A Brick Bearing Wall" by Walter A. Laska*)

FIGURE 21.6
A wood ledger attached with anchor bolts to a 4-inch loadbearing brick wall. Metal hangers on the ledger support joists for a structural panel subfloor, which will stabilize the walls. (*From* Masonry Concept Home *"4 Brick Bearing Wall" by Walter A. Laska*)

FIGURE 21.7
A typical section of an 8-inch CMU residential wall. The wall can also be insulated on the inside surface or within the hollow cores of the CMU.

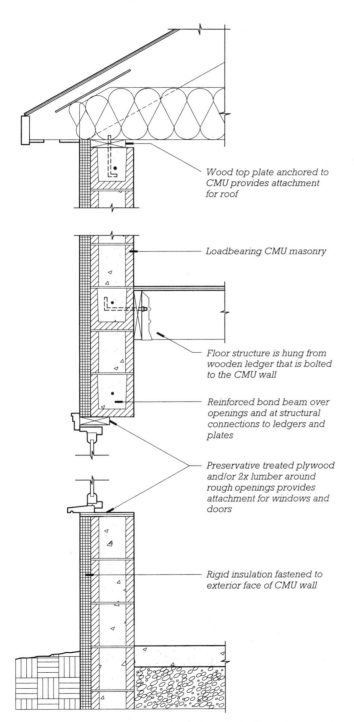

Wood top plate anchored to CMU provides attachment for roof

Loadbearing CMU masonry

Floor structure is hung from wooden ledger that is bolted to the CMU wall

Reinforced bond beam over openings and at structural connections to ledgers and plates

Preservative treated plywood and/or 2x lumber around rough openings provides attachment for windows and doors

Rigid insulation fastened to exterior face of CMU wall

FIGURE 21.8
The size of the masonry units allows CMU walls to be constructed efficiently and to be reinforced easily in both the vertical and the horizontal directions. Notice the vertical rebar projecting from the unfinished chimney structure. (*Photo by Charles Menefee*)

Thermal Insulation of Masonry Walls

A solid masonry wall is a good conductor of heat, which is another way of saying that it is a poor insulator. In hot, dry climates, the capacity of an uninsulated masonry wall to store heat is effective in keeping the inside of the building cool during the hot day and warm during the cold night, but in climates with sustained cold or hot seasons, measures must be taken to improve the insulating qualities of masonry walls. There are three general ways of insulating masonry walls: on the outside face, within the wall, and on the inside face.

Insulation on the outside face is a relatively recent development. It is usually accomplished by means of an exterior insulation and finish system (EIFS), which consists of panels of plastic foam that are adhered to the masonry and covered with a thin, continuous layer of polymeric stucco reinforced with glass fiber mesh (Figure 21.9). The appearance is that of a stucco building. The masonry is completely concealed and can be of inexpensive materials and workmanship. An advantage of an EIFS is that the masonry is protected from temperature extremes and can function effectively to stabilize the interior temperature of the building. Disadvantages are that the thin synthetic stucco coatings are usually not very resistant to denting or penetration damage and that the EIFS is combustible. Additionally, an EIFS that does not incorporate a fully drained and ventilated cavity between the insulation panels and the masonry is prone to water leakage.

Exterior insulation of a masonry wall can also be accomplished with rigid insulation set between furring strips (Figure 21.10). With this system, virtually any siding can be attached to the masonry walls while retaining the thermal advantages of interior masonry.

Insulation within the wall can take several forms. The hollow cores of a concrete block wall can be filled with loose granular insulation or with proprietary systems of molded-to-fit liners of foam plastic (Figure 21.11). Insulating the cores of concrete blocks allows both faces of the block to be exposed, simplifying the finishing process. Insulating the cores does not retard the passage of heat through the webs of the blocks, however, and is most effective when it is coupled with an unbroken layer of insulation on the interior or exterior face of the wall.

The inside surface of a masonry wall can be insulated by adhering slabs of plastic foam to the wall and applying plaster directly to the foam. More usual is the practice of attach-

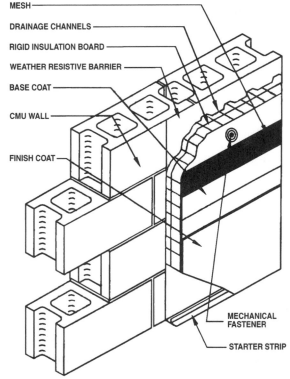

MESH
DRAINAGE CHANNELS
RIGID INSULATION BOARD
WEATHER RESISTIVE BARRIER
BASE COAT
CMU WALL
FINISH COAT
MECHANICAL FASTENER
STARTER STRIP

FIGURE 21.9
Water-managed EIFS applied to the exterior of CMU wall. (*From* Concrete Masonry Homes: Recommended Practices, *U.S. Department of Housing and Urban Development, Office of Policy Development and Research, Washington, D.C., 1999*)

FIGURE 21.10
Rigid insulation is being applied to the exterior of this masonry residence. Wooden furring strips between the insulating panels allow virtually any siding material to be attached to the wall. (*Photographs from National Concrete Masonry Association (NCMA) publication* Residential Technology: Volume 2, [*RESTEK 2*])

FIGURE 21.11
This proprietary concrete masonry system is designed with polystyrene foam inserts that provide a high degree of thermal insulation. The webs of the blocks are minimal in area to reduce thermal bridging where they pass through the foam. Vertical reinforcing bars can be placed as shown, and horizontal bars can be laid in the grooved webs. The cores must be grouted where reinforcing is used. (*Courtesy of Korfil, Inc., West Brookfield, Massachusetts*)

ing wood or metal furring strips to the inside of the wall with masonry nails or powder-driven metal fasteners (Figure 21.13). Furring strips are analogous to wood or metal studs and may be of any desired depth to house the necessary thickness of fibrous or foam insulation. Furring strips can also solve another chronic problem of masonry construction by creating a space in which electrical wiring and plumbing can easily be concealed.

Detailing Masonry Walls

Every masonry wall is porous to some degree. Some water will find its way through even the most carefully constructed new masonry if the wall is wetted for a sustained period. The most practical way to prevent the passage of water through exposed structural masonry, therefore, is to protect the surface of the masonry with wide roof overhangs. When single-wythe structural masonry must be exposed to the weather, it is important to provide high-quality flashing at parapets

FIGURE 21.12
Foamed-in-place insulation may be injected into the cavities of standard CMU walls to increase thermal performance. Excess foam is trimmed from the top of the wall before the top plate is attached. (*Photographs from National Concrete Masonry Association* (*NCMA*) *publication* Residential Technology: Volume 2, [*RESTEK 2*])

FIGURE 21.13
Foam plastic insulation on the interior of a concrete masonry wall with wooden furring strips between the insulating panels. There are also patented systems of metal furring strips in which only isolated clips contact the masonry wall so as to minimize thermal conduction through the metal. The furring strips serve as a base to which interior finish panels such as gypsum board can be attached. (*Photographs from National Concrete Masonry Association* (*NCMA*) *publication* Residential Technology: Volume 2, [*RESTEK 2*])

and wall openings and to protect the exposed surface of the wall with a moisture-resistant paint or sealer.

Paints and sealers can be very effective at keeping moisture from penetrating masonry walls, but most have a relatively short life expectancy and must be reapplied regularly. There are two basic categories for water-resistant masonry coatings: those that are opaque and mask the color of the masonry with a color of their own, and those that are transparent and allow the color and texture of the masonry to be seen. Transparent coatings generally have a shorter effective life (3 to 5 years) than opaque coatings (up to 10 years).

Loadbearing Masonry and the Building Codes

The International Residential Code contains a section that sets prescriptive standards for masonry walls of all types. By following these standards, the design of a masonry residence need not bear the seal of a licensed architect or engineer. The code specifies, for example, that loadbearing masonry walls for single-story buildings must be at least 6 inches (152 mm) thick, and walls more than one story high must be 8 inches (203 mm) thick. Anchoring of the roof and floor structures to the masonry walls and minimum reinforcing of the walls are also designated in the code.

INSULATING CONCRETE FORMS

Insulating concrete forms (ICFs) replace the wood or metal formwork required for conventional concrete walls and stay in place after the concrete is poured to provide an insulating layer on both sides of the reinforced concrete wall. ICFs are typically stacked like concrete masonry units and are held in place with adhesive or by interlocking with one another (Figures 21.14 and 21.15). Virtually all ICFs

are made of polystyrene (either EPS or XPS) and thus are very easily cut and carved with a simple hand saw, power saw, or hot wire to make openings and to accommodate electrical conduits and plumbing (Figure 21.17). Because polystyrene is so soft and degrades in ultraviolet light, ICF walls must be protected with other materials at both indoor and outdoor surfaces. Typically, ICF walls are finished with synthetic stucco on the exterior and covered with drywall on the interior, although most ICF units include *integral metal furring strips* to accept screws, thus allowing either side to be covered with virtually any finish material.

ICF systems have been commonly available in North America for only the

past 10 years, and at first were used primarily for foundations (Figure 21.18). Now ICF is commonly used for the above-grade walls of the building as well. Like loadbearing masonry, ICF systems tend to be used only for exterior walls, while interior walls, floors, and roof are usually framed using standard light frame construction (Figure 21.19).

Numerous companies manufacture and market ICF systems in a wide range of sizes and methods of connection. The smallest ICF units measure approximately 10 inches tall by 3 feet long (250 by 915 mm) and the largest are about 30 inches tall by 10 feet long (0.75 by 3.0 m). The thickness of the units typically ranges from

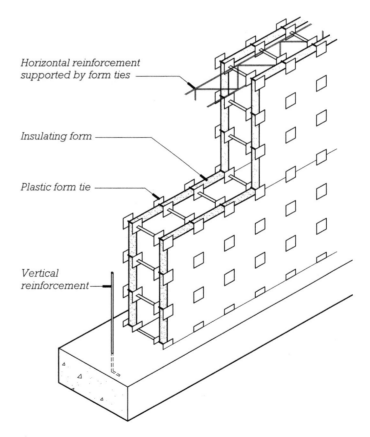

FIGURE 21.14

An example of insulating concrete forms on a concrete footing. The inner and outer form walls are spaced with plastic form ties. The ties resist the outward pressure of the wet concrete as it is poured and stay in place after the concrete is cured to provide screw attachment for finish surfaces. Forms can be quite large due to their very light weight.

FIGURE 21.16
The first two courses of ICF have been positioned on the footing. The forms will be aligned and braced when they reach their full height so that they remain in the proper position when the concrete is poured. (*Courtesy of Arxx Building Products*)

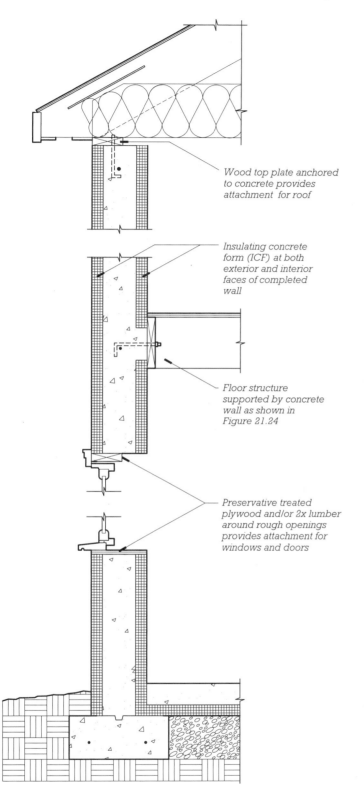

Wood top plate anchored to concrete provides attachment for roof

Insulating concrete form (ICF) at both exterior and interior faces of completed wall

Floor structure supported by concrete wall as shown in Figure 21.24

Preservative treated plywood and/or 2x lumber around rough openings provides attachment for windows and doors

FIGURE 21.17
Insulating concrete forms are easily adjusted at the site with simple hand tools. (*Photo by Rob Thallon*)

FIGURE 21.15
Section of a two-story wall made with insulating concrete forms. Finish materials have yet to be applied to either exterior or interior surfaces. The ground floor may be framed similar to the upper floor if a crawlspace or basement is desired.

FIGURE 21.18
Early experimental uses of ICF in North America were typically limited to foundations where insulation was required. (*Photo by Rob Thallon*)

FIGURE 21.19
As experience and confidence with ICF systems have increased, they have found acceptance for construction of exterior walls of new residences. Both the simple house on the left and the larger house on the right of this view are built with ICF systems. (*Photo by James McDonald*)

FIGURE 21.20
These forms illustrate two basic approaches taken by the manufacturers of ICF systems. The two forms on the left are monolithic, and different sized blocks are required to achieve a variety of form widths. The two forms on the right are made in different widths by using different length spacers between outer insulative panels that are identical. (*Photo by Rob Thallon*)

9 to 12 inches (230 to 300 mm), depending on the manufacturer. Some companies manufacture forms that can be varied in thickness by connecting the two sides with various lengths of rigid plastic ties (Figure 21.20).

The Insulating Concrete Form Construction Process

An ICF wall is constructed on a concrete footing (either a spread footing or a turned-down slab). Vertical rebar placed in the footing to align with openings in the forms ties the ICF wall to the footing (Figure 21.21). The first course of forms is held in place on the footing either with adhesive or by placing it in the wet concrete of the footing before it has set. Subsequent courses are stacked on top of one another in the same fashion as CMU until the top of the wall is reached.

Openings for doors and windows are made by omitting the ICF blocks in these areas and forming the perimeter of the rough opening with preservative-treated wood members. The wood form prevents concrete from flowing into the opening and is left in place to provide a nailing surface for the attachment of windows and doors (Figure 21.22). Corners are typically made with standardized form components, but some systems require mitering and gluing of corners.

After the forms have been stacked to the top of the wall, the next step is to complete the placement of the reinforcing steel within the forms. All ICF systems allow vertical reinforcing steel to be inserted into the forms from the top after the entire wall is in place, but the method and timing of placing horizontal reinforcement vary considerably from

manufacturer to manufacturer. Most ICF systems use forms with an open channel at the top of each course, allowing horizontal reinforcing bars to be laid in place as the forms are being set in place. Other systems have horizontal channels through the center of the forms, requiring that horizontal reinforcing be installed after the forms are in place.

Once the reinforcing steel is in place, the forms are aligned, braced sturdily with wood walers and diagonals, and filled with grout. The grout is typically a rich (3000 to 3500 psi) concrete mixture with a ⅜-inch pea gravel aggregate and a high (5 inch) slump. This mix is designed to flow horizontally without the need for vibration, completely fill the chambers in the forms, and totally encase all of the reinforcing steel. It is pumped into the open tops of forms up to 12

FIGURE 21.21
Two workers laid up this ICF work in about 2 hours. Much of the time was spent connecting vertical rebar to rebar projecting from the concrete footing. (*Photo by Rob Thallon*)

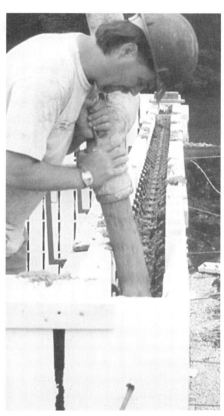

FIGURE 21.23
Concrete being pumped into reinforced ICF. (*Courtesy of Arxx Building Products*)

FIGURE 21.22
A wooden frame inserted into the ICF wall will form a window rough opening. (*Courtesy of Arxx Building Products*)

feet (3.6 m) tall, but it is allowed to accumulate in lifts of no more than 4 feet (1.2 m) in order to minimize the danger of blowouts from extreme fluid pressure at the base of the forms (Figure 21.23). Subsequent lifts can be added after the lower lift has begun to set (usually about a half-hour after it is poured). When the forms are completely filled, excess concrete is struck off to form a level surface even with the top of the forms, and anchor bolts are inserted into the soft concrete. A preservative-treated wooden sill for connecting to the roof structure will be later held in place by the anchor bolts. In a day or two, when the grout has fully set, the temporary braces are removed, and work crews can proceed with the framing of floors, interior walls, stairs, and roof.

For houses with one or more framed floors, ICF walls are typically formed only to the level of each floor before they are grouted. This requires two or three trips for the pumper truck, but allows the floor framing to stabilize the erected walls as well as form a platform for the erection of the ICF walls above floor level.

Framed floors are typically supported by a wood ledger connected to the inner side of the insulated concrete wall with common anchor bolts. The ledger cannot bear on the insulation, so holes are cut into the insulation at frequent intervals to allow the concrete to extend to the face of the ledger (Figure 21.24). A similar strategy is typically recommended for connecting interior partition walls to the ICF system.

Channels for electrical wiring and plumbing can be cut into the surface of the ICF walls. The insulation is sufficiently thick to accommodate the depth of most electrical boxes, which are glued in place after a pocket has been cut to receive them. The cables between boxes are pushed into grooves cut into the insulation. The holes and grooves can be filled with expandable liquid foam if necessary.

ICF walls are typically finished on the exterior with synthetic stucco because it can be applied directly to the surface of the insulating forms. The surface of the forms must usually be smoothed somewhat before stucco or siding material can be applied, and this is accomplished by "sanding" with a large open-mesh screen. Other siding materials may be attached by using furring strips adhered to the forms or connected to the concrete with powder-driven fasteners. Many ICF systems incorporate integral furring strips to which finishes can be screwed.

The interior surface of ICF walls is commonly covered with gypsum board so that all walls of interior rooms can be finished entirely with the same material. The gypsum board is attached to strips incorporated in the ICF system or to furring strips of wood or light-gauge metal that are fastened to the insulated wall with powder-driven fasteners.

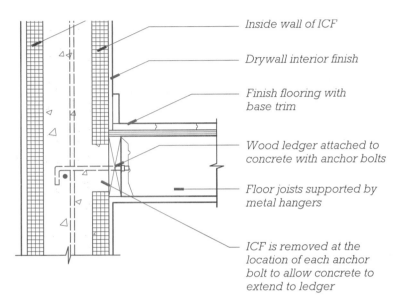

Inside wall of ICF

Drywall interior finish

Finish flooring with base trim

Wood ledger attached to concrete with anchor bolts

Floor joists supported by metal hangers

ICF is removed at the location of each anchor bolt to allow concrete to extend to ledger

FIGURE 21.24
A ledger to support floor joists is connected to an ICF wall with anchor bolts. The insulating form is cut away from around each bolt so that the poured concrete will extend to the surface of the ledger.

Insulating Concrete Forms and the Building Codes

ICF construction has grown so rapidly that it has outpaced the building codes, and many ICF houses are now built even though there is no reference to the system in the International Residential Code. Prescriptive tables prepared by the NAHB for seismic zones 0, 1, and 2 are accepted as standards by most code jurisdictions. In seismic zones 3 and 4, site-specific engineering for ICF residences is currently required. ICF prescriptive tables will likely be incorporated into the International Residential Code beginning in the 2001 edition.

FIGURE 21.25
From the street, an ICF house is indistinguishable from a typical wood-framed house. (*Courtesy of Arxx Building Products*)

```
C.S.I./C.S.C.
MasterFormat Section Numbers for
03300          CAST-IN-PLACE CONCRETE
03310          Structural Concrete
03330          Architectural Concrete
03350          Concrete Finishing
03390          Concrete Curing
04800          MASONRY ASSEMBLIES
04810          Unit Masonry Assemblies
04820          Reinforced Unit Masonry Assemblies
04830          Nonreinforced Unit Masonry Assemblies
```

SELECTED REFERENCES

1. There are several references listed in Chapter 5 that relate to the topic of masonry loadbearing walls.

2. Vanderwerf, Pieter A. et al. *Insulating Concrete Forms for Residential Design and Construction.* New York: McGraw-Hill, 1997.

KEY TERMS AND CONCEPTS

loadbearing masonry
unreinforced masonry wall
reinforced brick loadbearing wall
CMU loadbearing wall
framed floors and masonry walls
thermal insulation of masonry walls

exterior insulation
insulation within the wall
interior insulation
moisture protection of masonry walls
insulating concrete form (ICF)

integral furring strip
openings in ICF walls
reinforcement of ICF walls
framed floors and ICF walls
exterior finish of ICF walls
interior finish of ICF walls

REVIEW QUESTIONS

1. Where should flashings be installed in a masonry wall? What is the function of the flashing in each of these locations?

2. What are the various ways of insulating a loadbearing masonry wall? How do these ways depend on the type of masonry construction being employed?

3. Why is balloon framing used rather than platform framing in conjunction with masonry and concrete loadbearing construction?

4. What precautions should be taken when constructing masonry walls in Minneapolis in the winter?

5. What are the advantages and disadvantages of very large ICF units as compared to smaller units?

6. Why are ICF walls rarely formed and poured higher than one floor at a time?

7. How are utilities incorporated into ICF walls?

EXERCISES

1. Contact a local contractor, designer, or building inspector and inquire as to whether there has been any recent construction using loadbearing masonry or ICFs. If so, try to visit the site and interview the owner about his or her philosophical reasons for choosing this method of construction. Interview the builder (who is often the owner) about difficulties and delights with the construction process. What would be done differently if the project were repeated?

2. Design a small one-room vacation cottage using loadbearing masonry walls. Be sure to pay attention to the module of the masonry unit. How will the cottage be insulated?

3. Change the construction system of the cottage walls to ICFs. How will the dimensions of the cottage be affected? How will the wall finishes be affected?

TIMBER FRAME
CONSTRUCTION

Wood beams have been used to span the roofs and floors of buildings since the beginning of civilization. In earliest historic times, roof timbers were combined with masonry loadbearing walls to build houses and public buildings. In the Middle Ages, braced wall frames of timber were built for the first time (Figure 22.1). The carpenters from forested European countries who emigrated to North America in the 17th and 18th centuries brought with them a fully developed knowledge of how to build efficiently braced frames, and for two centuries North Americans lived and worked almost exclusively in buildings framed with hand-hewn wooden timbers joined by interlocking wood-to-wood connections (Figure 22.2). Nails were rare and expensive, so they were used only in door and window construction and, sometimes, for fastening siding boards to the frame.

Until 2 centuries ago, logs could be converted to boards and timbers only by human muscle power. To make timbers, axemen skillfully scored and hewed logs to reduce them to a rectangular profile. Boards were produced slowly and laboriously with a long, two-man pit saw, one man standing in a pit beneath the log, pulling the saw down, and the other standing above, pulling it back up. At the beginning of the 19th century, however, water-powered sawmills began to take over the work of transforming tree trunks into lumber, squaring timbers and slicing boards in a fraction of the time that it took to do the same work by hand. The house builders and barn raisers of the early 19th century switched from hand-hewn to sawn timbers as soon as they became available.

By the mid-19th century, the introduction of inexpensive wire nails combined with the availability of sawn lumber to spawn the wood light frame and mark the end of dominance by timber framing of residential construction. Many early timber-framed barns and houses still survive, however, and with them survives a rich tradition that continues to the present day. The fact that many of these structures are centuries old and still in service is a testament to the structural ingenuity of their design and the prowess of their builders.

FIGURE 22.1
Braced wall framing was not developed until the late Middle Ages and early Renaissance, when it was often exposed on the face of the building in the style of construction known as halftimbering. The space between the timbers was filled with brickwork or with wattle and daub, a crude plaster of sticks and mud, as seen here in Wythenshawe Hall, a 16th-century house near Manchester, England. (*Photo by James Austin, Cambridge, England*)

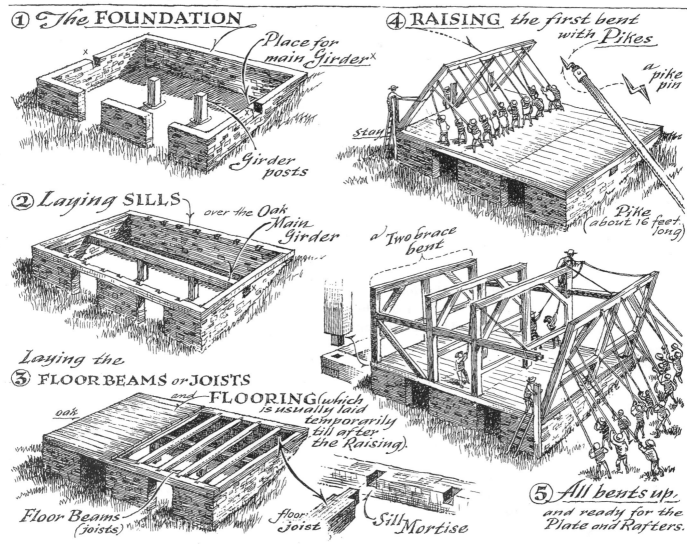

FIGURE 22.2
The European tradition of heavy timber framing was brought to North America by the earliest settlers and was used for houses and barns until well into the 19th century. (*Drawing by Eric Sloane, courtesy of the artist*)

THE UNIQUENESS OF THE TIMBER FRAME

Residential building systems must compete successfully on the basis of price and performance with other systems, particularly wood light frame construction. Timber framing is an anachronistic system that cannot compete on these grounds, but the timber frame has such aesthetic appeal that it is increasing in popularity despite its

higher cost and lack of energy or environmental advantage. To some degree, this appeal may stem from the color, grain figure, and warm feel of wood. In larger part, it may relate to the pleasure that people feel when they sense the structure at work and understand how the timbers collectively support and stabilize the building (Figure 22.3). Most people today live in dwellings where none of the framing is exposed, and even a few exposed ceiling beams in one's house have become a much-

desired amenity. The timber satisfies this desire and recalls the sturdy houses our ancestors erected from hand-hewn timbers only a few generations ago.

The Shape of the Timber Frame

The large timbers of the timber frame, rectangular in section, are most easily and effectively joined at right angles. Thus, the organization of the structural members tends to describe simple orthogonal spaces and simple building shapes. Connections are costly, so their number is minimized. Indeed, virtually all historic timber frame buildings of note are composed in a very straightforward way (Figure 22.4). Modern timber frame houses, whether they are tall, sprawling, or built on a hillside, generally follow these same guidelines because they employ the same structural vocabulary (Figure 22.5).

Enclosure of the Timber Frame

The main difference between historic timber frame structures and those constructed today is the method of enclosure. In the past, the frame itself was used as part of the enclosure system, and the spaces between members were filled in with stone, brick, and other materials to complete the walls and seal against the weather

(Figure 22.6a). This method of enclosure exposed the outside of the frame to the elements, which led to its deterioration. As the frame expanded and contracted with changes in moisture content, gaps were created around the infilled wall material, inviting moisture penetration and increasing air infiltration. The heavy infill materials were also poor thermal insulators. Prompted by more severe climatic conditions, early North American settlers improved the situation by cladding the entire outside surface of the building to protect it from the weather, but the thermal deficiencies of the system persisted (Figure 22.6b). Nowadays, we expect new construction to form both an effective weather barrier and an efficient thermal insulating layer, so timber frame buildings are usually wrapped on the exterior with well-insulated panels that are tightly sealed together. The position of the wall on the outside of the frame allows the frame to be fully exposed to the interior, unlike its historic counterparts (Figure 22.6c).

FIGURE 22.3
Architects Moore, Lyndon, Turnbull, and Whitaker and structural engineer Patrick Morreau designed a rough-sawn heavy timber frame both as a means of supporting this seaside condominium dwelling and as a major feature of its interior design. This project breaks from the tradition of all-wood joinery, but maintains the sense of a timber frame. For further photographs of this project, see Figures 22.5 and 22.21. (*Photo © Morley Baer*)

MATERIALS AND STRUCTURE OF THE TIMBER FRAME

Materials

As stands of old-growth trees have become scarce, wood for timber frame construction has become more precious and its quality has diminished. Some species of wood such as oak that were traditionally used for timbers are no longer commonly cut for building construction, so timber framers must cultivate relationships with small mills to obtain them. In fact, even the commercial species usually used for timber frames such as pine, spruce, and fir are seldom available from stock in the form of large timbers, so they are usually obtained through small, local, specialized mills attuned to the particular needs of timber framers.

Many timber framers have turned to recycled timbers as a

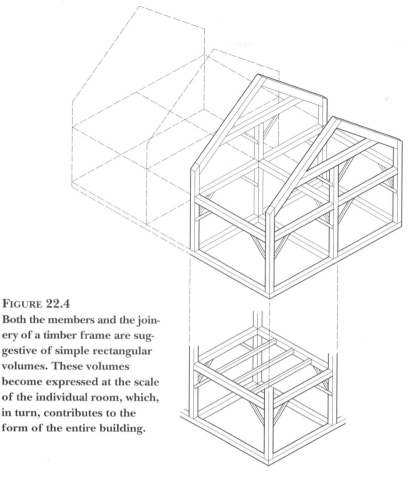

FIGURE 22.4
Both the members and the joinery of a timber frame are suggestive of simple rectangular volumes. These volumes become expressed at the scale of the individual room, which, in turn, contributes to the form of the entire building.

FIGURE 22.5
Each of the attached dwellings in Sea Ranch Condominium #1 in northern California, built in 1965, is framed with a simple cage of unplaned timbers sawn from trees taken from another portion of the site. The diagonal members are wind braces. (*Architects: Moore, Lyndon, Turnbull, and Whitaker. Photo by Edward Allen*)

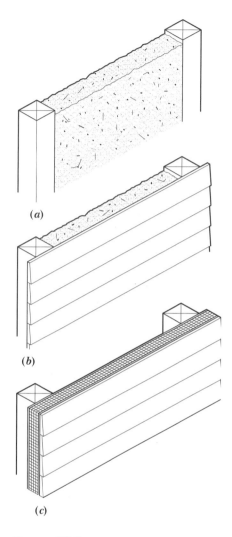

(*a*)

(*b*)

(*c*)

FIGURE 22.6
(*a*) Early timber frame walls were filled with masonry or wattle and daub. (*b*) North American builders covered timber frame walls with siding to protect them against the severe climate. (*c*) Timber frame buildings are now clad with insulated panels on both walls and roof. The panels provide excellent thermal insulation and a surface for the application of a weather envelope.

FIGURE 22.7
This historic timber frame barn is being enclosed with structural insulated panels. Because of the irregularity of the original frame, a decision was made to space the insulated panels away from the frame rather than fastening them directly to it. (*Courtesy of Thermal Shell Homes, Houston, Texas*)

source of material. Many mills and factories that were built with large timbers between the late 1800s and early 1900s have become obsolete, and when they are demolished, the straight and fine-grained timbers of which they were often made are sold as salvage. Timber framers, recognizing that the quality of this old-growth timber from virgin forests can no longer be equaled, are willing to expend extra effort to obtain it and have it sawn to the dimensions desired for their use.

Timber framers and clients for whom environmental responsibility is a priority often use glue-laminated timbers rather than sawn. Laminated beams are stronger and more dimensionally stable than sawn wood beams

and can be made in the exact sizes and shapes desired.

Joinery

Much of the beauty of a timber frame structure comes from the elegance of the joints that connect the members (Figure 22.8). These joints have evolved over time to respond to the physical properties of the wood and the nature of the forces imposed upon them. Rich traditions of wood joinery have developed independently in Japan, China, Europe, and North America. Modern timber framers, with the advantage of international communication, often combine details from several cultures to achieve the best overall result.

Joints in a timber frame structure typically must be designed to withstand the forces of compression, tension, shear, and torsion. These forces correspond respectively to pushing a beam into a column, pulling it away from the column, pushing it downward toward the ground, and rotating it (Figure 22.9). Not surprisingly, timber joints that can resist all of these forces tend to be complicated.

The traditional timber frame joint consists of a mortise and tenon held together with wooden dowels. When this joint is used to connect a post and a beam, a rectangular recess called a *mortise* is cut into the side of the post and a matching tongue known as a *tenon* is cut at the end of the beam (Figure 22.10). The

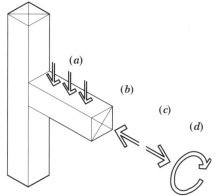

FIGURE 22.9
Basic forces on a timber frame joint: (*a*)
Shear at the joint is the result of gravity
loads on the beam, (*b*) compression at the
joint is the result of the beam being pushed
along its length into the post, (*c*) tension
at the joint results from horizontal forces
pulling the beam away from the post, (*d*)
torsion at the joint is the result of the rota-
tional forces on the beam (or on the post).

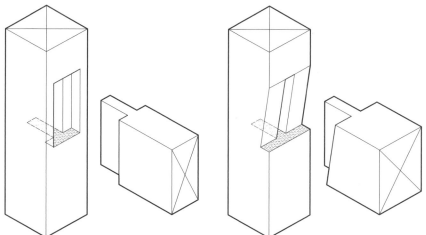

FIGURE 22.10
A typical mortise-and-tenon joint connecting beam to post. The shaded areas at
the bases of the mortises provide bearing surfaces for vertical loads imposed
by the beams.

beam bears on the horizontal surface at the base of the mortise, and it is prevented from rotating by the tenon. The beam cannot be pushed closer to the post without crushing wood fibers, and pegs inserted through the tenon prevent its withdrawal. The proportions of the joint are adjusted according to species of wood, size of members, and forces within the members. This basic joint is used in various versions throughout the frame to connect other members such as joists, rafters, and knee braces.

In the past, joints for timber frame buildings were made entirely with hand tools, but now power tools speed the joint-making process considerably without lowering quality. One tradition that unites most residential timber framers is the avoidance of metal in the connections. Metal would produce stronger and less expensive joints, but they would not be as beautiful.

Lateral Bracing of a Timber Frame

Horizontal forces such as wind and earthquake are resisted in a timber frame structure by means of diagonal *knee braces* located at the right-angle connections between principal members of the frame. Knee braces always occur in pairs so that when a force is applied to one side of the frame, one brace will be working in tension, and the other will be working in compression (Figure 22.11). The mortise-and-tenon joints used for knee braces are always stronger in compression than in tension, so the structure is designed to rely on only half of the knee braces at any one time. And even though the walls that are built around the frame to enclose it add some lateral strength to the structure, the timber frame is designed to stand on its own.

Wood Shrinkage in Timber Construction

Wood is subject to large amounts of expansion and contraction caused by

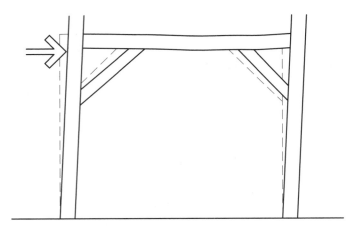

FIGURE 22.11
Paired knee braces in a single bay of a timber frame. A force applied to one end will put one brace in compression and the other in tension. A force on the other end will have the opposite effect.

seasonal changes in moisture content, particularly in the direction perpendicular to its grain. As if this were not enough to contend with, the problem is compounded for timber framers because they frequently work with green (unseasoned) timbers due to the cost and difficulty of seasoning such large pieces.

A timber frame is designed to minimize the cumulative effects of differential shrinkage by eliminating cross-grain wood from the principal lines of support. For example, columns are cut as a single piece running the entire height of the building, with the beams connected to the column by a mortise-and-tenon joint (Figure 22.12). The joints are also designed to minimize both the structural and the visual effects of shrinkage.

BUILDING A TIMBER FRAME HOUSE

The construction of a timber frame starts in the shop with the careful cutting and labeling of the timbers. These timbers are then transported to the site, where a foundation has been

prepared. Some timber framers start by framing a timber floor that is connected directly to the foundation. In this case, the timbers are fit together until the floor structure is complete and a deck is laid to make a subfloor. A timber-framed first floor will rarely, if ever, be seen and appreciated (from the underside) by the occupants of the house, so in many cases a less expensive, but equally solid, conventionally framed floor is used as a starting point (Figure 22.12).

With a level subfloor as a working surface, the frames are assembled on top of one another in the reverse order of their eventual installation (Figure 22.13). The beams and knee braces that will run perpendicular to the frames are also assembled in order to preassemble as many components as possible for efficient use of the crane that will lift them into place. The crane is used in place of the neighborly notion of a "barn raising" for reasons of safety and practicality. Once the principal parts of the frame are in place and temporarily braced, the smaller parts such as floor beams, rafters, and knee braces are installed to complete and stiffen the structure.

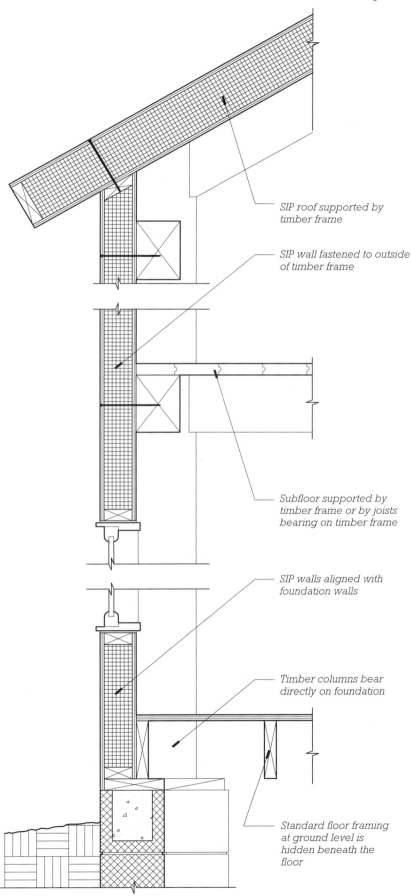

SIP roof supported by timber frame

SIP wall fastened to outside of timber frame

Subfloor supported by timber frame or by joists bearing on timber frame

SIP walls aligned with foundation walls

Timber columns bear directly on foundation

Standard floor framing at ground level is hidden beneath the floor

FIGURE 22.12
Typical timber frame construction. The ground floor can also be constructed of a concrete slab. Upper floors often incorporate extra layers of material for acoustic insulation or a secondary structure of joists that can house plumbing, ductwork, and other utilities.

(a)

(b)

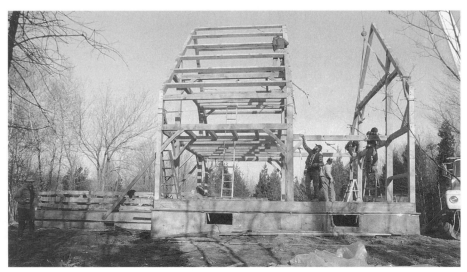

(c)

FIGURE 22.13

Constructing a timber frame: (*a*) Each plane (bent) of columns, beams, rafters, and braces is assembled. (*b*) The completed bents are laid out on the floor, ready for raising. (*c*) Raising the bents, using a truck-mounted crane, and installing floor framing and roof purlins. (*Courtesy of Benson Woodworking, Alstead, New Hampshire*)

FIGURE 22.14
The completed frame with the traditional roof tree in place at the peak of the gable. (*Courtesy of Benson Woodworking, Alstead, New Hampshire*)

Structural insulated panels (see Chapter 24) are typically used to enclose both walls and roof (Figure 22.15). The panels are screwed or nailed directly to the outside surfaces of the frame, where they form a continuous, insulated enclosure as well as an air infiltration barrier. During dry weather, panels with gypsum wallboard adhered to the inside face can be used to save the chore of applying an interior finish to the perimeter walls and ceiling. For a less expensive enclosure system, studs with batt insulation and sheathing can be substituted for insulated panels at the walls, and decking with rigid or batt insulation at the roof. Whichever system of enclosure is employed, interior partition walls are typically framed with 2×4 studs.

Timber frame houses present certain challenges with respect to planning for and installing the electrical wiring, plumbing, and heating/cooling ducts because the exterior walls,

FIGURE 22.15
Enclosing the timber-framed house with sandwich panels consisting of waferboard faces bonded to an insulating foam core. The panels are attached to the outside face of the frame, providing insulation and enclosure while revealing the frame on the inside. (*Courtesy of Benson Woodworking, Alstead, New Hampshire*)

the exposed beam ceilings, and the roof are often without accessible cavities. Even the interior partition walls are typically sandwiched between massive columns and beams, further complicating the running of wires, pipes, and ducts. All of these limitations on the installation of utilities point out the need for the detailed planning of these systems before construction begins (Figure 22.16).

Most conflicts between utilities and the timber frame can be resolved by routing the utilities through floors that are relatively independent of the frame because they are located above it (Figure 22.17). Electrical wiring located at exterior walls can be run in chases that are cast into insulating panels or applied to their interior surfaces (Figure 22.18).

FIGURE 22.16
Prewiring a timber frame house. Electric cables are placed in grooves routed into the top faces of timbers. The cables will be protected with metal plates before they are covered with finish flooring. Cables are also routed into the outside faces of walls, where they will be concealed with insulating panels. (*Courtesy of Benson Woodworking, Alstead, New Hampshire*)

FIGURE 22.17
Open-web joists allow large utilities such as plumbing drains, heating ducts, and electrical cables to be run through the floor structure in either direction with relative ease. (*Courtesy of Benson Woodworking, Alstead, New Hampshire*)

FIGURE 22.18
These insulating panels have been custom-made to provide a chase at their base through which electrical wiring can be run. The chase will be covered with base trim containing electrical outlets. (*Courtesy of Benson Woodworking, Alstead, New Hampshire*)

FIGURE 22.19
Architects Greene and Greene of Pasadena, California, were known for their carefully wrought timber frame houses such as this one, built for David B. Gamble in 1909. (*Photo © Wayne Andrews*)

FIGURE 22.20
This contemporary interpretation of the timber frame is made of laminated recycled redwood. Both contemporary and traditional timber frame components are often used in conjunction with other structural systems. (*Architect: Arne Bystrom. Photo by Rob Thallon*)

FIGURE 22.21
A view of the exposed timbers and connectors inside a dwelling in Sea Ranch Condominium #1. (*Photo © Morley Baer*)

TIMBER FRAMING AND THE BUILDING CODES

Timber frames are not described in the building code, so a structural engineer must analyze each timber frame building in order to obtain a building permit. Engineers do not have to start from scratch in this endeavor because certified structural tests have been performed and their results published with tables of values for all to use. A typical structural analysis of a timber frame structure includes a bent-by-bent computer examination of the forces with an emphasis on compression and tension at the braces. Wood-pinned joints in tension are often the weakest part of the system, but can usually be reinforced with metal straps at the outside or top of the frame where they will be covered with wall or roof panels. Uplift forces at the base of columns are frequently an issue also, and long drift pins or steel plate hold-downs set into the columns and anchored to the foundation are the typical resolution.

SELECTED REFERENCES

1. Benson, Tedd. *The Timber-Frame Home: Design, Construction, Finishing* (2nd ed.). Newtown, CT: Taunton Press, 1997.

A well-written and well-illustrated general description of timber frame construction by the acknowledged American master.

2. American Institute of Timber Construction. *Timber Construction Manual* (4th ed.). New York: John Wiley & Sons, 1994.

This is a comprehensive design handbook for timber structures, including detailed engineering procedures as well as general information on wood and its fasteners.

3. Goldstein, Eliot W. *Timber Construction for Architects and Builders.* New York,: McGraw–Hill, 1999.

A comprehensive guide to planning, engineering, detailing, and construction of heavy timber frames.

KEY TERMS AND CONCEPTS

timber frame
logical timber frame shapes
enclosure of the timber frame

sources of timber
timber frame joinery
mortise and tenon
knee brace

effects of timber shrinkage
structural insulated panel
utilities in the timber frame

REVIEW QUESTIONS

1. Why are the insulated enclosure walls for a modern timber frame structure generally located at the outside of the timber frame?

2. What are the important factors in detailing the junction of a major beam with a column in a timber frame structure? Draw several ways of making this joint.

3. Name several sources of quality timbers for timber frame buildings constructed today.

4. What are the typical problems in routing utilities through a timber frame structure? How are these problems usually resolved?

EXERCISES

1. Find an illustration of a particular timber frame connection and model it in wood. Make the model very large—half size or quarter size at the very least—and construct it of the same wood for which the connection is designed.

2. Design and model a timber frame house at ½ inch to the foot or larger. Can you figure out how to run all the utilities?

3. Find a barn or mill that was constructed in the 18th or 19th century and sketch some typical connection details. How is the structure stabilized against horizontal wind forces?

4. Obtain a book on traditional Japanese construction from the library and compare Japanese timber joint details with 18th- or 19th-century American practice.

23

LIGHT-GAUGE STEEL CONSTRUCTION

- **Light-Gauge Steel Frame Construction**

  Framing Procedures for Light-Gauge Steel

  Hybrid Uses of Light-Gauge Steel

  Advantages and Disadvantages of Light-Gauge Steel Framing

  Light-Gauge Steel Framing and the Building Codes

LIGHT-GAUGE STEEL FRAME CONSTRUCTION

Light-gauge steel frame construction is the noncombustible equivalent of wood light frame construction. The external dimensions of the standard sizes of light-gauge members correspond closely to the dimensions of the standard sizes of 2-inch (38-mm) framing lumber. These members are used in framing in much the same way as 2-inch wood members are used: as closely spaced studs, joists, and rafters. A light-gauge steel frame building may be sheathed, insulated, wired, and finished inside and out in the same manner as a wood light frame building.

The members used in light-gauge steel frame construction are manufactured by *cold forming:* Sheet steel is fed from continuous coils through machines that fold it at room temperature into long members whose shapes make them stiff and strong. The term light-gauge refers to the relative thinness (gauge) of the steel sheet from which the members are made.

For studs, joists, and rafters, the steel is formed into *C-shaped (cee) sections* that correspond to the 2 × 4 through 2 × 12 members used in wood light frame construction. The cee sections are designed to fit snugly inside channel sections called *tracks,* which are used for top and bottom wall plates and rim and header joists (Figure 23.1). The webs of cee members are punched at the factory to provide holes at 2-foot (600-mm) intervals; these are designed to allow wiring, piping, and bracing to pass through studs and joists without the necessity of drilling or punching holes at the construction site.

The strength and stiffness of a member depend on the shape and depth of the section and the *gauge* (thickness) of the steel sheet from which it is made. Studs in bearing walls, for instance, are usually made of 18-gauge steel, while those in non-bearing partition walls range from 20-gauge to 25-gauge. The gauge of joists can be adjusted according to the span, saving money and material without having to alter the depth of the joists. The designation of steel thickness is shifting from the old and confusing gauge system, in which increasing gauge numbers correspond to decreasing thickness, to a more logical system that describes thickness in terms of *mils,* which are thousandths of an inch (Figure 23.2).

For large projects, members may be manufactured precisely to the required lengths. Otherwise, they are furnished in standard lengths and are cut to length on the construction site with power saws (Figure 23.3). A variety of sheet metal angles, straps, plates, and miscellaneous shapes are manufactured as accessories for light-gauge steel construction (Figure 23.4).

Light-gauge steel members are usually joined with self-drilling, *self-tapping screws,* which drill their own holes and form helical threads in the holes as they are driven. Driven rapidly by handheld electric or pneumatic tools, these screws are plated with cadmium or zinc to resist corrosion, and are available in an assortment of diameters and lengths to suit a full range of connection situations. Some jobs such as fastening sheathing to framing are best done with pneumatically driven *pins* that penetrate the members and hold by friction. Welding is sometimes employed where particularly strong connections are needed.

Framing Procedures for Light-Gauge Steel

The sequence of construction for a building that is framed entirely with light-gauge steel members is essentially the same as that described in Chapter 9 for a building framed with nominal 2-inch wood members (Figure 23.5). Framing is usually constructed

FIGURE 23.1
Typical residential light-gauge steel framing members. To the left are the common sizes of the cee section, which is used for studs and joists. To the right is track section, which is primarily used for plates, rim joists, and headers. The sections correspond to 2 × 4 through 2 × 12 lumber sizes.

Thickness (in mils)	Corresponding Gauge
27	22
33	20
43	18
54	16
68	14

FIGURE 23.2
The thickness of light-gauge steel framing for years was described by the same gauge designation as is used for flashing and other thin metal work. Now the designation of thickness has been changed officially to mils, which are thousandths of an inch. The most common stud thicknesses are 43 mil for bearing walls and 33 mil for partition walls. One mil is equal to about 0.0254 mm.

FIGURE 23.3
A power saw with an abrasive blade cuts quickly and precisely through steel framing members. (*Courtesy of Unimast Incorporated—www.unimast.com*)

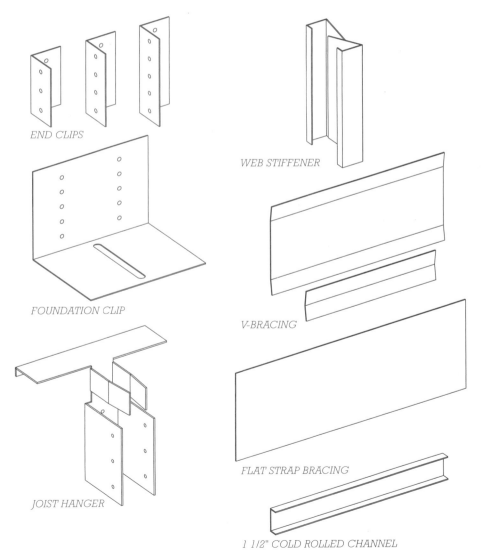

END CLIPS

FOUNDATION CLIP

JOIST HANGER

WEB STIFFENER

V-BRACING

FLAT STRAP BRACING

1 1/2" COLD ROLLED CHANNEL

FIGURE 23.4
Standard accessories for light-gauge steel framing: End clips are used to join members that meet at right angles. Foundation clips attach the ground-floor platform to anchor bolts embedded in the foundation. Joist hangers connect joists to headers and trimmers around openings. The web stiffener is a two-piece assembly that is inserted inside a joist and screwed to its vertical web to help transmit wall loads vertically through the joist. The remaining accessories are used for bracing.

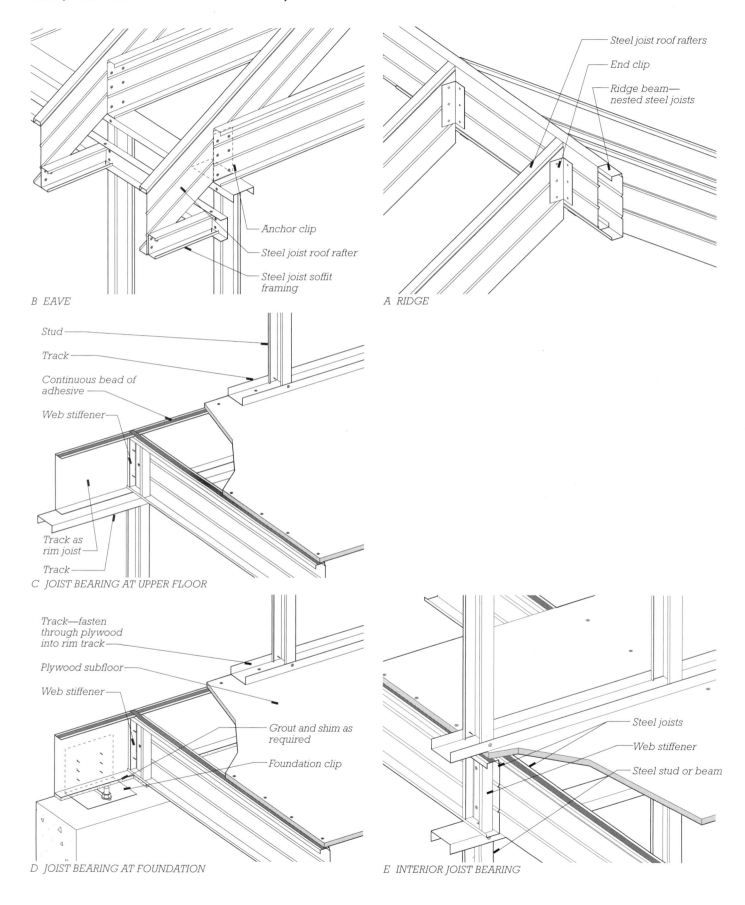

B EAVE

Anchor clip

Steel joist roof rafter

Steel joist soffit framing

A RIDGE

Steel joist roof rafters

End clip

Ridge beam— nested steel joists

Stud

Track

Continuous bead of adhesive

Web stiffener

Track as rim joist

Track

C JOIST BEARING AT UPPER FLOOR

Track—fasten through plywood into rim track

Plywood subfloor

Web stiffener

Grout and shim as required

Foundation clip

D JOIST BEARING AT FOUNDATION

Steel joists

Web stiffener

Steel stud or beam

E INTERIOR JOIST BEARING

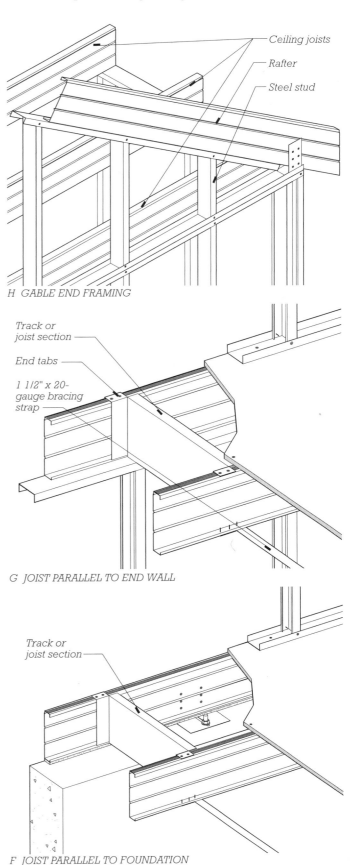

H GABLE END FRAMING

G JOIST PARALLEL TO END WALL

F JOIST PARALLEL TO FOUNDATION

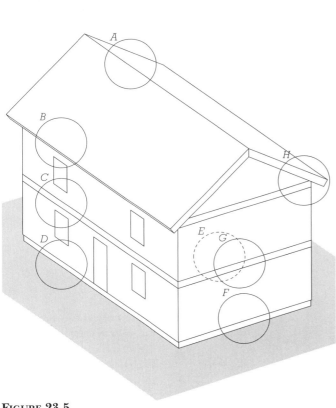

FIGURE 23.5

Typical light-gauge framing details. Each detail is keyed by letter to a circle on the whole-building diagram in the center to show its location in the frame. (*A*) A pair of nested joists makes a boxlike ridge board or ridge beam. (*B*) Anchor clips are sandwiched between the ceiling joists and rafters to hold the roof framing down to the wall. (*C*) A web stiffener helps transmit vertical forces from each stud through the end of the joist to the stud in the floor below. Mastic adhesive cushions the joint between the subfloor and the steel framing. (*D*) Foundation clips anchor the entire frame to the foundation. (*E*) At interior joist bearings, joists are overlapped back to back and a web stiffener is inserted. (*F, G*) Short crosspieces brace the last joist at the end of the building and help transmit stud forces through to the wall below. (*H*) Like all these details, the gable end framing is directly analogous to the corresponding detail for a wood light frame building as shown in Chapter 9.

platform fashion: The ground floor is framed with steel joists. Mastic adhesive is applied to the upper edges of the joists. Wood panel subflooring is laid down and fastened to the upper flanges of the joists with screws. Steel studs are laid flat on the subfloor and joined to make wall frames. Because the studs must be screwed to the top and bottom tracks from both sides, the wall frames must be flipped over before they are sheathed. Once sheathed or crossbraced with steel straps, the first-floor walls are tilted up and screwed to the floor frame (Figures 23.6 and 23.7). Next, the upper floor platform is framed, then the upper floor walls (Figure 23.8). Finally, the ceiling and roof are framed, all in much the same way as in a wood-framed house. Prefabricated trusses of light-gauge steel members that are screwed or welded together are often used to frame ceilings and roofs (Figures 23.9 and 23.10). It is possible, in fact, to frame any building with light-gauge steel members that can be framed with nominal 2-inch wood members.

FIGURE 23.6
Diagonal strap braces stabilize upper floor wall framing for an apartment building. (*Courtesy of United States Gypsum Company*)

FIGURE 23.7
Temporary braces support the walls at each level until the next floor platform has been completed. Small cold-rolled channels pass through the web openings of the studs; they are welded to each stud to help stabilize them against buckling. (*Courtesy of Unimast Incorporated— www.unimast.com*)

FIGURE 23.8
Ceiling joists in place for an apartment building. A brick veneer cladding has already been added to the ground floor. (*Courtesy of United States Gypsum Company*)

FIGURE 23.9
A worker tightens the last screws to complete a connection in a light-gauge steel roof truss. The truss members are held in alignment during assembly by a simple jig made of plywood and blocks of framing lumber. (*Courtesy of Unimast Incorporated—www.unimast.com*)

FIGURE 23.10
Installing steel roof trusses. (*Courtesy of Unimast Incorporated—www.unimast.com*)

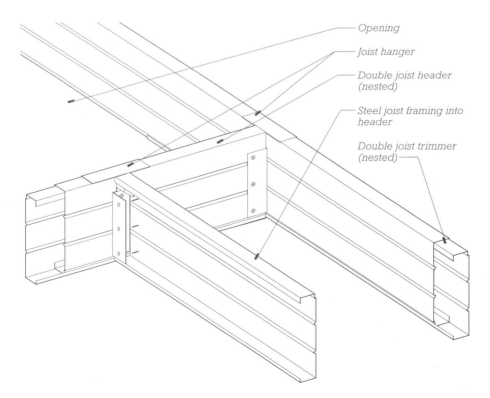

Opening

Joist hanger

Double joist header (nested)

Steel joist framing into header

Double joist trimmer (nested)

FIGURE 23.11
Headers and trimmers for floor openings are doubled and nested to create strong, stable box members. Only one vertical flange of the joist hanger is attached to the joist; the other flange would be used instead if the web of the joist were oriented to the left rather than the right.

FIGURE 23.12
A typical window or door head detail. The header is made of two joists placed with their open sides together. The top plate of the wall, which is a track, continues over the top of the header. Another track is cut and folded at each end to frame the top of the opening. Short studs are inserted between this track and the header to maintain the spacing of the studs in the wall.

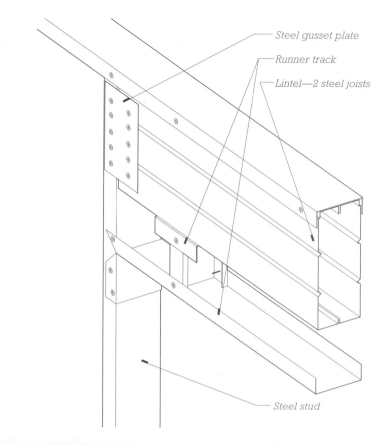

Steel gusset plate

Runner track

Lintel—2 steel joists

Steel stud

Openings in floors and walls are framed analogously to openings in wood light frame construction, with doubled members around each opening and strong headers over doors and windows (Figures 23.11 to 23.13). Joist hangers and *right-angle clips* of sheet steel are used to join members around openings. Light-gauge members are designed to nest together to form a tubular configuration that is especially strong and stiff when used for a ridge board or header (Figures 23.5A and 23.11).

Because light steel members are much more prone than their wood counterparts to buckle or twist under load, they must often be stiffened at locations where loads are great, and somewhat more attention must be paid to their bracing and bridging. In locations where large vertical forces must pass through floor joists (as occurs wherever load-bearing studs sit on a floor platform), steel *web stiffeners* (called *bearing stiffeners* in the code) are screwed to the webs of the joists to prevent the thin web from buckling (Figure 23.5C–E). If not sheathed with gypsum board on the inside and approved panel sheathing on the outside, studs must be braced horizontally at 4-foot (1200-mm) intervals, either with steel straps screwed to the

FIGURE 23.13
A detail of a window header. Because a supporting stud has been inserted under the end of the header, a large gusset plate such as the one shown in Figure 23.12 is not required. (*Courtesy of Unimast Incorporated—www.unimast.com*)

FIGURE 23.14
A detail of eave framing.
(*Courtesy of Unimast
Incorporated—
www.unimast.com*)

flanges of the studs or with 1½-inch (38-mm) cold-formed steel channels passed through the punched openings in the studs and screwed to an angle clip at each stud (Figure 23.14). Floor joists are bridged with solid blocking between and steel straps screwed to their top and bottom edges. Wall bracing consists of plywood or OSB panels or diagonal steel straps screwed to the studs (Figure 23.6). Permanent resistance to buckling, twisting, and lateral loads such as wind and earthquake is imparted largely and very effectively by subflooring, wall sheathing, and interior finish materials.

Hybrid Uses of Light-Gauge Steel

In residential construction, light-gauge steel and wood light framing are sometimes mixed in the same building. Some builders find it economical to use wood to frame exterior walls, floors, and roof, with steel framing for interior partitions. Sometimes all walls, interior and exterior, are framed with steel, and floors are framed with

wood. Steel trusses made of light-gauge members may be applied over wood frame walls. In such mixed uses, special care must be taken in the details to assure that wood shrinkage will not create unforeseen stresses or damage to finish materials.

In wood-framed dwellings, there are locations around fireplaces and flues where clearance to combustible framing is required by code. In these situations, the substitution of light-gauge steel for wood framing members can allow the finish materials to be applied adjacent to the flue without compromise.

Advantages and Disadvantages of Light-Gauge Steel Framing

Light-gauge steel framing shares most of the advantages of wood light framing: It is versatile; flexible; requires only simple, inexpensive tools; furnishes internal cavities for utilities and thermal insulation; and accepts an extremely wide range of exterior and interior finish materials. Additionally, steel framing may be used in buildings for which noncombustible con-

struction is required by code, thus extending its use to larger buildings and to those whose uses require a higher degree of resistance to fire.

Steel framing members are significantly lighter in weight than the wood members to which they are structurally equivalent, an advantage that is often enhanced by spacing steel studs, joists, and rafters at 24 inches o.c. (600 mm) rather than 16 inches (400 mm). Light-gauge steel joists and rafters can span slightly longer distances than nominal 2-inch wood members of the same depth. Steel members tend to be straighter and more uniform than wood members, and they are much more stable dimensionally because they are unaffected by changing humidity. Although they may corrode if exposed to moisture over an extended period of time, particularly in ocean-front locations, steel framing members cannot fall victim to termites or decay. Prices of steel framing members are relatively stable, while the price of wood, which is an agricultural commodity subject to fluctuations in supply, often varies over a wide range with changing market conditions.

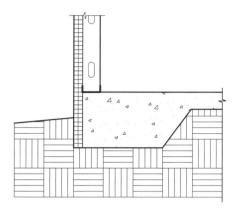

FIGURE 23.15
Detail of wall and floor framing at the foundation with thermal insulation. Rigid insulation is typically affixed to the exterior of the light-gauge steel frame to stem the flow of heat through the steel. Depending on lateral bracing and siding, a layer of sheathing may be introduced at the outside or inside face of the rigid insulation.

The thermal conductivity of light-gauge steel framing members is much higher than that of wood. In cold climates, steel framing members, unless detailed with thermal breaks such as foam plastic sheathing or insulating edge spacers between studs and sheathing, conduct heat rapidly enough to reduce substantially the thermal performance of a wall or roof. This can result in excessive energy loss and moisture condensation on interior building surfaces, with attendant mold and mildew growth and discoloration of surface finishes. A layer of rigid insulation must be added to the assembly in most climates to stem the flow of heat through the steel members (Figure 23.15). Special attention must be given to designing details to block excessive heat flow in every area of the frame. At the eave of a steel-framed house, for instance, the ceiling joists readily conduct heat from the warm interior ceiling along their lengths to the cold eave unless insulating edge spacers or foam insulation boards are used between the ceiling finish material and the joists.

Light-gauge steel framing takes longer than framing in wood for the following reasons: (1) Even though they are installed with power screwdrivers, each screw takes longer to install than does a nail into wood. (2) There are more screws to install because there are more angles and clips required. (3) Walls must be flipped over before they are sheathed. (4) Web stiffeners must be installed in the floor joists beneath each bearing wall. This extra framing labor usually more than offsets the slight material savings with steel and results in a frame that is more expensive overall than a comparable wood light frame.

FIGURE 23.16
Waferboard sheathes the walls of a house framed with light gauge steel studs, joists, and rafters.
(*Courtesy of Unimast Incorporated — www.unimast.com*)

Workers at the job site complain that light-gauge steel framing is noisy and creates considerable glare as compared with wood framing. Steel framing can also be difficult to handle because it can become extremely hot or cold, depending on the weather. On the positive side, framers like steel because of its precision and because mistakes can be easily corrected by unscrewing a piece and screwing it back in the correct location.

Any exterior or interior finish material that is used in wood light frame construction may be applied to a light-gauge steel frame. Whereas finish materials are most often fastened to a wood frame with nails, screws must be utilized instead with a steel frame. Wood trim components are applied with special finish screws that have very small heads; these are analogous to finish nails.

Light-Gauge Steel Framing and the Building Codes

The International Residential Code has adopted prescriptive requirements for framing light-gauge steel one- and two-family dwellings. This code, with its structural tables and standard details, allows for the design and construction of steel-framed houses without the need to employ an engineer or architect, just as with wood light frame construction (Figure 23.17).

Allowable Spans For Cold-Formed Steel Joists[a,b]

Nominal Joist Size	30 PSF Live Load Spacing (inches)		40 PSF Live Load Spacing (inches)	
	16	24	16	24
550S162-33	10'-7"	9'-1"	9'-7"	8'-1"
550S162-43	11'-6"	10'-0"	10'-5"	9'-1"
550S162-54	12'-4"	10'-9"	11'-2"	9'-9"
550S162-68	13'-2"	11'-6"	12'-0"	10'-6"
800S162-33	13'-3"	8'-10"	10'-7"	7'-1"
800S162-43	15'-6"	13'-7"	14'-1"	12'-3"
800S162-54	16'-8"	14'-7"	15'-2"	13'-3"
800S162-68	17'-11"	15'-7"	16'-3"	14'-2"
1000S162-43	18'-8"	15'-3"	16'-8"	13'-1"
1000S162-54	20'-1"	17'-6"	18'-3"	15'-11"
1000S162-68	21'-6"	18'-10"	19'-7"	17'-1"
1200S162-43	20'-3"	14'-1"	16'-10"	11'-3"
1200S162-54	23'-4"	19'-7"	21'-3"	17'-6"
1200S162-68	25'-1"	21'-11"	22'-10"	19'-11"

For SI: 1 inch = 25.4 mm, 1 foot = 304.8 mm, 1 pound per square foot = 0.0479 kN/m².
[a]Deflection criteria: $L/480$ for live loads, $L/360$ for total loads.
[b]Floor dead load = 10 psf.

FIGURE 23.17

A prescriptive table from the International Residential Code. The nominal joist size 550S162-33 corresponds to a 5.5 inch tall by 1.62 inch wide member in 20-gauge steel, which approximates the size of a 2 × 6 wood member. See Figure 23.2 for other gauge designations.

**C.S.I./C.S.C.
MasterFormat Section Numbers for
Light-Gauge Steel Construction**

05400	COLD-FORMED METAL FRAMING
05410	Loadbearing Metal Stud Systems
05420	Cold-Formed Metal Joist Systems

SELECTED REFERENCES

1. American Iron and Steel Institute. *AISI Cold-Formed Steel Design Manual*. American Iron and Steel Institute, Washington, D.C., 1996.

This is an engineering reference work that contains structural design tables and procedures for light gauge steel framing.

2. Scharpf, Robert. *Residential Steel Framing Handbook*. New York: McGraw–Hill, 1996.

The light-gauge steel equivalent of a wood framing manual. Material selection, tools, framing techniques, and attaching exterior and interior finish materials are discussed.

KEY TERMS AND CONCEPTS

light-gauge steel frame
cold-formed steel
cee section
track
gauge
mil

self drilling, self-tapping screw
pneumatically driven pin
light-gauge steel truss
right-angle clip
web stiffener

bearing stiffener
horizontal steel stud bracing
steel floor joist bridging
hybrid uses of light-gauge steel
thermal insulation of light-gauge steel

REVIEW QUESTIONS

1. How are light-gauge steel framing members manufactured?

2. How do the details of a house framed with light-gauge steel members differ from those of a similar house with wood platform framing?

3. What special precautions should be taken when detailing a steel-framed building to avoid excessive conduction of heat through the framing members?

4. What is the advantage of a prescriptive building code for light-gauge steel framing?

5. Compare the advantages and disadvantages of wood light frame construction and light-gauge steel frame construction.

EXERCISES

1. Convert a set of details for a wood light frame house to light-gauge steel framing.

2. Visit a construction site where light-gauge steel studs are being installed. Grasp the middle of an installed stud that has not yet been sheathed and twist it clockwise and counterclockwise. How resistant is the stud to twisting? How is this resistance increased as the building is completed?

3. On this same construction site, make sketches of how electrical wiring, electric fixture boxes, and pipes are installed in metal framing.

24

PANELIZED CONSTRUCTION

- **The Concept of Panelized Construction**
- **Types of Panels**
  Framed Panels
  Structural Insulated Panels
- **Framed Panel Systems**

- **Structural Insulated Panels**
  Today's Insulated Panels
  Performance of Structural Insulated Panel Systems
  Building with Structural Insulated Panels

The Concept of Panelized Construction

Panelized construction is almost as old as the wood light frame itself. Attempts to establish panelized housing businesses as early as the 1890s focused on the logic of building in the controlled environment of a factory. Panels could be built year-around free from the effects of the weather, and all parts of the house could be built on the ground without scaffolding. Although based on sound logic, early attempts at panelization met with limited success because transportation of finished panels to the site was difficult, and erection of the heavy panels above the first-floor level was awkward and dangerous.

In the early years of the 20th century, some companies capitalized on the benefits of factory manufacture without the troubles associated with large, cumbersome panels by offering complete house kits that were made up not of panels, but of individual pieces that were precut but unassembled. The most successful of these was Sears, Roebuck and Company, which sold more than 100,000 house kits between 1916 and 1933. The Sears kits included everything required to build a house down to sufficient paint for three coats, and each piece was labeled with instructions for easy assembly (Figure 24.1).

Since the era of the kit houses, lighter weight materials and the proliferation of small cranes and boom trucks have allowed panelizing companies to build larger and larger sections of houses in their factories. Today, there are 8-foot-tall by 8-foot-long panels that can be easily carried by two men, and panels as long as 40 feet or more that can be quickly set in place high on a building with a small crane (Figure 24.2). The NAHB reports that panelization is currently the fastest growing segment of the housing industry. Panel manufacturers cite as benefits of their systems the efficiency and

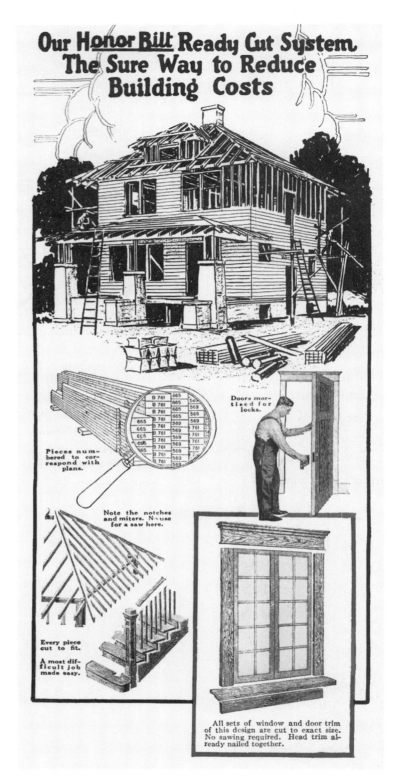

Figure 24.1
Sears, Roebuck and Company was the most successful of several companies offering kit houses early in the 20th century. All parts of an entire house were precut and labeled for "easy" assembly. (*Photo from page 9 from the* **1926 Book of Modern Homes,** *reprinted by arrangement with Sears, Roebuck and Co. and is protected under copyright. No duplication permitted*)

FIGURE 24.2
Versatile portable cranes and other hoisting machinery have allowed manufacturers to produce large sections of houses in the factory because these sections can now be easily assembled at the site. (*Courtesy of Thermal Shell Homes, Houston, Texas*)

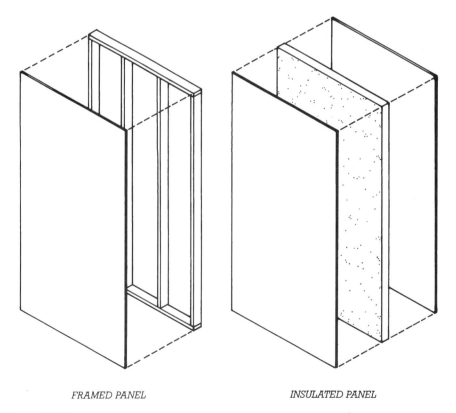

FRAMED PANEL   INSULATED PANEL

FIGURE 24.3
Two types of prefabricated panels. The framed panel is identical to a segment of a conventionally framed wall, floor, or roof. The facings on the insulated panel are adhesive bonded to a core of insulating foam to form a structural unit in which the facings carry the major stresses. Framed panels can be further completed at the factory by adding wiring, plumbing, thermal insulation, and interior finish, but this is unusual.

increased quality control of factory production as well as a shorter on-site construction schedule. The time saved at the site can translate to lower financing costs for the owner and/or more profit for the builder, who can produce more houses in a given year with the same amount of labor.

TYPES OF PANELS

There are fundamentally two types of panels currently being produced: *framed panels,* which are based on the standard materials and details of wood light frame construction, and *structural insulated panels,* which employ rigid insulation in conjunction with lightweight structural elements (Figure 24.3).

Framed Panels

Framed panels range from the simplest assemblies of structural framing and sheathing to fully finished panels containing insulation, wiring, plumbing, and both exterior and interior finishes. Framed panels are generally organized into two categories: open panels and closed panels. *Open panels* are essentially prefabricated sections of walls and (occasionally) floors and roofs with open cavities between the studs (and joists). Open panels may be finished on the exterior, but at the time they are incorporated into the shell of a house, they lack insulation, wiring, and interior finishes.

More complete *closed panels* include wiring, plumbing, insulation (usually glass fiber batts), a vapor retarder, and interior finishes. The production of closed panels is limited, however, because trade organizations have been successful in persuading code enforcement agencies not to grant manufacturing approval for custom designs. With closed panels, there are also thorny problems of joinery and difficulties in inclement weather of transporting and assembling panels with interior finishes.

FIGURE 24.4
Insulated panels are typically used only
at the exterior surfaces of the building—
the walls, the roof, and sometimes the
first floor. Interior walls and upper floors
are typically framed conventionally.

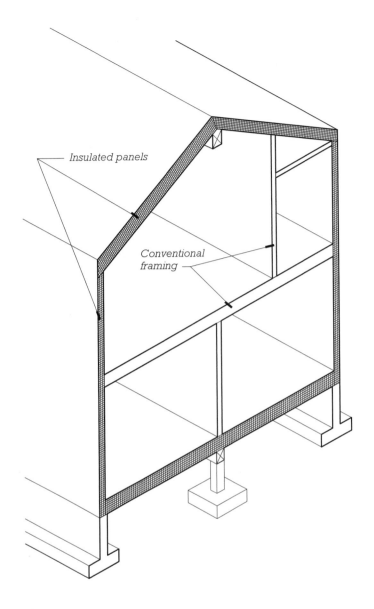

Structural Insulated Panels

Structural insulated panels (SIPs) are made of rigid insulation adhered in a factory to structural elements that work in concert with the insulation to provide the stiffness and strength required for use in bearing walls and roof assemblies. Structural insulated panels are typically used for the construction of exterior walls, roofs, and ground floors, while upper floors and interior walls are constructed with standard framing (Figure 24.4). Where SIPs are used for roofs, they are typically only employed for vaulted ceilings where the ceiling follows the slope of the roof. For the more common flat ceiling assemblies, insulated panels cannot compete economically with raised-heel trusses combined with fiberglass batt or fill-type insulation. This strategy recognizes that the panels are expensive and should therefore only be used at the exterior shell of the building where insulation is needed. Standard framing at the interior of the building provides wall and floor cavities in which electrical, plumbing, and heating systems can be easily installed.

FRAMED PANEL SYSTEMS

Open panels are noninsulated sections of a framed building that are manufactured in a factory and joined at the site. The panels are made with the same framing members, sheathing, and finish materials as standard wood light frame construction, and are joined to one another at the edges (Figure 24.5). Most open-panel manufacturers produce only walls, but some panelize floors as well. Roofs are not typically made of open panels because

time saving at the site is minimal, and the panels cannot compete economically with trusses or rafters. In order to avoid the problems associated with two separate framing contractors, panel manufacturers often train their crews both to set panels and to do conventional framing.

The first modern structural panels were built in 1935 by the Forest Products Laboratory of the Department of Agriculture (Figure 24.6). The panels were made of two sheets of plywood spaced apart and held rigid by 1 × 3 members. Three houses were constructed in which these uninsulated panels were used for both walls

and roof. The design of these houses depended strongly on the module of the plywood sheets, and the placement and sizes of windows and doors were coordinated with the panel dimensions.

Today, uninsulated panels are made with sheathing on the exterior side only in order to allow the installation of electrical wiring, plumbing, and insulation from the inside of the house. The framing members and sheathing are typically the same as for standard construction—2 × 4 or 2 × 6 studs, for example, with ½ inch (12 mm) sheathing. Wall panels typically have windows and doors installed at the factory. Interior walls are framed but

FIGURE 24.5
Open panels are used for both exterior and interior walls because they do not contain insulation. Exterior walls are typically installed first, utilizing the open sub-floor as a stage. The installed exterior walls will help to brace the interior walls. (*Courtesy of Soft Tech Building Systems, Springfield, Oregon*)

FIGURE 24.6
A prefabricated panel house developed in 1937 by the U.S. Forest Products Laboratory. (*Courtesy of USDA, Forest Products Laboratory*)

FIGURE 24.7
Open panels with exterior finish complete except for corner boards and bottom courses of siding. Windows and siding are often installed on open panels in the factory. (*Courtesy of Soft Tech Building Systems, Springfield, Oregon*)

not sheathed, so they are fitted with temporary diagonal braces to keep them square during transit. This level of completion of open panels corresponds to the rough framing of a conventionally framed residence, which, once the roof has been framed and dried in, allows crews to work simultaneously on the exterior and interior of the building.

Wall panels are often finished as completely as possible on the outside with siding, trim, and paint (Figure 24.7). This strategy takes advantage of the controlled environment of the factory for this work and yields a larger profit to the manufacturer. Panels finished on the exterior usually require additional exterior finish work (about 10 to 20 percent of the work) to be done at the site. This work entails the completion of siding and trim at the joints between and at the edges of

FIGURE 24.8
Exterior wall panels travel best in the vertical position if they have windows and/or siding installed because the window seals do not tend to break down and the siding is less likely to get scuffed. Interior wall panels may be transported flat and usually have temporary bracing since they do not have sheathing. (*Courtesy of Soft Tech Building Systems, Springfield, Oregon*)

FIGURE 24.9
**Open panels are secured
to the floor and to one
another in the same fash-
ion as stick frame compo-
nents. Note the bottom
plate, the double top
plate, and the standard
framing for structural
headers.** (*Courtesy of Soft
Tech Building Systems,
Springfield, Oregon*)

panels. Although finishing the exterior of panels in the factory is an
advantage to the manufacturer, no
calendar time is saved at the job site
by this practice because the exterior
of the house can be finished simultaneously with the work that must be
completed inside the house in any case.

With panelized systems, construction of the building shell at the site
takes just a few days. The foundation
will have been prepared in advance
so that it is ready for the installation of
the floor panels. The panels are organized on the delivery truck to facilitate their installation by the boom or
small crane that arrives with them. If
there are floor panels to span over a
basement, these are fastened to a
mudsill and strapped to the foundation
in a way similar to which joists are
connected in a conventionally framed
building. Wall panels are nailed
through their bottom plate to the
subfloor and their studs are strapped
to the foundation, also as in conventional framing (Figure 24.10). When
the first-floor walls are all in place,

FIGURE 24.10
**Upper floors can be
made of open panels
when spans do not exceed
the limits of what can be
transported. Accessibility
of the joist spaces from
the underside allows
plumbing, electrical
wiring, and other utilities
to be installed the same
as in a conventionally
framed house.** (*Photo by
Rob Thallon*)

FIGURE 24.11
**When complete, an open-panel house is
not distinguishable from a conventionally
framed house. This is the house shown
in Figures 24.7, 24.8, and 24.9.** (*Courtesy
of Soft Tech Building Systems, Springfield,
Oregon*)

they are tied together with a second top plate, and the roof or upper floor is built on top of this (Figure 24.9). When the shell is complete, the only significant difference between a panelized house and a conventionally framed one is that there are double framing members at the edges of joined panels of the panelized house.

Design limitations of open panels are related principally to shipping constraints and crane scheduling. According to panel manufacturers, the most practical house for their product is a single-story house with a hip roof. Wall panels can be as long as the truck used to transport them, but wall height is effectively limited to 10 feet (3 m) because walls must be shipped vertically to avoid breaking the seals on windows. This means that gable end walls or walls for rooms with vaulted ceilings must be made in two sections. A two-story house in a region where floor panelization is not practical will require two visits by the crane to place the panelized walls—one visit for the first floor, and one for the second. Because the crane rental company charges substantial fees for travel to and from the site and for setup once it arrives, the double scheduling of a crane can make panelization economically unfeasible.

STRUCTURAL INSULATED PANELS

In 1952, rigid insulation was introduced to panelized construction by the architect Alden B. Dow, who designed panels made of 1⅝-inch-thick (41-mm) polystyrene with ¼ inch (6.4 mm) plywood glued to each side (Figure 24.12). In Dow's system, studs and rafters were eliminated because the insulation aligned the two sheets of plywood in parallel planes to create a stiff assemblage that behaved structurally much like a steel I-beam. Lumber was used only at the edges of panels, which were joined to construct the walls and roofs of three experimental houses. One of these houses withstood the impact of a runaway car that went through the building and only damaged four panels—two on each side of the house.

Because of their composition and their structural behavior, these panels were called *sandwich panels* or *stressed skin panels*. The first serious attempt to mass-produce them was undertaken in 1959 by the Koppers Company, which produced enough panels to manufacture about 800 houses in 2 years. This endeavor ultimately failed because of slow production of the panels; low energy

costs, which meant that consumers saw little need for insulation, and some resistance from carpenters' unions that saw their jobs threatened. It took the energy crisis of 1973 to rekindle significant interest in insulated panels, and in the 1970s and 1980s, a number of panel companies were successfully established. Today, there are over 50 companies in North America that manufacture insulated panels, and it is estimated that the products of these companies are incorporated into the construction of between 400,000 and 800,000 houses per year.

Today's Insulated Panels

The insulated panels marketed today have a wide variety of designs that rely on a number of different strategies for their structural stability. Most have an OSB skin adhered to either side of a 3½-inch foam core of extruded polystyrene (EPS) or expanded polystyrene (XPS). These "skinned" panels perform structurally in the same way as Dow's experimental panels of 50 years earlier (Figure 24.14). Other panels do not have a structural skin but rather rely on spaced wood or metal members that are integrated with the rigid foam (Figure 24.15). Because of these differences, insulated panels are now referred to collectively as structural

FIGURE 24.12
The application of phenolic glue to the facing of Alden B. Dow's 1952 experimental polystyrene foam core panels. The panels were clamped in the press in the background. (*Photograph courtesy of the Alden B. Dow Archives*)

FIGURE 24.13
SIP production has become so auto-mated since Alden Dow's time that there is scarcely a person required on the floor of this high-volume plant. (*Courtesy of Winter Panel Corporation*)

insulated panels (SIPs) rather than stressed skin panels.

When compared with standard framing, SIP construction makes a superior thermal envelope and is faster to erect. However, the integration of electrical, plumbing, and heating systems is more complicated and less flexible because there are no open cavities in exterior walls in which to place wires, pipes, or ducts. The cost of the panels and the embodied energy in SIP construction are both generally higher than in standard construction, but life-cycle cost analysis typically shows that the initial investment can be recovered within 10 years through reduced heating and/or cooling bills.

SIP systems are faster to erect than standard framing because the panels are large (4×8 feet to 28×8 feet) and save most of the site labor it would take to construct a framed wall or roof section of comparable size. Panels are typically lifted into place with a small crane, and workers need only connect the panels to the foundation and to each other (Figure 24.16). The labor required to install insulation in a conventionally framed building is also saved with an SIP system.

The plumbing and heating systems in an SIP house can be located in conventionally framed interior walls, thus avoiding conflicts with insulated panels (Figure 24.17). Electrical outlets are always required in exterior walls, however, and usually some switches and light fixtures are needed as well, so panel manufacturers typically create 1-inch-diameter (25 mm) horizontal chases through wall panels. Some manufacturers locate vertical chases at panel ends as well. The various chases can be used by electricians to pull electrical cables through, much as they do through metallic or plastic conduit in commercial buildings. Boxes for receptacles and other electrical devices are cut into the inside face of the panel opposite the chases. This system allows for running electrical wires in exterior walls, but is not as flexible as a standard framed wall

FIGURE 24.14

SIPs are considerably stronger than conventional $2\times$ framing of equal dimension, but manufacturers and builders must resort to stunts to convince skeptical consumers of their strength. (*Courtesy of Fischer SIPS, Louisville, Kentucky*)

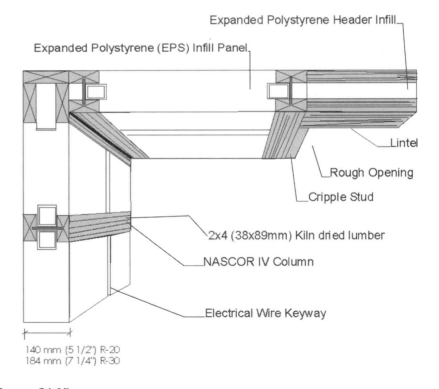

FIGURE 24.15

Some SIPs do not have a stressed skin but rely on other means for structural strength and stiffness. These panels can be used for walls only. (*Courtesy of NASCOR, Incorporated, Calgary, Alberta, Canada*)

FIGURE 24.16
With a crane to lift the panels into place, the construction of an SIP residential shell, even one as large as this multifamily building, can usually be completed within 2 or 3 days. (*Courtesy of NASCOR, Incorporated, Calgary, Alberta, Canada*)

FIGURE 24.17
Utilities in SIP houses are located within the cavities of conventionally framed interior walls whenever possible. In this house, plumbing for a sink is attached to the interior face of an SIP wall where it will later be covered by a cabinet. (*Courtesy of Energy Studies in Buildings Laboratory, Center for Housing Innovation, Department of Architecture, University of Oregon*)

(or ceiling) and requires somewhat more time for the electrician. Plumbing and heating can be installed on the (inside) surface of exterior walls when required, but the pipes and ducts must be covered for appearance and are usually unsightly even then.

Performance of Structural Insulated Panel Systems

Marketing for SIP systems stresses their performance as compared to standard framing. Potential customers are told that SIP systems are stronger, more energy efficient, and more environmentally responsible. Building scientists have studied all of these issues and have accumulated much data, but their work is complicated by differences in manufacturing and construction techniques, and as with all scientific work, preliminary results lead to more questions.

Structural Performance

SIPs satisfy the structural codes in a number of ways. Some rely almost entirely on an OSB skin that is laminated to the rigid insulation for structural integrity, while others have no skin and the edges of panels work collectively as a post-and-beam structure with the rigid insulation acting as diagonal bracing. In general, SIPs are stronger than standard framing, but there are isolated instances of delamination of facing from the foam plastic core that have led to structural failure. There have also been some reports of sagging in SIP roofs due to *creep* of the foam core, which is the long-term deformation of materials subjected to constant load. This issue is still being studied. To combat anxiety about panel failure, most SIP companies offer limited lifetime warranties on their products. With only 20 to 30 years of experience with SIPs in the field, however, a "lifetime" warranty is difficult to evaluate when compared with the lifetime of historic conventionally framed houses.

FIGURE 24.18
Violent winds toppled five 80-foot-tall hardwood trees onto this house built with SIP panels. The trees had 12- to 16-inch-diameter trunks and weighed up to 5000 pounds, but the house sustained no structural damage, and there was no cracking of drywall, wracking of windows, or separation of panels. Only siding and shingles needed to be repaired. (*House built in 1995 in Coopersville, MI, by Controlled Environment Structures*)

Energy Performance

A number of research reports indicate that houses built with SIP construction use less energy for heating and cooling than comparable wood light frame houses. In a comparison of 6-inch walls, the standard (EPS) SIP wall has an overall R-value of 21.6, whereas a wood stud wall has an overall value of 13.7. There are a number of reasons for the superior performance of the SIP wall:

1. The rigid insulation performs substantially better than fiberglass batts. EPS has an R-value approximately 20 percent higher than fiberglass batts.

2. There are fewer framing members in SIP walls to act as thermal bridging that conducts heat rapidly through the wall.

3. There are virtually no voids within an SIP wall.

4. The SIP panels, when installed carefully, create a more nearly airtight envelope than standard framing and thus reduce infiltration.

A detailed side-by-side comparison by the Florida Solar Energy Center of two otherwise identical houses, one built with SIPs and the other with conventional framing, showed that the SIP house was 38 percent tighter and reduced heating and cooling costs by 14 to 20 percent. Other side-by-side analyses have shown similar or more dramatic results.

Environmental Performance

Studies have shown that typical SIP houses use considerably less framing lumber and slightly less wood fiber overall than conventionally framed houses. The differences have to do with the substitution by SIP houses of interior OSB for studs and the reduction of wood waste by SIP systems. Proponents of SIPs point out that the use of less framing lumber is significant in itself because lumber is typically manufactured from high-grade trees while engineered OSB panels are made from waste wood products.

Because about 50 percent of the waste generated from the construction of a conventionally framed house consists of wood products, SIP houses have an advantage over conventional framing in this area. Construction waste volume differences depend more on management and house design than construction system, however, and SIP components are subject to rigid foam board waste both in the factory and at the site, so no clear advantage can be given to SIP houses for reduction of overall waste stream volume. Both wood scraps and polystyrene can be recycled.

As for embodied energy, SIP systems require three to five times as much energy to produce as the components for standard framing with comparable insulation, but it is estimated that initial energy investment in an SIP house should be reclaimed within a decade from saved heating and cooling energy.

Fire Resistance

Flame spread and smoke tests indicate that, overall, houses built with sandwich panels appear to be more resistant to fire than conventionally framed structures. The toxicity of smoke from burning insulation varies according to the insulation type: EPS fumes are more-or-less the same as wood smoke whereas polyurethane smoke contains cyanide. All foam plastics drip sticky, flaming goo during a fire, which is a serious hazard.

Building with Structural Insulated Panels

SIP components dovetail elegantly into the sequence of standard wood light frame construction, where they typically replace conventionally framed exterior walls and roofs. The SIP walls bear on the floor structure—either slab or framed—and the SIP roof bears on the walls (Figure 24.19). Once the panels are installed, the remainder of the construction proceeds as it would in standard construction except for the added labor involved in electrical wiring and sometimes plumbing and heating.

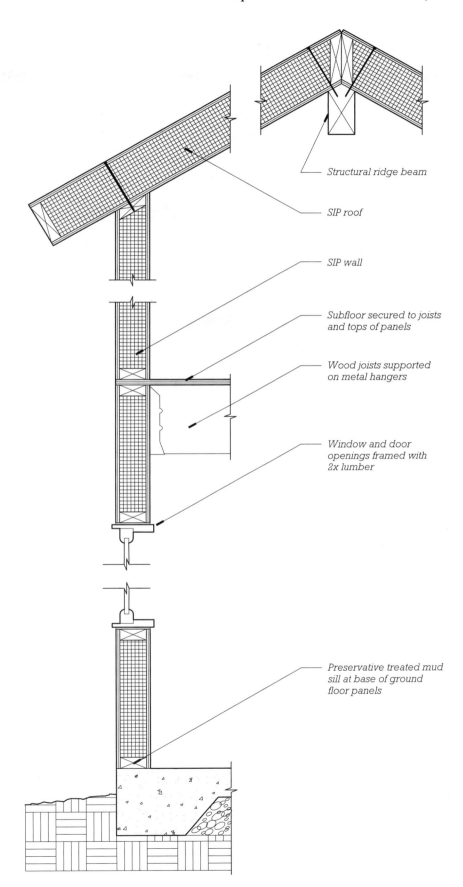

Structural ridge beam

SIP roof

SIP wall

Subfloor secured to joists and tops of panels

Wood joists supported on metal hangers

Window and door openings framed with 2x lumber

Preservative treated mud sill at base of ground floor panels

FIGURE 24.19
Building section showing typical SIP construction. The ground floor can also be conventionally framed with joists over a basement or crawlspace or made with SIPs as shown in Figures 24.20 and 24.21. The second floor can also be conventionally framed as shown in Figures 24.26 and 24.27. The roof, if not vaulted as shown here, is usually made of trusses.

FIGURE 24.20
SIPs used as a floor structure span between the beams of a simple post-and-beam foundation. The addition of a simple skirt between floor and ground will create a crawlspace. (*Courtesy of Energy Studies in Buildings Laboratory, Center for Housing Innovation, Department of Architecture, University of Oregon*)

FIGURE 24.21
A partially complete SIP floor structure. (*Courtesy of Thermal Shell Homes, Houston, Texas*)

FIGURE 24.22
Single-story SIP houses can be assembled without the assistance of a crane when 4- or 8-foot-long panels are used. (*Photo by Rob Thallon*)

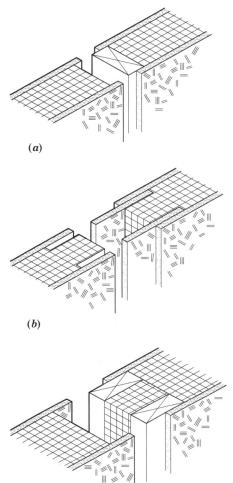

(*a*)

(*b*)

(*c*)

FIGURE 24.23
There are numerous ways to join insulated panels. Each is designed to provide structural integrity while simultaneously maintaining the continuity of thermal insulation.

Houses made with SIPs commonly are constructed on a basement or crawlspace foundation and a conventionally framed floor or a slab. However, the combined structural and insulating capacities of SIPs often inspire their use as a floor structure supported by a simple post-and-beam foundation (Figures 24.20 and 24.21). An SIP floor system provides the same subfloor platform on which to erect walls as a conventionally framed floor.

There is usually the need to add a skirt around the perimeter of a house with an SIP floor system.

The connection of wall panels to the floor structure is virtually identical to wood light framing except that the bottom plate is attached to the subfloor before the wall is erected and acts as a locating device (Figure 24.19). Typical 4-by-8-foot (1.2×2.4 m) panels weigh approximately 110 to 130 pounds (50 to 60 kg) and can

be positioned by two workers, although a light-duty crane can speed the process (Figure 24.22). (For two-story buildings, a crane would have to appear twice, as it does for open panels with conventional floor framing.) As they are being installed, panels are mechanically fastened edge to edge, and a number of details have been devised for this purpose (Figure 24.23). Wall panels are nailed or screwed to one another at corners, and all wall

FIGURE 24.24
Most SIP manufacturers compromise the thermal insulation slightly at the corner by introducing a thermal bridge in order to make a simple structural joint.

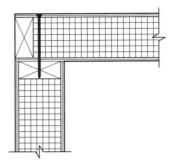

panels are tied together at the top with a continuous plate (Figure 24.24).

Upper floors are rarely made of insulated panels because there is no need for insulation at this location. Framing of upper floors with open-web joists or wooden I-joists provides cavities in which plumbing and other utilities can be easily located (Figure 24.26). The perimeter of framed upper floors must be insulated conventionally unless the SIP wall extends up to the level of the subfloor (Figure 24.27).

Roof panels are installed in a similar fashion to wall panels and are attached to the top plates of walls with long nails or screws that pass through the roof panels and penetrate the top

FIGURE 24.25
SIPs are relatively easy to modify at the site. Here a hot-wire cutting tool removes insulation from a floor panel so that a loadbearing rim joist can be inserted. Similar techniques can be employed to frame rough openings in walls. (*Courtesy of Energy Studies in Buildings Laboratory, Center for Housing Innovation, Department of Architecture, University of Oregon*)

plate of the wall (Figures 24.28 and 24.29). Eaves and overhanging rakes, where they exist, are commonly made by cantilevering the insulated panel, even though the insulation performs no thermal function where it is not located over a living space.

Openings for windows and doors are best manufactured into the panels at the factory but can also be cut on site with an oversized power saw and a router. All panels are caulked or sealed at all edges as they are installed in order to create as airtight an envelope as possible. Most SIP manufacturers suggest that the panels themselves are sufficient as an air barrier if properly caulked and sealed at top and bottom plates and at adjacent panels.

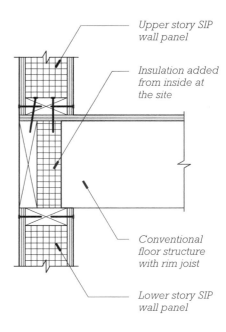

FIGURE 24.26
The common stacking of upper floor joists on SIP walls requires the addition of insulation across the height of the floor structure.

Upper story SIP wall panel

Insulation added from inside at the site

Conventional floor structure with rim joist

Lower story SIP wall panel

FIGURE 24.27
The use of open-web floor joists allows for easy routing of utilities through the floor structure. (*Courtesy of Thermal Shell Homes, Houston, Texas*)

FIGURE 24.28
These small roof panels were hoisted onto the second floor by hand and are being positioned by hand. The panels will be attached to the floor structure at their base and to each other at the ridge—forming a self-supporting structural triangle. (*Courtesy of Energy Studies in Buildings Laboratory, Center for Housing Innovation, Department of Architecture, University of Oregon*)

FIGURE 24.29
These large roof panels were placed with a crane and are supported at their top edge by a structural ridge beam. (*Courtesy of Thermal Shell Homes, Houston, Texas*)

FIGURE 24.30
The completed shell of the SIP house pictured in Figures 24.20 and 24.28. The extra electric meter bases are for energy consumption studies. (*Courtesy of Energy Studies in Buildings Laboratory, Center for Housing Innovation, Department of Architecture, University of Oregon*)

SELECTED REFERENCES

1. Andrews, Steve. *Foam-Core Panels and Building Systems: Principles, Practice, and Product Directory* (3rd ed.). Arlington, MA: Cutler Information Corporation, 1999.

A comprehensive survey of the residential SIP industry. Topics range from energy performance to the construction process to marketing. Updated frequently.

2. Morley, Michael. *Building with Structural Insulated Panels (SIPs)*. Newtown, CT: Taunton Press, 2000.

A comprehensive guide to the construction of SIP walls and roofs. Tools, panel selection, and the integration of mechanical systems are also discussed.

KEY TERMS AND CONCEPTS

panelized construction
framed panel
open panel
closed panel

open-panel construction
design of panels
sandwich panel
stressed-skin panel

structural insulated panel (SIP)
creep
thermal performance of SIP
SIP construction

REVIEW QUESTIONS

1. What are the significant differences between framed open-panel construction and conventional stick framing? What are the practical limitations of the open-panel system?

2. Why are framed closed panels seldom used?

3. Discuss the advantages and disadvantages of SIPs as compared to conventional stick framing.

4. Why does an SIP wall outperform a wood stud wall of equal thickness from a thermal point of view?

5. Describe the typical sequence of events required to build the shell of a house using SIP construction.

EXERCISES

1. Try to locate a local open-panel manufacturer. Visit one of its projects under construction if possible and compare what you see with conventional wood light frame construction. How much time does the panelized process save? Is a cost comparison available? What are the apparent limitations of the panelized process?

2. Locate two or three SIP manufacturers and have descriptive literature sent to you. Compare the systems and speculate about how the differences would make construction more or less complicated. What other effects would the differences between the systems appear to have?

3. Obtain a set of plans for a panelized house. Compare these plans to a conventional house of similar size. What are the major differences between the plans? Why do you suppose these differences occur? Can you think of alternative ways to design the details illustrated in the plans?

GLOSSARY

A

ABS pipe Plumbing pipe made from acrylonitrile butadiene styrene plastic.

AC *See* **Alternating current.**

ACC *See* **Autoclaved cellular concrete.**

Acrylic A transparent plastic material widely used in sheet form for glazing windows and skylights.

ADA *See* **Airtight drywall approach.**

Admixture A substance other than cement, water, and aggregates included in a concrete mixture for the purpose of altering one or more properties of the concrete.

Adobe Bricks made of clay and sand, dried but not fired. Walls of adobe bricks in mud mortar.

Aggregate Inert particles, such as sand, gravel, crushed stone, or expanded minerals, in a concrete or plaster mixture.

Air barrier A fabric, membrane, or continuous surface that is used to reduce infiltration of air through building assemblies, such as walls, ceilings, or floors.

Air conditioner A unit that takes in air, cools and dehumidifies it, and distributes it to the interior space of the building.

Air-conditioning system A system that achieves thermal comfort in a building by circulating filtered air that is either heated or cooled as conditions demand.

Air-entraining cement A portland cement with an admixture that causes a controlled quantity of stable, microscopic air bubbles to form in the concrete during mixing.

Air filter, electronic A device that removes particles from the air by applying an electric charge to the particles and a charge of opposite polarity to metal plates that attract the particles and hold them.

Airtight drywall approach (ADA) Restricting the passage of water vapor into the insulated cavities of a light frame building by eliminating passages around and through the gypsum board interior finish.

Air-to-air heat exchanger A device that exhausts air from a building while recovering much of the heat from the exhausted air and transferring it to the incoming air.

Air/vapor barrier (AVB) A sheet material used both as a vapor retarder and as an air barrier.

Air volume damper A flap or valve that may be adjusted to regulate the amount of air that passes through a duct.

All-in-one loan Money borrowed at interest that serves as both construction loan and permanent mortgage.

Allowable stress The tensile, compressive, or shear stress that may be borne safely by a structural element. Allowable stress is usually based on the breaking stress of the material, multiplied by a suitable factor of safety.

Alternating current (AC) Electrical current that oscillates in polarity and magnitude, so that it is capable of changing voltage with a transformer.

Ampere A measure of the rate of flow of electrical energy.

Amp Short for ampere.

Anchor bolt A bolt embedded in concrete for the purpose of fastening a building frame to a concrete or masonry foundation.

Angle A structural section of steel or aluminum whose profile resembles the letter *L*.

ANSI American National Standards Institute, an organization that fosters the establishment of voluntary industrial standards.

Appraised value The value of a building or piece of land as estimated by a professional who is expert in making such estimates.

Apron The finish piece that covers the joint between a window stool or sill and the wall finish below.

Arch A structural device that supports a vertical load by translating it into axial, inclined forces at its supports.

Architect One who is licensed by a state government as being legally entitled to prepare designs and plans for buildings of any size and scope.

Architectural drawings Construction drawings that show the general configuration and finishes of a building, as contrasted with electrical, structural, and mechanical drawings.

Ash dump A door in the underfire of a fireplace that permits ashes from a fire to be swept into a chamber beneath, from which they may be removed at a later time.

Ashlar Squared stonework.

Asphalt A tarry brown or black mixture of hydrocarbons.

Asphalt roll roofing A continuous sheet of the same roofing material used in asphalt shingles. *See* **Asphalt shingle.**

Asphalt-saturated felt A moisture-resistant sheet material, available in several different thicknesses, usually consisting of a heavy paper that has been impregnated with asphalt.

Asphalt shingle A roofing unit composed of a heavy organic or inorganic felt saturated with asphalt and faced with mineral granules.

ASTM American Society for Testing and Materials, an organization that promulgates standard methods of testing the performance of building materials and components.

Atactic polypropylene An amorphous form of polypropylene used as a modifier in modified-bitumen roofing.

Autoclaved cellular concrete (ACC) Concrete formulated so as to contain a large percentage of gas bubbles as a result of a chemical reaction that takes place in an atmosphere of steam.

AVB *See* **Air/vapor barrier.**

Awning window A window that pivots on an axis at or near the top edge of the sash and projects toward the outdoors.

Axial In a direction parallel to the long axis of a structural member.

B

Backer rod A flexible, compressible strip of plastic foam inserted into a joint to limit the depth to which sealant can penetrate.

Backfill Earth or earthen material used to fill the excavation around a foundation; the act of filling around a foundation.

Backsplash The upright curb at the rear of a countertop.

Balancing, system Adjusting a heating or cooling system so that it provides commensurate levels of comfort in all the rooms of a building.

Ballast A heavy material installed over a roof membrane to prevent wind uplift and shield the membrane from sunlight.

Balloon frame A wooden building frame composed of closely spaced members nominally 2 inches (51 mm) in thickness, in which the wall members are single pieces that run from the sill to the top plates at the eave.

Baluster A small, vertical member that serves to fill the opening between a handrail and a stair or floor.

Band joist A wooden joist running perpendicular to the primary direction of the joists in a floor and closing off the floor platform at the outside face of the building. Also called a rim joist.

Bar A small rolled steel shape, usually round or rectangular in cross section; a rolled steel shape used for reinforcing concrete.

Barge rafter A rafter in a rake overhang. Also known as a fly rafter.

Basalt A dark-gray to black, fine-grained, igneous rock.

Baseboard A strip of finish material placed at the junction of a floor and a wall

to create a neat intersection and to protect the wall against damage from feet, furniture, and floor-cleaning equipment.

Batten A strip of wood or metal used to cover the crack betwwen two adjoining boards or panels.

Batter board A temporary frame built just outside the corner of an excavation to carry marks that lie on the surface planes of the basement that will be built in the excavation.

Bay A rectangular area of a building defined by four adjacent columns; a portion of a building that projects from a facade.

Bead A narrow line of sealant; a strip of metal or wood used to hold a sheet of glass in place; a narrow, convex molding profile; a metal edge or corner accessory for plaster.

Beam A straight structural member, usually horizontal, that acts primarily to resist nonaxial loads.

Bearing A point at which one building element rests upon another.

Bearing capacity The relative ability of a soil to support a building, expressed as an allowable stress upon the soil.

Bearing stiffener Code terminology for web stiffener in light-gauge steel construction.

Bearing wall A wall that supports floors or roofs.

Bed joint The horizontal layer of mortar beneath a masonry unit.

Bedrock A solid stratum of rock.

Benchmark A fixed indication on a building site that may be used to establish grades and other vertical dimensions.

Bent A plane of framing consisting of beams and columns joined together, often with rigid joints.

Bentonite clay An absorptive, colloidal clay that swells to several times its dry volume when saturated with water.

Bevel An end or edge that is cut at an angle other than a right angle.

Bevel siding Wood cladding boards that taper in cross section.

Bifold door A door that consists of two leaves hinged together along their vertical edges and restrained by a guide wheel that runs in a track at the outer top corner. The door is opened by pulling the center, hinged joint outward into the room, which causes one edge of the door to pull away from its jamb and to move laterally.

Binder A substance that serves to hold a mass of fibers or particles together.

Bird's mouth The triangular notch in a rafter that gives it full bearing on a wall plate or beam.

Bitumen A tarry mixture of hydrocarbons, such as asphalt or coal tar.

Black pipe Iron pipe used for distribution of fuel gas in a building.

Blast furnace slag A byproduct of iron manufacture used as a concrete admixture.

Blind nailing Attaching boards to a frame, sheathing, or subflooring with toe nails driven through the edge of each piece so as to be completely concealed by the adjoining piece.

Blocking Pieces of wood inserted tightly between joists, studs, or rafters in a building frame to stabilize the structure, inhibit the passage of fire, provide a nailing surface for finish materials, or retain insulation.

Bluestone A sandstone that is gray to blue-gray in color and splits readily into thin slabs.

Board foot A unit of lumber volume, a rectangular solid nominally 12 square inches in cross-sectional area and 1 foot long.

Board siding Wood cladding made up of boards, as differentiated from shingles or manufactured wood panels.

BOCA Building Officials and Code Administrators International, Inc.

Bolster A long chair used to support reinforcing bars in a concrete slab.

Bolt A fastener consisting of a cylindrical metal body with a head at one end and a helical thread at the other, intended to be inserted through holes in adjoining pieces of material and closed with a threaded nut.

Bond In masonry, the adhesive force between mortar and masonry units, or the pattern in which masonry units are laid to tie two or more wythes together into a structural unit. In reinforced concrete, the adhesion between the surface of a reinforcing bar and the surrounding concrete. In construction law, a sum of money from which expenses are to be paid to do the work of a contractor if that contractor should fail to fulfill his or her contract.

Bond breaker A strip of material to which sealant does not adhere.

Bonded In construction law, covered by a bond. In concrete construction, adhered to concrete.

Boot fitting A transitional adaptor that connects a duct to a register.

Bottom bars The reinforcing bars that lie close to the bottom of a beam or slab.

Box beam A bending member of metal or plywood whose cross section resembles a closed rectangular box.

Box girder A major spanning member of concrete or steel whose cross section is a hollow rectangle or trapezoid.

Bracing Diagonal members, either temporary or permanent, installed to stabilize a structure against lateral loads.

Brad A small finish nail.

Branch line A secondary plumbing supply line.

Brick A rectangular building block made of clay or shale that has been fired in a kiln to convert it to a ceramic material. Bricks are sized to be held easily in one hand.

Bridging Bracing or blocking installed between steel or wood joists at midspan to stabilize them against buckling and, in some cases, to permit adjacent joists to share loads.

British thermal unit (Btu) The quantity of heat required to raise 1 pound of water 1 degree Fahrenheit.

Broom finish A skid-resistant texture imparted to an uncured concrete surface by dragging a stiff-bristled broom across it.

Brown coat The second coat of plaster in a three-coat application.

Brownstone A brownish or reddish sandstone.

Btu *See* **British thermal unit.**

Buckling Structure failure by gross lateral deflection of a slender element under compressive stress, such as the sideward buckling of a long, slender column or the upper edge of a long, thin floor joist.

Builder One who constructs houses or other buildings.

Builder's level An optical sighting device used to determine elevations and alignments on a building site.

Building code A set of legal restrictions intended to assure a minimum standard of health and safety in buildings.

Building designer One who undertakes the design of buildings of limited size and scope, usually at the scale of single-family residences.

Building drain The piping at the base of a DWV system that lies within the building.

Building sewer The piping that carries sewage from a building to the municipal sewer or a private sewer disposal system.

Built-up roof (BUR) A roof membrane laminated from layers of asphalt-saturated felt or other fabric, bonded together with bitumen or pitch.

BUR *See* **Built-up roof.**

Butt The thicker end, such as the lower edge of a wood shingle or the lower end of a tree trunk; a joint between square-edged pieces; a weld between square-edged pieces of metal that lie in the same plane; a type of door hinge that attaches to the edge of the door.

Butyl rubber A synthetic rubber compound.

Bypass door A door that opens by rolling laterally on a track at its top edge and that moves past another such door mounted parallel, offset, and to one side.

C

Cabinet door, full overlay A door without rabbeted edges that overlaps the face frame of a cabinet.

Cabinet door, inset A door that is fully recessed within the frame of the cabinet.

Cabinet door, lip A door with rabbeted edges whose lips overlap the face frame of a cabinet.

Cabinet, face frame A cabinet with a front frame that surrounds the doors and drawers.

Cabinet, frameless A cabinet on which only the doors and drawer fronts are visible.

Cabinet, modular A mass-produced cabinet that is furnished in standard sizes.

Cable, main service The large wire that brings electricity into a building.

Cable, metal clad Hot, neutral, and ground wires in a protective, flexible metal sheath.

Cable, nonmetallic Hot, neutral, and ground wires wrapped in plastic insulation as a single cable.

Calcining The driving off of the water of hydration from gypsum by the application of heat.

Camber A slight, intentional initial curvature in a beam or slab.

Cambium The thin layer beneath the bark of a tree that manufactures cells of wood and bark.

Cantilever A beam, truss, or slab that extends beyond its last point of support.

Cant strip A strip of material with a sloping face used to ease the transition from a horizontal to a vertical surface at the edge of a membrane roof.

Capillary action The pulling of water through a small orifice or fibrous material by the adhesive force between the water and the material.

Capillary break A slot or groove intended to create an opening too large to be bridged by a drop of water and, thereby, to prevent the passage of water by capillary action.

Carbide-tipped tools Drill bits, saws, and other tools with cutting edges made of an extremely hard alloy.

Carpenter One who makes things of wood.

Casement window A window that pivots on an axis at or near a vertical edge of the sash.

Casing The wood finish pieces surrounding the frame of a window or door; a cylindrical steel tube used to line a drilled or driven hole in foundation work.

Casting Pouring a liquid material or slurry into a mold whose form it will take as it solidifies.

Cast-in-place Concrete that is poured in its final location; sitecast.

Cast iron Iron with too high a carbon content to be classified as steel.

Caulk A low-range sealant.

Cavity wall A masonry wall that includes a continuous airspace between its outermost wythe and the remainder of the wall.

Cee section A metal framing member whose cross-sectional shape resembles the letter *C*.

Cellulose A complex polymeric carbohydrate of which the structural fibers in wood are composed.

Celsius A temperature scale on which the freezing point of water is established as 0 and the boiling point as 100 degrees.

Cement A substance used to adhere material together; in concrete work, the dry powder that, when it has combined chemically with the water in the mix, cements the particles of aggregate together to form concrete.

Cementitious Having cementing properties, usually with reference to inorganic substances, such as portland cement and lime.

Centering Temporary formwork for an arch, dome, or vault.

Central vacuum system A vacuum cleaning system with its fan, filter, and dust container mounted in a fixed location and connected to a system of pipes that make the vacuum available anywhere in the building.

Ceramic tile Small, flat, thin clay tiles intended for use as wall and floor facings.

Certified Professional Building Designer One who is licensed by a state government to design buildings of limited size and scope, usually at a residential scale.

Chair A device used to support reinforcing bars.

Chamfer A flattening of a longitudinal edge of a solid member on a plane that lies at an angle of 45 degrees to the adjoining planes.

Change order An agreement to add to, subtract from, or alter a work that is already under contract. The change order may include a provision for adjusting the contract price accordingly.

Channel A steel or aluminum section shaped like a rectangular box with one side missing.

Chimney A vertical tube to conduct smoke and combustion gases from a fuel-burning device to the atmosphere.

Chimney cap A concrete slab cast on top of a chimney and around the flue liners to keep water out of the interior construction of the chimney.

Chinking Any material that is inserted between logs in stacked log construction to seal out weather. The process of inserting chinking.

Chlorinated polyethylene A plastic material used in roof membranes.

Chlorosulfonated polyethylene A plastic material used in roof membranes.

Chord A top or bottom member of a truss.

Circuit A group of electrical outlets served by a single circuit breaker and cable.

Circuit breaker A device that shuts off electricity in a circuit if the capacity of the circuit is exceeded.

Cladding A material used as the exterior wall enclosure of a building.

Clamp A tool for holding two pieces of material together temporarily; unfired bricks piled in such a way that they can be fired without using a kiln.

Class A, B, C roofing Roof covering materials classified according to their resistance to fire when tested in accordance with ASTM E108. Class A is the highest, Class C the lowest.

Cleanout A removable plug or cover in a waste line for the purpose of allowing inspection and cleaning of the line.

Cleanout hole An opening at the base of a masonry wall through which mortar droppings and other debris can be removed prior to grouting the interior cavity of the wall.

Clear dimension, clear opening The dimension between opposing inside faces of an opening.

Clinker A fused mass that is an intermediate product of cement manufacture; a brick that is overburned.

Closer The last masonry unit laid in a course; a partial masonry unit used at the corner of a header course to adjust the joint spacing; a mechanical device for regulating the closing action of a door.

CLSM *See* **Controlled low-strength material.**

CMU *See* **Concrete masonry unit.**

Code *See* **Building code.**

Coefficient of performance (COP) A measure of the efficiency of a heating or cooling device, equal to the amount of energy that is produced by the device divided by the amount of energy introduced into it.

Cohesive soil A soil such as clay whose particles are able to adhere to one another by means of cohesive and adhesive forces.

Cold-formed steel construction Steel framing composed of members that were created by folding sheet steel at room temperature.

Collar joint The vertical mortar joint between wythes of masonry.

Collar tie A piece of wood nailed across two opposing rafters near the ridge to resist wind uplift.

Column An upright structural member acting primarily in compression.

Column footing A footing made to support the load of a column.

Combination door A door with interchangeable inserts of glass and insect screening, usually used as a second, exterior door and mounted in the same opening with a conventional door.

Combination window A sash that holds both insect screening and a retractable sheet of glass, mounted in the same frame with a window and used to increase its thermal resistance.

Combustion air Outdoor air that is taken into a building in order to provide oxygen for the combustion of fuel in a furnace, stove, or fireplace.

Common Bond Brickwork laid with each five courses of stretchers followed by one course of headers.

Competitive bidding A process of soliciting and comparing cost proposals from two or more business entities, usually for the purpose of making a business arrangement with the one that proposes the lowest cost.

Composite A material or assembly made up of two or more materials bonded together to act as a single structural unit.

Compression A squeezing force.

Compression gasket A synthetic rubber strip that seals around a sheet of glass or a wall panel by being squeezed tightly against it.

Compressive strength The ability of a structural material to withstand squeezing forces.

Concave joint A mortar joint tooled into a curved, indented profile.

Concrete A structural material produced by mixing predetermined amounts of portland cement, aggregates, and water, and allowing this mixture to cure under controlled conditions.

Concrete block A concrete masonry unit, usually hollow, that is larger than a brick.

Concrete masonry unit (CMU) A block of hardened concrete, with or without hollow cores, designed to be laid in the same manner as a brick or stone; a concrete block.

Concrete, stamped A concrete slab that has been impressed with a pattern to simulate brick or stone paving.

Condensate Water formed as a result of condensation.

Condensation The process of changing from a gaseous to a liquid state, especially as applied to water.

Conduit A steel or plastic tube through which electrical wiring is run.

Construction documents Construction drawings, specifications, and contracts for the construction of a building.

Construction drawings Drawings that show all the dimensional and technical information required to build a building.

Construction loan Money borrowed at interest to pay construction costs until a permanent mortgage is arranged after the completion of construction.

Construction observation The appearance of an architect on the building site at intervals during construction for the purposes of answering questions from contractors, resolving problems, identifying and correcting deficient work, certifying payments to contractors, etc.

Continuous ridge vent A screened, water-shielded ventilation opening that runs continuously along the ridge of a gable roof.

Contract builder One who undertakes a legal obligation to build houses that are owned by others.

Contract negotiations Discussions for the purpose of agreeing upon the terms of a contract.

Contractor A person or organization that undertakes a legal obligation to do construction work.

Control joint An intentional, linear discontinuity in a structure or component, designed to form a plane of weakness where cracking can occur in response to various forces so as to minimize or eliminate cracking elsewhere in the structure.

Controlled low-strength material (CLSM) A concrete that is purposely formulated to have a very low but known strength, used primarily as a backfill material.

Convection The transfer of heat by the circulation of air.

Convector A heat exchange device that uses the heat in steam, hot water, or an electric resistance element to warm the air in a room; often called, inaccurately, a radiator.

Cope The removal of a portion at the rear of a mitered joint in order to facilitate connection to another member.

Coped connection A joint in which the end of one member is cut to match the profile of the other member.

Coping A protective cap on the top of a masonry wall.

Coping saw A handsaw with a thin, very narrow blade, used for cutting detailed shapes in the ends of wood moldings and trim.

Corbel A spanning device in which masonry units in successive courses are cantilevered slightly over one another; a projecting bracket of masonry or concrete.

Corner bead A metal or plastic strip used to form a neat, durable edge at an outside corner of two walls of plaster or gypsum board.

Cornice The exterior detail at the meeting of a wall and a roof overhang; a decorative molding at the intersection of a wall and a ceiling.

Corrosion Oxidation, such as rust.

Corrosion inhibitor A concrete admixture intended to prevent oxidation of reinforcing bars.

Corrugated Pressed into a fluted or ribbed profile.

Cost-plus contract An agreement to perform work for payments equal to the direct cost of the work plus additional payments for the contractor's overhead and profit.

Counterflashing A flashing turned down from above to overlap another flashing turned up from below so as to shed water.

Countertop The horizontal working surface of a base cabinet, often furnished as a unit with a front edge and a rear backsplash.

Countertop, plastic laminate A level top surface for a counter, made of a thin plastic laminate glued to a particleboard backing.

Countertop, solid surface A level top surface for a counter, made of a synthetic sheet material without a facing.

Course A horizontal layer of masonry units one unit high; a horizontal line of shingles or siding.

Coursed Laid in courses with straight bed joints.

Cove base A flexible strip of plastic or synthetic rubber used to finish the junction between resilient flooring and a wall.

Cover plate A plate of plastic, metal, or wood that is fastened to the face of an electrical box to enclose its contents and give a finished appearance.

Crawlspace A space that is not tall enough to stand in, located beneath the lowest floor of a building.

Crawlspace plenum A crawlspace that is used in effect as a giant supply duct for heating and cooling.

Creep A permanent inelastic deformation in a material due to changes in the material caused by the prolonged application of a structural stress.

Cricket A small gable that diverts roof water from the upslope side of a chimney.

Cripple stud A wood wall-framing member that is shorter than full-length studs because it is interrupted by a header or sill.

Cross-grain wood Wood incorporated into a structure in such a way that its direction of grain is perpendicular to the direction of the principal loads on the structure.

Cruck A framing member cut from a bent tree so as to form one-half of a rigid frame.

Cup A curl in the cross section of a board or timber caused by unequal shrinkage or expansion between one side of the board and the other.

Curing The hardening of concrete, plaster, gunnable sealant, or other wet materials. Curing can occur through evaporation of water or a solvent, hydration, polymerization, or chemical reactions of various other types, depending on the formulation of the material.

Curing compound A liquid that, when sprayed on the surface of newly placed concrete, forms a water-resistant layer to prevent premature dehydration of the concrete.

Current The volume of electricity that flows through a wire, measured in amperes.

Custom-designed house A dwelling that has been designed wholly or in part for its specific owner.

D

Damper A flap to control or obstruct the flow of gases; specifically, a metal control flap in the throat of a fireplace or in an air duct.

Dampproofing A coating applied to the outside face of a basement wall as a barrier to the passage of water.

Dap A notch at the end of a piece of material.

Darby A stiff straightedge of wood or metal used to level the surface of wet plaster.

Daylighting Illuminating the interior of a building by natural means.

DC *See* **Direct current**

Deadbolt A sliding metal bar, activated by a key, that locks a door or window securely.

Dead load The weight of the building or building component itself.

Decking A material used to span across beams or joists to create a floor or roof surface.

Decking board, synthetic A manufactured, decay-resistant plank, usually made of wood fibers in a plastic binder.

Deed restriction A provision in the legal certificate of ownership that a specified activity may not occur, or a specified physical feature may not be built, on a piece of land.

Deformation A change in the shape of a structure or structural element caused by a load or force acting on the structure.

Dehumidifier A device that removes moisture vapor from the air.

Dehydration The removal of water from a substance.

Demand water heater A device that heats water instantaneously, as it is used.

Design/build Furnishing both design and construction services with a single business entity.

Design development A phase of the process of designing a building that starts with a schematic design and ends with a finished, detailed design.

Designer One who configures a building and prepares drawings for its construction.

Developer One who brings together money, land, dwelling designs, and contractors to create a group of buildings for sale or lease.

Dew point The temperature at which water will begin to condense from a mass of air of a given temperature and moisture content.

Diagonal bracing *See* **Bracing.**

Diamond saw A tool with a moving chain, belt, wire, straight blade, or circular blade whose cutting action is carried out by diamonds.

Diaphragm action A bracing action that derives from the stiffness of a thin plane of material when it is loaded in a direction parallel to the plane. Diaphragms in buildings are typically floor, wall, or roof surfaces of wood panels, reinforced masonry, steel decking, or reinforced concrete.

Differential settlement Subsidence of the various foundation elements of a building at differing rates.

Diffuser A louver shaped so as to distribute air about a room.

Dimension lumber Lengths of wood, rectangular in cross section, sawed directly from the log.

Dimension stone Building stone cut to a rectangular shape.

Direct current (DC) Electrical flow of constant magnitude and polarity.

Direct hiring Employing a contractor or builder for a project without competitive bidding.

Distribution box A small chamber that conducts the effluent from a septic tank into the various lines of a drain field.

Dormer A structure protruding through the plane of a sloping roof, usually containing a window and having its own smaller roof.

Dosing tank An underground vessel that houses a pump to pump sewage uphill to a septic tank.

Double glazing Two parallel sheets of glass with an airspace between.

Double-hung window A window with two overlapping sashes that slide vertically in tracks.

Double-strength glass Glass that is approximately ⅛ inch (3 mm) in thickness.

Dowel A short cylindrical rod of wood or steel; a steel reinforcing bar that projects from a foundation to tie it to a column or wall, or from one section of a concrete slab or wall to another.

Downflow furnace A furnace in which return air enters at the top and supply air exits at the bottom.

Downspout A vertical pipe for conducting water from a roof to a lower level.

Drafting service A business whose purpose is to prepare drawings for the construction of buildings.

Drainage Removal of water.

Drainage matting A thick, open-structured plastic fabric that is applied to the outside of a foundation wall to allow water that approaches the wall through the soil to fall freely to a drainage pipe at the base of the wall for removal, thus preventing the water from exerting pressure on the wall or passing through the wall.

Drain field An underground layer of porous material into which septic tank effluent is introduced in order to disperse it into the soil.

Drain piping Pipes that carry the waste from water closets.

Drain–waste–vent (DWV) system All the piping and accessories that work to remove waste water from a building.

Drawer guides Devices on which a drawer rests and moves in and out of the cabinet of which it is a part.

Drip A discontinuity formed in the underside of a window sill, soffit, or wall component to force adhering drops of water to fall free of the face of the building rather than to move farther toward the interior.

Dry press process A method of molding slightly damp clays and shales into bricks by forcing them into molds under high pressure.

"Dry" systems Systems of construction that use little or no water during construction, as differentiated from systems such as masonry, plastering, and ceramic tile work.

Drywall *See* **Gypsum board.**

Dry well An underground pit filled with broken stone or other porous material, from which rainwater from a roof drainage system can seep into the surrounding soil.

Duct A hollow conduit, commonly of sheet metal, through which air can be circulated; a tube used to establish a passage for a posttensioning tendon in a concrete structure.

Duct adaptor A sheet metal transition fitting that connects a duct to a furnace.

Duct board Rigid glass fiber board covered on the outside with metal foil, used to make air ducts.

Duct, branch An air duct that connects a trunk duct to one or more registers.

Duct, return An air duct that takes air from a room back to the furnace or air-handling unit.

Duct, supply An air duct that provides air to a room.

Duct tape Adhesive tape used to seal seams in ducts.

Duct, trunk A main air duct that supplies branch ducts.

DWV *See* **Drain–waste–vent system.**

E

Earthen construction Walls made of soil.

Eave The horizontal edge at the low side of a sloping roof.

Edge band A strip of wood veneer or other material that is applied to the edge of a panel material.

Edge bead A strip of metal or plastic used to make a neat, durable edge where plaster or gypsum board abuts another material.

Efflorescence A powdery deposit on the face of a surface of masonry or concrete, caused by the leaching of chemical salts by water migrating from within the structure to the surface.

Effluent The liquid that emerges from a septic tank or other sewage treatment facility or device.

Egress Emergency escape from a building.

Egress window A window that meets code-specified minimum dimensions and can be used for emergency escape from a building.

EIFS *See* **Exterior insulation and finish system.**

EIFS, barrier An exterior insulation and finish system that depends for watertightness on the integrity of its outer layer.

EIFS, water managed An exterior insulation and finish system that includes provisions for free drainage of any water that gets behind the exterior layer of the system.

Elastic Able to return to its original size and shape after removal of stress.

Elastomer A rubber or synthetic rubber.

Elastomeric Rubberlike.

Elbow A plumbing fitting that makes an angle (usually of 90°) in piping.

Electrical box A metal or plastic container that houses and supports a receptacle, switch, or fixture, and its connections to a circuit.

Electric baseboard heater An electric resistance coil in a sheet metal housing that runs along the wall–floor junction and heats the air in a room.

Elevation A drawing that views a building from any of its sides; a vertical height above a reference point such as sea level.

Elongation Stretching under load; growing longer because of temperature expansion.

Embodied energy The energy that is consumed in producing and transporting a material to the building site.

Enamel A glossy or semigloss paint.

End nail A nail driven through the side of one piece of lumber and into the end of another.

Engineered fill Earth compacted into place in such a way that it has predictable physical properties, based on laboratory tests and specified, supervised installation procedures.

English bond Brickwork laid with alternating courses, each consisting entirely of headers or stretchers.

EPDM Ethylene propylene diene monomer, a synthetic rubber material used in roofing membranes.

Erosion control plan A scheme for preventing the loss of soil from a site during a construction project.

Expanded metal lath A thin sheet of metal that has been slit and stretched to transform it into a mesh used as a base for the application of plaster.

Expansion joint A surface divider joint that provides space for the surface to expand. In common usage, a building separation joint.

Exposed aggregate finish A concrete surface in which the coarse aggregate is revealed.

Extended-life admixture A substance that retards the onset of the curing reaction in mortar so that the mortar may be used over a protracted period of time after mixing.

Exterior insulation and finish system (EIFS) A cladding system that consists of a thin layer of reinforced stucco applied directly to the surface of an insulating plastic foam board.

Extrusion The process of squeezing a material through a shaped orifice to produce a linear element with the desired cross section; an element produced by this process.

F

Face brick A brick selected on the basis of appearance and durability for use in the exposed surface of a wall.

Face nail A nail driven through the side of one wood member into the side of another.

Face shell The portion of a hollow concrete masonry unit that forms the face of the wall.

Fahrenheit A temperature scale on which the boiling point of water is fixed at 212 degrees and the freezing point at 32.

Fan A device for moving air. In buildings, it is invariably powered by electricity.

Fan-coil unit A heating device in which an electric fan circulates room air past a coil of pipe through which hot water is circulated.

Fanlight A semicircular or semielliptical window above an entrance door, often with radiating muntins that resemble a fan.

Fascia The exposed vertical face of an eave.

Felt A thin, flexible sheet material made of soft fibers pressed and bonded together. In building practice, a thick paper or a sheet of glass or plastic fibers.

Fiberboard A panel made with wood fibers that are bonded together with a resin adhesive under heat and pressure. Fiberboard resins typically emit toxic formaldehyde fumes, but formaldehyde-free fiberboard is available.

Fibrous admixture Short fibers of glass, steel, or polypropylene mixed into concrete to act as reinforcement against plastic shrinkage cracking.

Fieldstone Rough building stone gathered from river beds and fields.

Figure The surface pattern of the grain of a piece of smoothly finished wood or stone.

Fill, compacted Earth that has been deposited and tamped in accordance with an engineer's directions in order that it may have specified structural properties.

Filter fabric A textile used to prevent soil particles from entering an open drainage medium such as crushed stone.

Finger joint A glued end connection between two pieces of wood, using an interlocking pattern of deeply cut "fingers." A finger joint creates a large surface for the glue bond, allowing it to develop the full tensile strength of the wood that it connects.

Finial An ornament at the top of a roof or spire.

Finish Exposed to view; material that is exposed to view.

Finish carpenter One who does finish carpentry.

Finish carpentry The wood components exposed to view on the interior of a building, such as window and door casings, baseboards, bookshelves, and the like.

Finish coat The final coat of plaster.

Finish floor The floor material exposed to view, as differentiated from the subfloor, which is the loadbearing floor surface beneath.

Finish grade The final level of the topsoil around a building.

Firebox The chamber that houses the fire in a fireplace, stove, or furnace.

Firebrick A brick that is made of special clays and produced in such a manner that it is able to tolerate very high temperatures without being damaged.

Firebrick, refractory A brick made from special clays to withstand very high temperatures, as in a fireplace, furnace, or industrial chimney.

Firecut A sloping end cut on a wood beam or joist where it enters a masonry wall. The purpose of the firecut is to allow the wood member to rotate out of the wall without prying the wall apart, if the floor or roof structure should burn through in a fire.

Fireplace surround The decorative trim that closes the gap between the masonry front of a fireplace and the surrounding wall surface.

Fireproofing Material used around a steel structural element to insulate it against excessive temperatures in case of fire.

Fire resistance rating The time, in hours or fractions of an hour, that a material or assembly will resist fire exposure as determined by ASTM E119.

Fire resistant Noncombustible; slow to be damaged by fire; forming a barrier to the passage of fire.

Fire separation wall A wall required under a building code to divide two parts of a building from one another as a deterrent to the spread of fire.

Firestop A wood or masonry baffle used to close an opening between studs or joists in a balloon or platform frame in order to retard the spread of fire through the opening.

Firestopping A component or mastic installed in an opening through a floor or around the edge of a floor to retard the passage of fire.

Fire wall A wall extending from foundation to roof, required under a building code to separate two parts of a building from one another as a deterrent to the spread of fire.

Fire zone A legally designated area of a city in which construction must meet established standards of fire resistance and/or combustibility.

Firing The process of converting dry clay or shale into a ceramic material through the application of intense heat.

First cost The cost of construction.

Fixed window Glass that is immovably mounted in a wall.

Flagstone Flat stones used for paving or flooring.

Flange A projecting crosspiece of a wide-flange or channel profile; a projecting fin.

Flashing A thin, continuous sheet of metal, plastic, rubber, or waterproof paper used to prevent the passage of water through a joint in a wall, roof, or chimney.

Flat grain lumber Dimension lumber sawed in such a way that the annual rings run more or less parallel to the faces of each piece, creating a swirled, active pattern.

Flatwork Concrete slabs on grade.

Flemish Bond Brickwork laid with each course consisting of alternating headers and stretchers.

Flex duct Air ducting that is easily bent to change its direction.

Flitch-sliced veneer A thin sheet of wood cut by passing a block of wood vertically against a long, sharp knife.

Float A trowel with a slightly rough surface used in an intermediate stage of finishing a concrete slab; as a verb, to use a float for finishing concrete.

Floating floor A finished floor surface of wood and/or plastic laminate that is not attached directly to the subfloor, allowing the finished floor to be cushioned and to expand and contract.

Flue A passage for smoke and combustion products from a furnace, stove, water heater, or fireplace.

Flue liner Sections of the tube made of terra cotta or other material installed inside the masonry shell of a chimney to provide a tight interior surface that is resistant to heat and acids.

Fluid-applied roof membrane A roof membrane applied in one or more coats of a liquid that cure to form an impervious sheet.

Flush Smooth, lying in a single plane.

Flush door A door with smooth, planar faces.

Fly ash A waste product of coal-fired power plants, used as a concrete admixture.

Fly rafter A rafter in a rake overhang. Also known as a barge rafter.

Foamed-in-place insulation A plastic thermal insulating material that is sprayed, injected, or poured into a building cavity as a liquid that expands to become a lightweight foam.

Foil-backed gypsum board Gypsum board with aluminum foil laminated to its back surface to act as a vapor retarder and thermal insulator.

Footing The widened part of a foundation that spreads a load from the building across a broader area of soil.

Forced-air heating system A system that maintains thermal comfort in a building by distributing heated air through ducts to the various rooms.

Form-release compound A substance applied to concrete formwork to prevent concrete from adhering.

Form tie A steel or plastic rod with fasteners on each end, used to hold together the two surfaces of formwork for a concrete wall.

Formwork Temporary structures of wood, steel, or plastic that serve to give shape to poured concrete and to support it and keep it moist as it cures.

Foundation The portion of a building that has the sole purpose of transmitting structural loads from the building into the earth.

Framing plan A diagram showing the arrangement and sizes of the structural members in a floor or roof.

French door A symmetrical pair of glazed doors hinged to the jambs of a single frame and meeting at the center of the opening.

Frictional soil A soil such as sand that has little or no attraction between its particles and derives its strength from geometric interlocking of the particles; a noncohesive soil.

Frieze block A wood block fastened between rafters at the eave to prevent their rotation and to enclose the roof structure where a soffit is not employed.

Frieze vent A screened opening through or above a frieze block, used to allow air to flow into the attic or the space below the roof sheathing.

Frost line The depth in the earth to which the soil can be expected to freeze during a severe winter.

Furnace A device that burns fuel and transfers the heat of combustion to air for the purpose of heating a building.

Furring channel A formed sheet metal furring strip.

Furring strip A length of wood or metal attached to a masonry or concrete wall to permit finish materials to be fastened to the wall with nails or screws.

G

Gable The triangular wall beneath the end of a gable roof.

Gable-end vent A louvered, screened opening in the gable end of an attic for the circulation of air between the attic and the outdoors.

Gable roof A roof consisting of two oppositely sloping planes that intersect at a level ridge.

Gable vent *See* **Gable-end vent.**

Galvanizing The application of a zinc coating to steel as a means of preventing corrosion.

Gambrel A roof shape consisting of two superimposed levels of gable roofs with the lower level at a steeper pitch than the upper.

Gasket A dry, resilient material used to seal a joint between two rigid assemblies by being compressed between them.

Gauge A measure of the thickness of sheet material or of the diameter of a wire.

General contractor A contractor who has responsibility for the overall conduct of a construction project.

Geotextile A synthetic cloth used beneath the surface of the ground to stabilize soil or promote drainage.

GFI *See* **Ground-fault interrupter.**

GFRC *See* **Glass-fiber-reinforced concrete.**

Girder A horizontal beam that supports other beams; a very large beam, especially one that is built up from smaller elements.

Glass block A hollow masonry unit made of glass.

Glass fiber batt A thick, fluffy, nonwoven insulating blanket of filaments spun from glass.

Glass-fiber-reinforced concrete Concrete that is reinforced with short glass fibers mixed with the concrete.

Glaze A glassy finish on a brick or tile; as a verb, to install glass.

Glazed a. Fitted with sheets of glass. b. Coated with a vitrified layer.

Glazier One who installs glass.

Glazier's points Small pieces of metal driven into a wood sash to hold the glass in place.

Glazing The act of installing glass; installed glass; as an adjective, referring to materials used in installing glass.

Glazing compound Any of a number of types of mastic used to bed small lights of glass in a frame.

Glue-laminated timber A timber made up of a large number of small strips of wood glued together.

Glulam A short expression for glue-laminated timber.

Grade A classification of size, quality, or suitability for an intended purpose; to classify as to size or quality.

Grade The surface of the ground; to move earth for the purpose of bringing the surface of the ground to an intended level or profile.

Grade beam A reinforced concrete beam that transmits the load from a bearing wall into spaced foundations such as pile caps or caissons.

Grading, finish Final placement and smoothing of the soil around a building.

Grading, rough Bringing the soil up to its approximate final level on the site by means of power equipment.

Grain In wood, the direction of the longitudinal axes of the wood fibers, or the figure formed by the fibers.

Granite Igneous rock with visible crystals of quartz and feldspar.

Gravel base The cushion of gravel or crushed stone beneath a slab on grade.

Green building Building so as to satisfy the needs of the current generation without compromising the ability of future generations to satisfy their needs.

Ground A strip attached to a wall or ceiling to establish the level to which plaster should be applied.

Ground An electric wire that is connected to the earth; the earth itself.

Ground-fault interrupter (GFI) A device that shuts off the electricity in a circuit if any current is leaked to ground.

Grounding wire An uninsulated wire that is connected to the earth.

Grout A high-slump mixture of portland cement, aggregates, and water, which can be poured or pumped into cavities in concrete or masonry for the purpose of embedding reinforcing bars and/or increasing the amount of loadbearing material in a wall; a mortar used to fill joints between ceramic tiles or quarry tiles.

Gunnable sealant A sealant material that is extruded in liquid or mastic form from a sealant gun.

Gusset plate A flat steel plate used to connect the members of a truss; a stiffener plate.

Gutter A channel to collect rainwater and snowmelt at the eave of a roof.

Gutter, continuous metal A roof water channel made from sheet metal on the building site, which allows it to be made in lengths that are longer than prefabricated gutters.

Gypsum Hydrous calcium sulfate.

Gypsum backing board A lower cost gypsum panel intended for use as an interior layer in multilayer constructions of gypsum board.

Gypsum board An interior facing panel consisting of a gypsum core sandwiched between paper faces. Also called drywall, plasterboard.

Gypsum lath Small sheets of gypsum board manufactured specifically for use as a plaster base.

Gypsum plaster Plaster whose cementing substance is gypsum.

Gypsum sheathing panel A water-resistant, gypsum-based sheet material used for exterior sheathing.

Gypsum wallboard *See* **Gypsum board.**

H

Hardboard A very dense panel product, usually with at least one smooth face, made of highly compressed wood fibers.

Hawk A square piece of sheet metal with a perpendicular handle beneath, used by a plasterer to hold a small quantity of wet plaster and transfer it to a trowel for application to a wall or ceiling.

Header A band joist or a joist that supports other joists; a structural member that spans an opening in a framed wall; a brick or other masonry unit that is laid across two wythes with its end exposed in the face of the wall.

Head joint The vertical layer of mortar between ends of masonry units.

Hearth The noncombustible floor area outside a fireplace opening.

Heartwood The dead wood cells in the center region of a tree trunk.

Heating element A device that introduces heat into water or air, such as an electric resistance coil.

Heat of hydration The thermal energy given off by concrete or gypsum as it cures.

Heat pump A device that utilizes a refrigeration cycle either to heat or to cool a building by passing air or water over either the condensing coils or the evaporating coils, respectively.

Heaving The forcing upward of ground or buildings by the action of frost or pile driving.

Heavy timber construction Framing a building with timbers nominally 6 inches (150 mm) in thickness or greater.

High-range sealant A sealant that is capable of a high degree of elongation without rupture.

High-voltage line A wire that carries electricity at a very high potential in order to reduce transmission losses over long distances.

Hip The diagonal intersection of planes in a hip roof.

Hip roof A roof consisting of four sloping planes that intersect to form a pyramidal or elongated pyramid shape.

Hold-down A metal device that connects a building frame to its foundation in such a way that the frame cannot lift up due to external forces, such as wind or earthquake.

Hollow concrete masonry Concrete masonry units that are manufactured with open cores, such as ordinary concrete blocks.

Hollow-core door A door consisting of two face veneers separated by an airspace, with solid wood spacers around the four edges. The face veneers are usually connected by a grid of thin spacers within the airspace.

Hook A semicircular bend in the end of a reinforcing bar, made for the purpose of anchoring the end of the bar securely into the surrounding concrete.

Hopper window A window whose sash pivots on an axis along or near the sill and that opens by tilting toward the interior of the building.

Horizontal force A force whose direction of action is horizontal or nearly horizontal. *See also* **Lateral force.**

Housewrap A sheet material that does not allow the passage of air or liquid water but that permits the passage of water vapor. It is placed in the location of a weather barrier, where it is employed as both a weather barrier and an air barrier.

Humidifier A device that adds moisture vapor to air.

HVAC systems Heating, ventilating, and air-conditioning systems.

Hydrated lime Calcium hydroxide produced by burning calcium carbonate to form calcium oxide (quicklime), then allowing the calcium oxide to combine chemically with water.

Hydration A process of combining chemically with water to form molecules or crystals that include hydroxide radicals or water of crystallization.

Hydronic baseboard system A heating system in which warm water is circulated through baseboard fin-tube convectors that run along the wall–floor junction.

Hydronic heating system A system that circulates warm water through convectors to heat a building.

Hydronic tubing Plastic or metal tubing used to circulate warm water through a radiant floor or radiant panel.

Hydrostatic pressure Pressure exerted by standing water.

Hygroscopic Readily absorbing and retaining moisture.

I

Ice-and-water shield A modified-bitumen sheet material that is installed under shingles near the eaves of a roof or in other locations where ice damming may be a problem. The sticky material of which the sheet is made seals around nails that are driven through it and maintains the watertightness of the sheet.

Ice dam An obstruction along the eave of a roof caused by the refreezing of water emanating from melting snow on the roof surface above.

ICF *See* **Insulating concrete form.**

Igneous rock Rock formed by the solidification of magma.

I-joist A manufactured wood framing member whose cross-sectional shape resembles the letter I.

Infiltration The unintentional inward or outward leakage of air between indoors and outdoors through walls, floors, roofs, and around windows and doors.

Insulating concrete form (ICF) A hollow block made of polystyrene foam. Blocks are stacked into walls, after which reinforcing bars are placed in the cavities and the cavities are filled with concrete to produce a strong, well-insulated wall.

Insulating glass Double or triple glazing.

Insulation, thermal A material with a low thermal conductivity that is included in a building assembly for the purpose of reducing heat flow through the assembly.

Insured Covered by a contract or contracts that will pay for damages or expenses in the event that specified things go wrong.

Interior decorator One who designs rooms, including paint colors, wallpaper patterns, floorcoverings, and furnishings.

Interior designer One licensed by a state to practice interior decoration.

Interlayment A particular kind of roof underlayment that is placed between courses of wooden shakes.

Iron In pure form, a metallic element. In common usage, ferrous alloys other than steels, including cast iron and wrought iron.

Irrigation system A built-in arrangement of pipes and fittings designed to supply water to lawns and planting beds.

Isocyanurate foam A thermosetting plastic foam with thermal insulating properties.

J

Jack A device for exerting a large force over a short distance, usually by means of screw action or hydraulic pressure.

Jack rafter A shortened rafter that joins a hip or valley rafter.

Jamb The vertical side of a door or window.

Joint compound A paste that is applied to the edges and nail holes of gypsum wallboard to hide them and to create an apparently monolithic surface.

Joist One of a parallel array of light, closely spaced beams used to support a floor deck, ceiling, or low-slope roof.

Joist hanger A metal device, attached with nails, that connects a joist to a header or beam.

K

Key A slot formed in a concrete surface for the purpose of interlocking with a subsequent pour of concrete; a slot at the edge of a precast member into which grout will be poured to lock it to an adjacent member; a mechanical interlocking of plaster with lath.

Kiln A furnace for firing clay or glass products; a heated chamber for seasoning wood; a furnace for manufacturing quicklime, gypsum hemihydrate, or portland cement.

Knee brace A short diagonal brace between a column and a beam.

Knee wall A short wall under the slope of a roof.

L

Lacquer A coating that dries extremely quickly through evaporation of a volatile solvent.

Lag screw A large-diameter wood screw with a square or hexagonal head.

Laminate As a verb, to bond together in layers; as a noun, a material produced by bonding together layers of material.

Laminated glass A glazing material consisting of outer layers of glass bonded to an inner layer of transparent plastic.

Laminated veneer lumber (LVL) Wood members that are made up of thin wood veneers joined with glue.

Laminated wood *See* **Glue-laminated timber.**

Landing A platform in or at either end of a stair.

Land planner One who determines the physical layout of a housing development, including property lines, building placements, streets, drainage, utilities, and so forth.

Landscape architect One who prepares designs for outdoor plantings, pavings, and construction.

Lap joint A connection in which one piece of material is placed partially over another piece before the two are fastened together.

Laser level A device that uses a laser to establish level planes on a building site, from which measurements may be taken to locate floors, ceilings, and other portions of a building.

Lateral force A force acting generally in a horizontal direction, such as wind, earthquake, or soil pressure against a foundation wall.

Lateral thrust The horizontal component of the force produced by an arch, dome, vault, or rigid frame.

Latex caulk A low-range sealant based on a synthetic latex.

Lath (rhymes with "math") A base material to which plaster is applied.

Lathe (rhymes with "bathe") A machine in which a piece of material is rotated against a sharp cutting tool to produce a shape, all of whose cross sections are circles; a machine in which a log is rotated against a long knife to peel a continuous sheet of veneer.

Lather (rhymes with "rather") One who applies lath.

Leach line A perforated pipe that distributes septic tank effluent into a drain field.

Lead (rhymes with "bead") In masonry work, a corner or wall end accurately constructed with the aid of a spirit level to serve as a guide for placing the bricks in the remainder of the wall.

Leader (rhymes with "feeder") A vertical pipe for conducting water from a roof to a lower level. Also known as a downspout.

Leaf A separately movable division of a sliding or hinged door; one of a pair of doors.

Let-in bracing Diagonal bracing that is nailed into notches cut in the face of the studs so that it does not increase the thickness of the wall.

Level cut A saw cut that produces a level surface at the lower end of a sloping rafter when the rafter is in its final position.

Licensed Certified by a state government as being competent to practice a profession or trade.

Life-cycle cost A cost that takes into account both the first cost and the costs of maintenance, replacement, fuel consumed, monetary inflation, and interest over the lifetime of the object being evaluated.

Lift A horizontal layer of material that is all deposited at one time, such as a layer of earth deposited for compaction, or a layer of concrete in a form.

Light A sheet of glass.

Light-gauge steel stud A length of thin sheet metal folded into a stiff shape and used as a wall framing member.

Lighting, line voltage Bulbs that are powered by electricity at 110 to 120 volts.

Lighting, low voltage Bulbs that are powered by electricity at such a low voltage that there is no danger of electric shock.

Lignin The natural cementing substance that binds together the cellulose in wood.

Limestone A sedimentary rock consisting of calcium carbonate, magnesium carbonate, or both.

Linoleum A resilient floorcovering material composed primarily of ground cork and linseed oil on a burlap or canvas backing.

Lintel A beam that carries the load of a wall across a window or door opening.

Liquid sealant *See* **Gunnable sealant.**

Listing agreement A contract between a realtor and the owner of one or more buildings, for the purpose of selling the building(s).

Lite *See* **Light.**

Live load The weight of snow, people, furnishings, machines, vehicles, and goods in or on a building.

Load A weight or force acting on a structure.

Loadbearing Supporting a superimposed weight or force.

Local source heaters Space heating devices that are totally self-contained, needing no connection to hydronic piping or ducts.

Lockset A prefabricated hardware assembly for a door that includes a latch and a deadbolt.

Log construction Walls made of tree trunks laid horizontally one atop another, notched together at corners.

Log walls, settling of The large vertical shrinkage that occurs when stacked logs in a wall dry out.

Lookout A short rafter, running perpendicular to the other rafters in the roof, which supports a rake overhang.

Loose-fill insulation Thermal insulating material consisting of loose granules or fibers that are poured into cavities in the building, such as the spaces between the ceiling joists.

Louver An array of numerous sloping, closely spaced slats used to diffuse air or to prevent the entry of rainwater into a ventilating opening.

Low-e coating *See* **Low-emissivity (low-e) coating.**

Low-emissivity (low-e) coating A surface coating for glass that permits the passage of most shortwave electromagnetic radiation (light and heat), but reflects most longer wave radiation (heat).

Low-range sealant A sealant that is capable of only a slight degree of elongation prior to rupture.

Low-slope roof A roof that is pitched at an angle so near to horizontal that it must be made waterproof with a continuous membrane rather than shingles; commonly and inaccurately referred to as a flat roof.

LVL *See* **Laminated veneer lumber.**

M

Main disconnect A circuit breaker that can be tripped to shut off all flow of electricity to a building.

Main panel The group of circuit breakers, mounted in a steel box, from which electricity is distributed throughout a building.

Main, water A large, underground pipe that brings municipal water to a location just outside the boundaries of a piece of land.

Making-up a box Pulling cables into an electrical box, clamping them to the box, and stripping the sheath and insulation from the ends of the wires.

Mansard A roof shape consisting of two superimposed levels of hip roofs with the lower level at a steeper pitch than the upper.

Mantel A shelf above a fireplace opening.

Manufactured housing Dwellings that are assembled largely in a factory, with a relatively minor proportion of assembly work remaining to be done on the building site.

Manufactured stone Artificial stone, usually made from portland cement concrete.

Marble A metamorphic rock formed from limestone by heat and pressure.

Mason One who builds with bricks, stones, or concrete masonry units; one who works with concrete.

Masonry Brickwork, concrete blockwork, and stonework.

Masonry cement Portland cement with dry admixtures designed to increase the workability of mortar.

Masonry opening The clear dimension required in a masonry wall for the installation of a specific window or door unit.

Masonry unit A brick, stone, concrete block, glass block, or hollow clay tile intended to be laid in mortar.

Masonry veneer A single wythe of masonry used as a facing over a frame of wood or metal.

Masterformat The copyrighted title of a uniform indexing system for construction specifications, as created by the Construction Specifications Institute and Construction Specifications Canada.

Mastic A viscous, doughlike, adhesive substance; any of a large number of formulations for different purposes such as sealants, adhesives, glazing compounds, or roofing cements.

MDF Medium-density fiberboard.

Medium-range sealant A sealant material that is capable of a moderate degree of elongation before rupture.

Meeting rail The wood or metal bar along which one sash of a double-hung, single-hung, or sliding window seals against the other.

Member An element of a structure such as a beam, a girder, a column, a joist, a piece of decking, a stud, or a component of a truss.

Membrane A sheet material that is impervious to water or water vapor.

Metal lath A steel mesh used primarily as a base for the application of plaster or stucco.

Metamorphic rock A rock created by the action of heat or pressure on a sedimentary rock or soil.

Meter base A box on which an electric meter is mounted.

Mil One one-thousandth of an inch.

Mild steel Ordinary structural steel, containing less than three-tenths of 1 percent carbon.

Milling Shaping or planing by using a rotating cutting tool.

Millwork Wood interior finish components of a building, including moldings, windows, doors, cabinets, stairs, mantels, and the like.

Miter A diagonal cut at the end of a piece; the joint produced by joining two diagonally cut pieces at right angles.

Mobile home Euphemism for a portable house that is entirely factory built on a steel underframe supported by wheels.

Model building code A code that is offered by a recognized national organization as worthy of adoption by state or local governments.

Modified bitumen A natural bitumen with admixtures of synthetic compounds to enhance such properties as flexibility, plasticity, and durability.

Modular Conforming to a multiple of a fixed dimension.

Modular home Euphemism for a house assembled on the site from two or more boxlike factory-built sections.

Modulus of elasticity An index of the stiffness of a material, derived by measuring the elastic deformation of the material as it is placed under stress, and then dividing the stress by the deformation.

Moisture barrier A membrane used to prevent the migration of liquid water through a floor or wall.

Molding A strip of wood, plastic, or plaster with an ornamental profile.

Moment A twisting action; a torque; a force acting at a distance from a point in a structure so as to cause a tendency of the structure to rotate at that point.

Momentum The energy possessed by a moving body.

Monolithic Of a single massive piece.

Mortar A substance used to join masonry units, consisting of cementitious materials, fine aggregate, and water.

Mortar, refractory A mortar especially formulated to withstand high temperatures.

Mortise and tenon A joint in which a tonguelike protrusion (tenon) on the end of one piece is tightly fitted into a rectangular slot (mortise) in the side of the other piece.

Mudsill The wood strip that is bolted to the top of a foundation.

Mud slab A slab of weak concrete placed directly on the ground to provide a temporary working surface that is hard, level, and dry.

Mulch A granular or fibrous material spread over the ground to discourage weed growth and retain moisture in the soil.

Mullion A vertical or horizontal bar between adjacent window or door units. A framing member in a metal-and-glass curtain wall.

Muntin A small vertical or horizontal bar between small lights of glass in a sash.

Muriatic acid Hydrochloric acid.

N

Nail-base sheathing A sheathing material, such as wood boards or plywood, to which siding can be attached by nailing, as differentiated from one such as cane fiber board or plastic foam board that is too soft to hold nails.

Nailing fin A flange around the outside of a window or door unit that lies flat against the wall sheathing, through which nails are driven to fasten the unit to the wall.

Nailing plate A solid sheet metal plate through which common nails cannot be driven and which is used to protect electrical wiring and plumbing where it must pass near to the surface of a wood light frame.

Nail popping The loosening of nails holding gypsum board to a wall, caused by drying shrinkage of the studs.

Nail set A hardened steel punch used to drive the head of a nail to a level below the surface of the wood.

Natural gas Fuel gas that is obtained from wells in the earth, usually a mixture of short-chain hydrocarbons such as methane, ethane, propane, and butane.

Neoprene Polychloroprene, a synthetic rubber.

Net ventilation area The actual open area of a ventilation device, which is the gross area of the opening minus the total area that is occupied by louvers and/or by the wires or vanes of an insect screen.

Noise reduction coefficient (NRC) An index of the proportion of incident sound that is absorbed by a surface, expressed as a decimal fraction of 1.

Nominal dimension An approximate dimension assigned to a piece of material as a convenience in referring to the piece.

Nonaxial In a direction not parallel to the long axis of a structural member.

Nonbearing Not carrying a load.

Nosing The projecting forward edge of a stair tread; the projecting front edge of a countertop.

NRC *See* **Noise reduction coefficient.**

Nut A steel fastener with internal helical threads, used to close a bolt.

O

o.c. Abbreviation for "on center," meaning that the spacing of framing members is measured from the center of one member to the center of the next, rather than the clear spacing between members.

Ogee An S-shaped curve.

Oil-based caulk A low-range sealant made with linseed oil.

Organic soil Soil containing decayed vegetable and/or animal matter; topsoil.

Oriented strand board (OSB) A building panel composed of long shreds of wood fiber oriented in specific directions and bonded together under pressure.

Overhead Costs of doing business, such as office rental, salaries and wages of management and clerical personnel, expenses of vehicles, tools, and machines, etc.

Oxidation Corrosion; rusting; rust.

P

Package fireplace A factory-built fireplace that is installed as a unit.

Paint A heavily pigmented coating applied to a surface for decorative and/or protective purposes.

Painting, interior, first phase The painting of the plaster or gypsum board, which is usually done before the installation of trim, moldings, and woodwork.

Painting, interior, second phase The painting of trim, moldings, and woodwork.

Panel A broad, thin piece of wood; a sheet of building material such as plywood or particleboard; a prefabricated building component that is broad and thin, such as a curtain wall panel; a region of a truss bounded by two vertical interior members.

Panel, closed A prefabricated building component that is finished on the inside at the time it is installed.

Panel door A wood door in which one or more thin panels are held by stiles and rails.

Panel, framed A wall, floor, or roof component made of wood or steel framing members and applied sheathing. May also include insulation, wiring, and/or interior and exterior finishes.

Panelized construction Assembling a building from prefabricated panels.

Panel, open A prefabricated building component that is unfinished on the interior at the time it is installed.

Panel, sandwich A prefabricated building component that is made of wood and/or gypsum board facings adhered to a core of foam plastic.

Panel, stressed skin A prefabricated building component consisting of wood panel facings fastened or adhered to rigid spacers. In common usage, a sandwich panel.

Panel, structural insulated (SIP) A sandwich panel with structural capacity to carry more than its own weight.

Parallel strand lumber (PSL) Manufactured wood components that are made of wood shreds oriented parallel to the long axis of each piece and bonded together with adhesive.

Parapet The portion of an exterior wall that projects above the level of the roof.

Parge To apply portland cement plaster over masonry to make it less permeable to water.

Particleboard A building panel composed of small particles of wood bonded together under pressure.

Parting compound *See* **Form-release compound.**

Partition An interior nonloadbearing wall.

Partition wall An interior wall that subdivides space but does not support a floor or roof.

Patio door A large glass exterior door made of two panels, one of which is fixed and the other of which operates by sliding it horizontally.

Patterned glass Glass into which a texture has been rolled during manufacture.

Paver A half-thickness brick used as finish flooring.

Paving, asphalt A horizontal outdoor surface, compacted in place, consisting of fine and coarse aggregates in a binder made of heavy components of petroleum.

Paving, stone A horizontal outdoor surface made of stones.

Paving, unit A horizontal outdoor surface made of small units of brick, stone, or concrete masonry.

Percolation test A procedure for determining the porosity of a soil, used as a basis for designing a private sewage disposal system.

Performance grade A rating used to indicate the relative weather resistance of a window.

Perimeter drain A perforated pipe bedded in crushed stone or gravel that serves as a conduit for the removal of water from around a basement.

Periodic kiln A kiln that is loaded and fired in discrete batches, as differentiated from a tunnel kiln, which is operated continuously.

Perlite Expanded volcanic glass, used as a lightweight aggregate in concrete and plaster and as an insulating fill.

Perm A unit of vapor permeability.

Permanent mortgage. A long-term loan of money at interest, to be repaid during the occupancy of the building.

PEX tubing Flexible water piping made of cross-linked polyethylene, a very durable, stable material.

Pier block A precast foundation unit usually used as a column footing.

Pilaster A vertical, integral stiffening rib in a masonry or concrete wall.

Pitch The slope of a roof or other plane, often expressed as inches of rise per foot of run; a dark, viscous hydrocarbon distilled from coal tar; a viscous resin found in wood.

Pitched roof A sloping roof.

Pivoting window A window that opens by rotating around its vertical centerline.

Plainsawing Sawing a log into dimension lumber without regard to the direction of the annual rings.

Plain slicing Cutting a flitch into veneers without regard to the direction of the annual rings.

Planing Smoothing the surface of a piece of wood, stone, or steel with a cutting blade.

Planned unit development (PUD) A tract of houses that is planned as a coordinated group, that does not necessarily meet all zoning ordinances to the letter, but that offers compensatory features for those ordinances with which it does not comply.

Plaster A cementitious material, usually based on gypsum or portland cement, applied to lath or masonry in paste form, to harden into a finish surface.

Plasterboard *See* **Gypsum board.**

Plaster of Paris Pure calcined gypsum.

Plasticity The ability to retain a shape attained by pressure deformation.

Plastics Synthetically produced giant molecules.

Plate A horizontal top or bottom member in a platform frame wall structure.

Plate glass Glass of high optical quality produced by grinding and polishing both faces of a glass sheet.

Platform frame A wooden building frame composed of closely spaced members nominally 2 inches (51 mm) in thickness, in which the wall members do not run past the floor framing members.

Plenum The space between the ceiling of a room and the structural floor above, or a crawlspace, used as a passage for ductwork, piping, and wiring.

Plumb Vertical.

Plumb cut A saw cut that produces a vertical (plumb) surface at the lower end of a sloping rafter after the rafter is in its final position.

Plumbing fitting An elbow, tee, coupling, union, or other device for connecting pipes.

Plumbing up The process of making a building frame vertical and square.

Ply A layer, as in a layer of felt in a built-up roof membrane or a layer of veneer or veneers in plywood.

Plywood A wood panel composed of an odd number of layers of wood veneer bonded together under pressure.

Pocket door A door that opens by rolling laterally into a chamber in the interior of the adjacent partition.

Polycarbonate An extremely tough, strong, usually transparent plastic used for window and skylight glazing, light fixture globes, door sills, and other applications.

Polyethylene A thermoplastic widely used in sheet form for vapor retarders, moisture barriers, and temporary construction coverings.

Polymer A large molecule composed of many identical chemical units.

Polypropylene A plastic formed by the polymerization of propylene.

Polystyrene foam A thermoplastic foam with thermal insulating properties.

Polysulfide A high-range gunnable sealant.

Polyurethane Any of a large group of synthetic resins and synthetic rubber compounds used in sealants, varnishes, insulating foams, and roof membranes.

Polyurethane foam A thermosetting foam with thermal insulating properties.

Polyvinyl chloride (PVC) A thermoplastic material widely used in construction products, including plumbing pipes, floor tiles, wall coverings, and roof membranes. Called "vinyl" for short.

Portland cement The gray powder used as the binder in concrete, mortar, and stucco.

Potable water Water fit for human consumption.

Pour To cast concrete; an increment of concrete casting carried out without interruption.

Powder coating A coating produced by applying a powder consisting of thermosetting resins and pigments, adhering it to the substrate by electrostatic attraction, and fusing it into a continuous film in an oven.

Powder-driven Inserted by a gunlike tool using energy provided by an exploding charge of gunpowder.

Precast concrete Concrete cast and cured in a position other than its final position in the structure.

Predecorated gypsum board Gypsum board finished at the factory with a decorative layer of paint, paper, or plastic.

Prehung door A door that is hinged to its frame in a factory or shop.

Prescriptive building code A set of legal regulations that mandate specific construction details and practices rather than establish performance standards.

Pressure tank An airtight vessel in which a volume of compressed air maintains pressure to distribute water throughout a building.

Pressure-treated lumber Lumber that has been impregnated with chemicals under pressure, for the purpose of retarding either decay or fire.

Primer A coating that prepares a surface to accept another type of coating or a sealant.

Prime window A window unit that is made to be installed permanently in a building.

Propane A hydrocarbon gas commonly compressed for storage in tanks and used as a fuel for cooking, water heating, and space heating.

Property line The boundary of a tract of land.

Protected membrane roof A membrane roof assembly in which the thermal insulation lies above the membrane.

Protection board A panel material used to prevent damage to a waterproofing membrane on the outside of a foundation.

PSL *See* **Parallel strand lumber.**

P-trap A plumbing waste fitting that resembles the letter *P*, designed to hold a water seal to keep sewer gases from entering the building.

PUD *See* **Planned unit development.**

Pull A handle for operating a cabinet door or drawer.

Pump, submersible An electric water pump that is installed at the bottom of a well.

Purlins Beams that span across the slope of a steep roof to support the roof decking.

Putty A simple glazing compound used to seal around a small light.

PVC *See* **Polyvinyl chloride.**

PVC pipe Plumbing pipe made from polyvinyl chloride.

Q

Quarry An excavation from which building stone is obtained; the act of taking stone from the ground.

Quarry tile A large clay floor tile.

Quartersawn Lumber sawn in such a way that the annual rings run roughly perpendicular to the face of each piece.

Quartersliced Veneer sliced in such a way that the annual rings run roughly perpendicular to the face of each veneer.

Quoin (pronounced "coin") A corner reinforcing of cut stone or bricks in a masonry wall, usually done for decorative effect.

R

Rabbet A longitudinal groove cut at the edge of a member to receive another member; also called a rebate.

Radiant barrier A reflective foil placed adjacent to an airspace in roof or wall assemblies as a deterrent to the passage of infrared energy.

Radiant floor A floor that is heated to serve as a source of heat for the building.

Radiant floor, staple-up A radiant floor created by stapling hydronic tubing to the bottom of the subfloor.

Radiant floor, thin slab A radiant floor made of hydronic tubing embedded in a cementitious material.

Radiant panel A ceiling, wall, or floor panel that is heated to serve as a source of heat for the building.

Radiation The transfer of heat by electromagnetic waves.

Rafter A framing member that runs up and down the slope of a steep roof.

Rail A horizontal framing piece in a panel door; a handrail.

Rain cap A covering intended to minimize rainwater penetration of the top of a chimney flue.

Rainscreen principle A theory by which wall cladding is made watertight by providing wind-pressurized air chambers behind joints to eliminate air pressure differentials between the outside and the inside that might transport water through the joints.

Rake The sloping edge of a steep roof.

Rammed earth Wall construction that consists of damp earth compacted into formwork.

Ray A tubular cell that runs radially in a tree trunk.

RBM *See* **Reinforced brick masonry.**

Realtor One who acts as an agent to buy and sell buildings on behalf of others.

Rebar Building-industry slang for reinforcing bar.

Receptacle A place where an appliance can be connected to an electric circuit.

Recycled content The percentage of a product that has been recovered from waste rather than from original sources.

Reflective coated glass Glass onto which a thin layer of metal or metal oxide has been deposited to reflect light and/or heat.

Register The outlet into a room for conditioned air from a central heating or cooling system. A register is provided with built-in damper vanes that allow some control of air volume.

Reinforced brick masonry (RBM) Brickwork into which steel bars have been embedded to impart tensile strength to the construction.

Reinforced concrete Concrete work into which steel bars have been embedded to impart tensile strength to the construction.

Reinforcing bar A steel rod that is embedded in concrete in order to carry tensile forces and/or to prevent cracking of the concrete.

Relative humidity A percentage representing the ratio of the amount of water vapor contained in a mass of air to the amount of water it could contain under the existing conditions of temperature and pressure.

Relieved back A longitudinal groove or series of grooves cut from the back of a flat wood molding or flooring strip to minimize cupping forces and make the piece easier to fit to a flat surface.

Remodeling Modifying an existing building.

Removable glazing panel A framed sheet of glass that can be attached to a window sash to increase its thermal insulating properties.

Replacement window A window unit that is designed to install easily in an opening left in a wall by a deteriorated window unit that has been removed.

Resale value The price for which the owner of a residence can sell it.

Resilient clip A springy mounting device for plaster or gypsum board that helps reduce the transmission of sound vibrations through a wall or ceiling.

Resilient flooring A manufactured sheet or tile flooring made of asphalt, polyvinyl chloride, linoleum, rubber, or other elastic material.

Resin A natural or synthetic, solid or semisolid organic material of high molecular weight, used in the manufacture of paints, varnishes, and plastics.

Resistance Opposition to the flow of electricity, measured in ohms.

Retaining wall A wall that resists horizontal soil pressures at an abrupt change in ground elevation.

Return-air duct A tube through which air is sucked from an interior space and conducted back to the furnace or air-handling unit.

Ridge board The board against which the tips of rafters are fastened.

Ridge vent A device that allows air to circulate in and out along the ridge of a roof while preventing the passage of water and insects.

Rigid insulation A thermal insulating material that is furnished in the form of low-density boards or panels.

Rim joist *See* **Band joist.**

Rise A difference in elevation, such as the rise of a stair from one floor to the next or the rise per foot of run in a sloping roof.

Riser A single vertical increment of a stair; the vertical face between two treads in a stair; a vertical run of plumbing, wiring, or ductwork.

Rock wool An insulating material manufactured by forming fibers from molten rock.

Roof deck The structural surface to which roofing materials are attached.

Roofer One who installs roof coverings.

Roofing The material used to make a roof watertight, such as shingles, slate, tiles, sheet metal, or a roof membrane; the act of applying roofing.

Roof jack A bracket that supports a scaffold plank on the sloping surface of a roof.

Roof window An openable window designed to be installed in the sloping surface of a roof.

Rotary-sliced veneer A thin sheet of wood produced by rotating a log against a long, sharp knife blade in a lathe.

Rough arch An arch made from masonry units that are rectangular rather than wedge shaped.

Rough carpentry Framing carpentry, as distinguished from finish carpentry.

Roughing in The installation of mechanical, electrical, and plumbing components that will not be exposed to view in the finished building.

Rough-in plumbing Installation of pipes and fittings, but not of fixtures or trim.

Rough opening The clear dimensions of the opening that must be provided in a wall frame to accept a given door or window unit.

Rough sitework Clearing a lot for construction and bringing the soil to approximate grades.

Rowlock A brick laid on its long edge, with its end exposed in the face of the wall.

Rubble Unsquared stones.

Rumford fireplace A masonry fireplace designed to maximize radiation of heat by means of a shallow, tall firebox with widely splayed sides, as developed in the 18th century by Count Rumford.

Run Horizontal dimension in a stair or sloping roof.

Running Bond Brickwork consisting entirely of stretchers.

R-value A numerical measure of resistance to the flow of heat.

S

Sand filter system An underground layer of sand that is used to remove suspended solids from septic tank effluent.

Sand-molded brick A brick made in a mold that was wetted and then dusted with sand before the clay was placed in it.

Sandstone A sedimentary rock formed from sand.

Sandstruck brick *See* **Sand-molded brick.**

Sandwich panel A panel consisting of two outer faces of wood or gypsum bonded to a core of insulating foam.

Sapwood The living wood in the outer region of a tree trunk or branch.

Sash A frame that holds glass.

Scab A piece of framing lumber nailed to the face of another piece of lumber.

Scarf joint A glued end connection between two pieces of wood, using a sloping cut to create a large surface for the glue bond, to allow it to develop the full tensile strength of the wood that it connects.

Schematic design An early stage in designing a building, in which the configuration is established in a general way, without specifics or details.

Scratch coat The first coat in a three-coat application of plaster.

Screed A strip of wood, metal, or plaster that establishes the level to which concrete or plaster will be placed.

Scribe To mark the edge of a piece of material with the exact contour of an adjacent, irregular surface to which it will be mated. After scribing, the marked contour is cut with a scroll saw or saber saw and the piece is fitted to the surface.

Scupper An opening through a parapet through which water can drain over the edge of a flat roof.

Sealant gun A tool for injecting sealant into a joint.

Sealer A coating used to close the pores in a surface, usually in preparation for the application of a finish coating.

Sectional home A dwelling that is factory assembled in two or more fully finished, boxlike sections that are transported to the building site and connected to one another.

Sedimentary rock Rock formed from materials deposited as sediments, such as sand or sea shells, which form sandstone and limestone, respectively.

Segregation Separation of the constituents of wet concrete caused by excessive handling or vibration.

Seismic Relating to earthquakes.

Seismic load A load on a structure caused by movement of the earth relative to the structure during an earthquake.

Self-compacting fill A fill material that tamps itself into the minimum possible volume, such as a gravel made up of small, rounded stones.

Self-drilling Drills its own hole.

Self-furring lath Metal lath with dimples that space the lath away from the sheathing behind to allow plaster to penetrate the lath and key to it.

Self-tapping A screw that creates its own screw threads on the inside of a hole.

Self-weight The weight of a beam or slab.

Septic tank A large, underground vessel that accepts sewage, digests it by fermentation, and discharges liquid effluent.

Service entrance The assembly by which electricity is conducted from outdoor lines to the meter base.

Set To cure; to install; to recess the heads of nails; a punch for recessing the heads of nails.

Setback A legally required minimum distance between a building and a property boundary.

Setting bed A thin, wire-mesh reinforced concrete slab to which ceramic tile or stone is adhered.

Sewer, building *See* **Building sewer.**

Sewer, municipal An underground pipe that collects sewage from buildings and conducts it to a sewage treatment plant.

Shading coefficient The ratio of total solar heat passing through a given sheet of glass to that passing through a sheet of clear double-strength glass.

Shake A shingle split from a block of wood.

Shale A rock formed from the consolidation of clay or silt.

Shear A deformation in which planes of material slide with respect to one another.

Shear panel A wall, floor, or roof surface that acts as a deep beam to help stabilize a building against deformation by lateral forces.

Shear wall A vertical plane or assembly of material that acts to prevent lateral deflection of a building frame. A shear wall must be made of a material that is able to exert tensile force in diagonal direction within its plane and must be fastened to the foundation in such a way that it cannot lift up or slide.

Sheathing The rough covering applied to the outside of the roof, wall, or floor framing of a light frame structure.

Sheathing clip An H-shaped metal fitting that is used to keep the edges of adjacent sheets of roof sheathing material aligned with one another.

Shed A building or dormer with a single, sloping roof plane.

Sheeting A stiff material used to retain the soil around an excavation; a material such as polyethylene in the form of very thin, flexible sheets.

Sheet metal Flat rolled metal less than ¼ inch (6.35 mm) in thickness.

Shim A thin piece of material placed between two components of a building to adjust their relative positions as they are assembled; to insert shims.

Shingle A small unit of water-resistant material nailed in overlapping fashion with many other such units to render a wall or sloping roof watertight; to apply shingles.

Shiplap A board with edges rabbeted so as to overlap flush from one board to the next.

Shop drawings Detailed plans prepared by a fabricator or manufacturer to guide the shop production of such building components as cut stonework, steel or precast concrete framing, curtain wall panels, or cabinetwork.

Shoring Temporary vertical or sloping supports of steel or timber.

Short circuit An electric circuit that is accidentally created by an unintended connection through a metallic object or a human body.

Shotcrete A low-slump concrete mixture that is deposited by being blown from a nozzle at high speed with a stream of compressed air.

Shutoff valve A device that can close a pipe against the passage of water.

Sidelight A tall, narrow window alongside a door.

Siding The exterior wall finish material applied to a light frame wood structure.

Siding nail A nail with a small head, used to fasten siding to a building.

Silicone A polymer used for high-range sealants, roof membranes, and masonry water repellents.

Sill The strip of wood that lies immediately on top of a concrete or masonry foundation in wood frame construction; the horizontal bottom portion of a window or door; the exterior surface, usually sloped to shed water, below the bottom of a window or door.

Sill sealer A resilient, fibrous material placed between a foundation and a sill to reduce air infiltration between the outdoors and the indoors.

Single-hung window A window with two overlapping sashes, the lower of which can slide vertically in tracks, and the upper of which is fixed.

Single-ply membrane A sheet of plastic, synthetic rubber, or modified bitumen used as a roofing sheet for a low-slope roof.

Single-strength glass Glass approximately ³⁄₃₂ inch (2.5 mm) thick.

Sink, flush mounted A kitchen basin that is mounted flush with the surrounding countertop.

Sink, drop in A kitchen basin with an integral perimeter flange that is inserted into an opening in the countertop from above.

Sink, integral A kitchen basin that is formed continuously with the countertop.

Sink, undermount A kitchen basin that is mounted to the underside of the countertop.

SIP Structural insulated panel. *See* **Panel, structural insulated.**

Sitecast Concrete that is poured and cured in its final position in a building.

Site plan A drawing that shows the boundaries and features of a piece of land, usually including proposed physical improvements.

Site utilities Temporary electrical, water, telephone, and toilet services for use during construction.

Skylight A fixed window installed in a roof.

Slab on grade A concrete surface lying upon, and supported directly by, the ground beneath.

Slaked lime Calcium hydroxide.

Slate A metamorphic form of clay, easily split into thin sheets.

Sliding window A window with one fixed sash and another that moves horizontally in tracks.

Slope The inclination of a roof surface, a slab, or the surface of the ground.

Slump test A test in which wet concrete or plaster is placed in a cone-shaped metal mold of specified dimensions and allowed to sag under its own weight after the cone is removed. The vertical distance between the top of the mold and the top of the slumped mixture is an index of its working consistency.

Slurry A watery mixture of insoluble materials.

Smoke chamber The space that lies above the damper of a fireplace and below the flue.

Smoke shelf The horizontal area behind the damper of a fireplace.

Soffit The undersurface of a horizontal element of a building, especially the underside of a stair or a roof overhang.

Soffit vent An opening under the eave of a roof, used to allow air to flow into the attic or the space below the roof sheathing.

Soft mud process Making bricks by pressing wet clay into molds.

Soil, amended Earth whose properties have been adjusted with organic matter, fertilizer, lime, etc., to give it optimum properties for growing a lawn or garden.

Soil line A pipe that carries waste from a water closet.

Solar water heater A device that uses sunlight to heat domestic water.

Soldier A brick laid on its end, with its narrow face toward the outside of the wall.

Sole plate The horizontal piece of dimension lumber at the bottom of the studs in a wall in a light frame building.

Solid-core door A flush door with no internal cavities.

Solid sheathing A continuous roof sheathing material, such as plywood or OSB.

Solvent A liquid that dissolves another material.

Sound Transmission Class (STC) An index of the resistance of a partition to the passage of sound.

Span The distance between supports for a beam, girder, truss, vault, arch, or other horizontal structural device; to carry a load between supports.

Span rating The number stamped on a sheet of plywood or other wood building panel to indicate how far, in inches, it may span between supports.

Spark arrester A screen over the top of a chimney flue that is designed to prevent burning embers from passing.

Specifications The written instructions from an architect or engineer concerning the quality of materials and execution required for a building.

Speculative builder One who constructs dwellings to be sold.

Spirit level A tool in which a bubble in an upwardly curving, cylindrical glass vial indicates whether a building element is level or not level, plumb or not plumb.

Splash block A small precast block of concrete or plastic used to divert water at the bottom of a downspout.

Spline A thin strip inserted into grooves in two mating pieces of material to hold them in alignment; a ridge or strip of material intended to lock to a mating groove; an internal piece that serves to align two hollow components to one another.

Split jamb A door frame fabricated in two interlocking halves, to be installed from the opposite sides of an opening.

Sprinkler head A fitting that sprays water in a predetermined pattern, either for irrigation or for extinguishing fire.

Stain A coating intended primarily to change the color of wood or concrete, without forming an impervious film.

Stair jack *See* **Stringer.**

Staking out Driving stakes into the soil to mark the approximate future location of a building.

Standing and running trim Door and window casings and baseboards.

Standing seam A sheet metal roofing seam that projects at right angles to the plane of the roof.

STC *See* **Sound Transmission Class.**

Steel Iron with a controlled amount of carbon, generally less than 1.7 percent.

Steep roof A roof with sufficient slope that it may be made waterproof with shingles.

Stiff mud process A method of molding bricks in which a column of damp clay is extruded from a rectangular die and cut into bricks by fine wires.

Stile A vertical framing member in a panel door.

Stipulated sum contract An agreement that specified work is to be carried out for a given price.

Stirrup A vertical loop of steel bar used to reinforce a concrete beam against diagonal tension forces.

Stirrup tie A stirrup that forms a complete loop, as differentiated from a U-stirrup, which has an open top.

Stock plan A design for a house that is prepared without reference to any particular site, and is offered for sale.

Stool The interior horizontal plane at the sill of a window.

Storm drain A system for removing storm water from a site.

Storm sewer A large pipe that collects rainwater and/or snowmelt from storm drains and perimeter drains, and conducts it away from a site for disposal elsewhere.

Storm water Water that runs off a building or a site, having originated as rain or snow.

Storm window A sash added to the outside of a window in winter to increase its thermal resistance and decrease air infiltration.

Story pole A strip of wood marked with the exact course heights of masonry for a particular building, used to make sure that all the leads are identical in height and coursing.

Stove, radiant A fuel-burning appliance that heats a room by radiation.

Straightedge To strike off the surface of a concrete slab using screeds and a straight piece of lumber or metal.

Strain Deformation under stress.

Straw bale construction Walls made of stacked bales of straw, then stuccoed inside and out.

Stress Force per unit area.

Stressed-skin panel *See* **Panel, stressed-skin.**

Stretcher A brick or masonry unit laid in its most usual position, with the broadest surface of the unit horizontal and the length of the unit parallel to the surface of the wall.

Striated Textured with parallel scratches or grooves.

Stringer The sloping wood or steel member that supports the treads of a stair.

Strip flooring Wood finish flooring in the form of long, narrow tongue-and-groove boards.

Strip footing A continuous spread footing that is used to support a wall, essentially a thickening of the base of the wall to transmit the weight of a building into the soil at a stress equal to or below the bearing capacity of the soil.

Stripping Removing formwork from concrete; sealing around a roof flashing with layers of felt and bitumen.

Structural bond The interlocking pattern of masonry units used to tie two or more wythes together in a wall.

Structural engineer One who is licensed by a state government to prepare designs for loadbearing structures of buildings and bridges.

Structural tubing Hollow steel cylindrical or rectangular shapes made to be used as structural members.

Stucco Portland cement plaster used as an exterior cladding or siding material.

Stud One of an array of small, closely spaced, parallel wall framing members; a heavy steel pin.

Styrene–butadiene–styrene A copolymer of butadiene and styrene used as a modifier in polymer-modified bitumen roofing.

Subcontractor A contractor who specializes in one area of construction activity and who usually works under a general contractor or builder.

Subfloor The loadbearing surface beneath a finish floor.

Subgrade The soil beneath a slab on grade.

Subpanel A secondary group of circuit breakers in a metal box, provided with electricity via a cable from a main panel.

Substrate The base to which a coating or veneer is applied.

Sump A pit designed to collect water for removal from an excavation or basement.

Superplasticizer A concrete admixture that makes wet concrete extremely fluid without additional water.

Supply duct A tube through which conditioned air is distributed within a building.

Supply pipe A pipe that brings clean water to a plumbing fixture.

Supply system All the pipes and other components that distribute water through a building for drinking, bathing, washing, and other chores.

Supporting stud A wall framing member that extends from the sole plate to the underside of a header and that supports the header.

Surface bonding Bonding a concrete masonry wall together by applying a layer of glass-fiber-reinforced stucco to both its faces.

Surveyor, registered land One who is licensed by a state to measure land and mark property lines.

Surveyor's rod A measuring stick that is used as a target for sightings through a surveyor's transit or level, so as to establish vertical levels.

Sustainable yield forest A stand of trees that is managed in such a way that it will continue to produce logs for lumber in perpetuity.

Swale A shallow ditch around a building that conducts water away from the foundation.

Switch A device for completing or interrupting an electric circuit.

Switch, four-way A switch that allows a light to be turned on and off independently from three or more locations.

Switch, three-way A switch that allows a light to be turned on and off independently from two locations.

T

Tackless strip A wood strip with projecting points used to fasten a carpet around the edge of a room.

Tapered edge The longitudinal edge of a sheet of gypsum board, which is recessed to allow room for reinforcing tape and joint compound.

Tee A metal or precast concrete member or plumbing fitting with a cross section resembling the letter *T*.

Tempered glass Glass that has been heat treated to increase its toughness and its resistance to breakage.

Tensile strength The ability of a structural material to withstand stretching forces.

Tensile stress A stress caused by stretching of a material.

Tension A stretching force; to stretch.

Terrace door A double glass door, one leaf of which is fixed, and the other hinged to the fixed leaf at the centerline of the door.

Test pit A hole in the ground that is used to assess soil conditions and run percolation tests preparatory to constructing a private sewage disposal system.

Thatch A thick roof covering of reeds, straw, grasses, or leaves.

Thermal break A section of material with a low thermal conductivity, installed between metal components to retard the passage of heat through a wall or window assembly.

Thermal bridge A component of higher thermal conductivity that conducts heat more rapidly through an insulated building assembly, such as a steel stud in an insulated stud wall.

Thermal conductivity The rate at which a material conducts heat.

Thermal envelope The insulated exterior shell of a building.

Thermal insulation *See* **Insulation, thermal.**

Thermal resistance The resistance of a material or assembly to the conduction of heat.

Thermostat A device that utilizes a bimetallic strip to control temperature to a set value.

Thinbed A cementitious panel used as a setting bed for tiles.

Thincoat Veneer plaster.

Thrust A lateral force resulting from the structural action of an arch or inclined structure.

Thrust block A wooden block running perpendicular to the stringers at the bottom of a stair, whose function is to hold the stringers in place.

Tie A device for holding two parts of a construction together; a structural device that acts in tension.

Tie beam A reinforced concrete beam cast as part of a masonry wall, whose primary purpose is to hold the wall together, especially against seismic loads, or cast between a number of isolated foundation elements to maintain their relative positions.

Tie rod A steel rod that acts in tension.

Tile A fired clay product that is thin in cross section as compared to a brick, either a thin, flat element (ceramic tile or quarry tile), a thin, curved element (roofing tile), or a hollow element with thin walls (flue tile, tile pipe, structural clay tile); also a thin, flat element of another material, such as an acoustical ceiling unit or a resilient floor unit.

Tilt/turn window A window that may open either as a casement window or as a hopper window.

Timber Standing trees; a large piece of dimension lumber.

Toe nailing Fastening with nails driven at an angle.

Tongue and groove An interlocking edge detail for joining planks or panels.

Tooling The finishing of a mortar joint or sealant joint by pressing and compacting it to create a particular profile.

Top-hinged inswinging window A window that opens inward on hinges on or near its head.

Top plate The horizontal member at the top of a stud wall.

Topside vent A water-protected opening through a roof membrane to relieve pressure from water vapor that may accumulate beneath the membrane.

Torque Twisting action; moment.

Torsional stress Stress resulting from the twisting of a structural member.

Townhouse A dwelling that is part of a row of dwellings that share common walls between them.

Tracheids The longitudinal cells in a softwood.

Track A light-gauge steel channel that is used for the top and bottom members of a wall frame.

Tract A group of contiguous building lots that are being developed by one developer.

Traffic deck A special roof surface or assembly that is designed to allow people to walk above the roof without damaging its water-resistant components.

Transformer An electrical device that changes the voltage of alternating current.

Transit-mixed concrete Concrete mixed in a rotating drum on the back of a truck as it is transported to the building site.

Transom A small window directly above a door.

Tread One of the horizontal planes that make up a stair.

Trim accessories Casing beads, corner beads, expansion joints, and other devices used to finish edges and corners of a plaster wall or ceiling.

Trimmer A beam that supports a header around an opening in a floor or roof frame.

Trowel A thin, flat steel tool, either pointed or rectangular, provided with a handle and held in the hand, used to manipulate mastic, mortar, plaster, or concrete. Also, a machine whose rotating steel blades are used to finish concrete slabs; to use a trowel.

True divided-lite A window or door that is glazed with separate, small rectangles of glass, each mounted independently in muntins, as distinct from a window or door that is glazed with a single sheet of glass, over which a grid of imitation muntins is installed.

Truss A triangulated arrangement of structural members that reduces nonaxial external forces to a set of axial forces in its members.

Tunnel kiln A kiln through which clay products are passed on railroad cars.

Type X gypsum board A fiber-reinforced gypsum board, used where greater fire resistance is required.

U

Uncoursed Laid without continuous horizontal joints, random.

Undercourse A course of shingles laid beneath an exposed course of shingles at the lower edge of a wall or roof, in order to provide a waterproof layer behind the joints in the exposed course.

Underfire The floor of the firebox in a fireplace.

Underlayment A panel laid over a subfloor to create a smooth, stiff surface for the application of finish flooring. Also a layer of waterproof material such as building felt between roof sheathing and roofing.

Underpinning The process of placing a new foundation beneath an existing structure.

Uniform settlement Subsidence of the various foundation elements of a building at the same rate, resulting in no distress to the structure of the building.

Unreinforced Constructed without steel reinforcing bars or welded wire fabric.

Upflow furnace A furnace in which return air enters at the bottom and supply air exits at the top.

Upside-down roof A membrane roof assembly in which the thermal insulation lies above the membrane.

U-stirrup An open-top, U-shaped loop of steel bar, used as reinforcing against diagonal tension in a concrete beam.

Utilities, temporary Electrical, water, and telephone services that are brought to a site specifically for use during the construction of a building.

V

Valley A trough formed by the intersection of two roof slopes.

Valley rafter A diagonal rafter that supports a valley.

Valve, balancing A device for controlling the volume of water that flows through a pipe for the purpose of regulating a hydronic heating system.

Vapor barrier *See* **Vapor retarder.**

Vapor retarder A layer of material intended to obstruct the passage of water vapor through a building assembly. Also called, less accurately, vapor barrier.

Varnish A slow-drying transparent coating.

Vee joint A mortar joint whose profile resembles the letter *V.*

Veneer A thin layer, sheet, or facing.

Veneer-based lumber Linear wooden structural elements manufactured by bonding together strands or sheets of rotary-sliced wood veneer.

Veneer plaster A wall finish system in which a thin finish layer of plaster is applied over a special gypsum board base.

Veneer plaster base The special gypsum board over which veneer plaster is applied.

Vent-free A gas-burning fireplace that does not need a vent to the outdoors.

Venting, direct Removing the combustion gases from an appliance via a horizontal pipe through a wall.

Venting, natural Removing the combustion gases from an appliance by a vertical chimney.

Vent piping Pipes that maintain a system of drains and waste lines at atmospheric pressure by connecting them to the outdoor air.

Vent spacer A device used to maintain a free air passage above the thermal insulation in an attic or roof.

Vermiculite Expanded mica, used as an insulating fill or a lightweight aggregate.

Vertical bar An upright reinforcing bar in a concrete column.

Vertical grain lumber Dimension lumber sawed in such a way that the annual rings run more or less perpendicular to the faces of each piece.

Vinyl *See* **Polyvinyl chloride.**

Vitrification The process of transforming a material into a glassy substance by means of heat.

VOC *See* **Volatile organic compound.**

Volatile organic compound (VOC) A carbon-based chemical that evaporates into the atmosphere.

Volt A unit of electrical potential, equivalent to pressure in a pipe.

W

Waferboard A building panel made by bonding together large, flat flakes of wood.

Wainscoting A wall facing, usually of wood, cut stone, or ceramic tile, that is carried only partway up a wall.

Waler A horizontal beam used to support sheeting or concrete formwork.

Wallboard hanger One who mounts gypsum wallboard on framing.

Wane An irregular rounding of a long edge of a piece of dimension lumber caused by cutting the lumber from too near the outside surface of the log.

Washed aggregate A surface texture on a concrete slab created by washing away the fine particles of sand and cement to expose the more coarse particles of gravel.

Washer A steel disk with a hole in the middle, used to spread the load from a bolt, screw, or nail across a wider area of material.

Waste piping Pipes that carry wastewater that is free of solids.

Waste system The pipes and other components that remove wastewater from a building.

Water–cement ratio A numerical index of the relative proportions by weight of water and cement in a concrete mixture.

Water heater A device that heats water for domestic consumption.

Water-resistant gypsum board A gypsum board designed for use in locations where it may be exposed to occasional dampness.

Water smoking The process of applying heat to evaporate the last water from clay products before they are fired.

Waterstop A synthetic rubber or bentonite clay strip used to seal joints in concrete foundation walls.

Water-struck brick A brick made in a mold that was wetted before the clay was placed on it.

Water table The level at which the pressure of water in the soil is equal to the atmospheric pressure; effectively, the level to which groundwater will fill an excavation; a wood molding or shaped brick used to make a transition between a thicker foundation and the wall above.

Water vapor Water in its gaseous phase.

Wattle and daub Mud plaster (daub) applied to a primitive lath of woven twigs or reeds (wattle).

Weather barrier A moisture-resistant barrier, such as building felt or housewrap, that covers the sheathing and acts as a backup waterproofing layer to the siding that is applied over it.

Weathered joint A mortar joint finished in a sloping, planar profile that tends to shed water to the outside of the wall.

Weatherstripping A ribbon of resilient, brushlike, or springy material used to reduce air infiltration through the crack around a sash or door.

Web A cross-connecting piece such as the portion of a wood I-joist that is perpendicular to the flanges, or the portion of a concrete masonry unit that is perpendicular to the face shells.

Web stiffener A metal rib used to support the web of a light-gauge steel joist or a structural steel girder against buckling.

Weep hole A small opening whose purpose is to permit drainage of water that accumulates inside a building component or assembly.

Weld A joint between two pieces of metal formed by fusing the pieces together by the application of intense heat, usually with the aid of additional metal melted from a rod or electrode.

Welded wire fabric A grid of steel rods that are welded together, used to reinforce a concrete slab.

Welding The process of making a weld.

Well, cased A cylindrical hole in the ground, lined with a metal pipe, that reaches down to a water-bearing stratum.

"Wet" systems Construction systems that utilize considerable quantities of water on the construction site, such as masonry, plaster, and sitecast concrete.

Wind brace A diagonal structural member whose function is to stabilize a frame against lateral forces.

Winder (rhymes with "reminder") A stair tread that is wider at one end than at the other.

Wind load A load on a building caused by wind pressure and/or suction.

Window schedule A table that lists each window in a building, along with its dimensions, mode of operation, and other relevant information.

Wind uplift Upward forces on a structure caused by negative aerodynamic pressures that result from certain wind conditions.

Wire, grounding An uninsulated electric wire that connects electrical equipment to ground for the purpose of safety.

Wire, hot An electric wire that is not connected to ground.

Wire, neutral An electric wire that is connected to ground.

Wire nut A connecting device that is screwed onto the ends of two or more electrical wires, comprising a spiral coil of wire enclosed in a rigid insulating cap.

Wrought iron A form of iron that is soft, tough, and fibrous in structure, containing about 0.1 percent carbon and 1 to 2 percent slag.

Wythe (rhymes with "scythe" and "tithe") A vertical layer of masonry that is one masonry unit thick.

Y

Yield strength The stress at which a material ceases to deform in a fully elastic manner and begins to deform irreversibly.

Z

Z-brace door A door made of vertical planks held together and braced on the back by three pieces of wood whose configuration resembles the letter Z.

Z-flashing A strip of sheet metal that has been bent into a three-part zigzag profile, used to prevent the entry of water at window and door heads, at horizontal trim features, and at horizontal joints in plywood siding.

Zero-clearance fireplace A prefabricated fireplace that can be installed directly against wood framing without danger of igniting the framing.

Zero-slump concrete A concrete mixed with so little water that it does not sag when piled vertically.

Zoning ordinance A law that specifies in detail how land may be used in a municipality.

INDEX